QC794.6.C6 A836
Atomic physics, accelerators

3 4369 00123798 5

QC
794.6
C6
A836

D1561476

DATE DUE

DEC 15 1995 S

WITHDRAWN

Methods of Experimental Physics

VOLUME 17

ATOMIC PHYSICS: ACCELERATORS

METHODS OF EXPERIMENTAL PHYSICS:

L. Marton and C. Marton, *Editors-in-Chief*

Volume 17

ATOMIC PHYSICS
Accelerators

Edited by

PATRICK RICHARD

Department of Physics
Kansas State University
Manhattan, Kansas

1980

ACADEMIC PRESS
A Subsidiary of Harcourt Brace Jovanovich, Publishers
New York   London   Toronto   Sydney   San Francisco

COPYRIGHT © 1980, BY ACADEMIC PRESS, INC.
ALL RIGHTS RESERVED.
NO PART OF THIS PUBLICATION MAY BE REPRODUCED OR
TRANSMITTED IN ANY FORM OR BY ANY MEANS, ELECTRONIC
OR MECHANICAL, INCLUDING PHOTOCOPY, RECORDING, OR ANY
INFORMATION STORAGE AND RETRIEVAL SYSTEM, WITHOUT
PERMISSION IN WRITING FROM THE PUBLISHER.

ACADEMIC PRESS, INC.
111 Fifth Avenue, New York, New York 10003

United Kingdom Edition published by
ACADEMIC PRESS, INC. (LONDON) LTD.
24/28 Oval Road, London NW1 7DX

Library of Congress Cataloging in Publication Data
Main entry under title:

Atomic physics: Accelerators.

(Methods in experimental physics ; v. 17)
Includes bibliographical references and index.
1. Collisions (Nuclear physics)--Addresses, essays, lectures. 2. Ion bombardment--Addresses, essays, lectures. I. Marton, Ladislaus Laszlo. II. Marton, C. III. Richard, Patrick, Date IV. Series.
QC794.6.C6A25   539.7'3   80-15602
ISBN 0-12-475959-9 (v. 17)

PRINTED IN THE UNITED STATES OF AMERICA

80 81 82 83   9 8 7 6 5 4 3 2 1

This work is dedicated to the memory of James R. Macdonald, whose career was devoted to the study of atomic collisions. Jim made many valuable contributions through his diligent and uncompromising search for correct solutions to atomic problems. Probably his last written contribution to science is contained in this volume. He did not see a printed version of his contribution.

CONTENTS

CONTRIBUTORS xvii
FOREWORD xix
PREFACE xxi
VOLUMES IN SERIES xxiii

1. Introduction
by PATRICK RICHARD

    1.1. Atomic Collisions with Accelerated Ions 1

    1.2. Accelerated Ions for Atomic Collisions 3
        1.2.1. Impetus from Nuclear Physics for Accelerated Ions 3
        1.2.2. The Accelerated-Ion Method 4
        1.2.3. Use of Accelerated Ions for Atomic Collisions . 20

2. Energy Loss of High-Velocity Ions in Matter
by WEI-KAN CHU

    2.1. Introduction 25

    2.2. General Concepts of Energy Loss 26

    2.3. Energy Loss Theories: General Background 30
        2.3.1. Coulomb Scattering and Energy Loss 31
        2.3.2. Perturbation Method and Dielectric Description . 33
        2.3.3. Bethe Formula 34
        2.3.4. Mean Excitation and Ionization Energy I . . . 35

2.3.5.	Shell Correction.	36
2.3.6.	Dependence of Energy Loss on Z_1	37
2.3.7.	Low-Velocity Electronic Energy Loss	39
2.3.8.	Nuclear Energy Loss	41
2.3.9.	Electronic Energy Loss in the Medium-Velocity Region	45

2.4. Experimental Methods on Energy Loss 46
    2.4.1. Transmission Measurements on Thin Foils . . . 47
    2.4.2. Transmission Measurements on Gaseous Targets 51
    2.4.3. Transmission Measurements on Supported Films 53
    2.4.4. Transmission Measurements on a Thin Layer on a Solid-State Detector 54
    2.4.5. Backscattering Energy Loss 58
    2.4.6. Backscattering Thick-Target Yield. 61
    2.4.7. Measurement of Energy Loss by the Inverted Doppler Shift Attenuation (IDSA) Method . . . 65

2.5. Current Problems in Stopping Cross Sections 68
    2.5.1. Chemical Effect: Bragg's Rule 69
    2.5.2. Solid-State Effect 69
    2.5.3. Structure Effect. 70
    2.5.4. Energy Straggling 70
    2.5.5. Charge State of the Projectiles 71
    2.5.6. The Barkas Effect 71
    2.5.7. Ion Clusters Effect 71
    2.5.8. Nuclear Energy Loss 72

3. Charge Equilibration of High-Velocity Ions in Matter
by HANS D. BETZ

3.1. Introduction 73
    3.1.1. Historic Background. 73
    3.1.2. Present Situation 74
    3.1.3. Review Articles and Data Collections 76

3.2. Fundamental Charge Exchange Processes 78
    3.2.1. Definitions and Basic Processes 78
    3.2.2. Mathematical Description of Charge-State Fractions under Nonequilibrium and Equilibrium Conditions. 80

- 3.3. Experimental Techniques and Data Analysis. 83
  - 3.3.1. Measurement of Charge-State Fractions 83
  - 3.3.2. Auxiliary Methods. 87

- 3.4. Electron Capture. 89
  - 3.4.1. Theory of Electron Capture in Simple Collision Systems 89
  - 3.4.2. Electron Capture in Complex Systems 94
  - 3.4.3. Experimental Results on Electron Capture Cross Sections 97
  - 3.4.4. Multiple Electron Capture 104
  - 3.4.5. Very Large Charge Exchange Probabilities . . . 108
  - 3.4.6. Capture into Very High Quantum States 108
  - 3.4.7. Capture into Continuum States 109

- 3.5. Electron Loss 110
  - 3.5.1. Target Ionization Due to Light Particles 110
  - 3.5.2. Projectile Ionization 113
  - 3.5.3. Experimental Results on Electron Loss Cross Sections 114
  - 3.5.4. Multiple-Electron Loss. 116
  - 3.5.5. Small-Impact-Parameter Collisions and Multiple-Electron Loss. 117

- 3.6. Equilibrium Charge-State Distributions 118
  - 3.6.1. Experimental Results 118
  - 3.6.2. Semiempirical Descriptions of Equilibrium Charge-State Distributions 122

- 3.7. Average Equilibrium Charge States. 125
  - 3.7.1. On the Lamb–Bohr (LB) Criterion 125
  - 3.7.2. Experimental Average Equilibrium Charge States 127
  - 3.7.3. Semiempirical Relationships for the Average Equilibrium Charge 129

- 3.8. Density Effects and Excited States 131
  - 3.8.1. Density Effect in Gaseous Targets. 131
  - 3.8.2. Density Effect in Solid Targets 134

CONTENTS

3.9. Atomic Transitions and Charge Equilibration in Solids . 139
    3.9.1. Electron Capture and Inner-Shell Vacancies . . 139
    3.9.2. Equilibrium Excitation of Very Fast Ions in Solids 142

3.10. Radiative Electron Capture 144

4. Inelastic Energy-Loss Measurements in Single Collisions
by BENT FASTRUP

4.1. Introduction 149
    4.1.1. Introductory Remarks 149
    4.1.2. Definition of Inelastic Energy Loss 150
    4.1.3. Historical Remarks 151

4.2. Binary Encounters 159
    4.2.1. Derivation of Q 159
    4.2.2. Kinetic Energy of Scattered Particles 159
    4.2.3. Transformation from Laboratory System to CM System 161
    4.2.4. Interatomic Potentials and Distance of Closest Approach. 163
    4.2.5. Kinematical Effects in Cross Sections 167
    4.2.6. Inherent Broadening Effects 168

4.3. Experimental Method of Determining Q. 173
    4.3.1. Noncoincidence Methods. General 173
    4.3.2. Coincidence Methods. General. 184
    4.3.3. Error Analysis 186

4.4. Data Reduction and Comparison with Auger Electron/X-Ray Yields 188
    4.4.1. Data Reduction 188
    4.4.2. Spectroscopic Methods. 191

5. Target Ionization and X-Ray Production for Ions Incident upon Solid Targets
by Tom J. Gray

5.1. Introduction 193

5.2. Definitions of Parameters 195
    5.2.1. Relationship of Ionization and X-Ray Cross Sections 195
    5.2.2. Parameters Associated with Theoretical Considerations 197

5.3. Theoretical Models of Inner-Shell Ionization. 200
    5.3.1. Direct Coulomb Ionization 200
    5.3.2. Corrections to the Theory of Direct Coulomb Ionization. 203
    5.3.3. Additional Considerations 207
    5.3.4. Molecular Orbital Excitation 212

5.4. General Experimental Consideration 213
    5.4.1. Experimental Arrangement 213
    5.4.2. X-Ray Detector Efficiency 214
    5.4.3. Target Considerations for Light Ions. 218
    5.4.4. Target Considerations for Heavy Incident Ions . 221

5.5. X-Ray Production Cross-Section Measurements 228
    5.5.1. Introduction 228
    5.5.2. K-Shell X-Ray Measurements. 229
    5.5.3. L-Shell X-Ray Measurements 240
    5.5.4. M-Shell X-Ray Measurements 249
    5.5.5. Target K-Shell Cross Sections for Heavier Ions . 250
    5.5.6. Charge-State Fractions for Heavy Ions in Solids as Related to the Two-Component Model . . . 272

6. Charge Dependence of Atomic Inner-Shell Cross Sections
by James R. Macdonald

6.1. Introduction 279

6.2. Conditions for Measuring Charge-Dependent Collision Processes 280

6.3. Projectile Charge Dependence of Target Inner-Shell Cross Sections 284

6.4. Fluorescence Yield Effects in High-Charge Collisions . 289

6.5. Charge Dependence of Projectile Cross Sections 297

6.6. Summary 302

7. Coincidence Experiments for Studying Impact-Parameter-Dependent and Charge-Changing Processes
by C. Lewis Cocke

7.1. Introduction 303

7.2. Impact Parameter Dependence of Inner-Shell Vacancy Production 304
   7.2.1. Introduction and History 304
   7.2.2. The Single-Encounter Experiment 306
   7.2.3. Techniques 311
   7.2.4. Experimental Results 317

7.3. Coincidence Experiments Involving Charge-Analyzed Reaction Products 345
   7.3.1. Related to L-Shell Vacancy Production 345
   7.3.2. Coincident Charge State Analysis for Spectral Identification of X-Ray and Auger Electrons . . 347
   7.3.3. Electron Capture from the K Shell of Heavy Targets 350

8. Ion-Induced X-Ray Spectroscopy
by Forrest Hopkins

8.1. Introduction 355

8.2. Detectors and Targets. 356
    8.2.1. Proportional Counters 356
    8.2.2. Semiconductor Detectors. 358
    8.2.3. Single-Plane Crystal Spectrometers 361
    8.2.4. Single Cylindrical Curved-Crystal Spectrometers 365
    8.2.5. Doppler-Tuned Spectrometer 369
    8.2.6. Targets. 371

8.3. Multiple Inner-Shell Vacancy Production 374
    8.3.1. Multiple Vacancies and the K_α Satellite Structure 374
    8.3.2. Multiple-Ionization Formulation. 379
    8.3.3. Collision Mechanisms 382

8.4. Spectroscopy of Individual States 386
    8.4.1. Line Energies. 386
    8.4.2. Fluorescence Yields 392
    8.4.3. Multielectron Transitions. 395

8.5. Single-Collision Phenomena 401
    8.5.1. Multiple L-Shell Ionization Probabilities 401
    8.5.2. Hypersatellite Production. 408
    8.5.3. Neon as a Case Study 410

8.6. Chemical Effects 416
    8.6.1. Effects of Outer-Shell Relaxation on X-Ray Spectra. 416
    8.6.2. Dependence of Solid Target Spectra on Environment 418
    8.6.3. Outer-Shell Rearrangement in Gaseous Molecules 424

8.7.	Multiple-Collision Phenomena	426
	8.7.1. Collisional Quenching	426
	8.7.2. Ions Moving in Solids	428

9. Ion-Induced Auger Electron Spectroscopy
by Dennis L. Matthews

9.1.	Introduction	433
9.2.	Techniques for Production and Detection of Auger Electrons	436
	9.2.1. Chambers and Detectors	436
	9.2.2. Description of Electron Spectrometers	439
	9.2.3. Fast-Ion Accelerator Peculiarities	464
	9.2.4. Optimum Experimental Setup for a Given Type of Auger Electron Measurement	467
9.3.	Studies of Ion–Atom Collision Phenomena Using Auger Spectroscopy	468
	9.3.1. Theory of Auger Electron Emission	468
	9.3.2. Ion–Atom Collision Mechanisms	482
	9.3.3. Selected Experimental Topics	503

10. Radiative and Auger Beam–Foil Measurements
by David J. Pegg

10.1.	Introduction	529
10.2.	Experimental Arrangements	530
	10.2.1. Introduction	530
	10.2.2. Particle Accelerators and Ion Sources	533
	10.2.3. Beam Analysis and Source Purity	535
	10.2.4. Foil Targets for Excitation	536
	10.2.5. Normalization Procedures	537
	10.2.6. Filters, Spectrometers, Spectrographs, and Detectors	538

10.3.	Source Characteristics		548
	10.3.1.	General Properties	548
	10.3.2.	Comparison of the Beam–Foil Source with Other Spectroscopic Sources	548
10.4.	Beam–Foil Spectra		550
	10.4.1.	Introduction	550
	10.4.2.	Postfoil Charge Distributions	551
	10.4.3.	Charge State Identifications	552
	10.4.4.	Doppler Broadening and Shifts	554
	10.4.5.	Singly Excited States	557
	10.4.6.	Doubly Excited States	560
	10.4.7.	Hydrogenic (Rydberg) States	564
	10.4.8.	Auger-Emitting States	566
10.5.	Lifetimes		570
	10.5.1.	Introduction	570
	10.5.2.	Population Changes and Cascading	572
	10.5.3.	Postfoil Beam Velocity	575
	10.5.4.	Limits on Beam–Foil Lifetimes	578
	10.5.5.	Allowed Radiative Transitions	579
	10.5.6.	Forbidden Radiative Transitions	583
	10.5.7.	Forbidden Auger Transitions	586
	10.5.8.	Lamb Shift Determinations in Heavy Ions	587
10.6.	Quantum Beat Phenomena		590
	10.6.1.	Introduction	590
	10.6.2.	Alignment, Orientation, and Coherence	591
	10.6.3.	Quantum Beat Experiments	592
10.7.	Applications of Beam–Foil Results		595
	10.7.1.	Introduction	595
	10.7.2.	Atomic Structure Theory	596
	10.7.3.	Astrophysical and Laboratory Plasmas	599
	10.7.4.	Recent Additions	605

AUTHOR INDEX 607

SUBJECT INDEX 627

CONTRIBUTORS

Numbers in parentheses indicate the pages on which the authors' contributions begin.

HANS D. BETZ, *Sektion Physik, Universität München, 8046 Garching, Am Coulombwall 1, West Germany* (73)

WEI-KAN CHU, *IBM Data Systems Division, East Fishkill, Hopewell Junction, New York 12533* (25)

C. LEWIS COCKE, *Department of Physics, Kansas State University, Manhattan, Kansas 66506* (303)

BENT FASTRUP, *Institute of Physics, University of Aarhus, DK-8000 Aarhus C, Denmark* (149)

TOM J. GRAY, *Department of Physics, Kansas State University, Manhattan, Kansas 66506* (193)

FORREST HOPKINS, *Lawrence Livermore Laboratory, Livermore, California 94550* (355)

DENNIS L. MATTHEWS, *Physics Department, University of California, Lawrence Livermore Laboratory, Livermore, California 94550* (433)

JAMES R. MACDONALD,[*] *Department of Physics, Kansas State University, Manhattan, Kansas 66506* (279)

DAVID J. PEGG, *Department of Physics and Astronomy, University of Tennessee, Knoxville, Tennessee 37916 and Oak Ridge National Laboratory, Oak Ridge, Tennessee 37830* (529)

PATRICK RICHARD, *Department of Physics, Kansas State University, Manhattan, Kansas 66506* (1)

[*] Deceased

FOREWORD

The subject of experimental techniques in atomic physics has been treated in Volume 4, Part A: *Atomic Sources and Detectors,* Part B: *Free Atoms* (edited by Vernon W. Hughes and Howard L. Schultz), and Volume 7, *Atomic and Electron Physics—Atomic Interactions* (edited by Benjamin Bederson and Wade L. Fite).

The present volume deals, of course, with the use of accelerators applied to the study of atomic collisions. We are grateful to Professor Patrick Richard for bringing together a volume which complements the subject matter of the earlier volumes. Because of the usefulness of ion beams in many fields of physics, interest in the techniques covered here should extend to areas other than atomic physics.

We congratulate Professor Richard for a job well done.

L. MARTON
C. MARTON

PREFACE

The study of atomic collison physics has undergone tremendous growth in the past fifteen years. This growth is attributed to the development of several new types of measurement techniques that revealed unique information on the formation of excited states of ions in ion–atom collisions. In particular, the importance of inner-shell excitation processes over a very broad range of collision parameters and the ability to form discrete few electron–ion systems have been realized in many of these measurements. Conceptual models as well as detailed theoretical models appeared to give a good physical picture for many of the observations. As is the case in many a growing field, the models and theories have been subjected to severe tests with many of the theories shown to be adequate but not totally general. The feedback between theory and experiment has continued throughout this period and is projecting into the future.

The study of atomic collisions by the method of accelerated ions cuts through the lines of many traditional areas of physics, utilizing and extending such expertises as spectroscopy (e.g., x-ray, Auger electron, and ion spectroscopy), traditional nuclear collision physics, and accelerator physics. The study of atomic collisions using very highly charged ions has followed closely with the development of heavy-ion accelerators, which also dates back over the past two decades. This development continues to push the frontiers of this field of atomic collisions with few electron ions to higher Z projectiles. This frontier promises to provide more stringent tests of quantum electrodynamic theories of atomic structure and atomic collisions.

This volume is intended to provide a thorough review of the methods that have been employed in the study of atomic collisons using the accelerated ion method. The major subjects that are covered here are: energy loss of ions in matter, charge equilibration of ions in matter, inelastic energy loss in single collisions, target ionization in solid targets, charge dependence of atomic cross sections, coincidence measurements in atomic collisions, ion-induced x-ray spectroscopy, ion-induced Auger electron spectroscopy, and optical, x-ray, and Auger beam foil spectroscopy. The material is presented in a form and at a level where it can serve as an introduction to the field for a researcher in a different field of study, for a graduate student interested in obtaining a basic knowledge of atomic collision physics, and for the practitioner of ion–atom collision physics.

The application of the results of ion–atom collisions to other areas of physics as well as to industry is not specifically considered here. However, it is hoped that this volume will serve as a reference and study guide for future developments in atomic-collision related areas such as ion implantation, channeling, trace element analysis, plasma collision physics, laser–ion beam interactions, electron–ion beam interactions, and any future new developments using ion beam techniques.

PATRICK RICHARD

METHODS OF EXPERIMENTAL PHYSICS

Editors-in-Chief
L. Marton   C. Marton

Volume 1. Classical Methods
Edited by Immanuel Estermann

Volume 2. Electronic Methods. Second Edition (in two parts)
Edited by E. Bleuler and R. O. Haxby

Volume 3. Molecular Physics, Second Edition (in two parts)
Edited by Dudley Williams

Volume 4. Atomic and Electron Physics—Part A: Atomic Sources and Detectors, Part B: Free Atoms
Edited by Vernon W. Hughes and Howard L. Schultz

Volume 5. Nuclear Physics (in two parts)
Edited by Luke C. L. Yuan and Chien-Shiung Wu

Volume 6. Solid State Physics (in two parts)
Edited by K. Lark-Horovitz and Vivian A. Johnson

Volume 7. Atomic and Electron Physics—Atomic Interactions (in two parts)
Edited by Benjamin Bederson and Wade L. Fite

Volume 8. Problems and Solutions for Students
Edited by L. Marton and W. F. Hornyak

Volume 9. Plasma Physics (in two parts)
Edited by Hans R. Griem and Ralph H. Lovberg

Volume 10. Physical Principles of Far-Infrared Radiation
Edited by L. C. Robinson

Volume 11. Solid State Physics
Edited by R. V. Coleman

Volume 12. Astrophysics—Part A: Optical and Infrared
Edited by N. Carleton

Part B: Radio Telescopes, Part C: Radio Observations
Edited by M. L. Meeks

Volume 13. Spectroscopy (in two parts)
Edited by Dudley Williams

Volume 14. Vacuum Physics and Technology
Edited by G. L. Weissler and R. W. Carlson

Volume 15. Quantum Electronics (in two parts)
Edited by C. L. Tang

Volume 16. Polymers (in three parts)
Edited by R. A. Fava

Volume 17. Atomic Physics: Accelerators
Edited by P. Richard

Volume 18. Fluid Dynamics (in preparation)
Edited by R. J. Emrich

Volume 19. Ultrasonics (in preparation)
Edited by P. Edmonds

Methods of Experimental Physics

VOLUME 17

ATOMIC PHYSICS: ACCELERATORS

1. INTRODUCTION

By Patrick Richard

1.1. Atomic Collisions with Accelerated Ions

The methods of studying atomic collisions with accelerated ions are the basic subjects of this book. These subjects are enormous in terms of the atomic physics that can be extracted from the very simple atomic systems to the very exotic high-charge or heavy-ion systems. There are many parameters that must be considered in defining the methods of studying atomic collisions. The velocity of the incident ion dictates the degree to which the atoms can penetrate and thus the types of atomic excitations that one might expect to observe. The charge state q and the nuclear charge Z of the incident ion also play an important role in deciding the excited ions' final fate in the collision. Other parameters of importance that enter these considerations are the target atomic number, the binding energies of the electronic shells of the target and projectile, the chemical state of the target, and the physical state of the target.

In general it is safe to say that atomic physicists are interested in all measurable aspects of the atomic collision process, and in fact in some aspects that are not currently measurable. On the one hand, we are interested in the macroscopic phenomena of energy loss and charge state equilibrium of ions traversing solids and gases; on the other hand, we are interested in the microscopic phenomena that dictate specific excitation processes. The latter phenomena have been developed to a very high degree of sophistication and can be divided into several areas for detailed discussion. For example, the study of energy loss in single collisions has been developed in the last two decades and has shed tremendous light on the physics of atomic collisions. The excitation of particular states in ions can be studied in both gaseous and solid environments by observing the x-ray emission from the ions. A parallel study of the Auger emission from collisions in gaseous environments is also an area in which much information about the collision can be deduced. Observation of the x rays in high resolution using a Bragg spectrometer can lead to additional valuable detail about the competition among final atomic states, which might otherwise be unresolved. Several types of coincidence experiments have been developed that can yield information on the impact parameter dependence of the collision for a given type of atomic excitation process. The study of very heavy ions leads to new types of atomic collision phenomena that are of particular theoretical interest. The atomic spectroscopy of the excited projectile ions formed in ion–atom collisions is itself a very interesting area of work, which yields much information about the effects of configuration interaction, relativistic interactions, and quantum electrodynamics on the total lifetimes and branching ratios of atomic states. We could give many interesting examples of the developments in these various areas of atomic collisions, but these are the areas of atomic collisions that form the material for this book. We thus leave it to the authors of the remaining parts of this volume to guide the reader through these very interesting developments.

Atomic excitations can be studied by many methods other than the accelerated-ion method. Some of these methods can include ion–atom collisions as a subset of the collision phenomena. Examples are violent atomic systems such as spark discharges,[1,2] exploding wires,[3] tokamak plasmas,[4] laser-induced plasmas,[5] and the solar plasma.[2] These methods are not treated here, but are mentioned because of the close connection that exists between the various methods. For example, the violent atomic systems produce a very rich atomic spectroscopy (as does accelerated ion–atom collisions) that is of both fundamental and practical interest. The lifetimes of some of these states have been or can be studied by the beam foil method and are important in modeling the parameters, such as temperature and density, of violent atomic systems.

Accelerated ions are used in two other important atomic excitation methods. One is the interaction between accelerated ions and a photon field (e.g., crossed ion beam and laser beam[6]) and the other is the interaction between accelerated ions and electrons (crossed or merged beams[7]). The latter is of particular interest to the modeling of violent atomic systems due to the important role that electron capture plays particularly

---

[1] B. S. Fraenkel and J. L. Schwob, *Phys. Lett.* **40A**, 83 (1972); E. M. Reeves and A. K. Dupree, *in* "Beam Foil Spectroscopy," (I. A. Sellin and D. J. Pegg, eds.), p. 925. Plenum Press, New York, 1976.

[2] A. H. Gabriel and C. Jordan, "Case Studies in Atomic Collision Physics," Vol. II, pp. 209–291. North-Holland Publishing Co., Amsterdam, 1972.

[3] C. M. Dozier, P. G. Burkhalter, and D. J. Nagel, *Bull. Am. Phys. Soc.* **20**, 1303 (1975).

[4] K. M. Young, *Conf. Appl. Small Accelerators Res. Ind. IEEE Trans. Nucl. Sci.* **NS-26**, 1234 (1979).

[5] N. J. Peacock, "Beam Foil Spectroscopy," p. 925 Plenum Press, N.Y., 1976.

[6] H. W. Bryant, *XI Int. Conf. Phys. Electron. At. Collisions, Kyoto, Japan* (K. Takayanagi and N. Oda, eds.) North-Holland Publishing Co., Amsterdam, 1979.

[7] D. H. Crandall, R. A. Phaneuf, and P. O. Taylor, *Phys. Rev. A* **18**, 1911 (1978).

for highly charged ions. We expect these two areas will continue to see much development in the next few years. These subjects are not included in this book, however. The range of experiments possible depends on, among other things, the range of variables available by the accelerated ion method just as in the ion–atom collision case.

In the remainder of this short introduction, we concentrate on this latter point. How does atomic collision physics using the accelerated-ion method depend on the available ion variables such as range of ion velocity, range of ion atomic charge q, range of ion nuclear charge Z, and range of ion beam current? If all possible combinations of atomic variables were available, then the insatiable appetite of atomic physicists could be satisfied by merely giving them sufficient time to experimentally sift through the atomic parameters. If this were the case, then the proposed question would be of little interest. In the framework of reality, however, the question is: What are the ion variables that atomic physicists consider to be important and how do these variables overlap with present-day technology of accelerated ions? To continue these considerations we discuss in the Chapter 1.2 some of the past, present, and future anticipated developments in accelerating ions.

1.2. Accelerated Ions for Atomic Collisions

1.2.1. Impetus from Nuclear Physics for Accelerated Ions

The impetus[8] for generating accelerated ions very quickly followed the discovery of very large angle deflection of alpha particles from thin metallic foils by Rutherford[9] in 1906 and by Geiger and Marsden[10] in 1909 and the subsequent explanation of the results by Rutherford[11] in 1911 in terms of his nuclear model of the atom. The predictions of Rutherford's model were completely confirmed by Geiger and Marsden[12] in 1913. The search for the size of the nucleus was thus launched. At some distance of closest approach, which depends on the projectile energy, an incident alpha particle should be influenced by the nuclear forces as well as the nuclear Coulomb field. In 1919, Rutherford[13] himself showed deviations

---

[8] For a detail treatment of the use of accelerators in nuclear physics through 1963, the reader is referred to *Methods Exp. Phys.* **5A**, 580–689.
[9] E. Rutherford, *Phil. Mag.* **11**, 166 (1906); **12**, 134 (1906).
[10] H. Geiger and E. Marsden, *Proc. Roy. Soc.* (London) **A82**, 495 (1909).
[11] E. Rutherford, *Phil. Mag.* **21**, 669 (1911).
[12] H. Geiger and E. Marsden, *Phil. Mag.* **25**, 604 (1913).
[13] E. Rutherford, *Phil. Mag.* **37**, 537 (1919).

from the pure Rutherford Coulomb scattering in the lightest of elements, where internuclear separations of the order of 5×10^{-13} cm could be reached. However, as late as 1925, Rutherford and Chadwick were unable to detect any deviations from pure Rutherford Coulomb scattering for alpha particles on copper nuclei. They were able to probe the copper atom only to distances of $\sim 1.5 \times 10^{-12}$ cm.

The need for higher energy projectiles as a probe of the nucleus was clearly in demand. The experiments discussed above were all performed with alpha particles from radioactive sources such as radium, polonium, and thorium. The alpha-particle energies from these sources are typically in the 4–8 MeV range. The deviations from pure Rutherford Coulomb scattering could have been observed for alpha particles on copper if 40 MeV alpha particles, which can reach internuclear separations of $\sim 2 \times 10^{-13}$ cm, were available. Development of particle accelerators would also allow for the acceleration of other particles, particularly protons, which could penetrate the Coulomb field of the nucleus with greater ease. Not only was there a need for higher collision energies but also for higher fluxes of particles. Radioactive sources available to Rutherford and others were typically tens of milligrams of radium, which translates into usable collimated beams for collision studies of $\leq 10^{-12}$ A/mg radium. The development of accelerated beams of a microamp would thus represent a five to six order of magnitude increase in beam intensity over the radioactive-source method.

1.2.2. The Accelerated-Ion Method

In the next few years a surprisingly large number of unique ideas for accelerating ions was developed specifically in answer to the long need for probing the nucleus of the atom. It was around 1930, 20 years—or approximately one-half of a scientific career—following the pioneering work leading to the discovery of the nucleus that the four major types of accelerators were conceived; the Cockroft–Walton, the Van de Graaff generator, the cyclotron, and the linear accelerator. This is an interesting example of the significant time lag that so often occurs between the need for and the realization of a technology.

1.2.2.1. *Electrostatic Cockroft–Walton Accelerator.* The anticipation of using a direct potential drop to accelerate ions led to the extensive studies of various high-voltage devices around 1930. Even though high voltages could be reached, the main difficulty was to establish the voltage across a discharge tube without a low voltage breakdown. From 1930 to 1932 J. D. Cockroft and E. T. S. Walton[14] developed an accelera-

---

[14] J. D. Cockroft and E. T. S. Walton, *Proc. Roy. Soc. (London) A* **129**, 477 (1930); *A* **137**, 229 (1932); *Nature* **129**, 649 (1932).

tor named after them, which used a high-voltage multiplier system in the form of a cascade rectifier circuit. One of the major accomplishments of this accelerator was the successful attachment of an evacuated accelerating tube to a source of high voltage. They accelerated protons to 380 keV and were the first to artificially transmutate (or split) atoms by observing the ^{7}Li + p → ^{4}He + ^{4}He reaction at 150 keV.

1.2.2.2. Electrostatic Van de Graaff Accelerator. From 1929 to 1931 R. J. Van de Graaff[15] developed a high-voltage generator using the method of mechanically delivering charge to a conducting dome by means of a charged belt. He had attended the physics lectures of Marie Curie at the Sorbonne in 1924, where he learned of the use of alpha particles from radioactive sources.[16] While at Oxford as a Rhodes Scholar, he read the 1927 anniversary address of Rutherford to the Royal Society concerning the need for accelerated ions "to have available . . . a copious supply of atoms and electrons . . . transcending in energy the alpha and beta particles from radioactive substances." The first model of the Van de Graaff generator developed in 1929 reached a voltage of 80 kV.

Some of the first electrostatic accelerators to use the Van de Graaff generator were built at the Department of Terrestial Magnetism (DTM) of the Carnegie Institute of Washington, D.C. The DTM 1-MV Van de Graaff accelerator (see Fig. 1) completed in late 1933 by M. A. Tuve et al.[17] was the first megavolt accelerator to use the Van de Graaff generator. The first nuclear physics experiment with the Van de Graaff was not completed until 1935. Using this accelerator they first observed the energy dependence of a nuclear resonance, the one at 0.44 MeV in ^{7}Li(p, γ)^{8}Be, and Tuve et al.[18] first observed the anomalous (non-Mott) scattering of protons by hydrogen beginning at about 0.7 ± 0.1 MeV, which demonstrated the short-range attraction between protons.

Returning to the DTM 1-MV open-air Van de Graaff accelerator shown in Fig. 1, one can readily see the evacuated acceleration column, the large charged dome, and the mechanically driven belts used to carry the charge up to the dome. The Van de Graaff or electrostatic type of accelerator has been one of the most used of the accelerators developed for the pur-

---

[15] R. J. Van de Graaff, *Phys. Rev.* **38,** 1919A (1931).

[16] The discussion of R. J. Van de Graaff is taken from W. D. Bygrave, P. A. Treado, and J. M. Lambert, "Accelerator Nuclear Physics," High Voltage Engineering Corp., Burlington, Mass., 1970.

[17] M. A. Tuve, L. R. Hafstad, and O. Dahl, *Phys. Rev.* **48,** 315 (1935). This paper describes the DTM 1 MeV and a 400 keV model they built. These authors also reported observing 1 MeV H$^+$ particles from a Tesla coil discharge device in 1932, *Phys. Rev.* **39,** 384 (1932).

[18] M. A. Tuve, N. P. Heydenburg, and L. R. Hafstad, *Phys. Rev.* **49,** 402L (1936); **56,** 125 (1939).

FIG. 1. The first 1-MV DC accelerator to use the generator developed by R. J. Van de Graaff in 1929 was completed in 1933 by M. A. Tuve (in picture) at DTM. The accelerator, which operated for a quarter of a century, now resides in the Museum of History and Technology of the Smithsonian Institution in Washington, D.C. (reprinted with permission).

1.2. ACCELERATED IONS FOR ATOMIC COLLISIONS

FIG. 2. New electrostatic accelerator facility under construction in 1937. The huge air-insulated pressure tank Van de Graaff accelerator is located at DTM. Much smaller Van de Graaff accelerators of comparable and higher voltages have been subsequently constructed using insulating gases such as CO_2–N mixtures and SF_6 (reproduced courtesy of the Carnegie Institution of Washington).

pose of furthering the knowledge of the nucleus. Its properties of being a direct current device, easily variable in energy, and capable in principle of accelerating any mass positive ion of charge q to a final energy qV, where V is the terminal voltage of the accelerator, accounts for its versatility. In the continuing search for the nuclear potential it has been feasible to follow a course of continually upgrading the electrostatic accelerator.

Three rather distinct developments have occurred in making higher energy electrostatic accelerators. The first upgrading consisted of building pressurized tank accelerators. One of three giant air-insulated pressurized machines was constructed at DTM in 1937 and is shown in Fig. 2. The other machines were at Westinghouse and the University of Minnesota. The accelerator was completed in 1938 and on January 28, 1939, two days after the news that Hahn and Strassman[19] had discovered fis-

---

[19] O. Hahn and F. Strassmann, *Naturwissenschaften.* **27**, 11 (1939).

FIG. 3. Accelerator attracts a distinguished crowd. Left to right are R. C. Meyer, M. A. Tuve, E. Fermi, R. B. Roberts, L. Rosenfeld, E. Bohr, N. Bohr, G. Breit, and J. Fleming (director of DTM) in the target room of the pressure tank Van de Graaff (see Fig. 2) gathered for a counter experiment demonstration of fission only two days after the announcement of the radiochemical discovery of fission by Hahn and Strassman (reprinted courtesy of Carnegie Institution of Washington).

sion, Roberts et al.[20] demonstrated fission as a counter experiment to an illustrious group of physicists, shown in Fig. 3, in the target room of the pressure tank machine. The early pressure tank Van de Graaff generators were developed to a nominal 5-MV operation. This type of machine operating at approximately 15 atm of a CO_2–N mixture could be contained in a pressure vessel approximately 6 m long and 2.5 m in diameter. Sulfur hexafluoride, SF_6, is used in some accelerators due to its superior insulating properties. A machine operating with pure SF_6 can operate at ~4 atm with better insulating properties than CO_2–N mixtures.

The second major upgrade was the development of the "tandem" or two-stage accelerator, which was proposed in 1936 by Bennett and

[20] R. B. Roberts, R. C. Meyer, and L. R. Hafstad, *Phys. Rev.* **55**, 416L (1939).

Darby[21] and again in 1951 by Alvarez.[22] It was developed to practicality by High Voltage Engineering Corporation[23] in 1960. The principle of the tandem is to charge the terminal to some positive voltage just as in the conventional Van de Graaff, but to have two acceleration columns, one on either side of the terminal. Negative ions are accelerated toward the terminal and then stripped by a gas or thin foil to form positive ions, which are then accelerated away from the terminal in the second column just as in the regular one-stage Van de Graaff. For proton beams the energy is doubled that available with the one-stage accelerator of the same terminal voltage. In general, for any negative ion the final energy in millions of electron volts is $(1 + q)V$, where V is the terminal voltage in megavolts and q the charge state of the ion at the terminal. Negatively charged molecules such as CN^- can be accelerated in the first stage, broken into positively charged ions in the terminal, and accelerated to final energies of one of the ions (e.g., C^{q+} or N^{q+}) of $(M/m + q)V$, where M/m is the ratio of the mass of the final accelerated ion to that of the molecule. The tandem accelerator has been developed to terminal voltages of 6, 7.5, 10, 12, and 15 MV. The higher voltage machines get extremely large.

The third development in the Van de Graaff electrostatic accelerator[24] is the concept of the folded tandem.[22,25] The principle of the folded tandem is to place a 180° bending magnet in the terminal of a tandem and thereby use the same acceleration column for the two acceleration tubes. The negative ions travel up to the terminal, are bent 180°, stripped positive, and accelerated back down the column in a second acceleration tube. This essentially reduces the size of the pressure vessel by a factor of two. The largest electrostatic accelerator to date is the one at Oak Ridge National Laboratory at Oak Ridge, Tennessee (see Fig. 4), which is being built by National Electrostatics Corporation and has reached a DC voltage of 32 MV.[26] The folded tandem concept allows a 25 MV accelerator to be contained in a vertical pressure vessel 30 m long and 10 m in diameter and to be contained in a vertical tower approximately 50 m high and 14 m in diameter (an eight or nine story building).

---

[21] W. H. Bennett and P. F. Darby, *Phys. Rev.* **49,** 97, 422, 881 (1936).

[22] L. W. Alvarez, *Rev. Sci. Instr.* **22,** 705 (1951).

[23] R. J. Van de Graaff, *Nucl. Instr. Meth.* **8,** 195 (1960).

[24] The term Van de Graaff is now a trademark of the High Voltage Engineering Corporation and therefore the terminology electrostatic accelerator should be used as a general description of such accelerators.

[25] H. Naylor, *Nucl. Instr. Meth.* **63,** 61 (1968).

[26] This accelerator was developed by the National Electrostatics Corporation and is designed as a 25 MV accelerator.

FIG. 4. New electrostatic accelerator facility under construction, 1979. A 25-MV folded tandem electrostatic accelerator resides inside the huge tower adjacent to the Oak Ridge Isochronous Cyclotron Laboratory. This accelerator has recently produced the largest-ever laboratory-controlled DC voltage of 32 MV (reprinted courtesy of Oak Ridge National Laboratory).

It is interesting that the technology of electrostatic accelerators has continued to improve over the 50 years since its inception in 1929. Figure 5 is a graph that roughly outlines the historical development of the electrostatic accelerator in terms of the maximum available terminal voltages for acceleration of ions, the maximum available H^+ energy, the maximum available He^{2+} energy, and the maximum available oxygen beam energy for single-stage Van de Graaffs before 1960 and for tandem Van de

1.2. ACCELERATED IONS FOR ATOMIC COLLISIONS

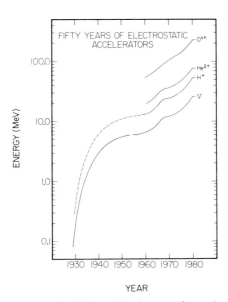

FIG. 5. The solid curve labeled V indicates the electrostatic accelerator DC terminal voltage. Prior to the tandem accelerator ($\sim$ 1960) that curve represents the energies of all singly ionized projectiles (H^+, $^2H^+$, He^+, etc.) and the dashed line represents the energies of doubly ionized projectiles (He^{2+}, Li^{2+}, etc.). The tandem accelerator energies for H^+, He^{2+}, and O^{q+} are also given as indicated.

Graaffs after 1960, not including double tandems, etc. (see Section 1.2.2.5).

The electrostatic accelerator as well as other accelerators have been continually improved and upgraded to higher energies for two main reasons. One reason is to increase the energy of light ion projectiles in order to probe the very depths of the nucleus, that is, to reach light ion energies exceeding the Coulomb barrier of even the heaviest of nuclei by several factors. The second reason is the growing interest in the nuclear physics of high-energy heavy-ion collisions. High-energy heavy ions, for example, allow for the possible production of a large number of new nuclei and possibly super heavy elements in islands of stability through heavy-ion fusion. New classes of phenomena are expected in heavy-ion–nuclear collisions.

It is useful to make a global plot of the performance of accelerators in terms of their energy per nucleon vs. ion mass. By comparing such a performance curve to the curve for the Coulomb barrier height for the collision of the incident ion with heavy ions, one can quickly see in which mass regions nuclear collisions can be studied with a given accelerator. Many such curves have been published in recent proposals for new accel-

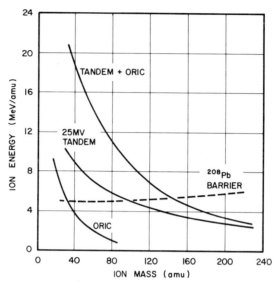

FIG. 6. Global plots of the maximum ion energy vs. ion mass for the 25-MV folded tandem, the Oak Ridge Isochronous Cyclotron, ORIC, and the 25-MV tandem injected into the ORIC are indicated by the solid curves. The Coulomb barrier height for incident ions on ^{208}Pb is indicated by the dashed curve.

erators. In Fig. 6 such a curve is given for a 25-MV tandem accelerator. The dashed curve represents the energy per nucleon of the Coulomb barrier height for the system of incident ion plus ^{208}Pb. From a comparison of the two curves we see that the ion beam energy can exceed the Coulomb barrier height of $A \leq 95$ in $A + {}^{208}$Pb collisions. For $A > 95$ the Coulomb barrier can still be reached for lighter mass targets. For example, for a projectile near $A \sim 235$, the Coulomb barrier can be reached for targets up to approximately mass 100.

1.2.2.3. Cyclotrons. All accelerators use the interaction between an external electric field and a charged particle to produce the acceleration. As discussed in the previous two sections, one of the early problems was attaching very high voltages across a discharge tube. From 1931 to 1932 E. O. Lawrence and his co-workers[27,28] discovered the principle of obtaining large accelerations by the repeated use of a relatively low accelerating voltage for particles moving in a uniform magnetic field. The mag-

---

[27] E. O. Lawrence and M. S. Livingston, *Phys. Rev.* **40**, 19 (1932). They reported on the acceleration of H$^+$ ions to 1.22 MeV in this paper; E. O. Lawrence, M. S. Livingston, and M. G. White, *Phys. Rev.* **42**, 150 (1932).

[28] E. O. Lawrence and M. S. Livingston, *Phys. Rev.* **45**, 608 (1934). They reported on the acceleration of H$^+$ ions to 5 MeV in this paper.

netic resonance principle they discovered for nonrelativistic particles is that the frequency f of revolution of a particle of charge e and mass m in a magnetic field B is independent of its radius (i.e., $f = qB/2\pi m$). By having the ions move inside two D-shaped electrodes, dees, separated by an acceleration gap, the ions that start in phase will remain in phase with an accelerating RF electric field at the resonance frequency f at the gap between the dees.

The first model of the cyclotron built in 1931 produced protons with energies of 80 keV and a second model in the next year produced 1.22-MeV protons. Constant-frequency cyclotrons are limited in energy due to the loss of resonance frequency of the ions from the relativistic increase in the mass with velocity, need for focusing of the beam in the r direction (this is accomplished in normal cyclotrons by shaping the pole faces to give a very slight decrease of field with radius) as the radius becomes large, and the practical considerations of magnets and RF systems. Since a fixed-frequency cyclotron can accelerate ions of a fixed q/m, it is not possible to accelerate H^+ and $^2H^+$ ions in the same cyclotron. A large cyclotron, which was used for many years, accelerates H_2^+, $^2H^+$, and $^4He^{2+}$ to energies of 10 MeV (per H^+), 20 MeV, and 40 MeV, respectively. The upper limit on the energy of the cyclotron is approximately 30 MeV for protons and 40 MeV for deuterons.[29]

It was suggested in 1938 by Thomas[30] that the two limiting problems with the cyclotron (see discussion above) could be overcome by a geometry including azimuthally varying magnetic fields. These accelerators are called AVF or sector-focusing cyclotrons and were developed in the late 1950s.[31,32] There are many varieties of AVF cyclotrons. They have in common the feature that they achieve strong axial focusing with a magnetic field that increases with radius in proportion to the particle's energy. This can be accomplished at constant frequency up to energies of several hundred million electron volts. These cyclotrons are variable in energy and can accelerate various mass particles by allowing the frequency to be variable. This is also possible in part because the main magnetic field coils can be varied separately from the shimming coils since they do not have permanent shimming as in a regular cyclotron. The Oak Ridge Isochronous Cyclotron, ORIC, is an example of this type of accelerator. It is a 76-in., spiral-ridge, variable-energy, variable-particle cyclotron,

[29] M. S. Livingston and J. P. Blewett, "Particle Accelerators," McGraw-Hill, New York, 1962.
[30] L. H. Thomas, *Phys. Rev.* **54**, 580 (1938).
[31] E. L. Kelly, R. V. Pyle, and R. L. Thornton, *Rev. Sci. Instr.* **27**, 493 (1956).
[32] K. R. Symon, D. W. Kerst, L. W. Jones, L. J. Laslett, and K. M. Terwilliger, *Phys. Rev.* **103**, 1837 (1956).

which accelerates H⁺ ions to 80 MeV. Its E/A vs. A capabilities are shown in the global plot given in Fig. 6. It reaches higher energies for H⁺ than does the 25-MV folded tandem, but it falls off more quickly with A. One of the other early machines is the 88-in. cyclotron at Berkeley, which can accelerate H⁺ ions to 50 MeV.

The most recent improvement in cyclotrons has followed the development of superconducting coil technology[33-36] in the 1960s and early 1970s. Superconducting cyclotrons are being built with a superconducting main coil combined with a room temperature cyclotron (trim coils, RF system, vacuum tank, ion source, dees, etc. are at room temperature). The cyclotron design is essentially the same as a normal spiral ridge or isochronous cyclotron but with a magnetic field approximately three times greater and a maximum energy approximately nine times greater for the same size machine. A very high energy superconducting cyclotron has been built by Blosser[37] at Michigan State University with an energy constant of $500(q^2/A)$ MeV. Figure 7 compares the performance of the accelerator labeled MSU Phase I 500 with the Oak Ridge 25-MV folded tandem labeled Oak Ridge Phase I 25 MV. This cyclotron has much higher energies for $A < 100$ than the 25-MV tandem and much lower energies for $A > 100$. Both accelerators cross the Coulomb barrier height curve for $A + U$ at $A = 100$. The MSU I 500 cyclotron is a 52 in. cyclotron and has a 10 ft overall diameter. Its performance can be compared to the room temperature ORIC cyclotron by noting the ORIC curve in Fig. 6. The ratio of energy constants of the two machines is $500/90 \simeq 5.5$ For example, for mass 80 the MSU I 500 curve falls at ~ 6 MeV/A and the ORIC curve falls at ~ 1 MeV/A.

1.2.2.4. Linear Accelerators (Linacs). The concept of the repeated application of a relatively small voltage to achieve a large acceleration was first demonstrated by Wideröe[38] in 1928 with a two-stage linear accelerator (linac). A linac consists of a string of conducting cylinders separated by an acceleration gap. An AC voltage is placed on the cylinders such that at each gap the particle sees a potential step that accelerates it and, while in each conducting cylinder, it drifts with constant velocity to the next gap. Wideröe accelerated positive ions to about 70 keV in what

---

[33] R. E. Berg, MSU report MSUCP-14 (1963).

[34] C. B. Bigham, J. S. Fraser, and H. R. Schneider, CRNL Report AECL-4654 (1973).

[35] D. J. Clark, *Proc. Quebec Heavy-Ion Accelerator Symp., McGill Univ.* (1973).

[36] Studio del progetto di un circlotrone superconduttore per ioni pesanti, Instituto nazionale di fisica nucleare (1976).

[37] H. G. Blosser, *VII International Conference on Cyclotrons*, Birkhauser Verlag, p. 584 (1975).

[38] R. Wideröe, *Arch. Elektrotech.* **21**, 387 (1928).

1.2. ACCELERATED IONS FOR ATOMIC COLLISIONS

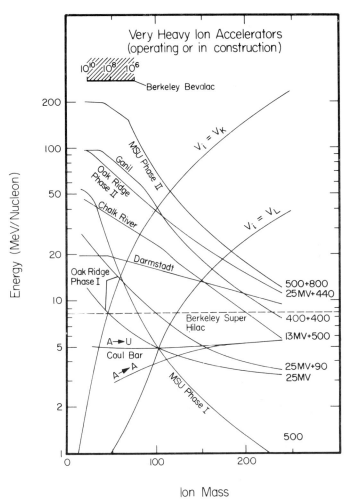

FIG. 7. Global plots of the energy per nucleon (MeV/A) vs. the ion mass (A) are given for several accelerators. The curves are labeled according to either their location or their accelerator name: MSU stands for Michigan State University, Ganil is the French project, etc. The numbers at the right indicate the type of accelerator (e.g., 13 MV refers to a 13-MV tandem; 400 refers to the energy constant of a cyclotron; the Berkeley Super Hilac and the Darmstadt accelerators are linear accelerators). The Coulomb barriers of $A + A$ and $A + U$ are indicated. The curves labeled $V_i \sim V_K$ and $V_i \sim V_L$ correspond to ion-beam energies for which the ion velocities are equal to the average K-shell and L-shell orbital velocities, respectively, for $A \to A$ collisions.

is perhaps the first particle accelerator. Sloan and Lawrence[39] in 1931 built a linac consisting of 30 acceleration gaps each with 42 kV. They accelerated Hg$^+$ ions to 1.2 MeV with this accelerator, but were unable to accelerate light ions to high velocities. This accelerator is the first heavy-ion accelerator. Sloan and Coates[40] in 1934 reported on a 36-gap, 79-kV linac in which Hg$^+$ ions up to 2.8 MeV were obtained. The success of the cyclotron in the early 1930s coupled with lack of high-power high-frequency RF sources prevented the exploitation of the linac for many years. The methods of producing linacs from large waveguides that have axial electric fields to provide the accelerating force were developed in the 1940s and 1950s (see L. Smith[41] for a review of these methods). Both proton and heavy-ion linacs have been built up to tens of millions of electron volts. Since protons and heavy ions travel much more slowly than the speed of light, it is necessary to have drift tubes to shield them from that part of the RF not suitable for acceleration. One of the first modern-day proton linacs is the 32-MeV accelerator Alvarez *et al.*[42] designed and built at the University of California. Heavy-ion linacs somewhat similar in design to proton accelerators are in operation at the Lawrence Radiation Laboratory, Berkeley, which recently achieved energies of 8.5 MeV/amu, and at the GSI laboratory in Darmstadt, West Germany, with energies ranging from 20 MeV/amu for light ions to 10 MeV/amu for very heavy ions. The performance curves of these accelerators are given in Fig. 7 and are labeled Berkeley Super Hilac and Darmstadt, respectively. Linac RF cavities are currently made from a vacuum tank structure of copper-covered steel and have a startling beauty, as can be witnessed by the interior view of the GSI accelerator given in Fig. 8.

The most recent success in linacs has been demonstrated by Bollinger[43] at Argonne National Laboratory, where he has developed a prototype precision heavy-ion superconducting linac. The accelerator consists of a series of split-ring resonators,[44] which are three-gap structures made of superconducting niobium. An average accelerating field of ~ 3.3 MV/m can be accomplished in a 35 cm resonator of less than 1 m diame-

---

[39] D. H. Sloane and E. O. Lawrence, *Phys. Rev.* **38**, 2021 (1931).

[40] D. H. Sloane and W. M. Coates, *Phys. Rev.* **46**, 542 (1934).

[41] L. Smith "Handbuch der Physik" (S. Flügge, ed.) Vol. 44, pp. 341–388. Springer, Berlin (1959).

[42] L. W. Alvarez *et al.*, *Rev. Sci. Instr.* **26**, 111 (1955).

[43] L. M. Bollinger *et al.*, *Proc. Proton Lin. Acc. Conf.* (AECL-5677 Chalk River, 1976) p. 95; L. M. Bollinger, *IEEE Trans. Nucl. Sci.* **NS-24**, 1076 (1977).

[44] The split-ring resonator was developed at Cal Tech. K. W. Shepard, J. E. Mercereou, and G. J. Dick, *IEEE Trans. Nucl. Sci.* **NS-22**, 1179 (1975); **NS-24**, 1147 (1977).

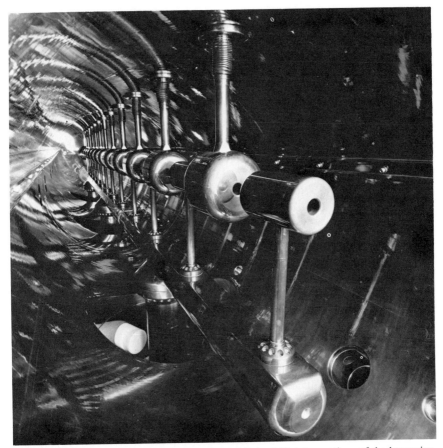

FIG. 8. The copper-clad steel tank together with the drift tube sections of the heavy-ion linear accelerator at GSI in Darmstadt are pictured in their cold beauty. The performance curve of this accelerator is given in Fig. 7 in comparison with other heavy-ion accelerators (reprinted with permission from GSI).

ter with a power consumption of ~ 4 W at 97 MHz. That compares with a room temperature proton linear accelerator that operates at typically 2 MV/m in a 1-m diam cavity with a power loss of 60 kW/MV at 200 MHz. The room temperature heavy-ion linacs are ~ 3 m in diameter and operate at ~ 70 MHz. The future use and development of superconducting linacs for heavy ions appear very promising since the accelerators can in principle be linearly stacked at will.

1.2.2.5. Multiaccelerators. For many years physicists have been designing and using multi-or coupled accelerators. Cockroft–Waltons have been injected into linacs, linacs injected into cyclotrons, cyclotrons injected into tandem Van de Graaffs, tandems injected into tandems, etc. The scheme of injecting one accelerator into another for obtaining higher-velocity heavy ions over the entire mass range is one of the largest efforts presently being pursued by nuclear physicists. Figure 7 contains the E/A vs. A curves for five of these facilities, which are presently in the construction process. These heavy-ion accelerators are attracting much attention from atomic physicists due to the areas of atomic collisions and atomic spectroscopy that become available for the first time with these ion velocities. This interest is further boosted by the success that has been obtained in using the Berkeley Super Hilac, the linac at GSI in Darmstadt, and the double MP tandem facility at Brookhaven National Laboratory in doing atomic collision studies (see Section I.2.3).

The heavy-ion facilities presented in Fig. 7 include (a) the MSU $500(q^2/A)$ MeV superconducting cyclotron injecting into a $800(q^2/A)$ MeV superconducting cyclotron, (b) the 25-MV Oak Ridge folded tandem injected into a $440(q^2/A)$ MeV cyclotron, (c) the French facility GANIL, with coupled $400(q/A^2)$ cyclotrons, (d) the Chalk River 13-MV tandem injected into a $500(q^2/A)$-MeV cyclotron, and (e) the Oak Ridge 25-MV folded tandem injected into the ORIC cyclotron $90(q^2/A)$. The relative merits of these machines in terms of the total energy per nucleon can be readily determined from the figure.

A very ambitious and interesting heavy-ion accelerator project[45] under study at the Lawrence Berkeley Laboratory at present is a system of two superconducting relativistic ion synchrotron accelerator storage rings. The accelerators can supply ion beams through uranium with energies from (a) 40 MeV/A to 20 GeV/A for the fixed-target-operation mode, and (b) 20 GeV/A to 2 TeV/A fixed-target equivalent energy for the colliding beam mode. This accelerator will use the Super Hilac as an injector. Figure 9 shows a global plot of the higher MeV/A facilities in comparison with some of the accelerators discussed previously. High-energy heavy-ion accelerators in Russia, Japan, France, and Germany are also included in this figure. The characteristic physics associated with the various energy regimes is indicated at the right.

The accelerators discussed in this section are the heavy-ion accelerators of the 1980s and perhaps the 1990s. Considering that the first models of all four original accelerators were in the tens of kilovolts to one million electron volt range with the researchers struggling to meet energies adequate to allow protons to penetrate at least medium-mass nuclei, the wishes of Rutherford to replace the α source with an artificial source of swift ions has been exceeded many times in the last 50 years.

---

[45] The Venus Project, LBL PUB-5025 (1979).

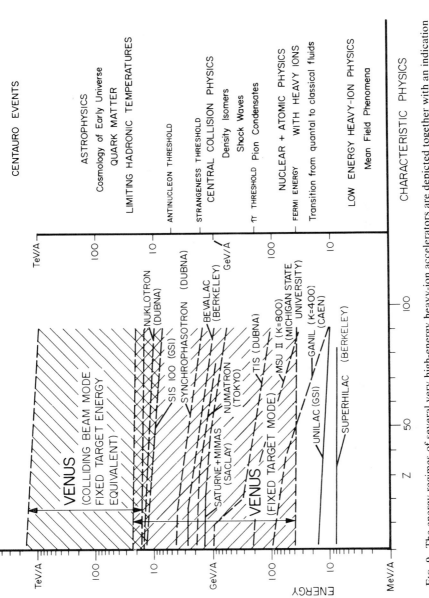

FIG. 9. The energy regimes of several very high-energy heavy-ion accelerators are depicted together with an indication of the characteristic physics in the different regimes. Venus is the acronym for the LBL variable energy nuclear synchrotron discussed in the text (reprinted with permission from Lawrence Berkeley Laboratory).

1.2.3. Use of Accelerated Ions for Atomic Collisions

The scattering of alpha particles from foils in the original experiments of Rutherford[9,11] and Geiger and Marsden[10,12] in the establishing of the nuclear model of the atom was as much atomic collision physics as it was nuclear physics. In fact it was only two years later that Bohr[46] in 1913 correctly predicted the one-electron spectra of ions by making Rutherford's nucleus a positive charge and allowing electrons to exist in quantized orbits around the nucleus. Most of the atomic spectroscopy of this era was done without the use of alpha rays, however. The fundamental problems of ionization of matter by charged particles, the electron loss (ionization) and electron gain (electron capture) of fast ions moving in matter, average charge distributions of and energy loss of ions after traversing thin materials were all undertaken with the use of alpha particles, particularly those of Po, RaC', and ThC'. Most of these studies were done between 1910 and 1935, with some of the pioneering work being done by Bohr,[47] Henderson,[48] Rutherford,[49] Briggs,[50] Kapitza,[51] and Blackett.[52] Quantum-mechanical treatments of these processes were undertaken in the early 1930s with the work of Bethe,[53] Bhadha,[54] Möller,[55] Bloch,[56] Mott,[57] Williams,[58] Livingston and Bethe,[59] and Henneberg.[60]

When the first accelerators were developed in the early 1930s, physicists were eager to explore the relatively unknown region of the atom, namely, the nucleus. Prior to this time atomic physics was the center of activity in physics. This situation changed in the late 1930s since many of the innovative atomic physicists turned their attention and talents toward

---

[46] N. Bohr, *Phil. Mag.* **26**, 1 (1913); **26**, 476 (1913).
[47] N. Bohr, *Phil. Mag.* **25**, 10 (1913); **30**, 581 (1915).
[48] G. H. Henderson, *Proc. Roy. Soc. (London) A* **102**, 496 (1922); *A* **109**, 157 (1925).
[49] E. Rutherford, *Phil. Mag.* **47**, 277 (1924).
[50] G. H. Briggs, *Proc. Roy. Soc. (London) A* **114**, 341 (1927).
[51] P. Kapitza, *Proc. Roy. Soc. (London) A* **106**, 602 (1924).
[52] P. M. S. Blackett, *Proc. Roy. Soc. (London) A* **135**, 132 (1932).
[53] H. A. Bethe, *Ann. Phys.* **5**, 325 (1930); *Z. Phys.* **76**, 293 (1932).
[54] H. J. Bhadha, *Proc. Roy. Soc. (London) A* **164**, 257 (1938).
[55] C. Möller, *Ann. Phys.* **14**, 531 (1932).
[56] F. Bloch, *Ann. Phys.* **16**, 285 (1933); *Z. Phys.* **81**, 363 (1933).
[57] N. F. Mott, Proc. Roy. Soc. *(London) A* **124**, 425 (1929); *A* **135**, 429 (1932); *Proc. Cambridge Phil. Soc.* **27**, 255 (1931).
[58] E. J. Williams, *Proc. Roy. Soc. (London) A* **135**, 108 (1932).
[59] M. S. Livingston and H. A. Bethe, *Rev. Mod. Phys.* **9**, 245 (1937).
[60] W. Henneberg, *Z. Phys.* **86**, 592 (1933).

the nucleus.[61] One of the exceptions to the early nuclear physics experiments performed on accelerators is the beautiful work by Coates,[62] who used the 2.8-MeV Hg$^+$ ions produced by the Berkeley linac[39,40] to study the atomic ionization of targets as a function of ion energy and target Z. This is the first heavy-ion atomic physics experiment using the accelerated-ion method.

During the 1940s atomic physics lost some momentum in the United States due to large efforts by many atomic physicists directed toward the development of electronics and radar. In addition many physicists felt that the major discoveries of atomic physics had already been made. It was not an unpopular belief that all of the electronic physics of atoms could be calculated by knowledge of the Schrödinger equation, the electromagnetic forces, and symmetry properties of electrons. The only remaining things were tedious calculations required to get a desired answer.

Many atomic physics studies in the area of collision physics or accelerated ion physics were kept somewhat vital due to the needs of the nuclear collision work with accelerators. Knowledge of methods for producing positive and negative ions as well as molecules was important in terms of making better ion sources for accelerators. This is still a very important area of research, in particular for the production of very high charge state ions. At present this is of interest from the atomic physics viewpoint. The interactions of highly charged, slowly moving ions with matter is of experimental, theoretical, and practical interest. The practical interest stems from the attempts to understand atomic processes in hot plasmas. One of the vanguard efforts at present is the development of the electron beam ion source (IBIS) and the CRYEBIS source, which hold the promise of creating low-velocity bare ions of medium-Z elements.[63,64]

Other accelerator-related atomic collision studies include the need for obtaining better detection devices for high-energy particles, the need for a better understanding of the energy loss of heavy ions in matter, and the need for studies of charge state equilibrium at high-ion velocities and for high-Z ions. Many of these studies became and are still closely related to solid-state studies with accelerated ions.

In the late 1950s and early 1960s many interesting developments oc-

---

[61] For a brief discussion of the history of atomic and molecular physics the reader is referred to B. Bederson *et al.*, "Atomic and Molecular Physics," Report to the National Academy of Sciences, Washington, D.C. (1971).

[62] W. M. Coates, *Phys. Rev.* **46**, 542 (1934).

[63] E. D. Donets and V. P. Ovsyannikov, JINR P7-10438 Dubna (1977).

[64] J. Arianer and Ch. Goldstein, IPNO-70-02 Orsay (1979).

curred, which seem to have caused a gradual but steady rise in the study of the accelerated-ion method and the type of physics one obtains from the strong interaction between ions and atoms of comparable nuclear charges. One of the first of these types of studies was performed interestingly enough with radioactive sources. Armbruster[65] and Specht[66] studied the X-ray emission from high-energy heavy-ion–atom collisions using the energetic fission fragments from fission sources. The perturbation techniques that describe the vacancy production by light ions are not successful in describing the results obtained from these experiments. At about the same time as these experiments were being performed, Afrosimov et al.[67] and Everhart and Kessel[68] developed the methods of studying inelastic collisions by energy loss measurements for single collisions. Fano and Lichten[69] interpreted the results of these experiments in 1965 in terms of inner-shell excitations created by a mechanism in which the colliding ions form a diatomic molecule whose internuclear separation varies during the collision. The inner-shell excitation occurs by a promotion of electrons from a lower molecular orbit to a higher molecular orbit at some internuclear distance. As the molecule breaks apart the promoted electron ends up in a higher atomic orbit of the separated atom. This mechanism predicts a large energy transfer to inner-shell electrons by relatively low velocity collisions. The third development taking place at this time is the bringing on-line of many heavy-ion accelerators with the capability of producing very highly charged ions for atomic spectroscopy, and a controlled source of accelerated ions for doing energy-dependent, target Z and projectile Z studies of collision processes. Processes penetrating to the inner shells of ions become more probable as the energy is increased. One immediate question is what type of global curve might define the high-energy region for atomic collisions. One criterion that can be used to explore this question is to assume that the interaction between a target electron and an incident ion maximizes when the average orbital velocity of that electron equals the velocity of the incident ion. The curve satisfying this criterion can be plotted on the global plots presented in the previous section. In Fig. 7, which contains the performance of many high-energy heavy-ion accelerators, and the nuclear Coulomb barrier for $A + A$ and $A + U$, is plotted the approximate curves for $A \rightarrow A$ colli-

---

[65] P. Armbruster, Z. Phys. **166**, 341 (1962); P. Armbruster, E. Roeckl, H. J. Specht, and A. Vollmer, Z. Naturforsh **19A**, 1301 (1964).
[66] H. J. Specht, Z. Phys. **185**, 301 (1965).
[67] V. V. Afrosimov, Y. S. Gordeev, M. N. Panov, and N. V. Fredorenko, Zh. Tekhn. Fiz. **34**, 1613; 1624; 1637 (1964) [Soviet Phys.-Tech. Phys. **9**, 1248; 1256; 1265 (1965)].
[68] E. Everhart and Q. C. Kessel, Phys. Rev. Lett. **14**, 247 (1965).
[69] U. Fano and W. Lichten, Phys. Rev. Lett. **14**, 627 (1965).

sions in which the velocity of the ion A is equal to the average orbital velocity of the K-shell and L-shell electrons. At first glance it is startling that even the highest energy accelerators cannot reach velocities sufficient to satisfy our proposed criterion for maximizing the ion–target K-shell interaction for $A > 100$. Likewise the ion–target L-shell interaction is not maximized for $A > 175$ using this criterion. This is not to say that K-shell excitations of the higher A masses cannot be produced. However, they will not be produced with the largest possible cross section. From what is known about systems much lighter than $A = 100$, we do expect a great enhancement in the K-shell excitation at the lower velocities from the Fano–Lichten model. These cross sections, however, do not exceed those corresponding to the maximum excitations due to a direct Coulomb excitation mechanism. What happens in these kinds of collisions is pure conjecture and is therefore not discussed in the remainder of this volume. However, at present it provides the impetus enabling atomic collision physicists to understand the lighter systems at lower velocities and from there to continue to push the frontiers of the accelerated-ion method to a global understanding of atomic collisions.

2. ENERGY LOSS OF HIGH-VELOCITY IONS IN MATTER

By Wei-Kan Chu

2.1. Introduction

When high-velocity ions penetrate a material, they interact with the target atoms along their trajectories and lose energy to them by several mechanisms. One mechanism transfers energy from the moving ions to the target electrons through excitation and ionization of the target atoms. Another mechanism transfers energy from the ions to the target nuclei through momentum transfer.

Energy loss and scattering have been an important source of information on the constitution of atoms. Theoretical and experimental work on this subject have been going on since the beginning of the century. Information on energy loss is needed in many experiments in atomic and nuclear physics: to determine the energy and mass of a given nuclear reaction product in a cloud chamber or an emulsion, to correct the energy of an ion beam that has passed through thin windows or materials in a given experiment, to design radiation shielding, etc.

More recently, the various applications of ion beams in material study (for example, range, range distribution, and radiation damage in ion implantation; ion sputtering; ion backscattering; and ion-induced excitation) have led to further needs for knowledge of the energy loss of ions in matter. Most experimental work with ion beams requires information on energy loss. Such information, then, is basic to experimental physics.

This part is addressed to experimentalists who (1) want a general background on energy loss, (2) need energy loss information for their experiments, or (3) want to measure energy loss. With these three objectives in mind, we start with the general concepts of energy loss, a few theories, some experimental verification of theories as background material, some rules for interpolation and extrapolation, and a few references on existing compilations of information for those who need to use specific values of energy loss. In the last half of this part we emphasize various experimental methods of measuring energy loss, the trade-offs among them, and the accuracy and applicable region of each.

There have been many publications on ion beam penetration. Sigmund[1] has written an excellent review article on the subject, with a list of 219 papers and a good extraction of current theories and formulas. Andersen[2] has made a compilation of papers on experimental range and energy loss. He has assembled a bibliography containing over 800 titles, cross-indexed for species as related to projectiles as well as targets so that publications relevant to any ion–target combination can be retrieved easily.

2.2. General Concepts of Energy Loss

There are several energy loss mechanisms for moving ions in a target material:

(1) *Excitation and ionization.* This is the principal mechanism for the energy loss of ions at high velocities. Moving ions transfer their energies to the target electrons and thus promote some of the target atoms into excited or ionized states. Energy loss due to excitation and ionization is also called electronic energy loss, or inelastic energy loss.

(2) *Nuclear collisions.* This is the major mechanism for energy loss of projectiles at low velocities. Projectiles transfer their energies to the target nuclei by elastic collisions, and consequently the target atom recoils. Energy loss due to a nuclear collision is also called nuclear energy loss or elastic energy loss.

(3) *Generation of photons.* This mechanism is significant only at relativistic velocities. Projectiles emit photons because of deceleration in the medium (bremsstrahlung).

(4) *Nuclear reactions.* For certain very specific combinations of projectile, energy, and target, a nuclear reaction can be induced.

Here we concentrate on the first two of these mechanisms.

Microscopically, energy loss due to excitation, ionization, or nuclear motion is a discrete process. Macroscopically, however, it is a good assumption that the moving ions lose energy continuously. All we are concerned with here is the total collective effect of the energy attenuation during the penetration of ions into a given material. Individual effects due to a single collision are treated in separate chapters.

---

[1] P. Sigmund, *in* "Radiation Damage Processes in Materials" (*Proc. Radiat. Damage Processes Mater., Aleria, Corsica, France, August 27 to September 9, 1973*), pp. 3–118. Noordhoff, Leyden, 1975.

[2] H. H. Andersen, "Bibliography and Index of Experimental Range and Stopping Power Data." Pergamon Press, New York, 1977.

2.2. GENERAL CONCEPTS OF ENERGY LOSS

A good assumption, which has been implied so far, is that electronic energy loss and nuclear energy loss are not correlated, and therefore can be treated separately and independently. To measure energy loss, we must determine two quantities: the distance Δx that the ions traverse in the target, and the energy loss ΔE in this distance. The mass density ρ or the atomic density N are frequently combined with the distance, in the form $\rho \, \Delta x$ or $N \, \Delta x$, to express the amount of material per unit area or the number of atoms per unit area that the projectiles have traversed in losing energy ΔE to the target material. The atomic density N is related to the mass density ρ by Avogadro's number N_0 and to the mass number of the target M_2 by

$$N \equiv N_0 \rho / M_2. \tag{2.2.1}$$

Energy loss can be expressed in several different ways. Some frequently used units are

$$\begin{aligned} dE/dx: &\quad \text{eV/Å, MeV/cm,} \\ dE/\rho \, dx: &\quad \text{eV/}(\mu\text{g/cm}^2), \text{keV/}(\mu\text{g/cm}^2), \text{MeV/}(\mu\text{g/cm}^2), \\ dE/N \, dx: &\quad \text{eV/(atoms/cm}^2), \text{eV-cm}^2. \end{aligned} \tag{2.2.2}$$

In the literature, and especially in some of the earlier experimental work, many different units derived from the above three forms have appeared. In many publications all three are called energy loss or dE/dx. One has to translate the units carefully when comparing one measurement to another.

Recently, most authors have adopted $dE/N \, dx$ (eV-cm^2) as the stopping cross section. Experimental physicists formerly used ϵ for this term, that is,

$$\epsilon \equiv \frac{1}{N} \frac{dE}{dx} \quad \text{(eV-cm}^2\text{).} \tag{2.2.3}$$

Theoretical physicists tend to prefer S, for example,

$$S \equiv \frac{\langle \Delta E \rangle}{N \, \Delta x} = \sum_i T_i P_i = \int T \, d\sigma, \tag{2.2.4}$$

where $\langle \Delta E \rangle$ is the average energy lost to a target of thickness $N \, \Delta x$, T_i the kinetic energy transferred to the ith electron with probability P_i, and the integral extends over all possible energy losses in individual collisions. The last term of Eq. (2.2.4) describes the cross section $d\sigma$ of the energy loss (stopping process), which therefore is called the stopping cross section. Thus Eqs. (2.2.3) and (2.2.4) have the same dimensions and the same meaning.

The advantage of using the stopping cross section (ϵ or S) rather than dE/dx is obvious. Especially in a systematic study, ϵ gives a description of energy loss on an atom-to-atom basis, which permits convenient extrapolation, whereas dE/dx changes for different materials, or even for the same material at different densities.

The earliest theoretical development on energy loss was done by Bohr.[3,4] His theory gives a quantitative account of the essential features of the process. Quantum-mechanical treatments later confirmed his analysis. Bethe[5-7] developed the theory further. Lindhard and co-workers[8-10] and Firsov[11,12] have contributed to the understanding of energy loss of ions at low velocity.

Figure 1 gives a schematic diagram of the proton-stopping cross section in silicon. The selection of projectile and target is arbitrary. Other selections will produce similar curves, with some differences in shape and scale. Some of the names and terms in Fig. 1 have not yet been defined or described here; they are presented in Fig. 1 for the purpose of relating each of the various theories to its region of applicability.

Figure 1 covers a very broad range of energies. It starts from a fraction of a kiloelectron volt to many gigaelectron volts. This curve has a peak at about 100 keV and a dip around 500 GeV. The increase in energy loss at relativistic velocities is due to photon emission, bremsstrahlung, and Cerenkov radiation. In this study we focus our attention on the nonrelativistic energy region.

In Fig. 1, the nuclear energy loss is small when compared to the electronic energy loss, even at very low energy. For example, according to Lindhard *et al.*[9,13] (LSS theory), nuclear stopping accounts for 2% of the total stopping for protons in silicon at 10 keV, and 16% for 1 keV. In studies of energy loss, therefore, nuclear energy loss can be ignored, especially for light ions at medium and high velocities. Nuclear collisions,

---

[3] N. Bohr, *Phil. Mag.* **25**, (6) 10 (1913).
[4] N. Bohr, *Mat. Fys. Medd. Dan. Vid. Selsk.* **18**, No. 8 (1948).
[5] H. A. Bethe, *Ann. Phys.* **5** (5), 325 (1930).
[6] H. A. Bethe, *Z. Phys.* **76**, 293 (1932).
[7] H. A. Bethe, *Phys. Rev.* **89**, 1256 (1953).
[8] J. Lindhard, *Mat. Fys. Medd. Dan. Vid. Selsk.* **28**, No. 8 (1954).
[9] J. Lindhard and M. Scharff, *Phys. Rev.* **124**, 128 (1961).
[10] J. Lindhard, *Mat. Fys. Medd. Dan. Vid. Selsk.* **34**, No. 4 (1965).
[11] O. B. Firsov, *Zh. Eksp. Teor. Fiz.* **32**, 1464 [Engl. transl., *Sov. Phys.—JETP* **5**, 1192 (1957)] O. B. Firsov, *Zh. Eksp. Teor. Fiz.* **33**, 696 [Engl. transl., *Sov. Phys.—JETP* **6**, 534 (1958)].
[12] O. B. Firsov, *Zh. Eksp. Teor. Fiz.* **36**, 1517 [Engl. transl., *Sov. Phys.—JETP* **9**, 1076 (1959)].
[13] J. Lindhard, M. Scharff, and H. E. Schiøtt, *Mat. Fys. Medd. Dan. Vid. Selsk.* **33**, No. 14 (1963).

2.2. GENERAL CONCEPTS OF ENERGY LOSS

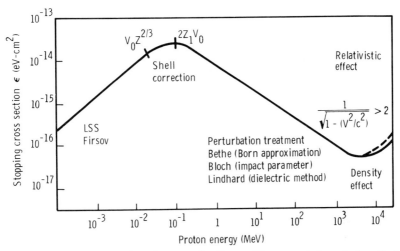

FIG. 1. Stopping cross section of protons in silicon. The general shape is described by various theories for various energy regions.

however, produce different effects in targets. In studying radiation damage, for example, one should focus on the nuclear stopping cross section. This part has a separate chapter on this subject.

For ions heavier than protons, the energy loss behavior is very similar to that of protons. In Fig. 2 we repeat the curve for proton energy loss in

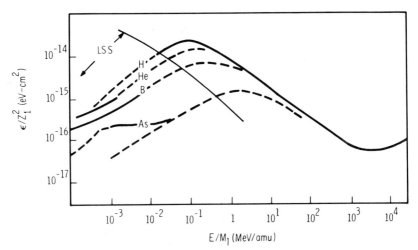

FIG. 2. Family of curves of stopping cross section for heavy ions in silicon. Based on Northcliffe and Schilling's tabulation (1970).[14]

silicon, simply changing the coordinates to ϵ/Z_1^2 for the normalized (or scaled) stopping cross section and to E/M_1 for the velocity parameter. For protons as projectiles, $Z_1 = M_1 = 1$; the curve is the same in Fig. 2 as in Fig. 1. One might object to the use of E/M_1 at relativistic velocities, but that is a minor detail. By presenting Fig. 2 with the normalized parameters, we can see several trends in the stopping cross section as a function of the projectile. The values of stopping cross sections in Figs. 1 and 2 are obtained from Northcliffe and Schilling.[14] The several features one can observe from Fig. 2 are:

(1) At high velocities, ϵ/Z_1^2 tends to converge to a single curve.

(2) For higher Z_1, the position of the peak in ϵ/Z_1^2 occurs at a higher velocity, and its location scales with Z_1.

(3) At lower velocities, the curves in the ϵ/Z_1^2 family are nearly parallel.

(4) At very low velocities, the nuclear stopping cross section begins to influence the curves. The larger the mass number of the projectile, the larger the influence.

These observations are purely empirical and phenomenological. However, various theories for various velocity regions give detailed descriptions of the behavior of stopping cross section as a function of the projectile, its velocity, and the target atoms.

We start our analysis with light ions at high velocities, where many of the ion beam analyses are applied. We then discuss the low-velocity region, where most of the ion implantation work is done and where nuclear stopping plays a role.

2.3. Energy Loss Theories: General Background

In this chapter, we briefly describe various theoretical treatments of energy loss. We keep the discussion at a very elementary level in order to provide the general background needed in extracting information on energy loss. Those interested in detailed theoretical treatments are referred to the excellent review articles by Sigmund.[1] When a charged particle penetrates a material and collides with a target nucleus, the projectile deflects and the target nucleus recoils. This deflection phenomenon is well described by Rutherford's scattering law, which provides an accurate description of large-angle scattering, but these so-called single Coulomb scatterings are very rare events, and are by no means the majority of the projectile scattering events. Most of the projectiles pass nearby multi-

---

[14] L. C. Northcliffe and R. F. Schilling, *Nucl. Data Tables* **7A**, 233 (1970).

2.3. ENERGY LOSS THEORIES: GENERAL BACKGROUND

tudes of atoms and interact with the circumnuclear electrons as well as with the atomic nucleus. At high velocities, as we shall see, almost all of the energy loss of the projectiles results from projectile–electron encounters, and almost all of the scattering deflection from projectile–nuclear encounters.

The theoretical treatments of inelastic collisions of charged particles with target atoms or molecules deal with fast collisions and/or with slow collisions. The criterion used in making this separation is the velocity of the projectile relative to the mean orbital velocity of the atomic or molecular electrons in the shell or subshell of a given target atom, for which the inelastic process is being considered. When the projectile velocity v is much greater than that of an orbital electron (fast-collision case), the influence of the incident particle on an atom may be regarded as a sudden, small external perturbation. This picture leads to Bohr's theory of stopping power.[3,4] The collision produces a sudden transfer of energy from the projectile to the target electron. The energy loss of a fast particle to a stationary nucleus or electron can be calculated from Rutherford scattering.

2.3.1. Coulomb Scattering and Energy Loss

When a projectile of mass M_1 with atomic number Z_1 at velocity v collides with a target of mass M_2 and atomic number Z_2, the projectile is deflected because of the Coulomb interaction, and the well-defined Kepler motion gives

$$\tan(\theta/2) = b/2p, \qquad (2.3.1)$$

where θ is the scattering angle in the center of mass system, p the impact parameter, and b the collision diameter, which is the distance of closest approach on a head-on collision:

$$Z_1 Z_2 e^2 / b = \tfrac{1}{2} M_0 v^2. \qquad (2.3.2)$$

Here M_0 is the reduced mass of the scattering system:

$$M_0 = M_1 M_2 / (M_1 + M_2). \qquad (2.3.3)$$

From conservation of energy and momentum, we derive maximum energy transfer to the target particle for a head-on collision as

$$T_{\max} = \frac{4 M_1 M_2}{(M_1 + M_2)^2} E \qquad (2.3.4)$$

and the energy transfer for a general case is

$$T = T_{\max} \sin^2(\theta/2). \qquad (2.3.5)$$

2. ENERGY LOSS OF HIGH-VELOCITY IONS IN MATTER

With Eqs. (2.3.1) and (2.3.5) the energy transfer becomes

$$T = \frac{T_{\max}}{1 + (2p/b)^2}. \qquad (2.3.6)$$

The differential scattering cross section can be written as

$$d\sigma = 2\pi p \, dp. \qquad (2.3.7)$$

Substituting Eqs. (2.3.6) and (2.3.7) into Eq. (2.2.4), we express the energy lost to the nuclei as

$$S_n = \int T \, d\sigma = T_{\max} 2\pi \int_{p_{\max}}^{p_{\max}} \frac{p \, dp}{1 + (2p/b)^2}. \qquad (2.3.8)$$

From Eq. (2.3.6), this equation can be written as

$$S_n = T_{\max}(b/2)^2 \pi \int_{T_{\min}}^{T_{\max}} dT/T \qquad (2.3.9)$$

and therefore

$$S_n = \frac{2\pi Z_1^2 Z_2^2 e^4}{M_2 v^2} \ln \frac{T_{\max}}{T_{\min}} \qquad (2.3.10)$$

The low-energy limit $T_{\min}$, which corresponds to $p_{\max}$, has been arbitrarily introduced to avoid divergence in the integration. At a very great distance, the nuclei are shielded by the electrons, and the interaction potential between the projectiles and the target nuclei is no longer $1/r^2$. For the interaction between the projectile and target electrons, no energy whatever will be transferred, if this energy is less than the ionization or excitation of the electron.

When the derivation is repeated for the electrons, the mass of the target electron becomes m rather than M_2 and the charge of the electron is 1 rather than Z_2; Eq. (2.3.10) then gives

$$S_e(\text{per electron}) = \frac{2\pi Z_1^2 e^4}{mv^2} \ln \frac{T_{\max}}{T_{\min}}. \qquad (2.3.11)$$

Because there are Z_2 electrons per target atom, the electronic energy loss per atom will be Z_2 times Eq. (2.3.11):

$$S_e = \frac{2\pi Z_1^2 Z_2 e^4}{mv^2} \ln \frac{T_{\max}}{T_{\min}}. \qquad (2.3.12)$$

The nuclear energy loss is much less than the electronic energy loss, because the ratio of Eq. (2.3.12) to Eq. (2.3.10) gives

$$\frac{S_n}{S_e} = \frac{Z_2 m}{M_2} \approx \frac{1}{2} \frac{\text{mass of electron}}{\text{mass of proton}} \approx \frac{1}{4000}. \qquad (2.3.13)$$

2.3. ENERGY LOSS THEORIES: GENERAL BACKGROUND

Equation (2.3.12) can be derived in many different ways. One necessary condition is that the projectiles must be moving much faster than the target electrons, which therefore can be treated as stationary. When the projectile velocity is compared with that of the most loosely bound target electrons, we get

$$v \gg v_0 = \frac{e^2}{\hbar} = \frac{c}{137}. \tag{2.3.14}$$

In the opposite case, if the projectiles are slow enough, all the electrons can adjust their orbital motion in accordance with the instantaneous positions of the projectiles, and therefore are expected to absorb comparatively little energy from them during a scattering event. Hence less electronic energy is lost at low energies (Figs. 1 and 2). This is a part of the adiabaticity argument of Bohr.[3] Other considerations—collision time, electron revolutional frequencies, and so on—complicate the matter further.

Rutherford scattering can provide only a crude estimate of electronic energy loss. It is not a sufficient basis for electron-stopping theories, whose evolution is briefly outlined in the next section.

2.3.2. Perturbation Method and Dielectric Description

A very good approximation is made for the case in which projectiles do not pass close to the target nucleus. One can assume that the direction of motion and the speed of the projectiles are essentially unchanged by the soft collision, and that neither the atomic nucleus nor its surrounding electrons move appreciably while the projectile is traversing the target atom. The momentum transferred from the projectile to the electron is perpendicular to the momentum of the projectile. From this transfer one can calculate the energy lost from the projectile to the target electrons; the result of such calculations leads to a formula similar to Eq. (2.3.12).

The energy loss problem can also be approached from a dielectric description that was suggested by Fermi[15] and developed by Kramers[16] and Lindhard.[10] The description starts with a point charge moving with a constant velocity, producing an electric field. The force $\mathbf{F}$ acting on the projectile when it is traversing a medium is dE/dx and is given by

$$\mathbf{F} = Z_1 e(\mathbf{E}_{\text{medium}} - \mathbf{E}_{\text{vacuum}}), \tag{2.3.15}$$

where $\mathbf{E}$ is the electric field of the moving point charge.

---

[15] E. Fermi, *Z. Phys.* **29**, 315 (1924).
[16] H. A. Kramers, *Physica* **13**, 401 (1947).

Both the perturbation treatment and the dielectric treatment lead to a formula very similar to Bethe's formula discussed in the next section.

2.3.3. Bethe Formula

Most of the classical treatment of energy loss is based on a well-defined physical length, e.g., impact parameter p or collision diameter b ($=2Z_1Z_2e^2/M_0v^2$). When the latter becomes of the order of the de Broglie wavelength λ ($=\hbar/M_0v$), quantum-mechanical treatment is required. Thus the criterion for the validity of the classical treatment is $b \gg \lambda$ or

$$x \equiv b/\lambda = 2Z_1Z_2e^2/\hbar v \gg 1. \quad (2.3.16)$$

When this condition is not fulfilled, Bethe's quantum theory of electronic stopping[5-7] is derived. It is based on a plane-wave Born approximation, which is a quantum-mechanical perturbation calculation. Such a treatment leads to a nonrelativistic stopping formula:

$$S_e = \frac{4\pi Z_1^2 Z_2 e^4}{mv^2} \log \frac{2mv^2}{I}, \quad (2.3.17)$$

where I is the mean ionization and excitation energy, defined by

$$\log I = \sum_n f_n \log(E_n - E_0), \quad (2.3.18)$$

where E_0 is the ground state of the target atom, E_n the excited state, and f_n the strength of the dipole oscillator.

The complete Bethe formula for electronic stopping of high-velocity charged particles is

$$\frac{1}{N}\frac{dE}{dx} = \frac{4\pi Z_1^2 Z_2 e^4}{mv^2}\left[\log \frac{2mv^2}{I} - \log\left(1 - \frac{v^2}{c^2}\right) - \frac{v^2}{c^2} - \frac{C}{Z_2} - \frac{\delta}{2}\right]. \quad (2.3.19)$$

The two terms containing v^2/c^2 are the relativistic correction terms. C/Z_2 is a velocity-dependent term, significant only at low velocities, which is included to correct the stopping cross section for the nonparticipation of inner-shell electrons in ionization and excitation of projectiles at low velocities. The $\delta/2$ term, another correction to stopping cross section, is important only at ultrahigh velocities; it corrects for the density effect, reducing energy loss from the dashed line to the solid line in the high-velocity region of Fig. 2. This subject has been reviewed by Crispin and Fowler.[17] Both C/Z_2 and $\delta/2$ are functions of the target atom as well as the projectile velocity.

---

[17] A. Crispin and G. N. Fowler, *Rev. Mod. Phys.* **42**, 290 (1970).

2.3. ENERGY LOSS THEORIES: GENERAL BACKGROUND

Bethe's theory gives a very accurate value for the energy lost by fast projectiles in elemental targets, provided that the input values I and C/Z_2 are accurate. At the high-velocity limit, C/Z_2 becomes less important; the energy-independent parameter I can be extracted from a careful measurement of energy loss. Such measurements, as a function of energy at low energies, provide combined information on $\log I$ and C/Z_2. That is, knowing I and C/Z_2, one can extract energy loss values by the Bethe formula for fast-moving ions with very high accuracy. We briefly discuss I and C/Z_2 next.

2.3.4. Mean Excitation and Ionization Energy I

In this section we focus our attention on the mean excitation and ionization energy I of Eqs. (2.3.17) and (2.3.19). I gives the stopping property of a given target Z_2, and is some energy that target electrons can obtain from a very fast projectile. This term is a weighted average of all the excitation and ionization processes possible for a given atom. For light target atoms each of which contains very few electrons, this value can be calculated by quantum-mechanical methods. Because of the complexity of this calculation, however, I is typically determined empirically from very accurate measurements of energy loss or range.

Bloch's relation[18,19] gives an approximation, based on the Thomas–Fermi model of atoms, for estimating I for heavy elements. This relation states

$$I = Z_2 I_0 \qquad (I_0 \simeq 10 \text{ eV}). \tag{2.3.20}$$

Fano[20] and Turner[21] have reviewed both theoretical and experimental determinations of I. Chu and Powers[22,23] made a statistical calculation of I, using Lindhard and Scharff's[9,24] approach with Hartree–Fock–Slater wave functions. Their results, plus some independent calculations for a few elements and some experimental measurements, are given in Fig. 3. Oscillation of I/Z_2 is observed, and the calculated results show excellent agreement with the experimental values, especially for the region studied by Andersen et al.,[25] $Z_2 = 20$–30.

---

[18] F. Bloch, *Ann. Phys.* **16** (5), 287 (1933).
[19] F. Bloch, *Z. Phys.* **81**, 363 (1933).
[20] U. Fano, in "Studies in Penetration of Charged Particles in Matter," Nat. Acad. Sci.—Nat. Res. Council, Publ. 1133, Washington, D.C., 1964.
[21] J. E. Turner, in "Studies in Penetration of Charged Particles in Matter," Nat. Acad. Sci.—Nat. Res. Council, Publ. 1133, Washington, D.C., 1964.
[22] W. K. Chu and D. Powers, *Phys. Lett.* **38A**, 267 (1972).
[23] W. K. Chu and D. Powers, *Phys. Lett.* **40A**, 23 (1972).
[24] J. Lindhard and M. Scharff, *Mat. Fys. Medd. Dan. Vid. Selsk.* **27**, No. 15 (1953).
[25] H. H. Andersen, H. Sørensen, and P. Vajda, *Phys. Rev.* **180**, 373 (1969).

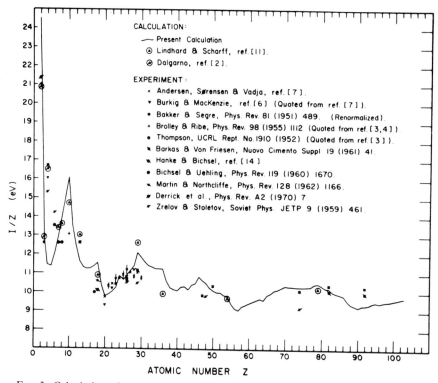

FIG. 3. Calculation of mean excitation energy by Lindhard and Scharff's theory with a Hartree–Fock–Slater charge distribution. The calculation I/Z vs. atomic number Z reveals structure, as was observed in many experimental measurements cited in the figure. From Chu and Powers (1972).[22,23]

The results given in Fig. 3 can be used as a guide for normalization and interpolation. For reliable values, I must always be determined empirically.

2.3.5. Shell Correction

Equation (2.3.19) contains a term C/Z_2 to correct for the nonparticipation of inner-shell electrons in the stopping power. When the projectile is extremely fast, all the target electrons contribute to the stopping power more predictably. When the projectile is not so fast, the inner-shell electrons contribute less to the stopping power, and the shell correction term enters in to account for the fact that a single variable I is insufficient to express the stopping problem. Since the correction is due to inner shells,

2.3. ENERGY LOSS THEORIES: GENERAL BACKGROUND

it can be expressed as

$$C/Z_2 = (C_K + C_L + \cdots)/Z_2, \qquad (2.3.21)$$

where C_K and C_L mean corrections to the K shell and the L shell stopping power. Shell corrections have been treated by Walske.[26] Empirically, a careful energy loss measurement should yield C/Z_2 and I. It is easy to see that if one ignores the density effect, a measurement of ϵ vs. E enables one to extract a single energy-dependent parameter X following Bichsel's treatment,[27] where

$$X = \log I + (C/Z_2). \qquad (2.3.22)$$

The mean excitation and ionization energy I is independent of energy; the shell correction C/Z_2 is a function of energy, whose value approaches zero at very high energies. In principle, these two boundaries make it possible to separate the two terms. In practice, any small error in experimental measurements will be amplified in the extraction of C/Z_2 and I. Once C/Z_2 and I are determined, values of ϵ vs. E can be obtained for light projectiles in a very broad energy region.

So much for the energy dependence of ϵ in the region of Bethe's formulation. In the next section we discuss the dependence of energy loss on the projectile.

2.3.6. Dependence of Energy Loss on Z_1

The basic result of the Bethe theory is that the energy loss is a property of the target medium and also a function of the projectile velocity. The nature of the projectile enters in only as a scaling factor Z_1^2 in Eqs. (2.3.17) and (2.3.19). Hence stopping power can be extrapolated from one projectile to another by the relation

$$\frac{1}{Z_A^2}\left(\frac{dE}{dx}\right)_{A,v} = \frac{1}{Z_B^2}\left(\frac{dE}{dx}\right)_{B,v}, \qquad (2.3.23)$$

where A and B are the atomic numbers of two different projectiles. Both projectiles are moving at a velocity v, which is sufficiently high that they are totally ionized, that is, are bare nuclei without electrons. Because of the Z_1^2 scaling, ϵ/Z_1^2 vs. velocity becomes a unique curve at high velocities—as we can see in Fig. 2, where all the curves merge into the proton-stopping curve. At not too high a velocity, a heavy projectile carries electrons such that the average net charge of the ion is no longer Z_1

---

[26] M. C. Walske, *Phys. Rev.* **88**, 1283 (1952); *ibid.* **101**, 940 (1956).
[27] H. Bichsel, *in* "Studies in Penetration of Charged Particles in Matter," Nat. Acad. Sci.—Nat. Res. Council, Publ. 1133, Washington, D.C., 1964.

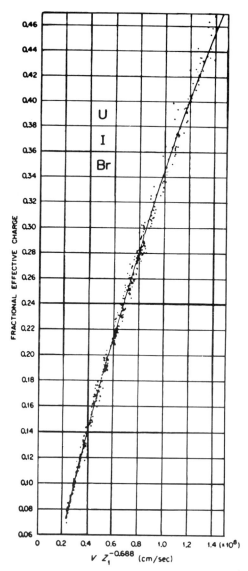

FIG. 4. Measured electronic stopping powers for uranium, iodine, and bromine ions in foils of carbon, aluminum, nickel, silver, and gold, plotted as Z^*/Z_1 vs. $Z_1^{-0.688}$, where Z^* is defined by Eq. (2.3.24), with the proton stopping power taken from Northcliffe and Schilling. The full-drawn line is a least-squares fit to Eq. (2.3.22), with $C = 1.034$ and $p = 0.688$. From Brown and Moak (1972).[28]

but Z_1^*. A generalized scaling that applies to a very broad velocity region becomes

$$\frac{1}{Z_A^{*2}} \left(\frac{dE}{dx}\right)_{A,v} = \frac{1}{Z_B^{*2}} \left(\frac{dE}{dx}\right)_{B,v}, \qquad (2.3.24)$$

where Z_A^* and Z_B^* are the effective charges of moving ions A and B traveling at velocity v, which could be considerably below the Bethe velocity region.

This type of scaling from one projectile to another has been demonstrated by Brown and Moak.[28] They measured the stopping powers for uranium, iodine, and bromine ions in foils of carbon, aluminum, nickel, silver, and gold in various velocity regions. They extracted the fractional effective charge Z_1^*/Z_1 from Eq. (2.3.24) by comparing their energy loss measurement to the published values of proton energy loss. This fractional effective charge is then plotted as a function of velocity times $Z_1^{-0.688}$. In their figure, reproduced here as Fig. 4, the solid curve is a least-squares fit to an equation representing Betz's[29] formula:

$$Z_1^*/Z_1 = 1 - C \exp(-v/v_0 Z_1 \gamma) \qquad (2.3.25)$$

where the fitting parameters $C = 1.034$ and $\gamma = 0.688$ are as obtained by Brown and Moak.[28] Recent measurements by Ward et al.[30] confirm the scaling of energy loss as given by Eq. (2.3.24). Ziegler[31] recently modified Eq. (2.3.25) by curve-fitting a large number of ion–target combinations over a wide energy region, and obtained consistent results among the measurements of energy lost by heavy ions.

2.3.7. Low-Velocity Electronic Energy Loss

At low velocities the Bethe formula does not apply to electronic stopping, because the inner-shell electrons contribute less to the stopping power. This reduction gives a very large correction. Also at very low velocities, the neutralization probability becomes so large that the collision between the projectile and the surrounding electrons is almost elastic in a reference frame moving with the ion. The energy loss then becomes proportional to the velocity of the projectile. Lindhard, Scharff, and Schiøtt[13] (abbreviated as LSS) and Firsov[11,12] gave theoretical descriptions for this energy region. The LSS expression is based on elastic scat-

---

[28] M. D. Brown and C. D. Moak, *Phys. Rev.* **B6**, 90 (1972).
[29] H. D. Betz, *Rev. Mod. Phys.* **44**, 465 (1972).
[30] D. Ward, J. S. Forster, H. R. Andrews, I. V. Mitchell, G. C. Ball, W. G. Davies, and G. J. Costa, unpublished, AECL-5313, Chalk River, Canada (1976).
[31] J. F. Ziegler, *Appl. Phys. Lett.* **31**, 544 (1977).

tering of free target electrons in the static field of a screened point charge. Firsov's expression is based on a simple geometric model of momentum exchanged between the projectile and the target atom during interpenetration of electron clouds. Both theories adequately describe the general behavior of the stopping power with regard to the energy dependence and magnitude of the stopping power.

The beauty of Firsov's approach[11,12] is its simplicity. The geometric model of the interaction of two atoms can be modified easily for more complicated atomic structures. Firsov's approach considers the transfer of momentum from projectile electrons to those of the target atom. An imaginary surface s is constructed between the two atoms at the middle or at the position of the minimum potential. As the projectile interpenetrates the target atom, electrons of one atom, upon reaching the surface s, are assumed to transfer a momentum mv to the other atom. The total momentum transfer per unit time is given by assuming an electron flux $\frac{1}{4}nv_e$, where n is the localized electron density and v_e the velocity of the electrons in the atom. The total energy loss in the collision is related to the impact parameter b by

$$T(b) = \tfrac{1}{4}mv \int_{-\infty}^{+\infty} dx \int_s ds\; nv_e, \qquad (2.3.26)$$

and the stopping cross section becomes

$$S_e = \frac{1}{N}\frac{dE}{dx} = \int_{b_0}^{\infty} 2\pi b\; db\; T(b). \qquad (2.3.27)$$

From here on, it is up to the user to determine what values of n and v_e to use and how to integrate over the plane s, the distance x, and the impact parameter b.

The beauty of the LSS approach is that by extensive use of the Thomas–Fermi model, similarities among different stopping systems can be obtained, and therefore scaling from one system to the other can be done with adequate accuracy. The reduction of energy, distance, and energy loss into a set of universal units, along with the proper treatment of nuclear energy loss and range study, has made the LSS theory one of the most influential theories on low-energy ion implantation. Because of the complexity of the problem and some necessary but crude approximations in the treatment, deviation between the theoretical calculations and the experimental measurements should not be taken as a surprise.

With respect to electronic energy loss, the major contribution to the difference between theoretical trend and experimental measurements is the charge distribution of a target atom. The smooth distribution described by the Thomas–Fermi model has great significance in producing the gen-

2.3. ENERGY LOSS THEORIES: GENERAL BACKGROUND

eral trend of the interaction of charged particles, and allows smooth scaling from one system to another. The Thomas–Fermi statistical treatment of LSS makes it possible to generalize the problems and reduce the parameters in order to form an overall picture of the stopping problem, whereas a Hartree–Fock–Slater charge distribution is very specific and irregular and causes stopping to fluctuate about the norm (LSS). Realizing the difference between the Thomas–Fermi and Hartree–Fock–Slater charge distributions enables one to adjust energy loss theory when necessary.

By the LSS theory, the electronic energy loss at low velocities can be written as

$$S_e = \frac{1}{N}\frac{dE}{dx} = \eta \frac{8\pi e^2 a_0}{N} \frac{Z_1 Z_2}{Z} \frac{v}{v_0} \tag{2.3.28}$$

where $\eta \simeq Z_1^{1/6}$. This relation is thought to be applicable to $v < v_0 Z^{2/3}$. The energy and depth, in LSS dimensionless units, are

$$\epsilon = E \frac{aM_2}{Z_1 Z_2 e^2 (M_1 + M_2)}, \tag{2.3.29}$$

$$\rho = xNM_2 4\pi a^2 \frac{M_1}{(M_1 + M_2)^2}, \tag{2.3.30}$$

where

$$a = 0.8853 a_0 Z^{-(1/3)} \tag{2.3.31}$$

$$Z = (Z_1^{2/3} + Z_2^{2/3})^{3/2}. \tag{2.3.32}$$

In dimensionless units, the electronic stopping cross section can be expressed as

$$(d\epsilon/d\rho)_e = k\epsilon^{1/2}, \tag{2.3.33}$$

where k is a constant depending on Z_1, Z_2, M_1, and M_2:

$$k = \frac{0.0793 Z_1^{2/3} Z_2^{1/2} (M_1 + M_2)^{3/2}}{Z^{1/2} M_1^{3/2} M_2^{1/2}}. \tag{2.3.34}$$

The values of k are from 0.1 to 0.2 unless the projectile is much lighter than the target atom. E/ϵ, x/ρ, and k for several systems are given in Table I.

2.3.8. Nuclear Energy Loss

Nuclear stopping is a relatively small effect in the Rutherford collision region, as was stated in Section 2.3.1. Its contribution to the total stopping cross section is significant only at low velocities. However, its sig-

TABLE I. LSS Conversion Factors

Target	Ion	E/ϵ (keV)	x/ρ (Å)	k
Si	H	1.163	1480	2.08
	He	2.674	492	0.45
	B	8.850	313	0.24
	As	209	590	0.12
	Sb	515	928	0.11
Ge	H	3.32	6730	5.47
	As	298	668	0.16
	Sb	656	835	0.14
Au	H	10.75	23540	15.1
	As	592	883	0.46
	Sb	1136	837	0.22

nificance in the theory of radiation effects, such as radiation damage, sputtering, and the relation between projected range and total range, makes the study of nuclear collision important. Scattering cross sections for heavy projectiles in the Thomas–Fermi and excessive-screening regions have been reviewed by Sigmund.[32-35]

By the Thomas–Fermi atomic model, the differential cross section can be written as

$$d\sigma = \pi a^2 \frac{dt^{1/2}}{t} f(t^{1/2}), \qquad (2.3.35)$$

where t is a reduced variable that contains both dimensionless energy ϵ and the scattering angle θ, that is,

$$t^{1/2} = \epsilon \sin(\theta/2). \qquad (2.3.36)$$

This type of cross section leads to a nuclear stopping cross section having the form

$$(d\epsilon/d\rho)_n = f(\epsilon), \qquad (2.3.37)$$

which is given in Fig. 5. Electronic stopping cross sections with various values of k are also given in this schematic. From Table I and Fig. 5, one can easily estimate at what energy region the nuclear stopping is compara-

---

[32] P. Sigmund, *Rev. Roum. Phys.* **17**, 823 (1972).
[33] P. Sigmund, *Rev. Roum. Phys.* **17**, 969 (1972).
[34] P. Sigmund, *Rev. Roum. Phys.* **17**, 1079 (1972).
[35] P. Sigmund, in "Physics of Ionized Gases" (M. Kurepa, ed.), p. 137. Inst. Phys., Belgrade, 1972.

2.3. ENERGY LOSS THEORIES: GENERAL BACKGROUND

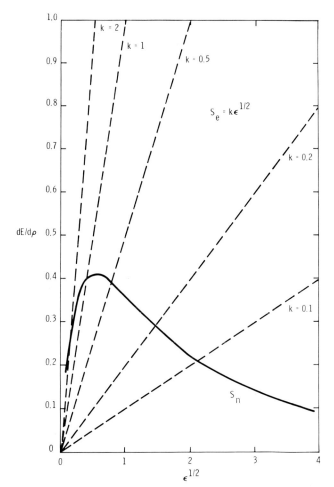

FIG. 5. Nuclear stopping (solid curve) and electronic stopping (dashed curves) as calculated by the LSS theory for various values of k. Light ions in a solid have a large k; heavy ions in a solid have k around 0.1.

ble to the electronic stopping. For example, for protons bombarding silicon, $k = 2.08$ and electronic stopping dominates the total energy loss, even down to the energy region below 1 keV. For heavy-ion implantation, nuclear stopping dominates in a very large energy region. For example, in arsenic stopping in silicon, Fig. 2 shows the nuclear stopping contribution at low energies.

It should be noted that the same symbol ϵ has been used in two different meanings. One is the LSS unitless energy, defined in Eq. (2.3.29) and

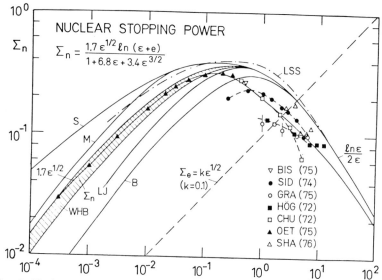

FIG. 6. Nuclear stopping power as a function of energy in LSS units. The points are experimental data; the theoretical curves are based on various potential calculations: S, Sommerfeld; M, Molière; LJ, Lenz–Jansen; B, Bohr; and LSS as in text. For details see Kalbitzer et al. (1976).[37]

used throughout the theory in the low-energy region. The other is the stopping cross section, defined by Eq. (2.2.3). Wilson et al.[36] and Kalbitzer et al.[37] have given a simple universal formula to fit the published data on nuclear stopping power. The version of Kalbitzer et al.[37] is

$$\left(\frac{d\epsilon}{d\rho}\right)_n = \frac{1.7\epsilon^{1/2} \ln(\epsilon + e)}{1 + 6.8\epsilon + 3.4\epsilon^{3/2}}, \qquad (2.3.38)$$

where here e refers to the natural base 2.718. This formula applies in the range $10^{-4} \leq \epsilon \leq 10^2$, and gives a value much lower than that obtained by the LSS theory.

The nuclear stopping study by Kalbitzer et al.[37] is summarized in Fig. 6. In their figure, nuclear stopping power $d\epsilon/d\rho$ is given as a function of energy ϵ, in LSS units. The points denote experimental data quoted in their paper. The curves are theoretical values of nuclear stopping power for various potentials (see the paper by Kalbitzer et al.[37] for references). The dashed line represents the electronic stopping power of Eq. (2.3.33) with $k = 0.1$. The solid curve fitted to the experimental points is that of Eq. (2.3.38), which is in good agreement with the calculation by Wilson, Haggmark, and Biersack[36] (WHB curve in Fig. 6).

[36] W. D. Wilson, L. G. Haggmark, and J. P. Biersack, *Phys. Rev. B* **15**, 2458 (1977).
[37] S. Kalbitzer, H. Oetzman, H. Grahmann, and A. Feuerstein, *Z. Phys.* **A278**, 223 (1976).

2.3.9. Electronic Energy Loss in the Medium-Velocity Region

The Bethe formula applies to the high-velocity region (Section 2.3.3), and the LSS and Firsov theories apply to the low-velocity region. For the medium-velocity region, which is the neighborhood of maximum energy loss, there is no adequate theory. The Bethe theory does not work because the charge of the projectile is partly neutralized and because the inner-shell electrons participate less in the stopping power. Accurate knowledge of shell corrections for the target atoms may push the applicability of the Bethe formula into the medium-velocity region, but when $2mv^2 \simeq I$, the value of Eq. (2.3.19) depends entirely on the value of $-C/Z_2$, and the Bethe formula no longer applies in this region. Accurate knowledge of energy loss in this region at the present time can be extracted only by semiempirical methods—which, of course, are also useful in other regions.

For the medium-velocity region, the electronic energy loss can be interpolated by the scaling, as follows:

(1) The scaling of energy loss from the charge state of the projectile (Section 2.3.6) is one of the methods that makes it possible to scale from one projectile to another for a given target at a given velocity.

(2) The three-parameter curve fit by Brice[38] enables one to interpolate an energy loss value from one energy region to another for a given projectile–target combination.

(3) For a given projectile at a given velocity, the energy loss from one target to another can be scaled from a semi-empirical relation that can be calculated by use of the Hartree–Fock–Slater charge distribution. Such a practice has been demonstrated by Ziegler and Chu.[39] As an example, Fig. 7 shows 2-MeV ^{4}He stopping cross section vs. Z_2. The dashed curve was calculated theoretically by the method of Lindhard and Winther,[40] here modified by the use of the Hartree–Fock–Slater atomic wave function. The calculation is similar to that made by Rousseau et al.[41] The interpolation suggested by Ziegler and Chu[39] is represented by the solid curve.

The accuracy of the interpolation and extrapolation depends on the amount and the accuracy of the data base used for interpolation. Each case will be different. In general, an accuracy of ± 10% can be reached

---

[38] D. K. Brice, *Phys. Rev. A* **6**, 1791 (1972).
[39] J. F. Ziegler and W. K. Chu, *At. Data Nucl. Data Tables* **13**, 463 (1974).
[40] J. Lindhard and A. Winther, *Mat. Fys. Medd. Dan. Vid. Selsk.* **34**, No. 4 (1964).
[41] C. C. Rousseau, W. K. Chu, and D. Powers, *Phys. Rev. A* **4**, 1066 (1971).

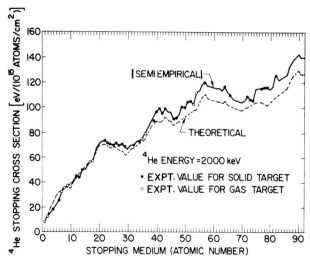

FIG. 7. Semiempirical values of the stopping cross section for a 2000 keV beam of ^{4}He. Also shown are the experimental values on which they are based as well as the theoretical values that were used in interpolation. From Ziegler and Chu (1974).[39]

by proper interpolation. Examples of the method of interpolation can be found in Northcliffe and Schilling,[14] Brown and Moak,[28] Ziegler and Chu,[39] Ward et al.,[30] Andersen and Ziegler,[42] and Ziegler.[43]

2.4. Experimental Methods on Energy Loss

Accurate values of energy loss always come from careful measurements. In this section we briefly discuss various experimental methods of measuring energy loss. Some of the methods are very accurate and straightforward and require very little description. Others, especially some of those recently developed, display some novelty. In all cases only one or two representative examples are given to illustrate the method.

There are several different methods of measuring energy loss. One is to prepare a thin foil or thin film, measure the film thickness, and measure the total energy loss of a beam transmitted through the film, to obtain the stopping cross section. Another is to make an indirect measurement of a physical quantity that is related to the stopping cross section in a predict-

---

[42] H. H. Andersen and J. F. Ziegler, "Hydrogen Stopping Powers and Ranges in All Elements," Pergamon Press, New York (1977).

[43] J. F. Ziegler, "Helium: Stopping Powers and Ranges in All Elemental Matter," Pergamon Press, New York (1977).

able way, e.g., range, backscattering yield, or Doppler shift of γ emission of compound nuclei in a target medium. All of these are related to energy loss, and therefore an accurate measurement of any one of them will yield information on energy loss. In energy loss measurements, a probable error of 2-4% is typical. A few experimenters have achieved probable error less than 1%.

2.4.1. Transmission Measurements on Thin Foils

The principle of this measurement is to prepare a uniform, self-supported thin foil of the target material and carefully measure the energy of the beam with and without transmission through the foil, to determine the amount of energy loss in the foil.

A thin foil is usually prepared by vacuum-evaporating the target material onto a plastic substrate, which is subsequently dissolved; an example of this method is given by Valenzuela and Eckardt.[44] Foils of some elements can be obtained commercially. The thickness of the foil is usually determined by measuring the mass of the foil on a microbalance or a quartz oscillator during the evaporation and measuring the area of the foil. The mass per unit area is equivalent to the density times the thickness of the foil. It can also be obtained by calibrating the observed energy loss against a given projectile whose stopping cross section in this element is known. Target mass per unit area can be expressed as $\rho \, \Delta s$ or $N \, \Delta x$, to an accuracy ranging from $\pm 0.1\%$ for thick foils to $\pm 2\%$ for thin foils. Thick foils, of the order of $1-10$ mg/cm², are suitable for protons with high energy (Bethe region), but are too thick for heavy projectiles, especially at medium or low energies. Foils on the order of fractions of 1 mg/cm² (e.g., 5-500 μg/cm²) are difficult to prepare but can also be made by vacuum evaporation.

The energy of the projectile can be measured by any of various instruments: solid-state detectors, electrostatic and magnetic analyzers, and so on. The various methods give measurements with various probable errors. Typically, a transmission experiment gives an accuracy of 2-5%. A few experiments give better accuracy.

As an example, Fig. 8 shows the experimental setup used by Ishiwari *et al.*[45] to measure the energy loss of 7.2-MeV protons in thin foils. The gold foil at the center of the chamber is not the sample; the elastically scattered beam from the gold foil is used in order to avoid damaging the silicon detector as an intense direct beam would do. Targets of alumi-

---

[44] A. Valenzuela and J. C. Eckardt, *Rev. Sci. Instrum.* **42**, 127 (1971).
[45] R. Ishiwari, N. Shiomi, S. Shirai, and Y. Uemura, *Bull. Inst. Chem. Res., Kyoto Univ.* **52**, 19 (1974).

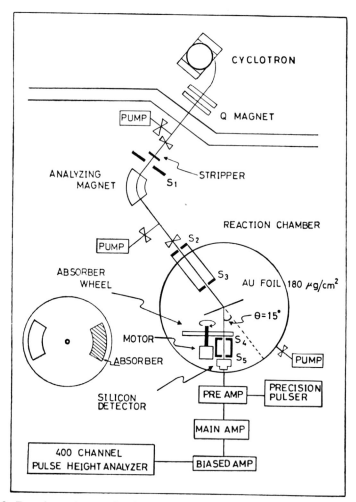

FIG. 8. Experimental setup for the energy loss measurement by the foil transmission method. From Ishiwari et al. (1974).[45]

num, titanium, iron, copper, molybdenum, silver, tin, tantalum, and gold foils, 10–20 mg/cm² thick, are mounted on the absorber wheel at lower left. This experimental setup is typical for transmission measurements of energy loss. Ishiwari's measurements are accurate to within ±0.3–0.5%.

Another very accurate method for measuring energy loss was developed by Andersen et al.[46] By this method, called the calorimetric

---

[46] H. H. Andersen, A. F. Garfinkel, C. C. Hanke and H. Sørensen, *Mat. Fys. Medd. Dan. Vid. Selsk.* **35**, No. 4 (1966).

2.4. EXPERIMENTAL METHODS ON ENERGY LOSS

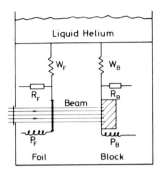

FIG. 9. Stopping power measuring system. W_F and W_B are thermal resistances, R_F and R_B thermometers, and P_F and P_B electrical heaters. From Andersen et al. (1966).[46]

method, the amount of heat that the projectiles give to the foil is measured. The principle of this method is illustrated in Fig. 9. A target foil of good heat conductivity is connected to a thermal resistance W_F, a thermometer R_F, and a heating coil P_F. Behind the target foil, a block of gold or any heat-conducting material whose thickness is greater than the range of the transmitted beam is connected to a thermal resistance W_B, a thermometer R_B, and a heating coil P_B. The measuring equipment is fastened to the bottom of a liquid-helium cryostat. The thermometer is an ordinary 0.1 W carbon resistor. The change in resistivity gives accurate temperature measurement at low temperatures.

The proton beam at energy E_0, as it passes through the foil and is stopped in the block, will cause a heating of the foil and block, raising the temperature of both as measured by R_F and R_B thermometers. The beam is then switched off, and electrical powers P_F and P_B are fed to heaters thermally connected to the foil and block, to produce the same temperature rise. The power is proportional to the amount of energy deposited in the foil and in the block; therefore the energy deposited in the foil is ΔE and will be related to the incident energy E_0 by

$$\Delta E/P_F = E_0/(P_F + P_B). \qquad (2.4.1)$$

The energy loss of 5–12-MeV protons and deuterons in various metals can be measured by this method with an accuracy of ±0.4%.

The above two examples (Figs. 8 and 9) represent two sets of the most accurate measurements of energy loss. Both types of experiments are reported with a probable error of less than 0.5%. However, the measurements by Ishiwari et al. are consistently 1.7–3.4% lower than those by Andersen et al. (Fig. 9). The difference is small, but outside the probable errors quoted.

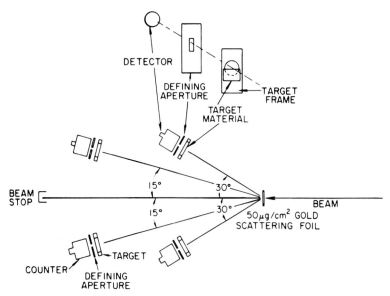

FIG. 10. The experimental arrangement for measuring ΔE. The gold scattering foil acts as a low intensity source of ions for the four target–counter arrays. This permits the measurement of ΔE values for four target materials simultaneously. As can be seen in the exploded view, each target frame is only half covered with the target material. Thus only half of the ions that reach the counter have suffered energy loss. From Ward et al. (1972).[47]

The energy loss of heavy ions can be measured by transmission only when the projectiles have enough energy to go through a self-supported foil. As an example, Ward et al.[47] have measured energy losses of approximately 1–3.5-MeV/amu ions transmitted through various metallic foils. Their experimental setup is given in Fig. 10. A high-energy ion beam from a tandem accelerator passes through a thin gold foil, which acts as a low-intensity source of ions for four to six target–counter arrays. Thus ΔE values for several target materials can be measured simultaneously. As can be seen from the enlarged view of the target setup in Fig. 10, each target frame is only half-covered with target foil; thus only half of the ions suffer energy loss. With this target arrangement, energies of projectiles can be measured with and without energy loss due to passing the foil. Some of the energy spectra are given in Fig. 11. The incident energies are calculated from the known scattering angles. The shift of the peak gives a measurement of ΔE, and independent measurement of $\rho \, \Delta x$ gives $dE/\rho \, dx$. An overall uncertainty of about 4% is claimed for these measurements.

[47] D. Ward, R. L. Graham, and J. S. Geiger, Can. J. Phys. **50**, 2302 (1972).

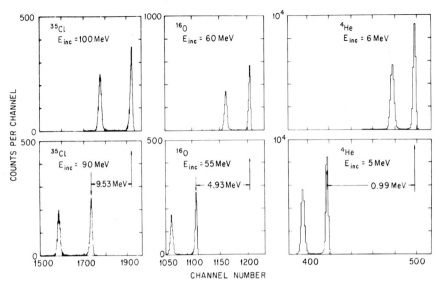

FIG. 11. Some of the spectra obtained with the apparatus shown in those ions which have lost energy in traversing the target material. The energy of the higher peak is calculated from the known incident beam energy E_{Inc} and the accurately known scattering angle (15° or 30°). The shifts of the upper peak position with incident-beam energy provide the energy calibration of the counting system. From Ward et al. (1972).[47]

High-precision transmission measurements can be made only when the foil is thick enough. For foils a fraction of 1 mg/cm² thick, the uncertainty in the foil thickness will be greater than ±1–2%. A recent measurement of energy loss and energy straggling of protons and helium ions in the energy region 20–260 keV on thin metal foils by Eckardt[48] has a probable error of 2–7%.

2.4.2. Transmission Measurements on Gaseous Targets

There are many measurements of energy loss of protons and α particles in gaseous targets. Typically a gaseous target is contained in a differentially pumped gas cell or a gas cell with thin windows. The thickness of the gas target $N \Delta x$ is related to the physical length of the gas cell Δx and the number of gas particles per unit volume N; by the ideal gas law,

$$N = 9.565 \times 10^{15}(P/T), \qquad (2.4.2)$$

where P is the pressure of the gas (Torr), and T the temperature of the gas (°K).

[48] J. C. Eckardt, unpublished (1976).

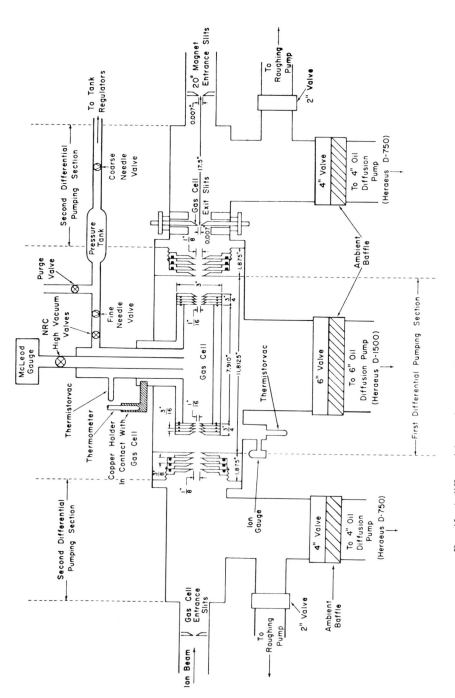

FIG. 12. A differential pumping gas cell system. From Bourland et al. (1971).[49]

2.4. EXPERIMENTAL METHODS ON ENERGY LOSS

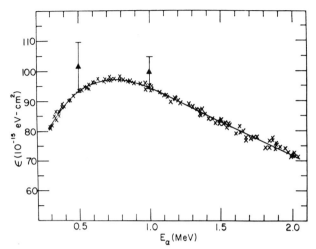

FIG. 13. Stopping cross section of α particle in oxygen gas. The smooth curve is an average-value curve drawn through the measurements, and the triangles are measurements by Rotondi. From Bourland et al. (1971).[49]

A typical experimental setup for the differentially pumped gas cell is given in Fig. 12. This example is extracted from Bourland et al.[49] Their gas cell has two-stage pumping and is 21.5 cm long. An end correction to this length for the pressure in the differential pumping section is less than 2.5%, except that for hydrogen gas the correction is 5%. A typical energy profile for measuring the energy loss of α particles in oxygen is given in Fig. 13. Each energy loss spectrum is made by use of a 60° magnetic spectrometer. A series of measurements at various gas pressures and various incident energies will produce an ϵ vs. E relation. In this experiment, many measurements produce a curve of ϵ vs. E that has a probable error of $\pm 1-2\%$. Many other elemental and compound gases are used in this setup (Fig. 12), in the study of the additivity rule for α-particle energy loss in gaseous compounds.

2.4.3. Transmission Measurements on Supported Films

Usually when a film is evaporated onto a substrate, the ion beam can be transmitted through the film but not through the substrate. The consequent difficulty in measuring transmission energy loss can be overcome by using a sharp nuclear resonance. For example, Leminen and Anttila[50] used the ^{27}Al (p, γ) reaction at various resonance energies to measure the

---

[49] P. D. Bourland, W. K. Chu, and D. Powers, *Phys. Rev. B* **3**, 3625 (1971).
[50] E. Leminen and A. Anttila, *Ann. Acad. Sci. Fenn., Ser. A6*, 370 (1971).

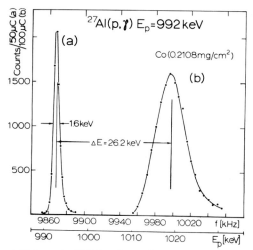

FIG. 14. Gamma yield from the ^{27}Al (p, γ) reaction at E_p = 991.8 keV, with and without stopping cobalt layer (0.211 mg/cm^2). The thickness of the reacting aluminum is 4 mg/cm^2. From Leminen and Anttila (1971).[50]

energy loss and energy straggling of 0.6–2-MeV protons in iron, cobalt, and antimony. A very thin aluminum layer, 4 μg/cm^2, was evaporated onto a tantalum backing. Figure 14 is an example from Leminen and Anttila.[50] Gamma yield from the ^{27}Al (p, γ) reaction at E_p = 991.8 keV is measured as a function of E_p, with and without a stopping cobalt layer. The reacting aluminum layer is 4 μg/cm^2 thick, the cobalt layer 211 μg/cm^2. The energy loss of the protons in the absorbing layer is determined by measuring the shift in the centroids of the γ-yield curve. The broadening of that curve gives the energy straggling.

This is a very useful method, but it is very specific. It works only for a projectile in an energy region in which known resonances exist. It is good for measuring the energy loss of protons when the (p, γ) reaction can be used over a broad energy region on several markers. It is not suitable for measuring the energy loss of helium ions or heavy ions at low energies, where no resonance or no sharp γ resonance exists.

2.4.4. Transmission Measurements on a Thin Layer on a Solid-State Detector

The scheme for measuring energy loss of a thin layer in front of a detector is very similar to that diagramed in Fig. 8, where a self-supported foil is placed in front of a solid-state detector. It differs in that since a

2.4. EXPERIMENTAL METHODS ON ENERGY LOSS

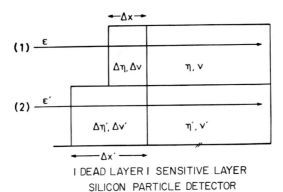

FIG. 15. Schematic for determining electronic (η, $\Delta\eta$) and nuclear (ν, $\Delta\nu$) stopping cross sections. From Grahmann and Kalbitzer (1976).[51]

layer on the detector can be made very thin (a few hundred angstroms), this method can be used for heavy ions at very low energies. These ions can penetrate a very thin film or dead layer on a detector. At very low energies the pulse height of a silicon detector corresponds not to the total energy of a particle that enters the detector, but rather to the electronic energies.

Grahmann and Kalbitzer[51] have used this fact in developing a simple but novel method by which electronic energy loss and total energy loss can be measured directly at very low energies (<60 keV). They use a silicon particle detector with a thin absorbing layer. Different effective thicknesses of this layer can be obtained by tilting the detector with respect to the incident particle beam. In Grahmann and Kalbitzer's experiment, the absorbing layer happens to be a dead layer of 400-Å silicon produced by ion implantation onto a silicon detector. The principle of the method is given schematically in Fig. 15, where the thickness of the dead layer is Δx for one case and $\Delta x'$ for another case, with the detector tilted. For all practical purposes the sensitive layer of the detector is infinitely thick. The total stopping cross section has two components, related by

$$\epsilon = \eta + \Delta\eta + \Delta\nu + \nu \quad \text{for} \quad \Delta x \text{ layer,} \qquad (2.4.3)$$

where ϵ is ion energy, $\Delta\eta$ and $\Delta\nu$ electronic and nuclear energy losses in the absorbing layer Δx, and ν the nuclear energy loss in the sensitive layer. The total ionization yield η is measurable and is the response of the solid-state detector. By tilting or evaporating to increase the effec-

---

[51] H. Grahmann and S. Kalbitzer, *Nucl. Instrum. Methods* **132**, 119 (1976).

tive thickness from Δx to $\Delta x'$, the second condition of Fig. 15 gives

$$\epsilon' = \eta' + \Delta\eta' + \Delta\nu' + \nu' \quad \text{for} \quad \Delta x' \text{ layer.} \quad (2.4.4)$$

By taking the difference between Eqs. (2.4.3) and (2.4.4), we obtain

$$\epsilon' - \epsilon = (\eta' - \eta) + (\Delta\eta' - \Delta\eta) + (\Delta\nu' - \Delta\nu) \\ + (\nu' - \nu) \quad \text{for} \quad \Delta x' - \Delta x \text{ layer,} \quad (2.4.5)$$

where η' and η are measurable quantities, and ϵ' and ϵ are adjustable and known incident energies.

One can adjust ϵ' so that the measured signals η' and η are equal ($\eta' = \eta$). This equality requires identical energies at the two interfaces of absorbing layers and sensitive layers of Fig. 15; therefore, $\nu' = \nu$ is also fulfilled, and Eq. (2.4.5) is reduced to

$$\epsilon' - \epsilon = (\Delta\eta' - \Delta\eta) + (\Delta\nu' - \Delta\nu) = N(\Delta x' - \Delta x)(S_e + S_n). \quad (2.4.6)$$

From this the total stopping cross section can be calculated.

If one makes the measurement using the same incident energy—that is, $\epsilon' = \epsilon$—Eq. (2.4.5) becomes

$$\eta - \eta' = (\Delta\eta' - \Delta\eta) + (\nu' + \Delta\nu' - \nu - \Delta\nu). \quad (2.4.7)$$

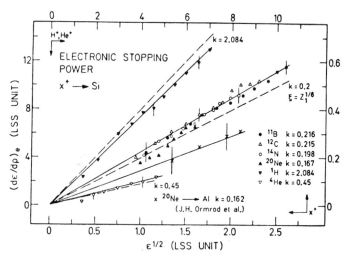

FIG. 16. Experimental values (points) and theoretical curves (dashed lines) of electronic stopping cross section for various ions in silicon and aluminum. From Grahmann and Kalbitzer (1976).[51]

2.4. EXPERIMENTAL METHODS ON ENERGY LOSS 57

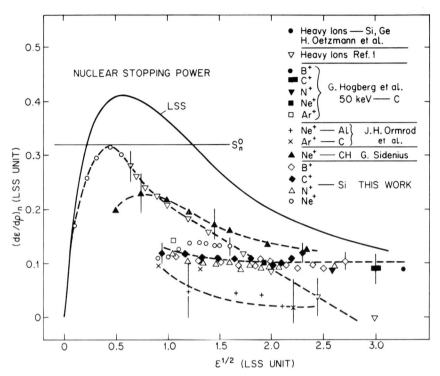

FIG. 17. Experimental values and theoretical curves of nuclear stopping cross sections. From Grahmann and Kalbitzer (1976).[51]

For this condition $\epsilon' = \epsilon$, however, the total amount of nuclear energy deposited in the target is conserved, that is,

$$\nu' + \Delta\nu' = \nu + \Delta\nu. \tag{2.4.8}$$

Therefore,

$$\eta - \eta' = \Delta\eta' - \Delta\eta = N(\Delta x' - \Delta x)S_e. \tag{2.4.9}$$

From this the electronic stopping cross section can be measured.

By this novel method, Grahmann and Kalbitzer are able to determine both electronic and total stopping cross sections. Their measurements indicate that the electronic stopping cross section obeys $S_e \propto \epsilon^{1/2}$ within ±5%. The electronic stopping power they measured is given in Fig. 16. The extracted measurements of nuclear stopping power are seen to be considerably lower than those obtained by use of the LSS theory. Figure 17 summarizes Grahmann and Kalbitzer's observations on $(d\epsilon/d\rho)_n$, which are consistent with their work presented in Fig. 6.

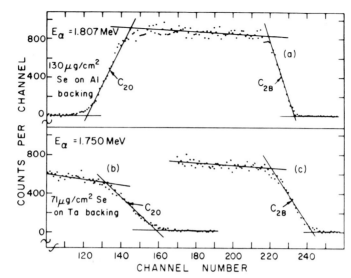

FIG. 18. Energy spectra of α particles elastically scattered at $\theta = 130°$. The target orientation was so fixed that the directions along both incident and detected α particles were 25° with respect to the normal to the target surface. Spectrum (a) describes α-particle scattering by a selenium target film prepared on an aluminum backing. The channel numbers C_{2B} and C_{20}, that is, the midpoints of the indicated steps, correspond to scattering from, respectively, the front and back surfaces of the target film. Spectra (b) and (c) were taken at the bombarding energy $E_\alpha = 1.750$ MeV; they represent scattering from a clean tantalum blank and from another onto which selenium has been uniformly deposited. In this case, however, both C_{2B} and C_{20} correspond to scattering by tantalum atoms on the front surfaces of the respective blanks. From Lin et al. (1973).[54]

2.4.5. Backscattering Energy Loss

Backscattering is one of the most often used methods of determining energy loss in solid targets. For a thin film on a thick substrate, the energy of projectiles backscattered from the thin film surface, or from the substrate surface if no thin film is present, will differ from that of projectiles scattered from the same element at the interface. The difference is attributed to the energy loss in the incident and outgoing paths.

This method was developed by Warters[52] to find the stopping cross section of protons in lithium. Several others have used it (see, for example, Chu and Powers[53] and Lin et al.[54]).

As an example, Fig. 18 from Lin et al.[54] shows energy spectra of helium

---
[52] W. D. Warters, unpublished Ph.D. thesis, Calif. Inst. Tech., Pasadena (1953).
[53] W. K. Chu and D. Powers, *Phys. Rev.* **187**, 478 (1969).
[54] W. K. Lin, H. G. Olson, and D. Powers, *Phys. Rev. B* **8**, 1881 (1973).

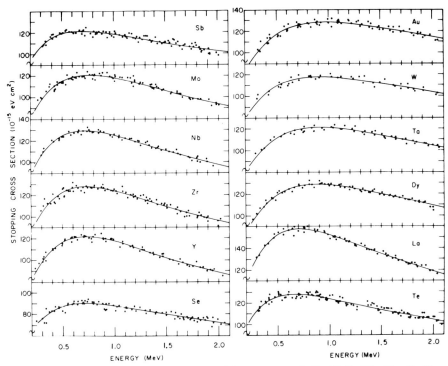

FIG. 19. Stopping cross sections of α particles vs. energy in Se, Y, Zr, Nb, Mo, Sb, Te, La, Dy, Ta, W, and Au. The experimental error ranges from ±2.2 to 3.3% of the average stopping cross section for the corresponding element. The solid curves were calculated from Brice's formula. For each element, these parameters were obtained individually by a least-squares fit of the data to the formula. From Lin et al. (1973).[54]

ions elastically scattered at $\theta = 130°$. Spectrum (a) describes helium ion scattering by a selenium target film prepared on an aluminum backing. The channel numbers C_{2B} and C_{20}, that is, the midpoints of the indicated steps, correspond to scattering from, respectively, the front and back surfaces of the selenium target film. Spectra (b) and (c) were taken at the bombarding energy $E_\alpha = 1.750$ MeV; they represent scattering from a clean tantalum substrate and from tantalum onto which selenium has been uniformly deposited. For both cases the difference between C_{2B} and C_{20} represents the energy loss in the incoming path, and that in the outgoing after backscattering at the interface. An independent measurement of film thickness $\rho \Delta x$ by weighing provides the information on stopping cross sections.

Stopping cross sections measured by this method have experimental errors from ±2.2 to 3.3%. As an example, Fig. 19 shows Lin et al. measurements[54] of stopping cross sections of α particles vs. energy in Se, Y, Zr, Nb, Mo, Sb, Te, La, Dy, Ta, W, and Au. The solid curves are calculated values based on three parameters obtained individually by a least-squares fit of the data to Brice's formula.[38]

Lin et al. present their measurements, in addition to the data obtained earlier, in the plot of ϵ vs. Z_2 given in Fig. 20. The solid irregular curves are taken from a calculation, based on Lindhard's statistical approach, carried out by Rousseau et al.[41] and by Chu and Powers.[22,23] Measurements were made at stopping powers of 0.8 MeV (open symbols) and 2.0 MeV (closed symbols); the sources of the references are given by Lin et al.[54] As Fig. 20 shows, the experimental results for the periodic dependence of the Z_2 structure in stopping cross section are in good agreement with theoretical calculations. Figure 20 is also in good agreement with the findings presented in Fig. 7.

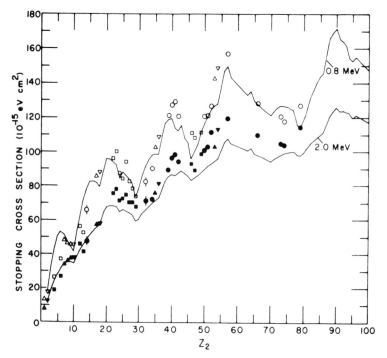

FIG. 20. α-Particle stopping cross section vs. the stopping-element atomic number Z_2 at 0.8 and 2.0 MeV. The solid curves are taken from the calculations, based on Lindhard's statistical approach, carried out by Rousseau et al. (1971).[41] Open and closed symbols, respectively, correspond to the 0.8- and 2.0-MeV data. From Lin et al.[54]

2.4.6. Backscattering Thick-Target Yield

All of the above-described methods of measuring energy loss involve measuring ΔE, an energy shift, and $N \Delta x$ or $\rho \Delta x$, the thickness of the target, which causes the energy shift. Therefore it is necessary to prepare a thin film and know its thickness. In this section, we describe a method that requires no thin-film target.

The thick-target yield method was first used by Wenzel and Whaling[55] in measuring the proton stopping cross section of ice. The method was later used in different manners, but the principle is always the same: the yield of elastic scattering of a projectile from a target is a function of the energy loss of the projectile in the target, before and after the scattering event.

As an example, Leminen[56] has measured the stopping power for protons in various metals by measuring the backscattering yield at the metal surface. Figure 21 shows his backscattering spectra of 500-keV protons elastically scattered from gold, tungsten, silver, molybdenum, copper, and titanium. The proton dose is 3 μC, and the scattering angle is 178°. The scattering yield is related to the scattering cross section, the solid angle of measurement, and the total number of protons incident on the target. The scattering yield at the surface is also related to the stopping cross sections at the incident energy and at scattered energies. A scattering yield measurement will produce energy loss information with probable error of about $\pm 3\%$. Leminen[56] used previously established stopping cross section data for copper, silver, and gold as his calibration standards in checking the internal consistency of his measurements. A similar principle has been applied by Feng et al.[57,58] to measure the relative stopping cross section ratio by measuring the spectrum height ratio at the interface of two layers.

These examples are measurements of backscattering yield at a well-defined energy, which corresponds to elastic scattering at the surface or the interface. By applying this method to the whole scattering yield spectrum, it should be possible to produce a curve of energy loss vs. energy. For example, Behrisch and Scherzer[59] assume the functional form of the stopping cross section to be $dE/dx = -AE^B$, where A and B are parameters. Since they also assume B to be 1/2, 0, or -1 for different energy regions, they can express energy loss analytically as a function of the

---

[55] W. A. Wenzel and W. Whaling, *Phys. Rev.* **87**, 499 (1952).
[56] E. Leminen, *Ann. Acad. Sci. Fenn.*, Ser. A6, 386 (1972).
[57] J. S.-Y. Feng, W. K. Chu, and M.-A. Nicolet, *Thin Solid Films* **19**, 227 (1973).
[58] J. S.-Y. Feng, W. K. Chu, and M.-A. Nicolet, *Phys. Rev. B* **10**, 3781 (1974).
[59] R. Behrisch and B. M. U. Scherzer, *Thin Solid Films* **19**, 247 (1973).

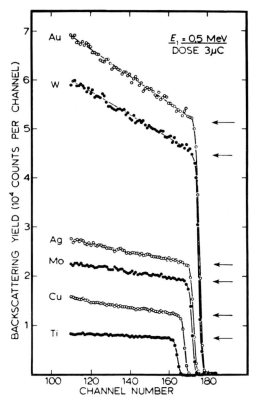

FIG. 21. Backscattering yield from Ti, Cu, Mo, Hg, W, and Au targets at a proton energy $E = 500$ keV. Proton dose is 3 μC. The arrows indicate the surface yield. From Leminen (1972).[56]

scattering yield at various energies. This enables them to translate a backscattering spectrum from a thick target into a curve of dE/dx vs. energy. Scherzer et al.[60] studied both methods, thick target and thin target, for energy loss of helium ions in gold. They list the approximations involved in both methods and the errors introduced by those approximations.

Lin et al.[61] have measured the stopping cross section of helium ions in gold and silver by the thick-target yield method. They use the stopping

---

[60] B. M. U. Scherzer, P. Børgesen, M.-A. Nicolet, and J. W. Mayer, in "Ion Beam Surface Layer Analysis" (O. Meyer, G. Linker, and F. Käppeler, eds.). Plenum Press, New York, 1976.

[61] W. K. Lin, S. Matteson, and D. Powers, Phys. Rev. B **10**, 3746 (1974).

2.4. EXPERIMENTAL METHODS ON ENERGY LOSS

cross section ϵ expressed by three parameters n, a, and z, following Brice's formula:[38]

$$\epsilon = \frac{4\hbar^2}{5m} \frac{Z_1 + Z_2}{1 + (av/v_0)^n}$$
$$\times \left[x^{1/2} \frac{30x^2 + 53x + 21}{3(x + 1)^2} + (10x + 1) \tan^{-1} x^{1/2} \right], \quad (2.4.10)$$

where $x = (v/2v_0 z)^2$, $v_0 = e^2/\hbar$, v is the velocity of the α particles, and m is the electron mass. Three parameters n, a, and z, are obtained by a least-squares method that minimizes the difference between the measured backscattering yield spectrum and the calculated backscattering spectrum by using Eq. (2.4.10) and its relation to the thick-target yield. As an example, Fig. 22 gives backscattering spectra for α particles from silver at three different incident energies. From these spectra (solid points) the three parameters are obtained by the variation method:

$$n = 3.10, \quad a = 0.352, \quad z = 2.32. \quad (2.4.11)$$

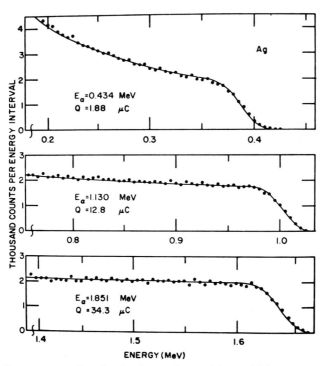

FIG. 22. Energy spectra from bombardment of α particles on thick mercury target. The curves were calculated from the stopping cross section of silver by use of Brice's formula. From Lin et al. (1974).[61]

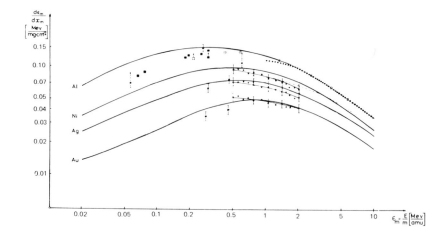
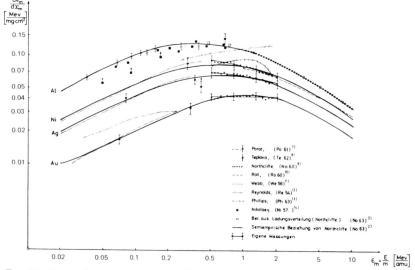

FIG. 23. Energy loss measurements of boron and nitrogen ions in metals by the thick-target yield method. From Bethge et al. (1966).[63]

By using this set of parameters, a calculated thick target yield curve can be obtained; it is shown as the solid curve in Fig. 21. The good agreement between the experimental spectra and the calculated ones provides a check of the thick-target yield method and shows that Brice's formula is an adequate expression of stopping cross section.

The thick-target yield method is not restricted to protons and helium

2.4. EXPERIMENTAL METHODS ON ENERGY LOSS

ions. Energy losses of boron, nitrogen, oxygen, and silicon ions in aluminum, nickel, silver, and gold have been measured by Bethge et al.[62,63] by a method based on the same principle. Their results for boron and nitrogen are illustrated in Fig. 23. Good agreement has been obtained between their measurements and earlier publications.

In general, the thick-target yield method is excellent for measuring energy loss. It avoids the difficulty of measuring the thickness of a thin film. However, the scattering cross section must be known; in all cases Rutherford scattering cross section is assumed to be valid in the application of the method. For an absolute measurement, accuracy in solid angle and current integration is crucial.

2.4.7. Measurement of Energy Loss by the Inverted Doppler Shift Attenuation (IDSA) Method

This experimental method for measuring energy loss is to some extent an inversion of the well-known Doppler shift attenuation (DSA) method for determining lifetime. For that method, dE/dx information is required. By inverting the problem, that is, knowing the lifetime of a γ-emitting reaction and measuring the Doppler shift attenuation of the γ ray, one can obtain information on dE/dx. This method has been described in several publications, such as Neuwirth et al.,[64] Hauser et al.,[65] and Latta and Scanlon.[66]

The γ-ray lifetime of an excited nuclear state can be measured by allowing a moving excited nucleus to slow down in some absorbing medium. The Doppler shift of the γ ray emitted is attenuated in a manner that depends on the lifetime of the γ-emitting state and on the slowing down (dE/dx) of the γ-emitting nuclei in the medium. Broude et al.[67] have found that apparently the lifetime value of a given ^{22}Ne state has an oscillatory dependence on the stopping medium Z_2. This dependence has been attributed to a nonsmooth dependence of the stopping power on Z_2. This Z_2 structure in the stopping cross section is predicted by Rousseau et al.,[41] on the basis of a Hartree–Fock–Slater atomic model that uses Lindhard and Winther's calculation. This demonstrates a potential for making accurate relative measurements of energy loss by inverting the DSA method.

---

[62] K. Bethege and P. Sandner, *Phys. Lett.* **19**, 241 (1965).
[63] K. Bethege, P. Sandner, and H. Schmidt, *Z. Naturforsch.* **21A**, 1052 (1966).
[64] W. Neuwirth, U. Hauser, and E. Kuehn, *Z. Phys.* **220**, 241 (1969).
[65] U. Hauser, W. Neuwirth, W. Pietsch, and K. Richter, *Z. Physik* **269**, 181 (1974).
[66] B. M. Latta and P. J. Scanlon, *Nucl. Instrum. Methods* **132**, 133 (1976).
[67] C. Broude, P. Englestein, M. Popp, and P. N. Tandon, *Phys. Lett.* B **39**, 185 (1972).

The stopping cross section ϵ is by definition connected with the differential energy loss along its path, that is,

$$\epsilon = \frac{1}{N}\frac{dE}{dx}, \qquad (2.4.12)$$

since

$$E = \tfrac{1}{2}M_1 v^2, \qquad (2.4.13)$$

$$\frac{dE}{dx} = \frac{dE}{dv}\frac{dv}{dx}, \qquad (2.4.14)$$

$$\frac{dE}{dx} = M_1 v \frac{dv}{dx}. \qquad (2.4.15)$$

Since $v = dx/dt$, Eq. (2.4.15) becomes

$$\frac{dE}{dx} = M_1 \frac{dv}{dt}, \qquad (2.4.16)$$

and

$$\epsilon = \frac{M_1}{N}\frac{dv}{dt}, \qquad (2.4.17)$$

if the atomic density N is known. This means that the energy loss, not ϵ, is related to the deceleration dv/dt, which can be derived from the slope of the Doppler spectrum. Therefore, the Doppler spectrum of an isotropically emitted γ ray is a direct measurement of the energy loss of the γ-emitting nucleus in its surrounding medium.

This method is very specific. It works only for a given nucleus that emits γ rays at a well-defined recoil energy. Therefore, it is not generally applicable to the problem of dE/dx. However, because it has high accuracy in measurements of relative stopping cross section, and because the study is for a specific ion at a well-defined energy region in various media, it is best applied when the target medium is the parameter. That means it is very powerful in applications of Bragg's rule—for example, in the work of Neuwirth et al.,[68] and also in the study of Z_2 structure in dE/dx by Pietsch et al.[69] as implied by Broude et al.[67]

As an example, we look at the study of energy loss of lithium in various media, using the IDSA method described by Hauser et al.[65] Nuclei of ^{7}Li* are produced from ^{10}B(n, α)^{7}Li* reaction. The asterisk indicates that ^{7}Li nuclei are in excited states and will emit γ rays for deexcitation. The

---

[68] W. Neuwirth, W. Pietsch, K. Richter, and U. Hauser, *Z. Phys.* A **275**, 209 (1975).
[69] W. Pietsch, U. Hauser, and W. Neuwirth, *Nucl. Instrum. Methods* **132**, 79 (1976).

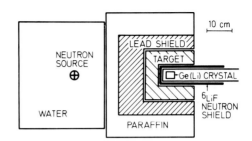

FIG. 24. Experimental setup for Doppler shift measurement. From Hauser et al. (1974).[65]

experimental setup is given in Fig. 24. The neutrons were produced by a 3-Ci americium–beryllium source. At this strength a given target material can be investigated over a period of one or two days. Water and/or paraffin are used as neutron moderators, so that the influx of neutrons into the target substance is isotropic. The target material has to contain ^{10}B in order to obtain ^{7}Li* from ^{10}B(n, α)^{7}Li reaction. Boron can be doped or implanted in a target of element Z_2 for the study of the ^{7}Li energy loss in Z_2. Boron can also be one of the elements in a compound for study of Bragg's rule, such as AlB_2, B_4C, B_2O_3, CrB_2, or B_4Si. The target medium can be in a solid, a liquid, or a gas.

In Fig. 24 a Ge(Li) detector is used in analyzing the energy of γ rays. A typical Ge(Li) detector will have a resolution of 1.3–1.6 keV (FWHM), which refers to the 477.55-keV γ rays emitted by ^{7}Li*. The γ-ray detector is shielded by ^{6}LiF to reduce neutron noise. The signal-to-noise ratio is proportional to the amount of ^{10}B content of the target. A typical γ-ray spectrum for the Doppler measurement is given in Fig. 25, where the target medium is CrB_2 and the detector resolution is 1.4 keV (FWHM). The reason for the broadening of the 477.55-keV γ ray is that the ^{7}Li is in motion, and so this is Doppler broadening. The shape of the broadening spectrum contains information on deceleration dv/dt, velocity, and the half-life of the nuclei emitting the γ rays. In other words, dv/dt can be written in terms of the half-life of the ^{7}Li*; the shape of γ-ray yield can be written as a function of the γ-ray energy and the velocity of ^{7}Li*. If we assume that the half-life of ^{7}Li* has been determined by another, independent method and by measuring the γ-ray yield vs. γ energy, the deceleration dv/dt vs. the velocity v of the ^{7}Li nuclei can be obtained, and energy loss vs. v for ^{7}Li is then obtained, in turn, by application of Eq. (2.4.16).

Neuwirth et al.[68] demonstrated this IDSA method in great detail in studies of the stopping cross section of ^{7}Li. A Z_2 oscillation found in their measurement can be predicted by calculations with a Hartree–Fock–Slater model. This is in good agreement with the Z_2 oscillation of ϵ for

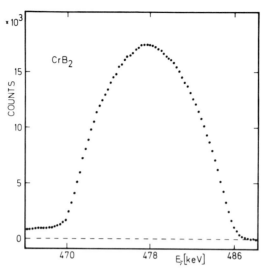

FIG. 25. Doppler spectrum of ⁷Li* stopped in CrB_2. Background is subtracted. From Hauser et al. (1974).[65]

⁴He ions such as given in Figs. 7 and 20. Z_2 oscillations of ϵ for ¹⁴N ions are measured and predicted in targets from carbon through molybdenum by Simons et al.,[70] and by Land and Brennan[71]; they have made range measurements of 800-keV ¹⁴N⁺ in targets, using protons as probing projectiles, to induce ¹⁴N(p, γ)¹⁵O nuclear interaction. They then extract energy loss information from the range measurement.

The IDSA method, then, is a very powerful method for studying the target dependence of the stopping cross section. Because of its high relative accuracy, it is particularly useful in the study of Bragg's rule, Z_2 oscillation, and the physical state effect of heavy ions' stopping cross sections in media. One disadvantage is that the method applies only to a few specific nuclei in a given energy region.

2.5. Current Problems* in Stopping Cross Sections

Energy loss is one of the macroscopic properties of a target material. Its description involves only the Coulomb interaction between the projec-

[70] D. G. Simons, D. J. Land, J. G. Brennan, and M. D. Brown, in "Ion Beam Surface Layer Analysis" (O. Meyer, G. Linker, and F. Käppeler, eds.). Plenum Press, New York 1976.

[71] D. J. Land and J. G. Brennan, Nucl. Instrum. Methods **132**, 89 (1976).

* See Note Added in Proof on p. 72.

2.5. CURRENT PROBLEMS IN STOPPING CROSS SECTIONS

tile and the target electrons and nuclei. The study of the energy loss process has been one of the major sources of information on atomic physics, and recent applications of ion beams in various disciplines have brought a need for further study. Accurate measurements of energy loss are always in demand for studies of various effects and deviations from theories. Comprehensive theories that provide fairly accurate information on energy loss have been developed. We conclude this part by listing a few current problems related to dE/dx.

2.5.1. Chemical Effect: Bragg's Rule

Bragg and Kleeman[72] first postulated the linear additivity of atomic stopping cross section. Because of its importance in radiation safety and health physics, the validity of this rule is constantly under test. For example, Powers et al.[73] have reviewed C–H and C–H–F compounds and found that there is a systematic deviation that depends on whether the compound is single-, double-, or triple-bonded. Neuwirth et al.[68,74] have shown that Bragg's rule is invalid for stopping cross sections of many metal–boron compounds for lithium ions of 80–800 keV. Deviations from Bragg's rule by up to 40% have been observed. Feng et al.[57,58] have found that Bragg's rule applies to metal alloys but not to metal oxides.

2.5.2. Solid-State Effect

A possible reason why Bragg's rule is invalid for a metal oxide is that the oxygen present in the metal oxide is in the solid phase rather than in the gaseous phase, according to Ziegler et al.,[75,76] who have given empirical corrections to the solid-state effect. That effect has also been calculated by Latta and Scanlon[77] and by Chu et al.[78] Both calculations indicate that the physical state of the stopping medium does exert a small effect on stopping cross sections. Recently Matteson et al.,[79] after reviewing and measuring the stopping cross section of helium ions in water vapor and in ice, observed that it is greater in vapor. Chu et al.[80] have measured

---

[72] W. H. Bragg and R. Kleeman, *Phil. Mag.* **10**, S318 (1905).

[73] D. Powers, A. S. Lodhi, W. K. Lin, and H. L. Cox, *Thin Solid Films* **19**, 205 (1973).

[74] W. Neuwirth, W. Pietsch, and R. Kreutz, *3rd Int. Conf. Ion Beam Anal.*, Washington, D.C. (1977).

[75] J. F. Ziegler, W. K. Chu, and J. S.-Y. Feng, *Appl. Phys. Lett.* **27**, 387 (1975).

[76] J. F. Ziegler and W. K. Chu, *J. Appl. Phys.* **47**, 2239 (1976).

[77] B. M. Latta and P. J. Scanlon, *Phys. Rev. A* **12**, 34 (1975).

[78] W. K. Chu, V. L. Moruzzi, and J. F. Ziegler, *J. Appl. Phys.* **46**, 2817 (1975).

[79] S. Matteson, D. Powers, and E. K. L. Chau, *Phys. Rev. A* **15**, 856 (1977).

[80] W. K. Chu, M. Braun, J. A. Davies, N. Matsunami, and D. A. Thompson, *3rd Int. Conf. Ion Beam Anal.*, Washington, D.C. (1977).

energy loss of helium ions in solidified oxygen, argon, and CO_2. Their results indicate that above 1 MeV, the energy loss differs only slightly in the gaseous phase and the solid phase; the maximum energy loss, however, occurs at somewhat higher energy in the solid phase. In the energy region below 1 MeV, the stopping cross sections of helium ions are about 5% lower than those reported for the corresponding gas.

2.5.3. Structure Effect

Softky[81] has found that the stopping cross section of protons of 1 MeV is 6% higher in graphite than in diamond. Matteson et al.[82] have found that the energy loss of 0.3–2-MeV α particles in graphite is 6–28% higher than the corresponding value for vapor-deposited carbon. They attributed the difference to an allotropic effect. For a single crystal, of course, the energy loss in a channeled direction is quite different from that in a random direction. Whether energy loss in a polycrystalline structure is or is not the same as that in an amorphous material of the same composition is not known; tests are being made to find out.

All the above three effects on the electronic stopping cross section are related to the slight change of outer-shell electronic configuration for a given element, either in a compound form or in a different phase or structure. They can therefore be considered basically the same, and an understanding of one will help in understanding the others.

2.5.4. Energy Straggling

Energy loss is a statistical process. The energy loss distribution is called energy straggling. A recent theory by Sigmund[83] indicates that correlation of target atoms in molecular gases causes an increase in straggling, and in addition to that, charge fluctuations of the projectile also give rise to an increase in straggling. Recent measurements by Besenbacher et al.[84] seem to confirm this theory. Measurements by Hvelplund[85] on low-energy helium ions having more energy straggling in N_2 gas than in neon can also be explained by this correlation effect proposed by Sigmund. Chu,[86] however, has interpreted Hvelplund's result with respect

---

[81] S. D. Softky, *Phys. Rev.* **123**, 1685 (1961).
[82] S. Matteson, E. K. L. Chau, and D. Powers, *Phys. Rev. A* **14**, 169 (1976).
[83] P. Sigmund, *Phys. Rev. A* **14**, 996 (1976).
[84] F. Besenbacher, J. Heinemeier, P. Hvelplund, and H. Knudsen, *Phys. Lett.* **61A**, 75 (1977).
[85] P. Hvelplund, *K. Dan. Vidensk. Selsk. Mat. Fys. Medd.* **38**, No. 4 (1971).
[86] W. K. Chu, *Phys. Rev. A* **13**, 2057 (1976).

to a Z_2 oscillation calculation of energy straggling. More straggling experiments are needed to verify some of the effects discussed here.

2.5.5. Charge State of the Projectiles

This subject is discussed in Part 6 by Macdonald. Because it is closely related to the energy loss of heavy ions, theoretical and experimental work on it will be important to information on the stopping cross section. So far there has been no direct experimental measurement of the charge state of a projectile *inside* a solid.

2.5.6. The Barkas Effect

The Barkas effect is the difference in the stopping of swift positive and negative particles. In Bethe's formula [Eq. (2.3.19)], the electronic stopping cross section derived by quantum perturbation theory to the first order gives the relation that the stopping cross section is proportional to Z_1^2. The Barkas effect could be explained by an extra term of order Z_1^3, which will yield a different stopping cross section according to whether Z_1 is positive or negative. Ashley et al.[87,88] have performed a classical perturbation calculation concerning a harmonic oscillator. The motion of the target electrons in the force field of the point charge projectile is treated up to quadratic terms in Z_1. Lindhard[89] has briefly reviewed the Barkas effect, and the higher-order effects. He approaches the problem from a simple classical Rutherford scattering from a screened potential. The Z_1^3 correction given by Lindhard is about twice that given by Ashley et al. The Z_1^3, Z_1^4 effect has been verified in a few experiments.

2.5.7. Ion Clusters Effect

When the projectiles are ion clusters rather than individual ions, the mode of propagation is influenced by the ions within the medium, by the ion–ion interaction in the cluster through the wakes trailing each ion, and by the Coulomb repulsion between the ions. Those correlations complicate the motion of the projectiles in condensed medium, and make the energy loss greater than for a single ion. The energy loss of the ion cluster depends on the *partition* of the energy loss between single electron collision and resonant excitations in the target. Careful measurements of the stopping cross sections of correlated clusters and individual ions provide

---

[87] J. C. Ashley, W. Brandt, and R. H. Ritchie, *Phys. Rev. B* **5**, 2393 (1972).
[88] J. C. Ashley, W. Brandt, and R. H. Ritchie, *Phys. Rev. A* **8**, 2402 (1973).
[89] J. Lindhard, *Nucl. Instrum. Methods* **132**, 1 (1976).

information on this partition relation. A recent review is given by Brandt and Ritchie.[90]

2.5.8. Nuclear Energy Loss

For very low-energy heavy ions the nuclear energy loss is totally dependent on the interacting potential. Recent experimental results in low-energy ion implantation indicate that projected ranges calculated by the LSS theory are off by as much as 100%. Nuclear stopping cross sections lower than those calculated by the LSS theory have been given recently by Kalbitzer et al.[37] and by Wilson et al.[36] (see Fig. 6). Interaction potentials need further experimental verification before accurate theory on nuclear energy loss for very low-energy heavy ions can be formulated.

Note Added in Proof

Much progress has been made recently in the topics covered in Chapter 2.5. For example, energy straggling, ion clusters effect, and the Barkas effect have received much theoretical and experimental attention. This detailed progress cannot be covered in this short part.

---

[90] W. Brandt and R. H. Ritchie, *Nucl. Instrum. Methods* **132**, 43 (1976).

3. CHARGE EQUILIBRATION OF HIGH-VELOCITY IONS IN MATTER

By Hans D. Betz

3.1. Introduction

3.1.1. Historic Background

The question of charge states of projectile ions during penetration through matter is a historic one, with interest dating back to the early part of this century. Initial interest arose from the observation that positive ions become neutralized when traveling through gaseous media. This observation caused intensive work on the free paths of protons,[1,2] helium ions,[3-5] and heavier ions of neon,[6,7] argon,[6,7] and krypton[7] with initial energies of up to ~22 keV in a variety of target gases such as helium, neon, and krypton. These and other early measurements were already detailed enough to include the intensity of projectile charge state fractions as a function of target thickness, projectile species, and collision energy.

Further impetus for charge state work came with the discovery of fission fragments. One problem concerned the velocity–range relations for the fragments in matter: in order to calculate the energy loss of the fragments it was necessary to determine the ionic charge states at all stages during the slowing down process. Attempts along these lines were relatively successful,[8,9] especially due to efforts of many outstanding investigators.[10-14] Bohr presented an extensive treatise[15] on the fundamental

---

[1] W. Wien, *Sitz. K. P. Akad. Wiss.*, p. 773 (July 1911).
[2] J. Koenigsberger and J. Kutschewski, *Ann. Phys.* **37**, 161(1912).
[3] E. Rutherford, *Phil. Mag.* **47**, 277(1924).
[4] G. H. Henderson, *Proc. Roy. Soc. A* **109**, 157(1925).
[5] P. Rudnick, *Phys. Rev.* **38**, 1342(1931).
[6] H. Kallmann and B. Rosen, *Z. Phys.* **61**, 61(1930).
[7] H. F. Batho, *Phys. Rev.* **42**, 753(1932).
[8] G. Beck and P. Havas, *Compt. Rend.* **208**, 1643(1939).
[9] P. Havas, *J. Phys. et Rad.* (8) **1**, 146(1940).
[10] W. E. Lamb, *Phys. Rev.* **58**, 696(1940).
[11] N. Bohr, *Phys. Rev.* **58**, 654(1940).
[12] N. Bohr, *Phys. Rev.* **59**, 270(1941).
[13] J. Knipp and E. Teller, *Phys. Rev.* **59**, 659(1941).

aspects of charge exchange collisions, parts of which are dealt with in Section 3.7.1. Refined work with fission fragments revealed a target density effect on heavy-ion charge states,[16] which was treated in a well-known paper by Bohr and Lindhard,[17] along with general considerations on electron capture and loss by heavy ions.

Considerable extension of early experimental work became feasible with the availability of heavy-ion accelerators. The primary interest was twofold: First, the possibility arose to undertake systematic studies of charge exchange phenomena with projectiles from the entire range of the periodic table and well-defined collision energies in a steadily increasing range. Efforts to construct heavy-ion accelerators of increasing capabilities have not yet ended and it appears reasonable to suspect that it will soon be possible to study the heaviest ions at collision velocities close to the speed of light. Second, optimized performance of a heavy-ion accelerator depends on effective charge stripping and it must be known in what way high degrees of projectile ionization can be obtained with sufficient particle intensity. A further problem is the conservation of the charge of an ion during the acceleration process and in the beam-handling system; interaction with the residual gas along the flight path of the ions can cause charge exchange and, thus, possibly loss from the ion beam. Of critical importance are extreme cases such as transport of (1) highly charged ions at relatively low velocity and (2) modestly charged ions at very high velocity. These situations are typical, for example, for the initial and final acceleration phases, respectively, in heavy-ion accelerating systems. Lately, particular interest is focused on case (2) in connection with heavy-ion-induced fusion research.[18]

3.1.2. Present Situation

Atomic collisions between heavy ions are complex in almost any aspect. Although the possibility of electron capture and loss by ions in matter was realized long ago,[3,19] we are far from having sufficiently complete experimental and theoretical information on these processes. Of course, much insight has been gained and the number of open basic questions on charge exchange phenomena is steadily shrinking. In very simple collision systems, electron capture and loss is understood fairly

---

[14] J. H. Brunings, J. K. Knipp, and E. Teller, *Phys. Rev.* **60**, 657(1941).
[15] N. Bohr, *Mat.-Fys. Medd. Danske Vid. Selsk.* **18**, No.8(1948).
[16] N. O. Lassen, *Mat.-Fys. Medd. Danske Vid. Selsk.* **26**, No.5(1951); **26**, No.12(1951).
[17] N. Bohr and J. Lindhard, *Mat.-Fys. Medd. Danske Vid. Selsk.* **28**, No.7(1954).
[18] W. D. Metz, *Science* **194**, 307(1976); R. Walgate, *New Scientist*, p. 156 (21 Oct. 1976).
[19] G. H. Henderson, *Phil. Mag.* **44**, 680(1922).

3.1. INTRODUCTION

well. However, when *heavy* ions and atoms become involved, complications arise primarily due to (1) the presence of many electrons with widely varying binding energies and (2) the large disturbance of the electron clouds experienced during a collision.

As regards electron loss processes, a fortunate situation has developed: in recent years, much progress has been achieved in the field of inner-shell ionization due to charged particles. At first glance, it may appear that this is of little importance for charge exchange processes that are assumed to occur mainly in outer shells, but there is a direct and deep connection between the two fields. First, results from studies of inner-shell ionization can be applied to outer shells as well, and second, fast ions can be stripped down to such an extent that any inner shell becomes the outermost shell remaining with the ion. As a result, estimates of loss cross sections can often be worked out with reasonable accuracy, although generally intricate problems remain to be solved. As regards electron capture in heavy-collision systems, the situation is somewhat less favorable and it must be admitted that present understanding, at least from a practical point of view, does not extend much beyond what was achieved by Oppenheimer.[20] The lack of comprehensive theoretical work may be taken as an indication of the high degree of complexity, although we may speculate that, in contrast to ionization phenomena, there has been less effective stimulation to tackle electron capture by heavy ions. Lately, however, it appears that particular activity is developing around the question of excited states produced by electron capture, with emphasis on the formation of very highly excited states, including capture into projectile continuum states, and on the role of very high angular momentum states.

Charge state distributions of ions penetrating through matter can be calculated from electron capture and loss cross sections. Since these cross sections are mostly hard to obtain, semiempirical techniques have been worked out that are of practical usefulness. Many attempts along these lines have been successful because the charge state fractions turn out to exhibit a surprisingly smooth dependence on important parameters such as ion species and ion velocity. General and satisfactory ab initio calculations on charge equilibration of heavy ions in matter remain a task for the future.

Passage of ions through *dense* media represents an interesting question that has received appreciable attention and continues to be discussed intensively. The question emerged from the observation of the density effect[16] and is focused on the role of excited states in heavy ion–atom collisions. Successful treatment of this problem required careful con-

---

[20] J. R. Oppenheimer, *Phys. Rev.* **31**, 349(1928).

sideration of the relative importance of collisional and collision-induced processes, which can affect the ionic charge of projectiles. Some qualitative understanding has been attained, but due to the difficulties of determining accurate charge exchange cross sections, quantitative improvements are very much desirable. In this connection, it should be mentioned that observation of collision-induced x rays has become a new and effective tool to collect information on projectile states.

Finally, it should be mentioned that the production of very highly stripped ions, especially of one- and two-electron ions, has become important for a variety of interesting experiments in atomic physics with high-Z ions. As an example, we mention Lamb shift experiments and high-precision spectroscopy of radiative transitions involving basic phenomena of quantum electrodynamics, relativistic effects, and electron–electron interaction.

3.1.3. Review Articles and Data Collections

Since the number of publications relevant for the topic of present interest is enormous and space limitations require to restrict discussion and quotation of references, it seems useful to present a list of some important and comprehensive review articles and data collections that are easily accessible and provide thorough information on detailed questions and experimental results related to charge exchange processes.

Allison[21] published an extensive survey dealing with mathematical descriptions of charge-changing collisions, experimental techniques, and results on charge state fractions and cross sections. An updated version of Allison's review was prepared by Allison and Garcia-Munoz,[22] giving further experimental details and containing data primarily on hydrogen and helium ions; results on heavier ions are briefly discussed.

Nikolaev discusses most of the fundamental questions of complex charge exchange processes in a thorough review[23] in which he outlines experimental techniques and data analysis. Electron loss and capture cross sections and equilibrium charge state distributions are treated on a theoretical and semiempirical basis with the help of statistical approaches and are compared with a number of experimental results obtained for relatively light ions; typical data concern beams of nitrogen in the range 2 to 20×10^8 cm/sec (0.3–30 MeV).

---

[21] S. K. Allison, *Rev. Mod. Phys.* **30**, 1137(1958).
[22] S. K. Allison and M. Garcia-Munoz, *in* "Atomic and Molecular Processes" (D. R. Bates, ed.), Chap. 19, p. 721. Academic Press, New York, 1962.
[23] V. S. Nikolaev, *Uspehi Fiz. Nauk* **85**, 679(1965) [*Sov. Phys. Usp.* **8**, 269, (1965)].

3.1. INTRODUCTION

This author compiled a review article[24] summarizing experimental methods, data analysis, and semiempirical and limited theoretical descriptions for capture, loss, and average equilibrium charge states, and also dealing with numerous problems such as the gas–solid density effect. In particular, it is shown that proper application of the simple Lamb[10]–Bohr[11,12] criterion allows fairly good ab initio calculations of equilibrium charge states. Most of the experimental data refer to ions with atomic numbers in the range $16 \leq Z \leq 92$.

Tawara and Russek presented a review[25] that gives detailed techniques of measurements, and experimental and theoretical results on electron capture and loss in gaseous targets. Although the paper is limited to hydrogen beams, most basic charge exchange problems are treated that are important for heavier systems.

Garcia et al.[26] reviewed inner-shell vacancy production in ion–atom collisions, and treated primarily collisional loss of inner-target electrons. However, basic theoretical concepts can be suitably applied to other electrons as well. The results and scaling procedures for electron loss are therefore of great importance to charge exchange phenomena.

Theoretical treatments of excitation, ionization, and electron capture can be found, for example, in works by McDowell and Coleman,[27] Mapleton,[28] and Basu et al.[29] Among basic concepts, they treat the OBK, the first and second Born, the distorted-wave, and the impulse approximations. Although the discussion remains limited to very simple collision systems, it exhibits all fundamental and important features that are essential for an approximate understanding of more complex systems.

Turning to some data collections, Lo and Fite[30] present electron capture and loss cross sections for fast, heavy particles passing through gases in graphical form for 18 different ions ranging from beryllium to uranium, stripped in five gaseous targets (N_2, O, O_2, Ne, and Ar). Only low collision energies are covered that, in all cases but one, do not exceed 2 MeV; initial projectile charge states are therefore limited to 0, 1+, and 2+.

A tabular compilation of equilibrium charge state distributions of ener-

---

[24] H. D. Betz, *Rev. Mod. Phys.* **44**, 465(1972).
[25] H. Tawara and A. Russek, *Rev. Mod. Phys.* **45**, 178(1973).
[26] J. D. Garcia, R. J. Fortner, and T. M. Kavanagh, *Rev. Mod. Phys.* **45**, 111(1973).
[27] M. R. McDowell and J. P. Coleman, "Introduction to the Theory of Ion-Atom Collisions," North-Holland, Amsterdam 1970.
[28] R. A. Mapleton, "Theory of Charge Exchange." Wiley (Interscience), New York, 1972.
[29] D. Basu, S. C. Mukherjee, and D. P. Sural, *Phys. Rep.* **42C**, 145(1978).
[30] H. H. Lo and W. L. Fite, *Atom. Data* **1**, 305(1970).

getic ions ($Z > 2$) in gaseous and solid media was published by Wittkower and Betz.[31] It lists 2525 equilibrium charge state distributions for 26 projectile ions, 50 target species, and 250 target–projectile combinations from ~100 sources. A significant part of the data originates from the work of Datz et al.[32] for bromine and iodine ions, which also shows interpolation graphs and nonequilibrium distributions for charge states of these ions.

Dehmel et al.[33] list, from some 100 references, experimental parameters (projectile, target, energy range, experimental technique) for cross-section measurements and present some 200 graphs of cross sections $\sigma(0 \to 1^+)$ and $\sigma(1^+ \to 2^+)$ as a function of projectile energy.

3.2. Fundamental Charge Exchange Processes

3.2.1. Definitions and Basic Processes

When an ion with charge q collides with a target atom, several processes may occur that change the charge into q'; these processes, which are schematically illustrated in Fig. 1, are the following ones:

(a) The ion captures one or more bound target electrons into any unoccupied bound state, resulting in either ground-state or excited ions. Capture can be nonradiative (Coulomb capture; see Chapter 3.4) or radiative (see Chapter 3.10).

(b) The ion loses one or more electrons due to Coulomb excitation into the continuum, leaving the ion in either a ground state or an excited state, depending on whether outer- or inner-shell electrons are lost, respectively.

(c) One or more electrons on the ion become excited to higher-lying bound projectile states.

(d) Excited states produced by mechanisms (a)–(c) decay either with or without emission of one or more electrons, depending on the initial degree of total excitation and selection rules for the possible transitions. In most cases, singly and multiply excited ions decay via x-ray and electron emission.

(e) When the ion is already in an excited state prior to the collision considered, processes (a)–(c) may also occur.

---

[31] A. B. Wittkower and H. D. Betz, *Atom. Data* **5**, 113(1973).

[32] S. Datz, C. D. Moak, H. O. Lutz, L. C. Northcliffe, and L. B. Bridwell, *Atom. Data* **2**, 273(1971).

[33] R. C. Dehmel, H. K. Chau, and H. H. Fleischmann, *Atom. Data* **5**, 231(1973).

3.2. FUNDAMENTAL CHARGE EXCHANGE PROCESSES

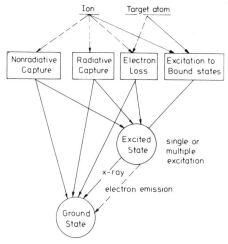

FIG. 1. Scheme for the basic collision-induced ion–atom processes, which lead to a change of the projectile charge state, either directly or via multiply excited states; the broken lines indicate charge-changing processes.

(f) Target atoms that are assumed to be neutral prior to the collision can undergo processes (b)–(d).

It becomes obvious that though only very few basic phenomena occur very complex situations will arise especially when multielectron atoms and ions participate in a collision. Moreover, the probabilities for these various processes are often very large, so that several of the processes can occur at the same time during a single collision.

It is customary to describe the net effect for charge-changing probabilities P for processes $q \to q'$ by a cross section $\sigma(q, q')$ with the understanding that only a single ion–atom collision has occurred ("single-collision conditions"). Cross sections are sometimes written in the form $\sigma(q, q \pm n)$, where positive and negative signs characterize electron loss and capture by the ions, respectively; and in the form $\sigma(q, q + n)$, with positive and negative values of n for electron loss and capture, respectively. Furthermore, subscripts are in use in order to distinguish cross sections for loss (σ_l) and capture (σ_c). Fortunately, collisions in which the ions capture or lose only a single electron ($|n| = 1$) are generally dominant, whereas multiple capture or loss collisions, i.e., those in which more than one electron is captured by or lost from the ion ($|n| > 1$), are less likely processes.

All of the processes (a)–(c) from above depend on the impact parameter b of the collision. In typical and normally important cases charge-changing probabilities are accentuated for values of b close to the relevant shell radii R; then, for not too slow collisions, scattering angles are small and ions that have undergone charge exchange can still be observed in forward direction. A total cross section

$$\sigma = \int P(b)\, d^2 b \qquad (3.2.1)$$

is measurable. In particular cases, relatively rare large-angle scattering events have also been observed and provide information on $P(b \simeq 0)$ rather than on σ from Eq. (3.2.1). Some of these cases are discussed in Section 3.5.5., but otherwise we always understand that cross sections are total ones.

For a given ion and target species with nuclear charges Z and Z_T, respectively, charge-changing cross sections depend not only on q and q' for, say, ground-state ions, but also on the collision velocity v. As long as v is not small compared with the orbital velocity v_e of bound electrons involved in the charge exchange process, cross sections can be easily interpreted in terms of processes (a)–(c) from above. In case of $v \ll v_e$, however, the collision is slow enough to allow formation of quasi-molecular orbitals, and other additional physical charge transfer concepts must be considered.[34] In the following, most cases of importance refer to the former, less-complicated situation where $v \gtrsim v_e$.

3.2.2. Mathematical Description of Charge-State Fractions under Nonequilibrium and Equilibrium Conditions

Let us assume that an ion beam passes through a target gas of total thickness x, where x is measured in units of atoms/cm². Depending on the actual charge q of an ion, various capture and loss processes become effective, and for increased target thickness the charge of any ion in the beam will fluctuate statistically. As a net effect, the charge composition of the ion beam penetrating through the target will vary and for each value of x certain relative charge state populations $Y_q(x)$ will be observed. The system of rate equations is given by

$$dY_q(x)/dx = \sum_{q' \neq q} \{\sigma(q', q)\, Y_{q'}(x) - \sigma(q, q')\, Y_q(x)\}, \qquad (3.2.2)$$

where q and q' extend over all possible charge states. For convenience, the fractions Y are normalized by $\Sigma_q Y_q = 1$. As long as v remains sharp,

---

[34] N. F. Mott and H. S. Massey, "The Theory of Atomic Collisions." Clarendon Press, Oxford, 1950; E. Gerjuoy, *Rev. Mod. Phys.* **33**, 544(1961); D. R. Bates, *in* "Atomic and Molecular Processes" (D. R. Bates, ed.), Chap. 14, p. 550. Academic Press, New York, 1962.

3.2. FUNDAMENTAL CHARGE EXCHANGE PROCESSES

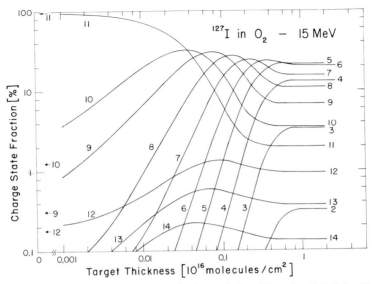

FIG. 2. Nonequilibrium charge-state distribution, calculated from Eq. (3.2.2) for 15-MeV iodine ions with initial charge 11+, passing through dilute oxygen. Charge states are indicated near each curve [H. D. Betz, *Rev. Mod. Phys.* **44**, 465 (1972)].

only a limited number of charge state fractions is influential, grouped around a certain most probable charge state. Nevertheless, a large number of cross sections is required to describe nonequilibrium charge state distributions. Figure 2 displays the typical behavior of $Y_q(x)$ as calculated from the set of Eqs. (3.2.2) for collisions between 15 MeV iodine ions and oxygen molecules.[24,35] It becomes obvious that the incident fraction decreases at the expense of neighboring fractions; for sufficiently thick targets no further changes occur and an equilibrium distribution will be reached that does not depend on the initial distribution of charge states in the beam incident on the target. Equilibrium conditions are attained when electron loss from and electron capture into any charge state component are balanced,

$$\Sigma_{q'} F(q')\sigma(q', q) = F(q) \Sigma_{q'} \sigma(q, q') \quad (q' \neq q), \quad (3.2.3)$$

where $F(q)$ symbolizes equilibrium fractions. For a given complete set of cross sections, $F(q)$ can be calculated either from full integration of Eq. (3.2.2) or from solving Eq. (3.2.3). In the example of Fig. 2, equilibrium is reached at a target thickness of only $x_\infty \simeq 10^{16}$ molecules/cm², corresponding to 0.53 µg/cm². This is an indication of the large magnitude of

[35] H. D. Betz and A. B. Wittkower, *Phys. Rev. A* **6**, 1485(1972).

charge-changing cross sections; since equilibrium requires more than one collision per ion to take place ($\bar{\sigma} x_\infty \gg 1$), relevant cross sections exceed 10^{-16} cm² and do indeed reach values as large as 7×10^{-15} cm²/molecule.[35]

The relation between charge-changing cross sections and charge states has been detailed frequently.[17,21,24] For this reason, we confine our considerations to the simplest possible case of a two-component system, which serves to illustrate some basic features. Equation (3.2.2) becomes in this particular case

$$dY_1/dx = -Y_1\sigma_l + Y_2\sigma_c, \qquad dY_2/dx = Y_1\sigma_l - Y_2\sigma_c, \qquad (3.2.4)$$

where $\sigma_l = \sigma(1, 2)$ and $\sigma_c = \sigma(2, 1)$ are the only two cross sections needed. Solution of Eq. (3.2.4) with the particular initial condition $Y_1(x = 0) = 1$ yields

$$Y_1(x) = F_1[1 + \exp(-\sigma_T x)], \qquad Y_2(x) = F_2[1 - \exp(-\sigma_T x)], \qquad (3.2.5)$$

where we use the abbreviation $\sigma_T = \sigma_c + \sigma_l$. Equilibrium fractions are given by

$$F_1 = \sigma_c/\sigma_T, \qquad F_2 = \sigma_l/\sigma_T. \qquad (3.2.6)$$

This example demonstrates that (1) charge equilibrium is reached for $\sigma_T x \gg 1$, where σ_T is the *sum* of capture and loss cross sections, and (2) equilibrium fractions depend only on *ratios* of cross sections. Qualitatively, this is no different in multicharge state systems.

Two remarks are in order, which pertain to the range of validity of Eq. (3.2.2). First, all ions must be in the ground state prior to a charge-changing collision, otherwise the cross sections are also functions of the degree of residual ion excitation and thus depend on x (see Section 3.8.1). Second, experimental equilibrium distributions are well defined only as long as x_∞ is a target thickness for which the average energy loss of the ions can be neglected. Normally, this condition is well fulfilled. In case of the example shown in Fig. 2, 15-MeV iodine ions lose only 10 keV energy in a 0.53 µg/cm² oxygen target. This trend is quite a general one and is mainly due to the relatively small energy loss per collision.

For a given equilibrium charge state distribution, we define the average charge by

$$\bar{q} = \Sigma_q\, qF(q), \qquad (3.2.7)$$

and a width parameter from the relation

$$d = [\Sigma_q\, (q - \bar{q})^2 F(q)]^{1/2}. \qquad (3.2.8)$$

These two quantities are often useful for an approximate characterization of $F(q)$, especially the average charge $\bar{q}$, which is discussed in further detail in Chapter 3.7. When $F(q)$ is a Gaussian distribution,

$$F(q) = d(2\pi)^{1/2} \exp[-(\bar{q} - q)^2/2d^2], \qquad (3.2.9)$$

the relation between width parameter d (standard deviation) and full half-width of the charge distribution H is given by $H = 2d(2 \ln 2)^{1/2}$.

3.3. Experimental Techniques and Data Analysis

3.3.1. Measurement of Charge-State Fractions

Experimental techniques for the measurement of charge-state fractions and charge-changing cross sections have been described by several authors.[21,22,23,24] Naturally, some of the early techniques have become outdated. Most of the experimental methods currently in use for the detection of few-component beams (hydrogen) have been detailed by Tawara and Russek.[25]

Figure 3 shows a typical arrangement for the measurement of charge fractions for light and heavy energetic ions. The three principal elements of such a setup are (a) the beam preparation system, (b) the target, which is either gaseous or solid, and (c) the charge-state detection systems, which are briefly discussed below.

At present, numerous ion accelerators are available for the production

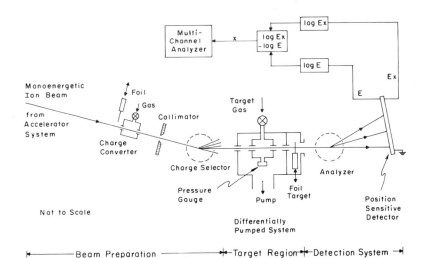

FIG. 3. Principal elements of experimental arrangements for the measurement of charge-state fractions with gaseous or solid targets.

of monoenergetic beams. Most of the data on heavy ions in the megaelectron volt range have been accumulated by utilizing tandem Van de Graaff accelerators[36] requiring injection of negative ions. Maximum terminal voltages of those machines reach and exceed 10 MeV and thus provide heavy-ion beams with energies E above some 100 MeV and DC intensities of up to $\sim 10^{11}$ particles/sec. The most important recent progress has been made in the development of versatile ion sources, in particular, sputtering sources,[37] which provide numerous negatively charged ions from the entire range of the periodic table. When very heavy ions are accelerated in any machine, it is essential that a vacuum can be maintained that does not attenuate the accelerated charge fraction due to charge exchange in the residual gas. A problem of this sort may arise for those ions that move with low velocities and thus have cross sections for charge exchange that are exceedingly large. The identity of the beam species (mass M) is normally established, especially in the case of large-intensity beams, by selection of particles with constant magnetic rigidity, $ME/q^2 = $ const. In this way, most mass impurities in the beam due to injection from the ion source, breakup of injected complex negative molecules in the terminal stripper, and ionized and accelerated components of the terminal stripper and of the residual gas are eliminated. It is obvious, however, that additional checks for the projectile mass may be necessary, especially when very low intensity beams are used. Then, additional electrostatic deflection, which is proportional to E/q, direct measurement of E with particle detectors, or observation of characteristic projectile x rays induced by collisions with any suitable target are helpful and, in the latter two cases, readily applied. Selection of the desired initial charge state is essential for comprehensive measurement of cross sections $\sigma(q, q')$. For this reason, a prestripper target followed by a beam deflector is needed.

Since charge-changing cross sections are mostly large, solid targets must always be relatively thin. When excessively large cross sections are dominant such as in the case of slow projectiles ($v \simeq v_0$, $\sigma_T \gtrsim 10^{-15}$ cm^2), target thicknesses near ~ 1 μg/cm^2 already produce charge state equilibrium. For faster ions, cross sections decrease and values of $x_\infty \gtrsim 100$ μg/cm^2 must be expected. Self-supporting foil targets are preferable to those with backings, but when backings are not avoidable the orientation of the target foil should be such that the backing faces the incident beam. As stated in Section 3.2.2. the energy loss of the beam in

---

[36] P. H. Rose and A. Galejs, in "Progress in Nuclear Techniques and Instrumentation" (F. J. Farley, ed.), p. 1–116. North-Holland, Amsterdam, 1967.

[37] R. Middleton and C. T. Adams, *Nuclear Instr. Meth.* **118**, 329(1974).

targets with minimum equilibrium thickness is mostly relatively small although not always negligible. In high $-Z_T$ targets, angular straggling of slow ions may give rise to problems with the charge state separation behind the target. Adequate estimates on small-angle multiple scattering of ions can be obtained, for example, from Sigmund and Winterbon[38] and references quoted therein.

Gaseous targets are used in the differentially pumped mode since any foil window would obscure the charge state distributions to be measured. When L (cm) and P (Torr) denote effective length and pressure, respectively, the total thickness of a gas target is given by

$$x = 9.66 \times 10^{18} LP/T \quad \text{molecules/cm}^2, \tag{3.3.1}$$

where T (K) is the temperature of the gas. Typical values are $P = 1$ Torr and $L = 10$ cm, which gives $x = 3.31 \times 10^{17}$ at room temperature (20°C). Such a maximum thickness suffices to produce charge equilibrium in beams that are not too fast. It should be pointed out that use of pressures in the Torr range does not guarantee that ions excited in one collision decay to the ground state prior to an average subsequent collision (compare Section 3.8.1).

High-density gas jet targets with mercury,[39] water,[40] carbon dioxide,[41] and other gases (CO_2, N_2, Ar)[42] have been used in the range of thicknesses between 4×10^{14} and 1.5×10^{17} molecules/cm². With nitrogen extreme densities near 10^{20} particles/cm³ and thicknesses up to ~ 1 mg/cm² have been reported,[43] and a liquid target (silicon oil droplets) has been constructed.[44] One of the strongest motivations for the development of dense gas or liquid targets lies in the hope of producing charge states as high as in a solid (see Section 3.8.2) for high-intensity heavy-ion beams without encountering the problem of shortlived foils. A thin carbon foil, for example, withstands a ~ 1-μA beam of ~ 5-MeV bromine

---

[38] P. Sigmund and K. B. Winterbon, *Nuclear Instr. Meth.* **119**, 541(1974).

[39] R. Beringer and W. Roll, *Rev. Sci. Instr.* **28**, 77(1957); I. M. Fogel, L. I. Krupnik, and V. A. Ankudinov, *Zh. Tekh. Fiz.* **26**, 1208(1956) [*Soviet Phys.-Tech. Phys.* **1**, 1181(1956)]; R. H. Dawton, *Nuclear Instr. Meth.* **11**, 326(1961); K. Bethge and G. Günther, *Z. Angew. Phys.* **17**, 548(1964).

[40] M. Roos, P. H. Rose, A. B. Wittkower, N. B. Brooks, and R. R. Bastide, *Rev. Sci. Instr.* **36**, 544(1965).

[41] B. Franzke, N. Angert, and Ch. Schmelzer, *IEEE* **NS-19**, No. 2, 266(1972).

[42] E. S. Borovik, F. I. Busol, V. B. Yuferov, and E. I. Sibenko, *Zh. Tekh. Fiz.* **33**, 973(1963) [*Soviet Phys.-Techn. Phys.* **8**, 724 (1964)]; F. I. Busol, V. B. Yuferov, and E. I. Sibenko, *Zh. Tekh. Fiz.* **34**, 2156(1964) [*Soviet Phys-Techn. Phys.* **9**, 1661(1965)].

[43] J. Ulbricht, G. Clausnitzer, and G. Graw, *Nuclear Instr. Meth.* **102**, 93(1972).

[44] A. Kardontchik and R. Kalish, *Nuclear Instr. Meth.* **131**, 399(1975).

ions only for minutes and particle intensities below ~ 100 nA must be chosen to attain foil lifetimes of hours.

The charge-state fractions produced in the target are spatially separated and detected (Fig. 3). When large beam intensities are available, the easiest detection method is to scan the fractions, one by one, with a simple Faraday cup, whereby fluctuating beam intensities must be monitored. It is more convenient, though, to detect the various charge state fractions simultaneously. For energetic ions several methods are available, including position-sensitive solid-state and proportional counters. Position-sensitive solid-state detectors are easily handled and can be operated with standard electronics; ions with energies above ~ 5 MeV can be detected, and count rates of up to $\sim 10^4$ particles/sec are feasible. Early results from this method were obtained by Moak et al.[45] for 10–180-MeV bromine and iodine ions. Excellent position resolution (fractions of 1 mm for a total length of 5 cm) can be obtained, i.e., more than ten adjacent states are readily recorded. The disadvantage lies in the short lifetime due to radiation damage. Usually, the tolerable maximum exposure amounts to $\sim 10^8$ particles/cm^2. This makes it difficult to detect relatively weak fractions in the presence of very strong ones.

Recently, many variations of position-sensitive, flow mode proportional counters have been constructed for efficient single-particle counting in connection with the needs of nuclear spectroscopy.[46] For example, versions employing a single wire,[47] two wires,[48] several (10) wires,[49] and thousands of wires[50] have been reported. For the purposes in charge state spectroscopy, the advantages of gas proportional counters are high count rate capability (up to $\sim 10^6$ particles/sec), almost arbitrary length while retaining good position resolution, and insensitivity to radiation damage. The entrance foil window is somewhat critical, but when the ions to be detected are sufficiently energetic, "safe" windows such as aluminized mylar with thicknesses in the 10-μm range can be used, which withstand the counter gas pressure and the exposure to the beam for a reasonably long time.

It should be mentioned that accelerators usually deliver ion beams with

---

[45] C. D. Moak, H. O. Lutz, L. B. Bridwell, L. C. Northcliffe, and S. Datz, *Phys. Rev.* **176**, 427(1968).

[46] D. L. Hendrie, *in* "Nuclear Spectroscopy and Reactions" (J. Cerny, ed.), Chap. III, C, p. 404. Academic Press, New York (1974).

[47] J. L. Ford, P. H. Stelson, and R. L. Robinson, *Nuclear Instr. Meth.* **98**, 199(1972).

[48] D. Shapira, R. M. Devries, H. W. Fulbright, J. Toke, and M. R. Clover, *Nuclear Instr. Meth.* **129**, 123(1975).

[49] R. G. Markham and R. G. Robertson, *Nuclear Instr. Meth.* **129**, 131(1975).

[50] P. Glässel, *Nuclear Instr. Meth.* **140**, 61(1977).

intensities in at least the nanoampere range (10^{10} particles/sec). When particle-counting techniques are used by means of solid-state detectors, beam intensities have to be reduced by six or seven orders of magnitude. This is no trivial task, and care must be taken to preserve a sharp beam energy by avoiding slit edge scattering and similar disturbing effects.

The measurement of charge-state fractions and equilibrium distributions for a given collision system is evident from the above. Further analysis of the charge fraction data is required when charge-changing cross sections are to be determined. The most straightforward method is to have a beam with a single charge state q_i impinge on a very thin gas target (single-collision condition) and to vary the target thickness. This results in an approximately linear increase of neighboring charge fractions and Eq. (3.2.2) reduces to

$$Y_q(x) \simeq \sigma(q_i, q)x. \qquad (3.3.2)$$

This procedure allows deduction of $\sigma(q_i, q)$ for fixed q_i and various q around q_i. A complete set of cross sections requires variation of q_i, i.e., a number of nonequilibrium distributions must be measured. A very accurate analysis of cross sections must take into account deviations from the linear approximation Eq. (3.3.2) and should be based on full consideration of the complete Eq. (3.2.2) by, say, employing standard least-squares fitting techniques.[24,51] Regarding the interpretation of experimentally determined charge-changing cross sections, we refer to the comments in Sections 3.2.1, 3.4.5, and 3.8.1.

3.3.2. Auxiliary Methods

The state of an ion after a collision is not necessarily specified by giving the ionic charge of the projectile; we have already mentioned that considerable excitation may occur that does not decay fast enough to allow it to be ignored. This situation is typical for dense targets and gives rise to research fields such as beam–foil spectroscopy.[52,53] Observed charge state distributions can therefore be influenced by excitation and in case of equi-

---

[51] S. Datz, H. O. Lutz, L. B. Bridwell, C. D. Moak, H. D. Betz, and L. D. Ellsworth, *Phys. Rev. A* **2**, 430(1970).

[52] *Proc. First Int. Conf. Beam-Foil Spectroscopy* (S. Bashkin, ed.). Gordon and Breach, New York, 1968; *Proc. Second Int. Conf. Beam-Foil Spectrosc.* (I. Martinson, J. Bromander, H. G. Berry, eds.) *Nuclear Instr. Meth.* **90**(1970); *Proc. Third Int. Conf. on Beam-Foil Spectrosc.* (S. Bashkin, ed.) *Nuclear Instr. Meth.* **110**(1972); *Proc. Fourth Int. Conf. Beam-Foil Spectrosc.* (I. A. Sellin and D. J. Pegg, eds.), Plenum Press, New York, 1976.

[53] D. J. Pegg, this volume, Part 10.

librium distributions we have to reckon with steady-state excitation. Thus, it is appropriate to point to methods for the detection of excited states.

The most convenient method for the measurement of excited states is to observe the decay products, electromagnetic radiation and electrons. In case of radiative transitions we distinguish two cases. When outer shells with low binding energies are involved, as is the case for very light and any slow projectile, transitions fall into or near the visible range; then well-developed techniques of beam–foil spectroscopy[52,53] can be applied to provide information on projectile excitation inside a gas or just outside a solid foil target. For not too light and highly stripped ions, shells with larger binding energies will be important and deexcitation produces x rays. Figure 4 shows a particular experimental setup for the measurement of x rays with intermediate [Si(Li) detector] and high resolution (crystal spectrometer), which is referred to in Chapters 3.9 and 3.10. Very thorough and detailed descriptions of x-ray techniques are given by Kauffman and Richard,[54] Hopkins,[55] and Bertin.[56]

Techniques for the measurement of Auger electrons are treated, for example, by Matthews.[57] Due to the limited penetration depth of low-

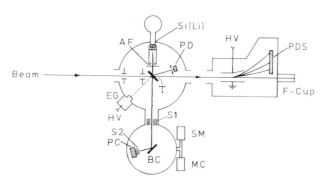

FIG. 4. Schematic setup for the measurement of collision-induced x rays with a solid state Si(Li) detector and a high-resolution system. AF, absorber foil; EG, electron gun; HV, high voltage; PD, particle detector; T, target; S1, S2, collimating soller slits; BC, Bragg crystal; PC, proportional counter; SM, stepping motor; MC, motor control; PDS, particle detection system.

[54] R. L. Kaufman and P. Richard, in "Methods of Experimental Physics" (D. Williams, ed.), Vol. 13, Chap. 3.2, p. 148. Academic Press, New York, 1976.
[55] F. Hopkins, this volume, Part 8.
[56] E. P. Bertin, "Introduction to X-Ray Spectrometric Analysis," Plenum, New York, 1978.
[57] D. L. Matthews, this volume, Part 9.

energy electrons in matter, decays cannot be observed from deep inside solids.

Metastable states such as the 2s state in hydrogenlike ions can be detected by quenching techniques, which are common in, say, Lamb shift measurements.[58] The metastable ions move into an electric or magnetic field region where Stark-mixing induces prompt 2p → 1s radiative transitions. A less direct technique is to utilize the difference of charge exchange cross sections for ground-state and excited ions[59] (see also Section 3.8.1). Very highly excited, long-lived states with quantum numbers $n \gg 1$ are detectable by means of direct field ionization.[60]

3.4. Electron Capture

3.4.1. Theory of Electron Capture in Simple Collision Systems

Nonradiative electron capture by a proton colliding with a hydrogen atom represents the simplest case that must be understood and solved before more complex capture collisions can be treated. For more than about 50 years, the collision p + H → H + p received appreciable theoretical attention and continues to stimulate efforts. One reason for the intricacy of the capture process lies in the fact that one deals, in contrast to treatment of excitation phenomena, with a kind of three-body problem, although the large electron/nucleon mass ratio allows easy determination of the ion trajectory. In the following, we briefly cite some of the important contributions, mostly in chronological order, and then describe attempts to treat heavier systems.

Oppenheimer[20] used the first Born approximation in a simplified version by considering only the interaction between the proton and the incident electron and neglecting the one between the two nuclei. He notes the use of *almost* orthogonal wavefunctions and derives an analytical expression for 1s → ns capture at large velocities, yielding the well-known approximate relation

$$\sigma_c(1s \longrightarrow ns) \propto n^{-3}. \tag{3.4.1}$$

---

[58] H. W. Kugel and D. E. Murnick, *Rep. Prog. Phys.* **40**, 297(1977).
[59] H. B. Gilbody, R. Browning, and G. Levy, *J. Phys. B* **1**, 230(1968); H. B. Gilbody, R. Browning, G. Levy, A. I. McIntosh, and K. F. Dunn, *J. Phys. B* **1**, 863(1968); H. B. Gilbody, K. F. Dunn, and B. J. Gilmore, *J. Phys. B* **7**, L187(1974).
[60] J. E. Bayfield, G. A. Khayrallah, and P. M. Koch, *Phys. Rev. A* **9**, 209(1974).

Brinkman and Kramers[61] use essentially the same technique and obtain in close analogy with Oppenheimer's work the analytical result (OBK)

$$\sigma_c(n_i s \longrightarrow ns)$$

$$= \frac{2^{18} v^8 Z_1^5 Z_2^5}{5 n^3 n_i^5 \left[v^4 + 2v^2 \left(\frac{Z_2^2}{n_i^2} + \frac{Z_1^2}{n^2} \right) + \left(\frac{Z_2^2}{n_i^2} - \frac{Z_1^2}{n^2} \right)^2 \right]^5} (\pi a_0^2), \quad (3.4.2)$$

where Z_1 and Z_2 refer to the nuclear charge of projectile and target, respectively, and a_0 denotes the Bohr radius. Similar work was also carried out by Massey and Smith.[62] It must be pointed out, though, that use of a sum rule for the Fourier transforms of hydrogenic wavefunctions[63,64] reveals that Eq. (3.4.2) does, in fact, represent $\sigma_c(n_i \rightarrow n)$, i.e., capture *averaged* over all initial (l, m) states in a target shell with principal quantum number n_i and *summed* over all final states in an initially empty projectile shell with principal quantum number n, and is then valid in a broader velocity range $(v \gtrsim v_0)$ and not just in the high-velocity limit.

Considerable and decisive improvement of the OBK approximation was achieved by taking into account the full interaction potential. Calculations of that kind were carried out by Bates and Dalgarno[65] for 1s → 1s transitions, Jackson and Schiff[66] for 1s → 1s, 2s, 2p capture in closed analytical form, Bates and Dalgarno[67] for transitions into all subshells with $n \leq 4$, and by Omidvar.[68] Dalgarno and Yadav[69] verified that the results from the revised OBK approximation agree with the ones from the perturbed stationary-state method. Pradhan[70] and Bassel and Gerjuoy[71] pointed out in detail that, although the inclusion of the internuclear Coulomb interaction in the OBK formalism (where approximately orthogonal wavefunctions are used) yields improved results, such an additional interaction is unphysical and can be circumvented by using exactly orthogonal wavefunctions, and he suggests the distorted-wave method. In particular, Pradhan[70] employs the impulse approximation and succeeds in

---

[61] H. C. Brinkman and H. A. Kramers, *Proc. Acad. Sci.*, Amsterdam **33**, 973(1930).
[62] H. S. Massey and R. A. Smith, *Proc. Roy. Soc. A* **142**, 142(1933).
[63] B. Podolsky and L. Pauling, *Phys. Rev.* **34**, 109(1929); V. Fock, *Z. Phys.* **98**, 145(1935).
[64] R. M. May, *Phys. Rev. A* **136**, 669(1964).
[65] D. R. Bates and A. Dalgarno, *Proc. Phys. Soc. A* **65**, 919(1952).
[66] J. D. Jackson and H. Schiff, *Phys. Rev.* **89**, 359(1953).
[67] D. R. Bates and A. Dalgarno, *Proc. Phys. Soc. A* **66**, 972(1953).
[68] K. Omidvar, *Phys. Rev. A* **12**, 911(1975).
[69] A. Dalgarno and H. N. Yadav, *Proc. Phys. Soc. A* **66**, 173(1953).
[70] T. Pradhan, *Phys. Rev.* **105**, 1250(1957); comments and corrections see T. Pradhan and D. N. Tripathy, *Phys. Rev.* **130**, 2317(1963); N. Urata and T. Watanabe, *J. Phys. B* **4**, L121(1971) and **5**, L49(1972); J. P. Coleman, *J. Phys. B* **5**, L19(1972).
[71] R. H. Bassel and E. Gerjuoy, *Phys. Rev.* **117**, 749(1960).

3.4. ELECTRON CAPTURE

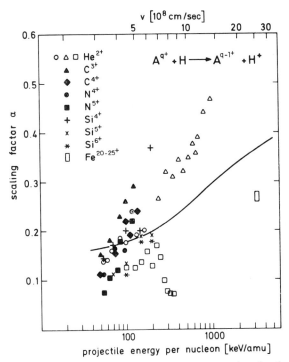

FIG. 5. Scaling factor obtained by dividing capture cross sections by the corresponding OBK cross sections, as a function of projectile velocity; broken line, eikonal approximation [F. T. Chan and J. Eichler, *Phys. Rev. Lett.* **42**, 58 (1979)].

employing orthogonal wavefunctions. A detailed presentation of the perturbation theories for electron capture is given by McDowell.[72] Born cross sections were calculated numerically for capture into various final states,[73] and Cheshire[74] obtained 1s → 1s cross sections in the framework of the impulse approximation. In all of these refined calculations capture cross sections are obtained that for obvious reasons are significantly smaller than the OBK values. The reduction factor ranges from ~8 at low velocities down to ~1.5 or, depending on the theory used, unity in the high-velocity limit, in fair agreement with abundant experimental data. Chan and Eichler[75] employ an eikonal treatment and show that capture cross sections scale to the OBK values with a Z-independent factor between 0.1 and 0.4. Their results are displayed in Fig. 5.

[72] M. R. McDowell, *Proc. Roy. Soc. A* **264**, 277(1961).
[73] R. A. Mapleton, *Phys. Rev.* **126**, 1477(1962).
[74] I. M. Cheshire, *Proc. Phys. Soc.* **82**, 113(1963).
[75] F. T. Chan and J. Eichler, *Phys. Rev. Lett.* **42**, 58(1979).

The second Born approximation has also been considered by several authors and we refer the reader to the discussions in Refs. 27, 28, and 29 and the references quoted therein. Although no detailed calculations have been carried out, Drisko[76] obtained an interesting result in the high-velocity limit,

$$\sigma_{B2} \propto (0.3 + 5\pi v/2^{12})\sigma_{OBK}, \quad v \ggg v_0,$$

which shows that the second Born term dominates only at very large collision velocities and is then proportional to v^{-11}. We note that it does not appear to be justified to reduce too large OBK values by a factor of 0.3 on the basis of the second Born term in the range of practically encountered, modest velocities.

Relativistic effects in charge transfer have been calculated by Mittleman[77] and the asymptotic form of capture has been treated by Mapleton[78] in first Born and distorted-wave approximations and by Coleman and McDowell[79] in the impulse approximation. Kleber and Nagarajan[80] used the distorted-wave Born approximation to derive a simple analytical correction factor for the OBK cross section in the high-velocity range. Results on the impact parameter dependence of the capture process have been given by Oppenheimer,[20] Schiff,[81] and Mittleman.[82]

The methods from above appear to be quite successful in a wide range of collision velocities, although the situation is still far from satisfactory. The case of extremely high velocities seems to remain controversial; for low velocities $v < v_0$ the Born approximation can become less successful, whereas distorted-wave techniques remain powerful, especially when the wavefunctions are expanded in terms of molecular wave functions.[34] Finally, it may be pointed out that a further problem concerns the post–prior discrepancy,[25,34,62] which can become particularly important in more complex, highly asymmetric collisions.

In the nonrelativistic regime, Coulomb capture can be worked out with high accuracy when one accepts elaborate numerical computations and does not insist on simple, analytical relations. This has been demonstrated for selected cases. A different approach to solving electron capture problems quite rigorously consists of numerical solutions of Hamilton's or Schrödinger's equation. The main advantage of such a pro-

---

[76] R. M. Drisko, Carnegie Inst. Tech. (1955), unpublished.
[77] M. H. Mittleman, *Proc. Phys. Soc.* **84**, 453(1964).
[78] R. A. Mapleton, *Proc. Phys. Soc.* **83**, 895(1964).
[79] J. Coleman and M. R. McDowell, *Proc. Phys. Soc.* **83**, 907(1964).
[80] M. Kleber and M. A. Nagarajan, *J. Phys. B* **8**, 643(1975).
[81] H. Schiff, *Can. J. Phys.* **32**, 393(1954).
[82] M. H. Mittleman, *Phys. Rev.* **122**, 499(1961).

3.4. ELECTRON CAPTURE

cedure is to circumvent the problems of choosing suitable basis sets of orthogonal wave functions throughout the collision process. Abrines and Percival[83] and Salop[84] use a Monte Carlo classical trajectory method, whereby it is assumed that the particles obey Newtonian laws during the collision. Maruhn-Rezwani et al.[85] proceed more rigorously and solve the time-dependent Schrödinger equation for the motion of the electron in a proton–hydrogen collision. Except for possible difficulties with numerical accuracy, the method works for any (nonrelativistic) collision velocity. For each impact parameter a separate calculation must be performed. Figure 6 shows results for the electron density in a head-on 20-keV p–H collision and demonstrates nicely the time-development of probabilities for charge transfer, target excitation, and ionization, which are necessarily obtained simultaneously.

The experimental situation for electron capture by protons in hydrogen and other target gases has been thoroughly reviewed.[25] Data are avail-

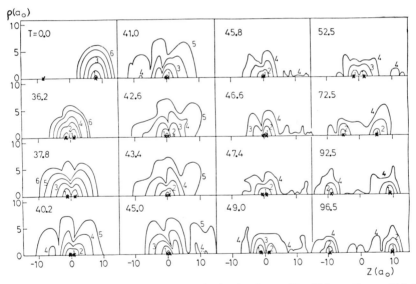

FIG. 6. Contour maps of the electron density in a head-on p–H collision at E_{Lab} = 20 keV in the center-of-mass system, for various elapsed times (given on the upper left of each frame in units of 10^{-17} sec). From one contour line to the next, the density changes by a factor of 10, indicated by the numbers for the negative exponents [V. Maruhn-Rezwani, N. Grün, and W. Scheid, Phys. Rev. Lett. **43**, 512 (1979)].

---

[83] R. Abrines and I. C. Percival, Proc. Phys. Soc. **88**, 861(1966).
[84] A. Salop, J. Phys. B **12**, 919(1979).
[85] V. Maruhn-Rezwani, N. Grün, and W. Scheid, Phys. Rev. Lett. **43**, 512(1979).

able from the lowest energies up to 38 MeV[86] and 600 MeV.[87] In the high-velocity range, radiative electron capture becomes important and probably dominates over Coulomb capture (Chapter 3.10.).

3.4.2. Electron Capture in Complex Systems

3.4.2.1. Theoretical Treatments.

Electron capture by protons in atomic hydrogen is not readily treated in a satisfactory manner, and thus it is not surprising that consideration of molecular hydrogen as a target results in additional complications[88] at least when accuracies of better than a factor of ~2 are desired. Considerable efforts have been devoted to electron capture by protons from helium,[89,90] helium ions from helium,[81,91] protons from heavy atoms,[92-97] and for more general systems.[25,27,98,99] Some of these authors have evaluated capture processes in view of the formation of multiplet states.[91,97] One of the interesting results on capture from heavy ions, which is easily understandable from Eq. (3.4.2), concerns the dominance of capture from inner target shells for sufficiently large projectile velocities. Calculations of the contributions from various shells of heavy targets seem to be in reasonable agreement with experimental data. Here, it is worthwhile to point to the consequence that capture of an inner target electron leaves an inner-shell vacancy and thus represents an often significant contribution to target excitation.

Rigorous treatments of very complex cases in which the incident projectile is also a multielectron system are not available. Bohr[15] and Bohr and Lindhard[17] derive several rough estimates for electron capture by fast fission fragments passing through light and heavy gases, which are in-

---

[86] E. Acerbi, M. Castiglioni, G. Dutto, F. Resmini, C. Succi, and G. Tagliaferri, *Nuovo Cimento* **50B**, 176(1967) and **64A**, 1068(1969).

[87] G. Raisbeck and F. Yiou, *Phys. Rev. A* **4**, 1858(1971).

[88] T. F. Tuan and E. Gerjuoy, *Phys. Rev.* **117**, 756(1960); G. Chatterjee and M. R. McDowell *J. Phys. B* **4**, 896(1971).

[89] B. H. Bransden and I. M. Cheshire, *Proc. Phys. Soc.* **81**, 820(1963); R. A. Mapleton, *Phys. Rev.* **130**, 1839(1963); R. A. Mapleton, *J. Phys. B* **1**, 529(1968).

[90] B. J. Szuster and K. E. Banyard, *Phys. Rev. A* **15**, 435(1977).

[91] T. G. Winter and C. C. Lin, *Phys. Rev. A* **12**, 434(1975).

[92] R. A. Mapleton, *Phys. Rev.* **130**, 1829(1963).

[93] R. A. Mapleton, *Phys. Rev.* **145**, 25(1966).

[94] V. S. Nikolaev, *Zh. Eksperim. i Teor. Fiz.* **51**, 1263(1966) [*Soviet Phys. JETP* **24**, 847(1967)].

[95] D. R. Bates and R. A. Mapleton, *Proc. Phys. Soc.* **90**, 909(1967).

[96] A. V. Vinogradov and V. P. Shevelko, *Zh. Eksperim. i Teor. Fiz.* **59**, 593(1970) [*Soviet Phys. JETP* **32**, 323(1971)].

[97] D. N. Tripathi and D. K. Rai, *Phys. Rev. A* **6**, 1063(1972).

[98] A. Dalgarno, in "Atomic Collision Processes" (M. R. McDowell, ed.), p. 609. North-Holland, Amsterdam, 1964.

[99] D. R. Bates and R. A. Mapleton, *Proc. Phys. Soc.* **87**, 657(1966).

tended to serve as guidelines but seem to be of little general usefulness. They realize that a projectile with a very high ionic charge can disturb the target electrons to be captured to such a large extent that there is no more theoretical justification for perturbation techniques such as the Born approximation. This is one of their motivations for arguing on a classical basis. Along these lines, Bell[100] outlines a method for calculating capture cross sections for highly charged heavy ions ($Z = 40, 50; q \leq 20$).

It is extremely difficult to specify the range of validity of particular theoretical techniques and most authors avoid confrontation with this important question. A sufficient condition for applicability of perturbation techniques is to require that the projectile ion does not appreciably alter the wavefunction of the target electron. Small perturbation during a collision may be expressed by $\Delta p \, \Delta b < \hbar$, where Δp is the momentum transfer in the most important region of impact parameters and Δb represents the associated uncertainty of the electron's location. When we take into account that the electron orbits with velocity v_e, simple though nontrivial evaluation reveals approximate equivalence of the above condition with

$$q < \begin{cases} Z_T^*, & v < v_e \\ v/v_0, & v \geq v_e \end{cases} \quad (3.4.3)$$

where q and Z_T^* may be understood as effective (screened) charges of projectile and target with respect to the considered target electron. For practical purposes, it is also useful to check whether average electron capture probabilities that result from perturbation techniques are small[27] ($\bar{P}_c \ll 1$) for the system of interest as well as for the reversed collision system (post–prior discrepancy). We note that Eq. (3.4.3) implies that OBK cross sections cannot be obtained for slow heavy-ion collisions in which the target electron moves along transiently formed quasi-molecular orbitals.

Of course, numerical solution of the time-dependent Schrödinger equation[85] is expected to represent a quite rigorous technique also for complex collision systems whereby wavefunctions are computed by Hartree–Fock methods. However, computational efforts involved are tremendous and no results are as yet available.

3.4.2.2. Semiempirical Procedures. There have been several attempts to modify basic results for electron capture in simple collision systems to almost any ion–atom collisions. According to the characteristics of its development, semiempirical procedures work quite well as

[100] G. I. Bell, *Phys. Rev.* **90**, 548(1953).

long as experimental data are considered in a restricted range of parameters, but they become less useful when some generality is demanded. Only a few of the more successful procedures are mentioned: Gluckstern[101] modified Bell's treatment[100] for ions in the range $8 \leq Z \leq 18$ and $3 \leq v/v_0 \leq 7$, and Nikolaev varies Bohr's[15] approximation for fast nitrogen ions in various gases[102] and presents a number of cross section formulas for a variety of cases involving modestly heavy, fast ions.[23] He also describes a modified OBK version for protons in any target atom.[94]

Generalization of the Brinkman–Kramers cross section Eq. (3.4.3) for heavy systems can be attempted by separating Z_1^5 into $q^2 Z_1^3$ and introducing binding energies $E_{i,f}$ instead of quantum numbers $n_{i,f}$. This yields

$$\sigma_c = \frac{8.53 \times 10^{-16} q^2 E_i^{5/2} E_f^{3/2} E_K^4}{[E_K^2 + 2E_K(E_i + E_f) + (E_i - E_f)^2]^5} \quad \text{cm}^2, \qquad (3.4.4)$$

where E_K denotes the kinetic energy of an electron traveling with projectile velocity v, E_i and E_f stand for the binding energy of the transferred electron before and after capture, respectively, and all energies (E_K, E_i, E_f) are to be inserted in units of kilo-electron volts. For one-electron systems, Eqs. (3.4.2) and (3.4.4) are identical. As in Eq. (3.4.2), initial and final principal shells are singly occupied and empty, respectively. According to the results indicated in Section 3.4.1, one must expect that Eq. (3.4.4) overestimates the true capture cross section by a factor that may be as large as ~ 8 but will be much less, especially in fast collisions.* Otherwise, Eq. (3.4.4) is expected to reproduce cross section trends fairly well on a physically sound basis. The availability of binding energies from Hartree–Fock calculations, tabulated for neutral atoms[104] (target) and calculable for any desired projectile electron configuration,[105] renders Eq. (3.4.4) particularly useful and versatile.

Several difficulties must be pointed out: in practice, capture must often be estimated for partially filled shells. Then, subshell transfer cross sec-

---

[101] R. L. Gluckstern, *Phys. Rev.* **98**, 1817(1955).

[102] V. S. Nikolaev, *Zh. Eksperim. i Teor. Fiz.* **33**, 534(1957) [*Soviet Phys. JETP* **6**, 417(1957)].

[103] J. R. Macdonald, M.D. Brown, S. J. Czuchlewski, L. M. Winters, R. Laubert, I. A. Sellin, and J. R. Mowat, *Phys. Rev. A* **14**, 1997(1976).

[104] C. C. Lu, T. A. Carlson, F. B. Malik, T. C. Tucker, and C. W. Nestor, *Atom. Data* **3**, 1(1971).

[105] J. P. Desclaux, *Comp. Phys. Comm.* **9**, 31(1975).

---

* There have been claims that this factor exceeds ~ 10 in, for example, 132-MeV Cl^{17+} collisions with argon.[103] Our estimate from Eq. (3.4.4) with realistic binding energies yields a factor of no more than 4.

tions must be known, but there is no easy way to obtain these without elaborate calculations. Nikolaev[94] worked out an analytical approximation for capture by protons from a few specific target subshells using the OBK approximation and generalizing results for electron capture obtained by May and Lodge.[106] In a most general case one would apply the OBK procedure, on a numerical basis, for each angular-momentum state without using the sum rules,[63,64] which allow for the simple analytical description Eq. (3.4.2). In typical heavy-ion–atom collisions, the velocity v will not be large enough to render s states dominant and one must reckon with important and even dominant contributions from p and higher states. Furthermore, one must take into account varying degrees of screening, especially during close collisions such as, for example, additional screening of large projectile charges due to outer target electrons when inner target electrons are captured. Finally, we mention that the OBK formalism represents only a single, direct capture process and ignores interaction between the target and an electron already captured. These postcollision effects may be influential in case of inner-shell capture from very heavy target atoms and may also cause a reduction of capture contributions into projectile states with large values of n_f, where binding energies E_f are relatively low, i.e., depending on v and Z_T the summation over final states in a calculation of the total capture cross section should be altered when E_f becomes too small.

3.4.3. Experimental Results on Electron Capture Cross Sections

In Section 3.1.3 we mentioned review articles and data collections[23–25,30] that cover experimental results on electron capture cross sections by protons,[25] various slow ions with charge states $q \leq 2^+$,[30] several ion species in the range $2 \leq Z \leq 18$ with velocities up to $5v_0$ and charge states $q \leq 7^+$,[23] and heavy ions ($16 \leq Z \leq 92$) with charge states that extend beyond 20^+.[24]

Nikolaev et al.[107] present capture cross sections in nitrogen for nitrogen ions with $v \leq 5v_0$ and $q \leq 4^+$ and find an empirical formula for $\sigma_c(v, q, Z_T)$. Extension of that work yields results on more ions (Z = 2, 3, 5, 7, 10, 15, 18) with $v \leq 5v_0$ and $q \leq 7^+$ passing through helium, nitrogen, argon, and krypton.[108] They find approximately the same dependence of

---

[106] R. M. May and J. Lodge, *Phys. Rev.* **137A**, 699(1965).

[107] V. S. Nikolaev, L. N. Fateeva, I. S. Dmitriev, and Y. A. Teplova, *Zh. Eksperim. i Teor. Fiz.* **33**, 306(1957) [*Soviet Phys. JETP* **6**, 239(1957)].

[108] V. S. Nikolaev, I. S. Dmitriev, L. N. Fateeva, and Y. A. Teplova, *Zh. Eksperim. i Teor. Fiz.* **40**, 989(1961) [*Soviet Phys. JETP* **13**, 695(1961)].

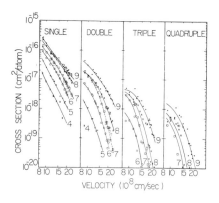

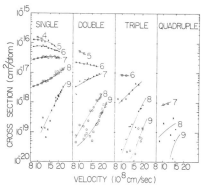

FIG. 7. Cross sections for single- and multiple-electron capture (top) and loss (bottom) of fluorine ions passing through argon, as a function of ion velocity. Parameter is the initial charge state indicated near each curve [S. M. Ferguson, J. R. Macdonald, T. Chiao, L. D. Ellsworth, and S. A. Savoy, *Phys. Rev. A* **8**, 2417 (1973)].

$\sigma_c(v)$ for all Z, v, and Z_T combinations. Pivovar et al.[109] measured $\sigma(1, 0)$ of He$^+$ in gases ($Z_T = 2, 7, 18, 36$) for velocities $v \leq 4v_0$, Ryding et al.[110] determined cross sections of aluminum in nitrogen for $v \leq 2.4v_0$ and $q \leq 6^+$, and Lo et al.[111] obtain results for nine different singly charged ions in atomic oxygen in the velocity range below $2v_0$.

In the range of higher velocities, Reynolds et al.[112] studied capture by nitrogen ions ($v = 8.6v_0$) in thin Zapon foils, Macdonald and Martin[113]

[109] L. I. Pivovar, V. M. Tubaev, and M. T. Novikov, *Zh. Eksperim. i Teor. Fiz.* **41**, 26(1961) [*Soviet Phys. JETP* **14**, 20(1962)].

[110] G. Ryding, A. Wittkower, G. Nussbaum, A. Saxman, and P. H. Rose, *Phys. Rev. A* **2**, 1382(1970).

[111] H. H. Lo, L. Kurzweg, R. T. Brackmann, and W. L. Fite, *Phys. Rev. A* **4**, 1462(1971).

[112] H. L. Reynolds, L. D. Wyly, and A. Zucker, *Phys. Rev.* **98**, 1825(1955).

[113] J. R. Macdonald and F. W. Martin, *Phys. Rev. A* **4**, 1965(1971).

3.4. ELECTRON CAPTURE

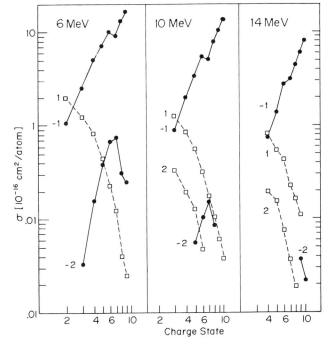

FIG. 8. Cross sections $\sigma(q, q + n)$ for electron capture (solid lines) and electron loss (dashed lines) by bromine ions passing through helium with energies of 6, 10, and 14 MeV, as a function of the initial ionic charge state q. The values of n are indicated near each curve [H. D. Betz, G. Ryding, and A. B. Wittkower, *Phys. Rev. A* **3**, 197 (1971)].

examine oxygen ($v \leq 10v_0$) in helium, nitrogen, and argon for charge states from 2+ to 8+. Similar work was performed for fluorine ($v \leq 10.7v_0$) in nitrogen[114] and argon.[115] Results from the latter work are shown in Fig. 7.

As regards highly charged heavy ions, there are numerous results for bromine and iodine. Angert *et al.* report capture by iodine in nitrogen[116] and argon[117] ($1.2 \leq v/v_0 \leq 4.2$ and $4^+ \leq q < 19^+$). Further studies of iodine ions have been carried out by Wittkower and Betz[118] in complex mol-

---

[114] J. R. Macdonald, S. M. Ferguson, T. Chiao, L. D. Ellsworth, and S. A. Savoy, *Phys. Rev. A* **5**, 1188(1972).

[115] S. M. Ferguson, J. R. Macdonald, T. Chiao, L. D. Ellsworth, and S. A. Savoy, *Phys. Rev. A* **8**, 2417(1973).

[116] N. Angert, B. Franzke, A. Möller, and Ch. Schmelzer, *Phys. Lett.* **27A**, 28(1968).

[117] N. Angert, B. Franzke, U. Grundinger, W. Kneis, A. Möller, and Ch. Schmelzer, *IEEE Trans. Nuclear Sci.* **NS-19**, No. 2, 263(1972).

[118] A. B. Wittkower and H. D. Betz, *J. Phys. B* **4**, 1173(1971).

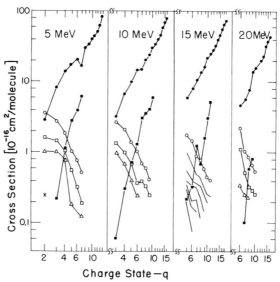

FIG. 9. Cross sections $\sigma(q, q + n)$ for iodine ions passing through oxygen with energies of 5, 10, 15, and 20 MeV, as a function of the initial charge state q. The full symbols refer to electron capture, $n = -1$ (●), $n = -2$ (■), and the open symbols refer to electron loss, $n = 1$ (○), $n = 2$ (□), $n = 3$ (△), and $n = 4$ (×). The location of the values of the multiple-electron loss cross sections at 15 MeV is indicated by solid lines for $2 \leq n \leq 7$ [H. D. Betz and A. B. Wittkower, *Phys. Rev. A* **6**, 1485 (1972)].

ecules ($v = 1.9v_0$; $q = 5^+$), Betz et al.[119] in helium ($1.4 \leq v/v_0 \leq 2.2$; $2^+ \leq q \leq 10^+$), Betz and Wittkower[35] in hydrogen and oxygen ($1.2 \leq v/v_0 \leq 2.8$; $2^+ \leq q \leq 15^+$), and Franzke et al.[41] in carbon dioxide ($1.8 \leq v/v_0 \leq 4.9$; $3^+ \leq q \leq 15^+$). Results for bromine ions have been reported by Datz et al.[51] in hydrogen, helium, and argon ($v/v_0 = 2.6$ and 3.6; $3^+ \leq q \leq 11^+$) and Betz et al.[119] in hydrogen and helium ($1.7 \leq v/v_0 \leq 2.7$; $2^+ \leq q \leq 10^+$).

Most of the trends obtained for $\sigma_c(Z, q, v, Z_T)$ look very similar, although the magnitude of the cross sections varies considerably. We display some typical results for $\sigma_c(q)$ with bromine ions in helium in Fig. 8, with iodine ions in oxygen in Fig. 9, and for $\sigma_c(v)$ with bromine and iodine in helium in Fig. 10.

In the low-velocity range $v \simeq v_0$, one finds a broad maximum of $\sigma_c(v)$ for all ion and target species,[23,24,30] whereas a more or less steep decrease is observed for higher collision velocities.[113–115] Compared with p → H collisions we note two additional effects in the behavior of $\sigma(v)$. First, in heavy targets, capture from inner shells (E_i) contributes significantly

---

[119] H. D. Betz, G. Ryding, and A. B. Wittkower, *Phys. Rev. A* **3**, 197(1971).

when v becomes sufficiently large and, second, since heavier ions become more highly stripped when v increases, additional final states (E_f) for electron capture are provided. For these reasons, $\sigma_c(v)$ in heavy-ion–atom encounters will not exhibit the strongest velocity dependence $\sigma_c \propto v^{-12}$ until v exceeds the K-shell velocity of the target electrons [$E_K \gg \max(E_f)$]. In the intermediate velocity range, therefore, descriptions of the data in the form $\sigma_c \propto v^{-m}$ will result in widely varying values of m. In essence, these features of $\sigma_c(v)$ are at least in qualitative agreement with the OBK approximation Eq. (3.4.4), although the limitations such as $P \ll 1$ (Section 3.4.1.) should not be ignored.

For decreasing velocity in the range $v \lesssim v_0$, however, it will be necessary to give attention to the fact that the collisions become increasingly adiabatic so that Eq. (3.4.4) is no longer justified. Details on the location

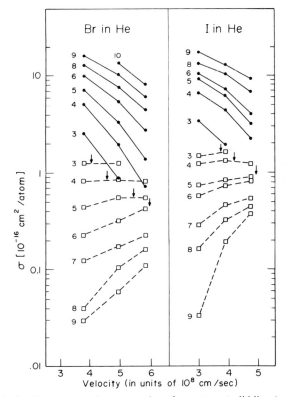

FIG. 10. Velocity dependence of cross sections for capture (solid lines) and loss (dashed lines) of a single electron by bromine and iodine ions passing through helium. The initial ionic charge state of the projectiles is indicated near each curve [H. D. Betz, G. Ryding, and A. B. Wittkower, *Phys. Rev. A* **3**, 197 (1971)].

of the maximum of $\sigma_c(v)$ have been of particular interest both theoretically[120-122] and experimentally.[123-126]

When a larger number of projectile charge states is available, it becomes interesting to investigate the trend $\sigma_c(q)$ for constant parameters Z, v, and Z_T. Again, the gross features are well described by Eq. (3.4.4); it is obvious that the q^2 dependence will mostly be strong though one should not expect it to become pure until v becomes large enough. It is not surprising, therefore, that empirical attempts to reproduce experimental data in the form $\sigma_c \propto q^p$ yield very different powers p, which are mostly larger than 2 and may still not be independent of q in a large range of charge states.[23,35,107,108,119] The situation is nicely illustrated in Fig. 11; one finds experimentally that p is large when v or q is low, while p seems to approach the value of 2 when v and q becomes sufficiently large. Moreover, when $p \simeq 2$, the absolute magnitude of σ_c becomes independent of the ion species. These results can be understood fairly well from Eq. (3.4.4): the q dependence lies primarily in the term q^2 and, to a varying extent, in E_f. For low q, the possible values of E_f are limited, but when q increases more final states become available and capture proceeds with maximum probability into a range of projectile states that can be estimated from Eq. (3.4.4). Further stripping will no longer affect σ_c/q^2 and the structure of the ion will not enter very specifically. Formation of excited states due to electron capture is then very likely.

The occurrence of shell effects has also been observed in both $\sigma(q)$[35,44,55,108,117,119] and $\sigma(Z_T)$.[23,127] These effects are relatively small and, in most cases reported, appear especially in slow collisions, which are more difficult to describe. Very recently, more pronounced oscillations of $\sigma_c(q)$ have been observed[128] in collisions between heavy ions (Ta, W, Au; $v < 2v_0$) and hydrogen (Fig. 12). The oscillatory behavior is attrib-

[120] G. F. Drukarev, *Zh. Eksperim. i Teor. Fiz.* **37**, 847(1959) [*Soviet Phys. JETP* **37**, 603(1960)].

[121] H. S. Massey, *Rep. Prog. Phys.* **12**, 248(1948).

[122] G. F. Drukarev, *Zh. Eksperim. i Teor. Fiz.* **52**, 498(1967) [*Soviet Phys. JETP* **25**, 326(1967)].

[123] I. M. Fogel, V. A. Ankudinov, and D. V. Pilipenko, *Zh. Eksperim. i Teor. Fiz.* **35**, 868(1958) [*Soviet Phys. JETP* **35**, 601(1959)].

[124] S. K. Allison, J. Cuevas, and M. Garcia-Munoz, *Phys. Rev.* **120**, 1266(1960).

[125] J. K. Layton, R. F. Stebbings, R. T. Brackmann, W. L. Fite, W. R. Ott, C. E. Carlston, A. R. Comeaux, G. D. Magnuson, and P. Mahadevan, *Phys. Rev.* **161**, 73(1967).

[126] L. I. Pivovar, L. I. Nikolaichuk, and A. N. Grigorev, *Zh. Eksperim. i Teor. Fiz.* **57**, 432(1969) [*Soviet Phys. JETP* **30**, 236(1970)].

[127] Y. S. Volodyagin, I. S. Dmitriev, V. S. Nikolaev, Y. A. Tashaev, and Y. A. Teplova, *J. Phys. B* **6**, L171(1973).

[128] H. J. Kim, P. Hvelplund, F. W. Meyer, R. A. Phaneuf, P. H. Stelson, and C. Bottcher, *Phys. Rev. Lett.* **40**, 1635 (1978); F. W. Meyer, R. A. Phaneuf, H. J. Kim, P. Hvelplund, and P. H. Stelson, *Phys. Rev. A* **19**, 515(1979).

3.4. ELECTRON CAPTURE

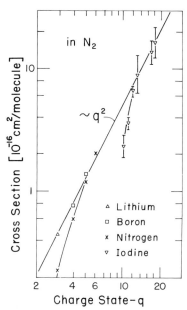

FIG. 11. Cross sections $\sigma(q, q - 1)$ for capture of a single electron by various ions of the same velocity $v = 4.17v_0$ in collision with nitrogen [N. Angert, B. Franzke, A. Möller, and Ch. Schmelzer, *Phys. Lett.* **27A**, 28 (1968)].

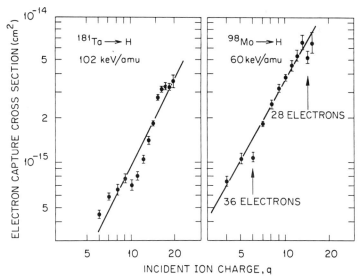

FIG. 12. Experimental electron capture cross sections for tantalum and molybdenum ions in atomic hydrogen, as a function of initial charge state [H. J. Kim *et al.*, *Phys. Rev. Lett.* **40**, 1635 (1978)].

uted to interference effects caused by the long-range Coulomb and a short-range, screened Coulomb force acting on the captured electron. The authors estimate these interference terms, but it remains to be seen whether their explanation withstands a more thorough investigation. It is often found that electron capture cross sections for slow ions of low charge are much smaller in helium than in any other target.[129] Qualitatively, this decrease can be understood from Eq. (3.4.4) as a result of the energetic mismatch of low E_f and large E_i. Electron capture has also been investigated for specific final states. Abundant data are available for capture into 2s states[59] and 3s states[130] at low collision velocities.

3.4.4. Multiple Electron Capture

In single collisions between ions and atoms, the simultaneous capture of several electrons can take place. Double-electron capture has been reported by Allison[131] for α particles moving through molecular hydrogen and helium ($1.2 \leq v/v_0 \leq 2.1$). Nikolaev et al.[132] measured cross sections for simultaneous capture of two, three, and four electrons in single collisions between 11 different ions ($2 \leq Z \leq 36$) in several gases for velocities from $1.2v_0$ to $5.5v_0$. Further results for multiple capture by fast ions ($v > v_0$) are available for oxygen,[113] fluorine,[114,115] aluminum,[110] bromine,[119] and iodine[35,119] in a variety of target gases. In the low-velocity range ($v < v_0$) multiple capture has been observed, for example, for α particles,[133] ions of carbon, nitrogen, oxygen,[134] and argon.[135] Some of the results for heavy and highly charged ions are contained in Figs. 8, 9, 13, and 14.

Cross sections for double-electron capture can amount to an appreciable fraction of the corresponding cross section for capture of a single electron. For example, in collisions between 5-MeV iodine ions and oxygen, ratios $k_n = \sigma(q, q + n)/\sigma(q, q - 1)$ as large as 0.36 result for $n = -2$ and $q = 7+$ (Fig. 9). Values of $k_{-2} \simeq 0.2$ are found in many other collision systems. By contrast, multiple-capture events can also

---

[129] A. B. Wittkower and H. B. Gilbody, *Proc. Phys. Soc.* **90**, 353(1967).

[130] B. M. Doughty, F. L. Brandon, C. W. Bray, R. W. Cernosek, and M. L. Goad, *Phys. Rev. A* **17**, 59(1978).

[131] S. K. Allison, *Phys. Rev.* **109**, 76(1958).

[132] V. S. Nikolaev, L. N. Fateeva, I. S. Dmitriev, and Y. A. Teplova, *Zh. Eksperim. i Teor. Fiz.* **41**, 89(1961) [*Soviet Phys. JETP* **14**, 67(1962)].

[133] R. A. Baragiola and I. B. Nemirovsky, *Nuclear Instr. Meth.* **110**, 511(1973).

[134] D. H. Crandall, M. L. Mallory, and D. C. Kocher, *Phys. Rev. A* **15**, 61(1977).

[135] A. Müller, H. Klinger, and E. Salzborn, *Phys. Lett.* **55A**, 11(1975); H. Klinger, A. Müller, and E. Salzborn, *J. Phys. B* **8**, 230(1975); E. Salzborn, *IEEE Trans. Nuclear Sci.* **NS-23**, No. 2, 947(1976); H. Klinger, A. Müller, and E. Salzborn, *J. Chem. Phys.* **65**, 3427(1976).

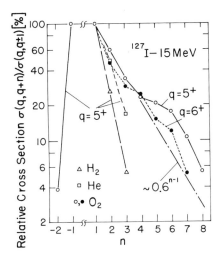

FIG. 13. Cross section ratios $\sigma(q, q + n)/\sigma(q, q \pm 1)$, in percent, for multiple-electron capture and loss of 15-MeV iodine ions with initial charge $q = 5+$ passing through hydrogen ($\triangle$), helium ($\square$), and oxygen ($\bigcirc$), and with initial charge $q = 6+$ passing through oxygen ($\bullet$) [H. D. Betz, *Rev. Mod. Phys.* **44**, 465 (1972)].

become exceedingly small. For 10-MeV iodine with $q = 4^+$ in helium, a ratio $k_{-2} = 0.006$ has been found,[35] and still smaller relative values occur but are not easily detected in experiments. Cross sections for capture of more than two electrons are usually identified only for $k_n \geqslant 0.01$; the reason for this lies in the difficulty of extracting very small cross sections in the presence of very large ones from the system of rate equations [Eqs. (3.2.2)].

When one considers m individual equivalent target electrons as being independent of each other with respect to capture by an ion, one obtains the approximate relation[23,136]

$$k_{-2} \simeq \frac{\int P_{cT}^2(b)\,db^2}{\int P_{cT}(b)\,db^2} \simeq \frac{0.5m(m-1)\bar{P}_c^2 \pi \bar{R}^2}{m\bar{P}_c \pi \bar{R}^2}$$

$$= \frac{m-1}{2}\bar{P}_c, \quad \bar{P}_c \ll 1, \quad (3.4.5)$$

i.e., the ratio k_{-2} essentially reflects the average probability for capture of a single electron in a single collision. Of course, effects due to different energies of initial and final states for the two captured electrons are ig-

[136] F. W. Martin and J. R. Macdonald, *Phys. Rev. A* **4**, 1974(1971).

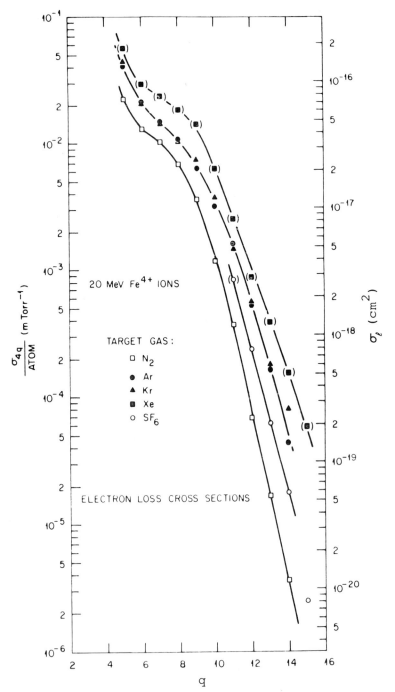

FIG. 14. Single- and multiple-electron loss cross sections for 20-MeV iron ions with initial charge 4+ passing through various target gases, as a function of final charge q [from Knudsen *et al.*, *Phys. Rev. A* **19**, 1029 (1979)].

3.4. ELECTRON CAPTURE

nored in Eq. (3.4.5). The important region of impact parameters for electron capture follows immediately from Eq. (3.4.5),

$$\pi \bar{R}^2 \simeq [(m - 1)\sigma(q, q - 1)]/2mk_{-2}. \tag{3.4.6}$$

Although it must be remembered that the capture probability $P_c(b)$ exhibits a rather weak dependence on b,[61,81] it is an interesting and important result that simple measurement of capture cross sections allows a crude, but direct estimate on $\bar{P}_c$ and $\bar{R}$. The ratio k_{-n} has also been examined as a function of q, n, v, and Z_T. It has been found, for example, that $k_{-2}(v)$ exhibits a maximum,[114,115,136] but a consistent explanation is not readily at hand.

To the above, a decisive argument has to be added. It has been demonstrated before that capture proceeds often into excited states. According to Eq. (3.4.4) this situation arises especially when the ion is highly stripped; fully stripped heavy ions, for example, may capture from light target atoms with higher probability into the L shell than into the empty K shell. When capture into excited states is more likely to occur than capture into the ground state, double- or multiple-electron capture will with high probability result in the formation of doubly excited projectile states. Then, deexcitation via emission of Auger electrons will be a possible consequence and will cause original double capture to appear as a single-capture process. For this reason, many of the reported cross sections for multiple-electron capture may have to be reinterpreted.

Incidentally, the shell effect in $\sigma(q, q - 2)$ for bromine and iodine ions visible in Figs. 8 and 9 may be explained in terms of multiple excitation. Iodine with charge 7+ has a closed N shell and both single and double capture proceed into the O shell (and into higher shells). When we consider 8+ ions, it may be argued that capture into the O and higher shells is still very likely; double capture into these shells, however, leaves a vacancy in the iodine N shell, which may be filled by an Auger transition and increases the charge state of the ion. Contributions from such processes will cause a decrease of effective double capture, and just this behavior is observed in the experiments for $q = 8+$. Similar arguments can be applied for 8+ bromine ions.

In very slow collisions, very large multiple capture cross sections are found[135] and it may occur that double capture exceeds single capture.[137] This is an indication of electron transfers between quasi-molecular orbitals, i.e., both projectile and target electrons are highly disturbed and molecular-potential curve models must be invoked.

---

[137] J. E. Bayfield and G. A. Khayrallah, *Phys. Rev. A* **11**, 920(1975); D. H Crandall, R. E. Olsen, E. J. Shipsey, and J. C. Browne, *Phys. Rev. Lett.* **36**, 858(1976).

3.4.5. Very Large Charge Exchange Probabilities

Experimentally determined total cross sections for capture or loss of target electrons by ions are directly and uniquely interpreted in terms of a single physical interaction only when all probabilities P for charge-changing processes in the collision in question are small compared to unity. Measurements have revealed that actual cross sections can become extraordinarily large, values of $\sim 10^{-15}$ cm² are not rare, and exceed simple geometrical ones ($\pi a_0^2 = 8.8 \times 10^{-17}$ cm²). In such cases, the relation $P \ll 1$ will not necessarily remain fulfilled and the so-called single-collision condition tends to lose significance in a single ion–atom collision. In other words, multiple interactions are likely to take place and actually observed cross sections reflect the combined effect of these interactions.

As an example, we consider such a collision in which a very large loss cross section $\sigma_l \simeq 10^{-15}$ cm² and a small capture cross section $\sigma_c \simeq 10^{-17}$ cm² have been found experimentally. Interpretation of σ_l is then difficult because of $\overline{P}_l \lesssim 1$, but σ_c is also not easily related to a pure capture process: the actually measured probability is more closely represented by $\overline{P}_c(1 - \overline{P}_l)$ than by $\overline{P}_c$ alone. A more precise description requires an impact-parameter formulation that will be quantitatively different for capture and loss. It must be concluded that results for small cross sections obtained from collisions in which exceedingly large cross sections with $P \lesssim 1$ are also effective may have to be reinterpreted when comparison with theoretical models is intended. Practical significance of the measured values as effective cross sections remains unaffected.

3.4.6. Capture into Very High Quantum States

As early as in 1928 Oppenheimer[20] predicted that electron capture proceeds into final principal shells according to $\sigma_c(n) \propto n^{-3}$ (see Section 3.4.1.), provided that the collision velocity is not too small. All subsequent calculations confirmed this early prediction. It cannot be claimed, however, that the n^{-3} hypothesis has been put to a stringent experimental test since all investigations suffer due to experimental uncertainties, low collision velocity, complex systems, or restricted range of observed states. As of today, it appears that the available data are not at variance with theory. For example, Bayfield et al.[60] investigated the production of final states with $13 \leq n \leq 28$ for 7–60 keV protons on atomic hydrogen and argon. They do not detail $\sigma_c(n)$ and find approximate agreement with OBK theory for the sum of capture cross sections $\Sigma_n \sigma_c(n)$.

The situation becomes much more complicated when the formation of l

3.4. ELECTRON CAPTURE

substates is differentiated. For this case, theoretical results come from the first Born approximation, but different investigators present somewhat different results. For example, some authors represent the cross sections by $\sigma_c(n, l) = \sigma_{OBK}(n) * f(l)$ with $\sigma_{OBK}(n) \propto n^{-3}$, while others do not factor n- and l-dependent terms. May[138] and Omidvar[68] arrive at leading terms ($v \gg v_0$)

$$f(l) = 5 \cdot 2^{2l}[(5 + l)(2l + 1)!!(2l - 1)!!(v/v_0)^{2l}]^{-1} \quad (3.4.7)$$

where $\Sigma_l f(l) = 1$. Vinogradov et al.[139] and Shevelko[140] use

$$f(l) \propto (e/2l)^{-2l}, \quad (3.4.8)$$

whereas Hiskes[141] indicates as dominant coefficient

$$\sigma_c(n, l) \propto [2^{2l} l!(n - l - 1)!/(n + l)!]^2. \quad (3.4.9)$$

Basu et al.[29] discuss a more complicated relation for $\sigma_c(n, l)$, which we do not reproduce here.

All of the above results are approximations of the same problem and are in agreement with Oppenheimer's conjecture that capture into states with higher l becomes increasingly unlikely. Especially at large collision velocities, capture into ns states dominates. This result can be qualitatively understood on the basis of overlap of initial and final wavefunctions in momentum space when one takes into account that wavefunctions become less broad when l increases.

Very recent experimental evidence[142] suggests that both the predicted n and especially l dependence are measurably different compared to theoretical expectations. We expect that this important question can be elucidated in the near future. It is not yet clear to what extent the l distribution is affected by Stark mixing due to Coulomb interaction with the ionized target atom during the outgoing part of the collision.

3.4.7. Capture into Continuum States

In ion–atom collisions one can observe that a substantial number of electrons is ejected in forward direction with velocities sharply centered

---

[138] R. M. May, *Nuclear Fusion* **4**, 207(1964).
[139] A. V. Vinogradov, A. M. Urnov, and V. P. Shevelko, *Zh. Eksperim i Teor. Fiz.* **60**, 2060(1971) [*Soviet Phys. JETP* **33**, 1110(1971)].
[140] V. P. Shevelko, *Z. Phys. A* **287**, 19(1978).
[141] J. R. Hiskes, *Phys. Rev.* **137**, A361(1965).
[142] H. D. Betz, J. Rothermel, and F. Bell, *Nuclear Instr. Meth.* **170**, 243(1980); R. W. Hasse, H. D. Betz, and F. Bell, *J. Phys. B* **12**, L 711(1979); H. D. Betz, to be published.

around the exact beam velocity.[143] This phenomenon is found not only in ion–solid interaction but also in single ion–atom collisions and has been interpreted in terms of charge exchange or capture to continuum states of the projectile ions. Its occurrence is not more surprising than ionization of target atoms, where one finds a maximum of ejected electrons just at the ionization limit in the continuum of the target atom. Theoretical explanations have been worked out in some detail[144]; characteristic features of the phenomena are understood but some aspects such as possible differences due to use of solid and gaseous targets remain to be clarified. Further unresolved questions refer to possible capture of continuum electrons into Rydberg states of projectile ions.[142]

3.5. Electron Loss

3.5.1. Target Ionization Due to Light Particles

In recent years significant progress has been achieved in the understanding of collisional electron loss. The field of charge-changing processes where one is primarily though not exclusively concerned with the outermost shell of projectile ions has greatly benefited from the advancement of the treatments of inner-shell excitation phenomena. Numerous and comprehensive articles on Coulomb excitation of target electrons due to impact of light particles are available, and a few of them are mentioned for further reference.

The Born approximation has been used to calculate cross sections for excitation and ionization in hydrogen–hydrogen collisions,[145] and for K- and L-shell ionization of atoms.[146] K-shell ionization of atoms with inclusion of the deflection of the bombarding particle in the Coulomb field of the target nucleus has been obtained on a semiclassical, time-dependent perturbation method (SCA).[147] This SCA model has been applied to adiabatic as well as the nonadiabatic velocity region[148] and yielded important results on the impact-parameter dependence of excitation processes. Ex-

---

[143] W. Meckbach, K. C. Chiu, H. H. Brongersma, and J. W. McGowan, *J. Phys. B* **10**, 3255(1977); C. R. Vane, I. A. Sellin, M. Suter, G. D. Alton, S. B. Elston, P. M. Griffin, and R. S. Thoe, *Phys. Rev. Lett.* **40**, 1020(1978).

[144] K. Dettmann, K. G. Harrison, and M. W. Lucas, *J. Phys. B* **7**, 269(1974); W. Brandt, and R. H. Ritchie, *Phys. Lett.* **62A**, 374(1977).

[145] D. R. Bates and G. Griffing, *Proc. Phys. Soc. A* **66**, 961(1953).

[146] E. Merzbacher and H. W. Lewis, "Handbuch der Physik," Vol. 34, p. 166. Springer-Verlag, Berlin, 1958.

[147] J. Bang and J. M. Hansteen, *Mat.-Fys. Medd. Danske Vid. Selsk.* **31**, No. 13 (1959).

[148] J. M. Hansteen and O. P. Mosebekk, *Nuclear Phys. A* **201**, 541(1973).

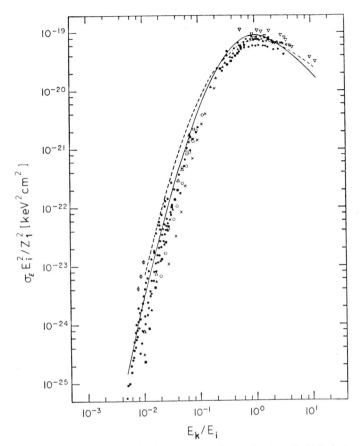

FIG. 15. Universal plot of the ionization cross section σ_i for two K-shell electrons with binding energy E_i due to impact of low-Z_1 particles with velocity v, as a function of reduced impact energy $E_K = m_e v^2/2$ [taken from J. D. Garcia, R. J. Fortner, and T. M. Kavanagh, *Rev. Mod. Phys.* **45**, 111 (1973)].

citation and ionization have been treated with the first Born and higher approximations,[34] for slow collisions,[34] and in terms of the generalized oscillator strength for fast collisions.[149] Inner-shell vacancy production in atoms has been reviewed,[26] and corrections due to Coulomb deflection* and binding effects have been derived in the plane-wave Born[150] and

[149] M. Inokuti, *Rev. Mod. Phys.* **43**, 297(1971).
[150] G. Basbas, W. Brandt, and R. Laubert, *Phys. Rev. A* **7**, 983(1973).

* We note that appreciable corrections do not arise due to the actual deviation from a straight line trajectory but due to the change of collision velocity during Rutherford scattering.

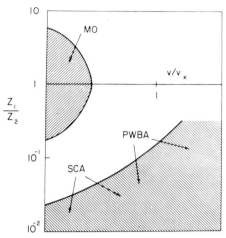

FIG. 16. Estimated regions of validity for various approximation schemes, as a function of Z_1/Z_2 and v/v_K (with velocity v_K of the electron to be ionized). PWBA, Born approximation; SCA, semiclassical approximation; MO, quasi-molecular electron promotion model [from D. H. Madison and E. Merzbacher, in "Atomic Inner-Shell Processes" (B. Crasemann, ed.), Vol. 1, p. 1. Academic Press, New York, 1975].

perturbed-stationary-state approximation.[151] A comprehensive treatment of Coulomb excitation of inner shells is available[152] and contains explicit analytical results, which can be used to compute K- and L-shell excitation by numerical integration.

The typical and well-known behavior of the ionization cross sections is illustrated in Fig. 15. The curve shown is a quite universal function $f(E_K/E_i)$ with binding energy E_i of the target electron to be removed, defined by

$$\sigma_I = Z_1^2 f(E_K/E_i)/E_i^2, \qquad (3.5.1)$$

and is relatively independent of whether the Born, SCA, or binary encounter approximation is used. A similar function f holds for other subshells with no radial node ($n - l = 1$), provided that occupation numbers are taken into account; in cases $n - l > 1$, the curve f exhibits a pronounced knee.[152] For rough estimates, however, f and Eq. (3.5.1) may be used for any shell. A closer comparison of these theoretical results with experiments reveals systematic deviations for $E_K/E_i < 0.1$, especially for

[151] G. Basbas, W. Brandt, and R. H. Ritchie, *Phys. Rev. A* **7**, 1971 (1973).
[152] D. H. Madison and E. Merzbacher, in "Atomic Inner-Shell Processes" (B. Crasemann, ed.), Vol. 1, p. 1. Academic Press, New York, 1975.

low-Z_2 targets. This effect has been explained as follows[150]: the major contributions to ionization arise from impact parameters[147]

$$b_l \simeq \hbar v/E_i. \qquad (3.5.2)$$

In case of low velocities the projectile must penetrate to distances $b_l \ll R_K$ deep within the K shell radius R_K to cause ionization. Since the collision is slow, $v \ll v_i$ ($E_i = mv_i^2/2$), the target electron has enough time to adjust to the presence of the projectile, its binding energy increases to $E_i + \Delta E_i$ and therefore reduces the probability for ionization. As long as ΔE_i is small compared to E_i, which is the case for $Z_1 \ll Z_T$, perturbation techniques can be applied and do indeed result in improved agreement with experiment.[150,151] This binding correction may be viewed as a special case of qualitatively distinct though quantitatively weak formation of molecular orbitals in the limit of very small collisional disturbances. An estimate of the range of validity for several approximation schemes is shown in Fig. 16. The situation is not principally different from the one for electron capture expressed in Eq. (3.4.3).

3.5.2. Projectile Ionization

The results for ionization of target electrons that have been sketched in Section 3.5.1 can be principally applied to the reversed case, ionization of projectile electrons in collisions with target atoms. Unfortunately, several complications arise: first, heavier systems have to be treated and one is thus subjected to the limitations sketched in Fig. 16; second, the ionizing particle is an initially neutral atom and one must find the effective charge q_T that acts on the projectile electron to be removed and enters in Eq. (3.5.1) instead of Z_1. Regarding the latter problem, q_T varies during the collision, which makes it difficult to obtain an accurate and universal description. A crude estimate, however, is readily determined from consideration of impact parameters: the major contribution to ionization will arise from impact parameters b_l' as given by Eq. (3.5.2) and q_T is found as the charge experienced at a distance $R = b_l'$ from the target nucleus. As seen in Section 3.7.1, electrons to be removed in typical collisions have velocities $v_i \simeq v$ and one obtains the simple relation

$$b_l' \simeq R_i, \quad v_i \simeq v, \qquad (3.5.3)$$

i.e., the most important impact parameters are of the order of the shell radius R_i associated with the particular projectile electron. In the majority of cases R_i is large enough to give relatively small values of effective charges q_T. This, in turn, validates the condition expressed in Eq. (3.4.3) or in Fig. 15 for the use of the Born approximation according to Eq. (3.5.1). Of course, polarization and ionization of the target electrons due to the approaching charged projectile deserve consideration, although the response of electrons with velocities near v_0 will be relatively slow for $v > v_0$. Since description of charge-changing processes deals mostly with loss of such projectile electrons for which we have $v_i \simeq v$, several complications can be often, though not always, avoided that arise in the treatment of close collisions where large effective charges q_T, large disturbances, and adiabatic effects occur.

Projectile ions are often highly charged and the possibility arises that projectile electrons become excited into higher-lying bound states. This mode of excitation is not contained in Eqs. (3.4.7)–(3.4.9) and in the results shown in Fig. 14, where only excitation to continuum states is considered, but it may add a contribution of equal order of magnitude.[145,149]

Several special cases of projectile ionization have been treated in the literature. Bohr[15] deals with collisions between fast light ions and light target atoms and arrives at a loss cross section that should be quite reasonable for $v \gg v_i$. Refined and more rigorous treatments for ionization of light ions in very light targets have been reported[153] and electron loss and excitation cross sections have been described in terms of elastic and inelastic x-ray form factors,[154] always on the basis of the Born approximation. Relatively slow collisions must be treated very differently[26,155,156] and need not be discussed here.

3.5.3. Experimental Results on Electron Loss Cross Sections

Abundant results from measurement of electron loss cross sections are available and have been summarized in reviews for very light projectiles,[21,22,25] ions with intermediate atomic numbers,[23] and heavier projectiles,[24] and in data collections for relatively slow and low-charge ions.[30,33] In the low-Z range it is often possible to compare results with theoretical

---

[153] I. S. Dmitriev and V. S. Nikolaev, *Zh. Eksperim. i Teor. Fiz.* **44**, 660(1963) [*Sov. Phys. JETP* **17**, 447(1963)]; I. S. Dmitriev, Y. M. Zhileikin, and V. S. Nikolaev, *Zh. Eksperim. i Teor. Fiz.* **49**, 500(1965) [*Sov. Phys. JETP* **22**, 352(1966)]; V. S. Senashenko, V. S. Nikolaev, and I. S. Dmitriev, *Zh. Eksperim. i Teor. Fiz.* **54**, 1203(1968) [*Sov. Phys. JETP* **27**, 643(1968)]. K. L. Bell, V. Dose, and A. E. Kingston, *J. Phys. B* **2**, 831(1969).

[154] H. Levy, *Phys. Rev.* **185**, 7(1969).

[155] Q. C. Kessel and B. Fastrup, in "Case Studies in Atomic Physics" (M. R. McDowell and E. W. Daniel, eds.), Vol. 3, No. 3, p. 137. North-Holland, Amsterdam, 1973.

[156] K. Smith, in "Scattering Theory" (A. O. Barut, ed.), p. 249. Gordon and Breach, New York, 1969.

3.5. ELECTRON LOSS

models,[157–159] whereas relatively few attempts have been reported to reproduce loss cross sections of highly charged ions.[23,107,160] Among the data for very energetic ions in the velocity range $v \gtrsim v_0$ we mention loss cross sections for oxygen,[113] fluorine,[114,115] aluminum,[110] chlorine,[161] iron,[162] bromine,[51,119] and iodine[35,41,117–119,161,163] ions in various gaseous targets, some of which are contained in Figs. 8–10.

Experimentally, total cross sections for the loss of *any* projectile electron are normally measured and thus represent the sum of cross sections for each individual electron on the ion. As a rule, the most weakly bound electrons from, say, the two outermost principal shells contribute most to the total loss and inner electrons need not always be considered. The loss cross section depends on the binding energy E_i, i.e., q and Z are not principal parameters. As is to be expected from Eq. (3.5.1), σ_l decreases with increasing E_i (increasing q for a given Z); a quantitative analysis reveals that the trend $\sigma_l(q)$ is very well described by the proportionality $\sigma_l(q) \propto E_{i,mq}^{-2}$, provided that (1) the summation extends over all projectile electrons and (2) binding energies $E_{i,mq}$ are accurately calculated for each combination of q and m equivalent electrons.

The dependence $\sigma_l(v)$ should primarily reflect the excitation function f in Eq. (3.5.1). For example, a maximum is expected near $v = v_i(q)$; for a multielectron projectile, the observed cross section is a sum from contributions that peak at somewhat different velocities. This results in a broadening of the peak in $\sigma_l(v)$ and the actually observed peak will be found at a velocity larger than the one of the most weakly bound electron on the ion, in agreement with experimental findings.[23,35,114,115,119]

Dependence of σ_l on the target species is mostly weak. As long as large-impact-parameter collisions are giving the dominant contributions to electron loss, q_T will be small, quite independently of Z_T. A noticeable increase in $\sigma_l(Z_T)$ occurs in the range $Z_T \lesssim 18$, but cross sections, for example, in argon and krypton do not differ by much.[109,160]

[157] K. H. Berkner, S. N. Kaplan, and R. V. Pyle, *Phys. Rev.* **134**, A1461(1964).
[158] R. Smythe and J. W. Toevs, *Phys. Rev.* **139**, A15(1965).
[159] H. B. Gilbody, K. F. Dunn, R. Browning, and C. J. Latimer, *J. Phys. B* **3**, 1105(1970); E. H. Pedersen and P. Hvelplund, *J. Phys. B* **6**, 1277(1973); E. H. Pedersen and P. Hvelplund, *J. Phys. B* **7**, 132(1974).
[160] I. S. Dmitriev, V. S. Nikolaev, L. N. Fateeva, and Y. A. Teplova, *Zh. Eksperim. i Teor. Fiz.* **42**, 16(1962) [*Soviet Phys. JETP* **15**, 11(1962)].
[161] H. A. Scott, L. B. Bridwell, C. D. Moak, G. D. Alton, C. M. Jones, P. D. Miller, R. O. Sayer, Q. C. Kessel, and A. Antar, *Phys. Rev. A* **18**, 2459(1978).
[162] H. Knudson, C. D. Moak, C. M. Jones, P. D. Miller, R. O. Sayer, and G. D. Alton, *Phys. Rev. A* **19**, 1029(1979).
[163] A. Möller, N. Angert, B. Franzke, and Ch. Schmelzer, *Phys. Lett.* **27A**, 621(1968).

3.5.4. Multiple-Electron Loss

In single ion–atom encounters, a projectile may lose more than one electron. Observations of multiple-electron loss have been reported for helium,[131] negative deuterium,[157] and hydrogen,[158] low-charge heavy ions,[30,33,111] fast light ions,[23,164] oxygen,[113] fluorine,[114,115] aluminum,[110] chlorine,[161] iron,[162] bromine,[51,119] and iodine,[35,41,51,118,161,163,165] passing through various gaseous targets. Some of the results for heavier ions are shown in Figs. 8, 9, 13, and 14. It is found that multiple-electron loss $\sigma(q, q + n)$ is, in general, more likely to occur than multiple-electron capture, especially when n becomes large. Very often, ratios $k = \sigma(q, q + n)/\sigma(q, q + 1)$ as large as ~ 0.5 are found for $n = 2$ and a decrease of k_n slower than $k_n \propto 2^{n-1}$ is observed.[162,163] This is nicely demonstrated for iron and iodine ions over a wide range of n values (Figs. 13 and 14).

Multiple-electron loss is caused in two different ways. First, direct ionization of more than one electron in the outermost shell occurs just as for single loss. Then, in analogy to multiple-electron capture, loss probabilities per electron and average impact parameters can be estimated from relations of the kind in Eqs. (3.4.5) and (3.4.6), provided that P is small compared to unity. Second, an electron in the next-inner shell may become excited. This is a likely process in heavy ions: iodine with $q = 4^+$, for example, contains 3 and 18 electrons in the outermost (O) and next-inner (N) shell and it is obvious that N electrons contribute more than O electrons to the total electron-loss cross section. In case of production of a vacancy in an inner shell, we must reckon with subsequent Auger relaxation, i.e., additional electron loss occurs shortly after the collision. Likewise, primary excitation of electrons in still deeper shells results in subsequent Auger cascades that are likely to be responsible for multiple loss with $n \gg 1$. As long as excitation of an electron considered is well described by a direct and single Coulomb interaction the associated multiple-electron loss can be estimated on well-established grounds. However, when inner electrons are removed at small impact parameters in collisions with high-Z_T target atoms, quasi-molecular effects usually become important and are more difficult to treat.[26,155] In a particular range of collision parameters, multiple loss is connected with large scattering angles; these cases are briefly discussed in Section 3.5.5.

---

[164] I. S. Dmitriev, V. S. Nikolaev, L. N. Fateeva, and Y. A. Teplova, *Zh. Eksperim. i Teor. Fiz.* **43**, 259(1962) [*Soviet Phys. JETP* **16**, 259(1963)].

[165] L. B. Bridwell, J. A. Biggerstaff, G. D. Alton, P. D. Miller, Q. C. Kessel, and B. W. Wehring, in "Beam Foil Spectroscopy" (I. Sellin and D. J. Pegg, eds.), Vol. 2, p. 657. Plenum Press, New York, 1976.

3.5.5. Small-Impact-Parameter Collisions and Multiple-Electron Loss

Excitation of an inner-shell projectile electron with $v_i \gg v$ due to collisions with target atoms occurs mainly at impact parameters smaller than the largest projectile shell radius. When Z_T is low, the process may be described in terms of Coulomb excitation, in line with Eq. (3.4.3) when interpreted for projectile instead of target excitation. For low Z_T, however, the probability for removal of an inner electron will be relatively small and large scattering angles are not accentuated for the projectile ions. By contrast, large-Z_T targets cause increased inner-shell excitation and scattering to large angles with respect to the beam axis. For this reason, experimental data deal mostly with collisions between heavy ions and atoms. Electron loss at large scattering angles has been reported, for example, for nitrogen,[166] neon,[166,167] argon,[166,168] krypton,[169] and iodine[165,170] ions, and has also been treated theoretically.[171–173] Fairly comprehensive summaries and reviews on the extensive work in this field are available.[26,155,174] Recently, charge-state distributions of fast oxygen ions ($5 \leq v/v_0 \leq 8.7$) in single collisions with neon, argon, krypton, and xenon have been measured for scattering angles between 5° and 8°.[175]

As a general result, extremely high charge states are found as soon as the observation angle becomes sufficiently large. Then, impact parameters are small enough to allow a certain inner-shell penetration, which leads to extensive electron excitation. In 60 MeV iodine on xenon,[165] for example, net loss of 16 electrons ($I^{10+} \rightarrow I^{26+}$) in a single collision can be observed with fairly large probability for modest scattering angles; a total loss cross section $\sigma(10,24) \simeq 10^{-18}$ cm² has been measured and the domi-

---

[166] L. I. Pivovar, M. T. Novikov, and A. S. Dolgov, *Zh. Eksperim. i Teor. Fiz.* **50**, 537(1966) [*Soviet Phys. JETP* **23**, 357(1966)].

[167] Q. C. Kessel, M. P. McCaughey, and E. Everhart, *Phys. Rev.* **153**, 57(1967).

[168] L. I. Pivovar, M. T. Novikov, and V. M. Tubaev, *Zh. Eksperim. i Teor. Fiz.* **46**, 471(1963) [*Soviet Phys. JETP* **19**, 318(1964)]; Q. C. Kessel and E. Everhart, *Phys. Rev.* **146**, 16(1966).

[169] L. I. Pivovar, M. T. Novikov, and A. S. Dolgov, *Zh. Eksperim. i Teor. Fiz.* **49**, 734(1965) [*Soviet Phys. JETP* **22**, 508(1966)].

[170] Q. C. Kessel, *Phys. Rev. A* **2**, 1881(1970).

[171] A. Russek, *Phys. Rev.* **132**, 246(1963).

[172] E. Everhart and Q. C. Kessel, *Phys. Rev.* **146**, 27(1966).

[173] U. Fano and W. Lichten, *Phys. Rev. Lett.* **14**, 627(1965); W. Lichten, *Phys. Rev.* **164**, 131(1967).

[174] Q. C. Kessel, in "Case Studies in Atomic Physics" (E. W. McDaniel and M. R. McDowell, eds.), Vol. 1, Chap. 7, p. 399. North-Holland, Amsterdam, 1969.

[175] B. Rosner and D. Gur, *Phys. Rev. A* **15**, 70(1977); B. Rosner, D. Gur, J. Alessi, K. C. Chan, and L. Shabason, *Phys. Lett.* **57A**, 320(1976).

nant contributions arise from laboratory angles smaller than $\sim 0.5°$. This demonstrates that multiple electron loss in small-impact-parameter collisions can represent an important process that contributes quantitatively to the production of very high projectile charge states. It is worthwhile to add that the probability for some charge exchange in these collisions is essentially unity; projectile electrons can even be affected strongly enough to give sort of near-equilibrium charge-state distributions in single collisions,[175] i.e., both certain electron loss and capture probabilities can approach unity.

3.6. Equilibrium Charge-State Distributions

3.6.1. Experimental Results

Measurements of equilibrium charge-state distributions are very abundant and the resulting data have been reviewed in great detail. Summaries can be found for ions with $Z \le 2$,[21,22] $Z \le 18$,[23] $Z \le 92$,[24] and a fairly complete tabulation of equilibrium charge state fractions ($2 < Z \le 92$) published prior to 1973 is available[31] and contains data from ~ 100 references, which will not be listed here (see Chapter 3.1). More recent data for heavy ions contain results on $Z \le 18$ for $6.3v_0$ and $12.9v_0$,[176-178] iodine at $2.5v_0$,[179] and krypton at velocities up to $16.2v_0$.[180]

Early work on charge fractions by Batho[7] ($Z \le 36$, $E \le 22$ keV) was extended considerably when fission fragments became available.[8,9,16,181-188] Due to the broad mass and energy distribution of the uranium fragments it was difficult to obtain well-defined charge-state distributions until sophisticated separation techniques were used.[188] Nuclear reactions

---

[176] U. Scharfer, C. Henrichs, J. D. Fox, P. von Brentano, L. Degener, J. C. Sens, and A. Pape, *Nuclear Instr. Meth.* **146**, 573(1977).

[177] R. B. Clark, I. S. Grant, R. King, D. A. Easthan, and T. Joy, *Nuclear Instr. Meth.* **133**, 17(1976); see also R. B. Clark, I. S. Grant, and R. King, preprint Nuclear Phys. Lab. Daresbury, Manchester, Sept. 1973.

[178] H. Münzer, *Sitz. Bericht Österr. Akad. Wiss. Math.-Naturw. II*, **183**, 229(1974).

[179] G. D. Alton, J. A. Biggerstaff, L. Bridwell, C. M. Jones, Q. Kessel, P. D. Miller, C. D. Moak, and B. Wehring, *IEEE Trans. Nuclear Sci.* **NS-22**, 1685(1975).

[180] E. Baron and B. Delaunay, *Phys. Rev. A* **12**, 40(1975).

[181] N. A. Perfilov, *Dokl. Akad. Nauk. SSSR* **28**, 5(1940).

[182] N. O. Lassen, *Phys. Rev.* **68**, 142(1945).

[183] N. O. Lassen, *Phys. Rev.* **69**, 137(1946).

[184] N. O. Lassen, *Mat.-Fys. Medd. Danske Vid. Selsk.* **30**, No. 8(1955).

[185] A. Papineau, *Compt. Rend.* **242**, 2933(1956).

[186] C. B. Fulmer and B. L. Cohen, *Phys. Rev.* **109**, 94(1958).

[187] H. Opower, E. Konecny, and G. Siegert, *Z. Nat. Forsch.* **20 A**,131(1965).

[188] E. Konecny and G. Siegert, *Z. Nat. Forsch.* **21 A**, 192(1966).

3.6. EQUILIBRIUM CHARGE-STATE DISTRIBUTIONS

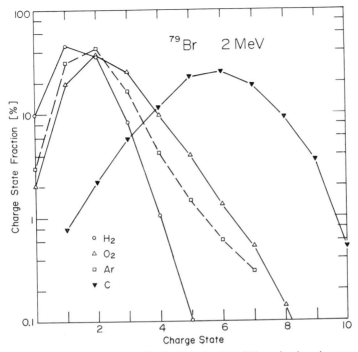

FIG. 17. Equilibrium charge-state distributions for 2-MeV bromine ions in several target species [from A. B. Wittkower and G. Ryding, *Phys. Rev. A* **4**, 226 (1971)].

have occasionally been sources for energetic ions,[189,190] but most data on charge-state distributions were obtained with the superior technique of direct acceleration.

Typical equilibrium charge-state distributions $F(q)$ are shown in Figs. 17–20 for bromine, iodine, xenon, and uranium ions. Evidently, the various charge states center around an average $\bar{q}$ and form a Gaussian-like distribution at least near the maximum. In case of targets that are not too light, pronounced asymmetries occur: the fractions on the high-charge-state side are relatively larger than the ones on the low-charge-state side. This effect results from the influence of cross sections for multiple-electron loss,[24] which are usually influential in heavy targets (Section 3.5.4.). In particular, fractions F_q of charge states q far above $\bar{q}$ will then be primarily formed by multiple-loss collisions.

Shell effects are sometimes found to alter the generally smooth trend of

---

[189] E. Preischkat and R. Vandenbosch, *Nuclear Instr. Meth.* **46**, 333(1967).
[190] R. L. Wolke, R. A. Sorbo, M. A. Volkar, S. Gangadharan, and E. V. Mason, *Phys. Rev. Lett.* **27**, 1449(1971).

3. CHARGE EQUILIBRATION OF HIGH-VELOCITY IONS

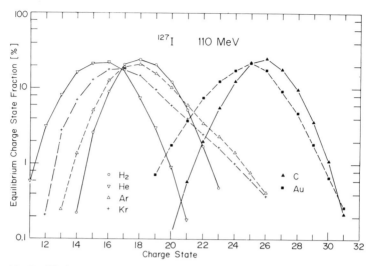

FIG. 18. Equilibrium charge-state distributions for 110-MeV iodine ions in several target species [from S. Datz. C. D. Moak, H. O. Lutz, L. C. Northcliffe, and L. B. Bridwell, *Atom. Data* **2**, 273 (1971)].

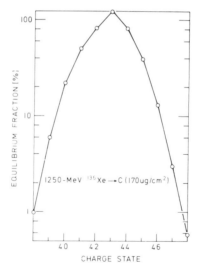

FIG. 19. Equilibrium charge-state distribution of 1250-MeV xenon ions in a carbon foil (from K. Blasche, R. Steiner, and B. Franzke, *GSI Nachrichten,* Darmstadt, No. 12–76, 1976, unpublished).

3.6. EQUILIBRIUM CHARGE-STATE DISTRIBUTIONS

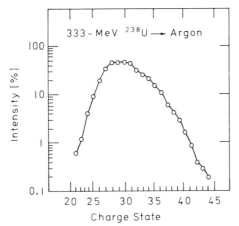

FIG. 20. Near equilibrium charge-state distribution of 333-MeV uranium ions with primary charge 42+ in argon of thickness 5×10^{16} atoms/cm^2 (from B. Franzke, *GSI Nachrichten*, Darmstadt, **No. 9-76**, 1976, unpublished).

$F(q)$. For example, when an ion with charge q_1 has neonlike configuration, formation of fractions below and above q_1 involves major contributions from M- and L-shell ionization, respectively. Due to the discontinuity of M- and L-shell binding energies, fractions F_q will be smaller for $q > q_1$ than one would expect from mere extrapolation of F_q for $q < q_1$.[32] A particular anomaly occurs for low velocities ($v \leq v_0$) and low charge states ($q \leq 1^+$) in helium targets and, to a lesser extent, in other rare gases. It is found that fractions for $q \leq 1^+$ are unusually large.[129,191-195] The reason lies in a sharply reduced electron capture cross section,[129] as can be understood from Eq. (3.4.4) in view of the mismatch between large and small binding energies E_i and E_f, respectively.

Equilibrium charge-state distributions in solids are usually measured for a random orientation. In channeling experiments, the range of impact parameters is restricted and this may result in pronounced changes of effective capture and loss cross sections. Certain charge-exchange pro-

---

[191] P. M. Stier, C. F. Barnett, and G. E. Evans, *Phys. Rev.* **96**, 973(1954).

[192] L. I. Pivovar, V. M. Tubaev, and M. T. Novikov, *Zh. Eksperim. i Teor. Fiz.* **48**, 1022(1965) [*Soviet Phys. JETP* **21**, 681(1965)].

[193] L. I. Pivovar and L. I. Nikolaichuk, *Zh. Eksperim. i Teor. Fiz.* **58**, 97(1970) [*Soviet Phys. JETP* **31**, 55(1970)].

[194] P. Hvelplund, E. Laegsgaard, and E. H. Pedersen, *Nuclear Instr. Meth.* **101**, 497(1972).

[195] A. B. Wittkower and H. D. Betz, *Phys. Rev. A* **7**, 159(1973).

cesses may then become inhibited to such an extent that no true charge state equilibrium can be achieved.[196]

3.6.2. Semiempirical Descriptions of Equilibrium Charge-State Distributions

In view of the difficulty of obtaining sufficiently complete and accurate sets of cross sections $\sigma(q, q', Z, v, Z_T)$ for the computation of charge-state distributions $F(q; Z, v, Z_T)$ a number of more direct, semiempirical descriptions of $F(q)$ have been attempted.[23,32,197–201] The usefulness of these techniques is unquestionable although only gross features of F are sometimes predicted. For practical needs, two different methods turned out to be quite successful. First, experimental equilibrium distributions for a given ion and target species are taken for several velocities and are simply interpolated and extrapolated to give $F(q, v)$ in a fairly wide range of q and v.[32,45] Second, observed regularities and trends of the shape of $F(q)$ and the average $\bar{q}$ can be utilized to predict charge distributions for almost arbitrary cases. The former method has the advantage that no input is needed other than the original data, irregularities such as shell effects and distribution asymmetries are partially included, but the description cannot be extended to any other collision system. The latter method, by contrast, is far more general but uses simplifying assumptions that may lead to significant discrepancies between prediction and experiment; assumptions are usually (1) Gaussian distributions, and simple estimates of (2) the width parameter d and (3) the average charge $\bar{q}$ (see Chapter 3.7). With regard to (1) we have pointed out before that strong asymmetries are to be expected for distributions in most target species.[202,203] Close estimate of d in relatively light targets was found, for example, in the form[204]

$$d = 0.5\{\bar{q}[1 - (\bar{q}/Z)^{1.67}]\}^{1/2}. \tag{3.6.1}$$

---

[196] S. Datz, F. W. Martin, C. D. Moak, B. P. Appleton, and L. B. Bridwell, *Radiation Effects* **12**, 163(1972).

[197] I. S. Dmitriev, *Zh. Eksperim. i Teor. Fiz.* **32**, 570(1957) [*Soviet Phys. JETP* **5**, 473(1957)]; I. S. Dmitriev and V. S. Nikolaev, *Zh. Eksperim. i Teor. Fiz.* **47**, 615(1964) [*Soviet Phys. JETP* **20**, 409(1965)].

[198] D. Nir, S. Gershon, and A. Mann, *Nuclear Instr. Meth.* **155**, 183(1978).

[199] B. Delaunay and J. Delaunay, preprint C.E.N. Saclay, April 1974 and July 1974.

[200] E. Baron, *IEEE Trans. Nuclear Sci.* **NS-19**, No. 2, 256(1972).

[201] E. Veje, *Phys. Rev. A* **14**, 2077(1976).

[202] G. Ryding, A. B. Wittkower, and P. H. Rose, *Phys. Rev.* **185**, 129(1969).

[203] M. D. Brown, *Phys. Rev. A* **6**, 229(1972).

[204] V. S. Nikolaev and I. S. Dmitriev, *Phys. Lett.* **28 A**, 277(1968).

3.6. EQUILIBRIUM CHARGE-STATE DISTRIBUTIONS

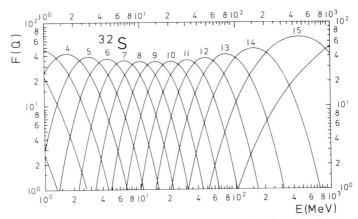

FIG. 21. Calculated graph for equilibrium charge-state distributions of sulfur ions in carbon, as a function of the projectile energy; parameter is the charge state (B. Delaunay and J. Delaunay, preprint C. E. N. Saclay, 1974).

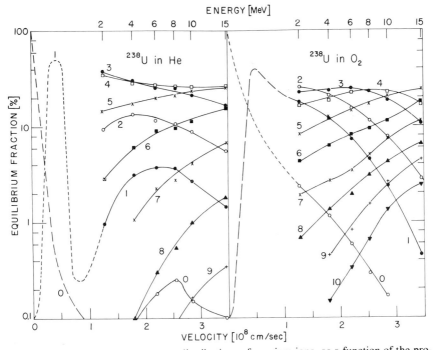

FIG. 22. Equilibrium charge-state distributions of uranium ions, as a function of the projectile velocity; parameter is the charge. The dashed lines are estimates only [From A. B. Wittkower and H. D. Betz, *Phys. Rev. A* **7**, 159 (1973)].

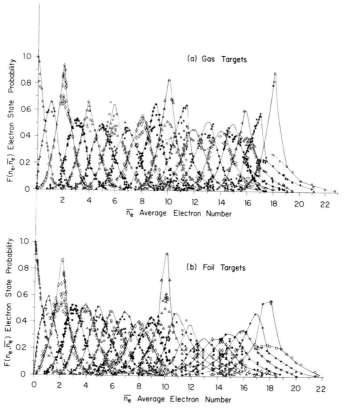

FIG. 23. Probability of electron states (number of unstripped electrons n_e) as a function of the average $\bar{n}_e = Z - \bar{q}$ for equilibrium distributions of ions in gases (a) and solids (b) for electron numbers in the range $18 \leq n_e \leq 39$ [from D. Nir, S. Gershon, and A. Mann, *Nuclear Instr. Meth.* **155**, 183 (1978)].

Delaunay and Delaunay[199] took into account more recent data on heavy ions, refined semiempirical generalizations of d and $\bar{q}$ for symmetric distributions, and obtained very reasonable estimates on $F(q, v)$ for all Z in typical gaseous and solid targets. Figure 21 shows their result for sulfur in carbon. Fortunately, extreme irregularities due to shell effects such as the ones shown in Fig. 22 for relatively slow uranium ions are relatively rare and occur, in general, only when $\bar{q}$ is either near zero or Z. Interestingly, a plot of electron state probabilities as a function of $Z - \bar{q}$ (Fig. 23) yields distributions that turn out to be relatively insensitive to Z and v (Ref. 198) and exhibit shell effects especially for closed-shell configurations.

3.7. Average Equilibrium Charge States

3.7.1. On the Lamb–Bohr (LB) Criterion

Lamb[10] and Bohr[11,12,15,17] were the first to present a relatively sound criterion for the determination of average equilibrium charges $\bar{q}(z, v)$ in dilute gaseous targets. In brief, they argue that equilibrium ionization of a projectile produced by electron loss and capture processes is such that the velocity v_e of the most weakly bound remaining electron equals v. Figure 24 demonstrates the usefulness of this criterion; the various theoretical curves shown differ mainly in the approximations employed to obtain v_e as needed for application of the LB criterion. It could be proven much later[24] that Lamb's formula[10] for calculating v_e from measurable binding energies E_e, or ionization potentials from Hartree–Fock calculations,

$$v_e = (2E_e/m)^{1/2}, \tag{3.7.1}$$

gives indeed surprising accuracy for any ion and velocity; this can be seen from inspection of Fig. 25 for the case of iodine projectiles.

Although the LB criterion sounds reasonable and yields general and fairly accurate results, it should not be overlooked that no precise theoretical justification has been given and that the explanatory arguments by

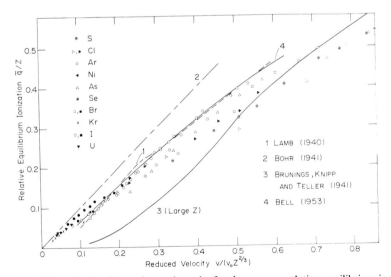

FIG. 24. Theoretical and experimental results for the average relative equilibrium ionization of heavy ions passing through gaseous targets, as a function of the reduced ion velocity $v/(v_0 Z^{2/3})$; see text for explanation of details [from H. D. Betz, *Rev. Mod. Phys.* **44**, 465 (1972)].

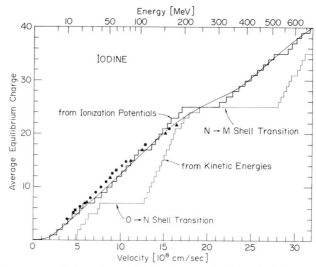

FIG. 25. Average equilibrium charge of iodine ions passing through dilute gases, as a function of the ion velocity; the step functions have been calculated from the Lamb–Bohr criterion [from H. D. Betz, *Rev. Mod. Phys.* **44**, 465 (1972)].

Lamb and Bohr are, at best, incomplete if not partially wrong. Bohr, for example, argues[15] that ionization of projectile electrons occurs as long as $E_e < E_K$ and subsides drastically as soon as E_e exceeds E_K, whereas just the opposite behavior is assumed for electron capture. It is now known that there is no such drastic change in the tendencies of $\sigma_l(q)$ and $\sigma_c(q)$ when q changes in the range near $\bar{q}$ (compare Figs. 8 and 9). The really decisive point is that one must show that $\sigma_l(q)$ is approximately equal to $\sigma_c(q)$ just when $E_e(q) \simeq E_K$; no author has succeeded in doing that on the basis of simple theoretical arguments although numerical evaluation of Eqs. (3.4.4) and (3.5.1) does indeed provide an approximate verification.

Limitations on the general applicability of the LB criterion have not been specified, but several shortcomings are obvious: First, the nature of the target does not enter and it is essentially assumed that Z_T is large enough to provide all those initial states that maximize the capture cross section for any given final projectile states. Second, the target density is assumed to be low enough so that only ground-state ions are to be dealt with; as a consequence, solid targets are excluded. Third, effects of multiple-electron loss are omitted, which are known to have a noticeable influence on $\bar{q}$. Altogether it can be stated that for reasons not fully understood the LB criterion allows estimates on $\bar{q}$ in some gaseous target that are surprisingly accurate and do not deviate further from experi-

3.7. AVERAGE EQUILIBRIUM CHARGE STATES

mental results than the experimental values scatter for all the different target gases.

3.7.2. Experimental Average Equilibrium Charge States

Experimental results on average equilibrium charge states $\bar{q}$ are contained in all the references on charge-state distributions mentioned in Chapter 3.6. In particular, we refer to data compilations[31,32] and some subsequent publications.[177-180,195]

Typical tendencies $\bar{q}(v)$ are presented for iodine and uranium ions in Figs. 26 and 27, respectively. As is expected from the LB criterion, $\bar{q}$ increases with v almost linearly for $\bar{q} \leq Z/2$ and less rapidly as $\bar{q}$ approaches Z. The large difference of $\bar{q}$ in gases ($\bar{q}_G$) and solids ($\bar{q}_s$) is evident and typical for heavy ions as long as $\bar{q}$ is not too close to zero or Z. Figure 27 exhibits the influence of the target species. Obviously, there is

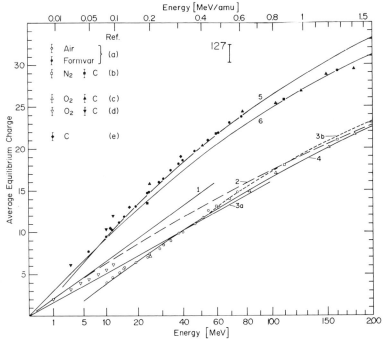

FIG. 26. Average equilibrium charge of iodine ions in gaseous and solid targets, as a function of the projectile energy. Theoretical and semiempirical curves: 1, Eq. (3.7.2); 2, 4, 5, Eq. (3.7.4); 3a, Eq. (3.7.3); 3b, Eq. (3.7.6); 6, Eq. (3.7.5) [from H. D. Betz, *Rev. Mod. Phys.* **44**, 465 (1972)].

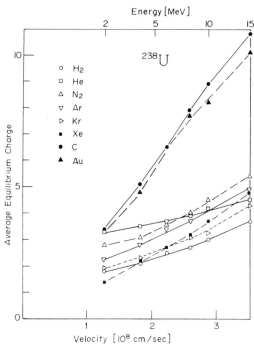

FIG. 27. Average equilibrium charge of uranium ions in gaseous and solid targets, as a function of ion velocity [from A. B. Wittkower and H. D. Betz, *Phys. Rev. A* **7**, 159 (1973)].

a definite scatter of $\bar{q}$ among gaseous and solid targets, which is mostly much smaller than the difference $\bar{q}_s - \bar{q}_G$. The anomaly for low-velocity ions in helium has been indicated before (Section 3.6.1.) and can become pronounced enough to render helium a more effective stripper than carbon (Fig. 27). A similar unfamiliar trend is observable for high-velocity ions in hydrogen (Fig. 18), where more effective stripping is achieved than in all other gases. The reason for these effects may be seen in unusually small capture cross sections resulting from the energetic mismatch $E_i \gg E_f$ and $E_i \ll E_f$ in Eq. (3.4.4) for the former and latter case, respectively.

Nonmonotonous trends have also been observed in $\bar{q}(Z)$ for given v and Z_T.[194,205] These shell effects can be largely reproduced by utilizing the LB criterion, and should also become understandable from consideration of loss and capture cross sections for the relevant projectile states.

[205] L. A. Petrov, V. A. Karnaukhov, and D. D. Bogdanov, *Zh. Eksperim. i Teor. Fiz.* **59**, 1926(1970) [*Soviet Phys. JETP* **32**, 1042(1971)].

3.7.3. Semiempirical Relationships for the Average Equilibrium Charge

The regular and smooth behavior of experimental results on $\bar{q}(v, Z)$ invited attempts to search for semiempirical and purely empirical relationships, which are useful in the entire range of parameters v and Z. One of the starting points was Bohr's evaluation[11,12] of v_e on the basis of the Thomas–Fermi statistical model, which yields

$$\bar{q}/Z = v/(v_0 Z^{2/3}), \quad \bar{q} < Z/2. \quad (3.7.2)$$

Several generalizations of the form $\bar{q}/Z = f[v/(v_0 Z^{2/3})]$ have been reported that were surprisingly successful.[185,206,207,208] Of practical importance are analytical expressions; it turns out that very limited data on $\bar{q}$ can be utilized to develop relations $\bar{q}(Z, v)$ that are reasonably accurate far outside the range of construction.[204,208,209] Widely used expressions are

$$\bar{q}/Z = A v Z^{-1/2}/v_0, \quad \bar{q} \lesssim 0.3 Z, \quad (3.7.3)$$

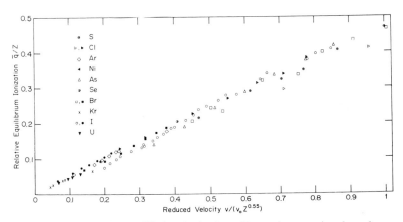

FIG. 28. Average relative equilibrium ionization of heavy ions passing through gaseous targets (N_2, O_2, air), as a function of the reduced ion velocity $v/(v_0 Z^{0.55})$ [from H. D. Betz, *Rev. Mod. Phys.* **44**, 465 (1972)].

[206] H. H. Heckman, B. L. Perkins, W. G. Simon, F. M. Smith, and W. H. Barkas, *Phys. Rev.* **117**, 544(1960).
[207] H. H. Heckman, E. L. Hubbard, and W. G. Simon, *Phys. Rev.* **129**, 1240(1963).
[208] H.-D. Betz, G. Hortig, E. Leischner, Ch. Schmelzer, B. Stadler, and J. Weihrauch, *Phys. Lett.* **22**, 643(1966).
[209] I. S. Dmitriev and V. S. Nikolaev, *Zh. Eksperim. i. Teor. Fiz.* **47**, 615(1964) [*Soviet Phys. JETP* **20**, 409(1965)].

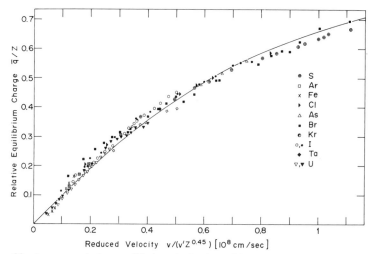

FIG. 29. Average relative equilibrium ionization of heavy ions passing through solid targets (mostly carbon), as a function of the reduced ion velocity $v/(1.65v_0 Z^{0.45})$ [from H. D. Betz, *Rev. Mod. Phys.* **44**, 465 (1972)].

where A is equal to approximately 0.39 in gaseous targets[192,195,209,210] and, for very heavy ions, 0.72 in solid targets. With use of two parameters, the range of higher charge states can be covered:

$$\bar{q}/Z = 1 - C \exp(-vZ^{-\gamma}/v_0), \quad v \geq v_0, \quad (3.7.4)$$

where, when gaseous targets are used, C and γ are close to 1 and 2/3, respectively, and somewhat different for solids.[208] With three parameters, data in solids are well represented by[204]

$$\bar{q}/Z = [1 + (Z^{-\alpha}v/v')^{-1/k}]^{-k}, \quad Z \geq 16, \quad (3.7.5)$$

where $\alpha = 0.45$, $k = 0.6$, and $v' = 1.65v_0$. A four-parameter formula* for gaseous and solid targets has also been constructed,[209]

$$\bar{q}/Z = \log[v/(m_1 Z^{\alpha_1})]/\log[n_1 Z^{\alpha_2}], \quad 0.3 \leq \bar{q}/Z \leq 0.9, \quad (3.7.6)$$

which is expected to hold for any ion. Evaluation of these prescriptions and comparison with data has been discussed in detail[24,180,210-212] and is partly shown in Fig. 26 for the case of iodine ions. A universal plot of data from ions in the range $16 \leq Z \leq 92$ is shown for gaseous targets in

[210] A. B. Wittkower and G. Ryding, *Phys. Rev, A* **4**, 226(1971).
[211] H. D. Betz, *IEEE Trans. Nuclear Sci.* **NS-18**, No. 3, 1110(1971).
[212] H. D. Betz, *IEEE Trans. Nuclear Sci.* **NS-19**, No. 2, 249(1972).

---

* Formula (8) in the translation of Ref. 209 is incorrectly printed.

Fig. 28 with use of the description $\bar{q}/Z = f(vZ^{-0.55}/v_0)$ and for solids in Fig. 29 according to Eq. (3.7.5). It is evident that there is relatively little spread in wide ranges of v and Z. It may be worthwhile to point out that in the dependence of $\bar{q}/Z$ in terms involving Z^γ, the values of γ should, on the basis of the LB criterion, approach unity when $\bar{q}$ comes close to Z for velocities $v \simeq Zv_0$.

3.8. Density Effects and Excited States

3.8.1. Density Effect in Gaseous Targets

When the density of a gaseous target is increased, the mean time Δt_c, between two successive charge-changing collisions decreases and one must reckon with the possibility that Δt_c becomes comparable or even larger than lifetimes τ of excited states, which are inevitably formed in these collisions. Then, it is not necessarily guaranteed that ions are in the ground state prior to a collision and an effect on charge-changing probabilities must be expected. When we assume a gas pressure of $P = 1$ Torr, a typical cross section $\sigma = 10^{-16}\text{cm}^2$, and an ion velocity of $\sim 5v_0$, we obtain

$$\Delta t_c = (3.35 \times 10^{16} P\sigma v)^{-1} \longrightarrow 3 \times 10^{-10} \text{ sec.} \quad (3.8.1)$$

Singly excited ions decay mostly by radiative dipole transitions,[213]

$$\tau = 0.375(2l + 1)[\max(l, l') \Delta E^3 R_{if}^2]^{-1}, \quad 10^{-9} \text{ sec,} \quad (3.8.2)$$

where the transition energy ΔE is to be inserted in Rydberg units and R_{if}^2 represents the square of the dipole moment in units of a_0^2. In outer shells where ΔE is below ~ 100 eV, Eq. (3.8.2) yields lifetimes in the order of 10^{-9}sec, i.e., $\tau \gtrsim \Delta t_c$ is no unusual condition. Nevertheless, occurrence of density effects has been underestimated and even ignored for a long time, despite hints by Lamb[10] and direct evidence from the work of Lassen.[16] Further evidence for the density effect was accumulated,[214-218]

---

[213] H. Bethe and E. E. Salpeter, in "Quantum Mechanics of One- and Two-Electron Atoms" (S. Flügge, ed.), p. 320. Springer-Verlag, Berlin, 1957.
[214] V. S. Nikolaev, I. S. Dmitriev, L. N. Fateeva, and Y. A. Teplova, *Izv. Acad. Nauk SSR, Ser. Fiz.* **26**, 1430(1962) [*Bull. Acad. Sci. USSR, Phys. Ser.* **26**, 1455(1962)].
[215] I. S. Nikolaev, I. S. Dmitriev, Y. A. Teplova, and L. N. Fateeva, *Izv. Acad. Nauk USSR, Ser. Fiz.* **27**, 1078(1963) [*Bull. Acad. Sci. USSR, Phys. Ser.* **27**, 1049(1963)].
[216] L. I. Pivovar, L. I. Nikolaichuk, and F. M. Trubchaninov, *Zh. Eksperim. i Teor. Fiz.* **52**, 1160(1967) [*Soviet Phys. JETP* **25**, 770(1967)].
[217] B. Franzke, A. Angert, A. Möller, and C. Schmelzer, *Phys. Lett.* **25 A**, 769(1967).
[218] G. Ryding, A. B. Wittkower, and P. H. Rose, *Phys. Rev.* **184**, 93(1969).

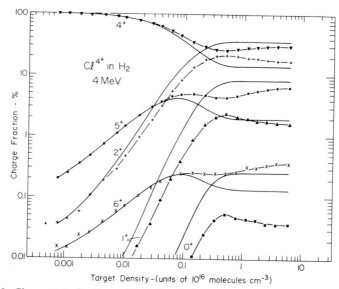

FIG. 30. Charge-state fractions of 4-MeV chlorine ions as a function of hydrogen target density. The solid lines are computed from ground-state cross sections, and the deviations from the measured fractions are indicative for the density effect [from H. D. Betz, L. Grodzins, A. Wittkower, and G. Ryding, *Phys. Rev. Lett.* **26**, 871 (1971)].

but the first direct elucidation of the mechanism was not obtained until 1970.[219–221]

Ryding et al.[219] measured charge-state distributions for 4 MeV chlorine ions in hydrogen, incident with varying initial charge states, and obtained sufficiently complete sets of cross sections that are effective at low and higher gas densities (see Fig. 30). The result can be summarized as follows:

(1) Electron loss cross sections differ little for singly excited ions and ground-state ions.

(2) Electron capture cross sections can be drastically reduced when the ions are in excited states prior to the capture event.

At first, these findings were surprising since it is just the contrary of what Bohr and Lindhard[17] had predicted, but a satisfactory explanation is immediately evident from the discussions of capture and loss in Chapters

---

[219] G. Ryding, H. D. Betz, and A. B. Wittkower, *Phys. Rev. Lett.* **24**, 123(1970).
[220] H. D. Betz, *Phys. Rev. Lett.* **25**, 903(1970).
[221] H. D. Betz, L. Grodzins, A. B. Wittkower, and G. Ryding, *Phys. Rev. Lett.* **26**, 871(1971).

3.8. DENSITY EFFECTS AND EXCITED STATES

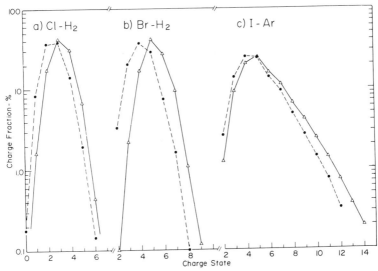

FIG. 31. Equilibrium charge-state fractions of 4-MeV chlorine, 14-MeV bromine, and 12-MeV iodine ions in dilute (dashed lines) and dense gases (solid lines) [from H. D. Betz, *Phys. Rev. Lett.* **25**, 903 (1970)].

3.4 and 3.5. Result (1) is understandable for many-electron ions. According to Eq. (3.5.1), ionization of an excited electron is enhanced when its binding energy is reduced, but this represents only a small increase of the total loss cross section since the dominant contributions are due to many electrons from, say, the two outermost shells. Regarding (2), it is sufficient to point out that capture often proceeds into excited states so that an ion already excited will most likely undergo an Auger transition; normally, higher shells with $n > 1$ will be involved where doubly excited ions are not likely to decay radiatively. The net result of such capture events is then no change of charge and, due to these contributions, the total observed capture cross section is reduced. Further experiments with iodine ions[41] confirmed the above observations, as well as comparison between equilibrium charge-state distributions in dilute and dense-gas targets (Fig. 31).[220,222]

Density-dependent distributions of the kind shown in Fig. 27 can be used to extract average effective lifetimes $\tau(q)$.[221] In a reasonably close approximation, Eq. (3.2.2) is utilized with loss cross sections, which are relevant for ground-state ions, and capture cross sections, which depend

[222] C. D. Moak, L. B. Bridwell, H. A. Scott, G. D. Alton, C. M. Jones, P. D. Miller, R. O. Sayer, Q. C. Kessel, and A. Antar, *Nuclear Instr. Meth.* **150**, 529(1978).

on the degree of residual excitation in the beam, i.e., on the density

$$\sigma_c^*(q) = \sigma_c(q)\{1 - g(q)\exp[-\rho v \sigma_t(q)\tau(q)]^{-1}\}, \qquad (3.8.3)$$

where σ_t is the total cross section for charge-changing events in the ground state and $g(q)$ signifies the initial degrees of excitation, i.e., those fractions of ions with charge q the electron capture processes of which are followed by immediate emission of an Auger electron. Application of this technique yielded lifetimes of excited chlorine ions in the range of $\sim 1-100$ nsec,[221] in close agreement with estimates from Eq. (3.8.2).

In addition to promptly decaying states one must reckon with the formation of longer-lived, metastable states.[52,223] In general, however, these states will not be strongly populated. With increasing q the relevant transition energies ΔE in Eq. (3.8.2) will increase and shorten the lifetimes of dominantly formed excited states. One would conclude that the density effect in gases becomes less important when $\bar{q}$ grows large.

3.8.2. Density Effect in Solid Targets

3.8.2.1. Passage of Protons through Solids.
Charge states of ions emerging from solids are well-defined observables, but no such conditions prevail inside solids. The electrons of the medium can respond to the disturbance set up by the projectile and screen the moving charge essentially by the formation of a polarization cloud around the projectile. Theoretical descriptions of this kind of screening and of the charge-changing processes that take place at the exit surface are still the subject of intensive discussions.[224-226] It became quite clear that screening causes all energy levels of protons to shift away from their unscreened values toward the continuum and completely unbound states result for[224]

$$D < 0.84 a_0, \qquad Z = 1, \qquad (3.8.4)$$

where D is the screening length. D can be expressed by

$$D^{-1} = \begin{cases} \omega_p \sqrt{3}(1 - 0.1 v^2/v_F^2)/v_F, & v < v_F, \\ \omega_p \pi/(2v), & v > v_F, \end{cases} \qquad (3.8.5)$$

where ω_p and v_F denote plasma frequency and Fermi velocity, respectively. As a consequence a proton has no bound state when it moves

---

[223] I. S. Dmitriev, V. S. Nikolaev, and Y. A. Teplova, *Phys. Lett.* **26 A**, 122(1968).

[224] F. J. Rogers, H. C. Graboske, and D. J. Harwood, *Phys. Rev. A* **1**, 1577(1970).

[225] W. Brandt and R. Sizmann, *Phys. Lett.* **37 A**, 115(1971); V. N. Neelavathi, R. H. Ritchie, and W. Brandt, *Phys. Rev. Lett.* **33**, 302(1974); W. Brandt, in "Atomic Collisions in Solids" (S. Datz, B. R. Appleton, and C. D. Moak, eds.), Vol. 1, p. 261. Plenum Press, New York, 1975.

[226] M. C. Cross, *Phys. Rev. B* **15**, 602(1977).

3.8. DENSITY EFFECTS AND EXCITED STATES

through a solid with a velocity below $\sim v_0$ and considerations of electron capture and loss inside the solid become simply irrelevant; the formation of charge states (neutral fraction) will then occur when the proton exits from the solid. For larger velocities, however, certain bound states exist inside the solid although they are weakly populated, and it is meaningful to define a ionic charge as well as cross sections for electron capture and loss.[226] The question arises as to how these results can be generalized for heavy ions. Since only the ion velocity and the binding energy $mv_e^2/2$ of a projectile electron are influential, an approximate evaluation for any Z reveals that Eq. (3.8.4) is almost equivalent to the condition $v_e v < c_1 v_F^2$, where c_1 is of the order of unity. It is concluded that very lightly bound states of ions may lose significance inside solids, but it becomes also clear that screening effects will be relatively unimportant for large v and v_e. In typical cases of present interest ($v > v_0$, $v_e \gtrsim v_0$), we assume that ionic charge states remain well defined inside solids, at least as long as electronic orbitals do not reach dimensions larger than, say, the lattice spacing of the target material.

3.8.2.2. Multiple-Excitation Model. The charge-state density effect in solids is clearly visible in Figs. 17, 18, 26, and 27: higher charge states of ions are measured *behind* a solid target compared with a gas target for otherwise identical conditions. A consistent explanation of this effect is provided by a multiple-excitation model,[227] which is briefly described below.

Due to the high atom density in solids, the mean time between two successive collisions of an ion is very short,

$$\Delta t_c = (\rho v \sigma)^{-1}. \quad (3.8.6)$$

A typical value is $\Delta t_c \simeq 10^{-16}$ sec, which results for $\rho = 10^{23}$ atoms/cm^3, $v = 10^9$ cm/sec, and $\sigma = 10^{-16}$ cm^2. Total lifetimes of almost any singly excited state in higher shells (we may ignore L → K transitions in heavy ions) are usually much longer. Consequently, equilibrium conditions inside solids involve balance not only between capture and loss of electrons, but also of excitation and deexcitation processes. In case of multiple-electron ions it is assumed that, under equilibrium conditions, *many* electrons from, say, the two outermost shells of the ground-state ion with charge $q \simeq \bar{q}_G$ are excited to higher shells. As a result, multiple vacancies are present that predominantly decay by Auger emission as soon as the ion emerges from the solid. Thus, ejection of several electrons per ion increases the ionic charge *behind* the solid to the observed value $\bar{q}_s$ (Fig. 32).

[227] H. D. Betz and L. Grodzins, *Phys. Rev. Lett.* **25**, 211(1970).

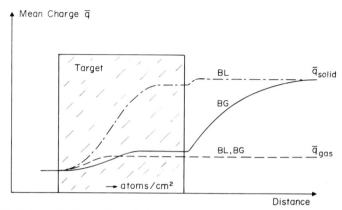

FIG. 32. Schematic illustration of the average charge of multielectron ions passing through gaseous and solid targets; BL: Bohr–Lindhard, BG: Betz–Grodzins model [from H. D. Betz, *Rev. Mod. Phys.* **44**, 465 (1972)].

Multiple excitation can be maintained inside the solid by the following processes: on the one side, by direct excitation to a higher lying unoccupied bound state and capture of target electrons into excited states; on the other side, by loss of excited electrons due to ionization and, in some cases, deexcitation via Auger and radiative transitions to lower-lying states. Quantitative analysis of this model is quite complex[228] but reveals

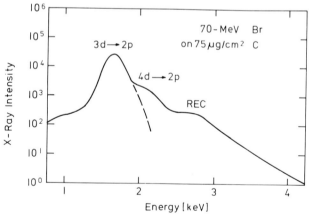

FIG. 33. L x-ray spectrum observed in collisions between 70-MeV bromine ions and a 75-μg/cm^2 carbon target. The dominant part of x rays is emitted by the ions during passage through the foil and reflects excitation conditions inside the solid. REC: radiative capture (see Chapter 3.10).

[228] H. D. Betz, *Nuclear Instr. Meth.* **132**, 19(1976).

3.8. DENSITY EFFECTS AND EXCITED STATES

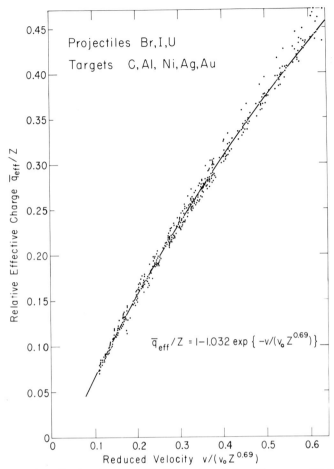

FIG. 34. Relative effective charge $\bar{q}_{\text{eff}}/Z$ of heavy ions deduced from measurements of the energy loss of these ions in various solid targets, as a function of the reduced ion velocity $v/(v_0 Z^{0.69})$. The solid line closely represents the locus of average relative equilibrium charge states obtained in gaseous targets [from M. Brown and C. D. Moak, *Phys. Rev. B* **6**, 90 (1972)].

that many electrons can indeed be stabilized in excited states; unfortunately, precise numbers are hard to come by because the relevant charge-changing cross sections are not known with sufficient accuracy. Nevertheless it results that the older single-excitation model[17] cannot be supported, which predicted that the higher charge states $\bar{q}_s$ are essentially produced inside solids. The problematic has been dealt with in great detail[24,227,228] and is partially evident from Fig. 32.

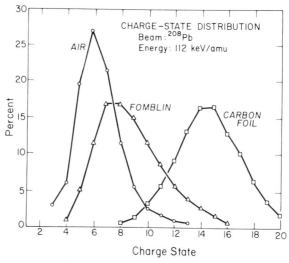

FIG. 35. Charge-state distribution for 23-MeV lead ions passing through air, fluorocarbon vapor (Fomblin), and a carbon foil [*Res. Accelerators* **3**, No. 4, 21 (1978)].

A direct indication for steady-state population of rather high projectile states inside solids may be seen in the occurrence of strong radiative transitions from these states. Figure 33 shows an x-ray spectrum obtained in collisions between 70-MeV bromine ions and carbon. Remarkably, both 4p → 1s and 4d → 2p transitions can be found with relatively large intensity although the ground-state population of equilibrated 70-MeV bromine ions ($\bar{q} \cong 19^+$) fills levels only up to 3p; in fact, the 4d level is not even populated in neutral bromine atoms. Closer examination reveals that the observed ratios of line intensities can be converted into population ratios. It results[229] that the latter ratios are indeed close to unity, i.e., several electrons are simultaneously in excited states at any time. Multiple steady-state excitation thus represents not a rare but the typical situation of multielectron ions moving through solids.

Further strong evidence for the validity of the multiple-excitation model may be seen in results on the energy loss S_e of fast ions in solid and gaseous targets. Since $S_e \propto q^2$, very different charge states of ions inside gases ($\bar{q}_G$) and solids ($\bar{q}_S$) would cause different values of S_e, contrary to experimental observations. Moreover, when average effective charge states $\bar{q}_{\text{eff}}$ are derived from comparison of S_e in *solids* with energy loss S_p of protons in the same solid target,

$$\bar{q}_{\text{eff}}^2 = S_e(v)/S_p(v), \qquad (3.8.7)$$

[229] H. D. Betz, unpublished.

it turns out that $\bar{q}_{eff}$ is surprisingly close to average equilibrium charges $\bar{q}_G$ obtained from charge state measurements in *gases*[24,230] (Fig. 34). In addition, evidence for ionic charge states that are lower inside than behind solids was found from the observation of x rays.[231]

For practical purposes, it would be important to find targets that have the high densities of solids but last longer under heavy-ion bombardment. Stripping in large (gaseous) molecules[232,233] indeed revealed some simulation of the density effect but did not yield the relatively high charge states obtained from solids (Fig. 35).

3.9. Atomic Transitions and Charge Equilibration in Solids

3.9.1. Electron Capture and Inner-Shell Vacancies

Ion–atom collisions produce abundant excited projectile states, which influence the formation of ionic charges in solids and in not too dilute gases. As long as predominantly produced excited states have lifetimes that are long compared to mean collision times $\tau \gg \Delta t_c$, decay of excited states by atomic transitions cannot affect the balance between electron capture and loss. Depending on the various parameters of the collision system and the projectile shell considered, the reverse condition $\tau \leqslant \Delta t_c$ can also become effective. Figure 36 illustrates the most important processes for production and destruction of projectile vacancies. In particular, it is possible that radiative and Auger transitions compete with electron capture.

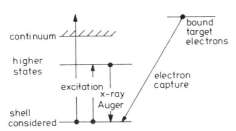

FIG. 36. Schematic illustration for production and destruction of an inner-shell vacancy in an ion moving through a solid target.

---

[230] M. D. Brown and C. D. Moak, *Phys. Rev. B* **6**, 90(1972).

[231] S. Datz, B. R. Appleton, J. R. Mowat, R. Laubert, R. S. Peterson, R. S. Thoe, and I. A. Sellin, *Phys. Rev. Lett.* **33**, 733(1974).

[232] G. Ryding, A. B. Wittkower, and P. H. Rose, *Particle Accelerators* **2**, 13 (1971).

[233] A. Kardontchik and R. Kalish, *Phys. Lett.* **54 A**, 219(1975); *Res. Accelerators* **3** (4), 21(1978).

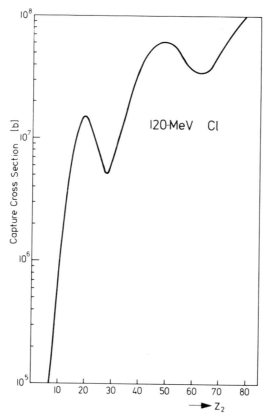

FIG. 37. Cross section for electron capture into the K shell of 120-MeV chlorine ions, as a function of the target atomic number Z_2, calculated from Eq. (3.4.4) [from F. Bell and H. D. Betz, *J. Phys. B* **10**, 483 (1977)].

Results in Chapter 3.4 imply that a fast ion can capture electrons from inner target shells and into excited projectile states. As a consequence, atomic transitions occur in the target and/or the projectile. Surprisingly, this well-known electron capture mode[94-96] has been ignored in the field of inner-shell excitation until very recently.[103,234,235] A further connection between electron capture and atomic transitions shows up when, for example, K-shell x-ray yields from fast projectiles are studied as a function of Z_T and are found to exhibit an oscillatory behavior.[236] It could be

[234] A. M. Halpern and J. Law, *Phys. Rev. Lett.* **31**, 4(1973).
[235] M. D. Brown, L. D. Ellsworth, J. A. Guffey, T. Chiao, E. W. Pettus, L. M. Winters, and J. R. Macdonald, *Phys. Rev. A* **10**, 1255(1974).
[236] H. D. Betz, F. Bell, and E. Spindler, in "Atomic Physics V" (R. Marrus, ed.), p. 493. Plenum Press, New York, 1977.

3.9. ATOMIC TRANSITIONS AND CHARGE EQUILIBRATION

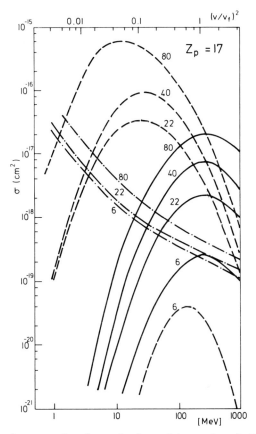

FIG. 38. Estimated cross sections for production and destruction of a K-shell vacancy in chlorine ions passing through solids, as a function of the ion energy. The target atomic number is indicated near each curve; —, electron loss; – – –, electron capture; – · –, effective cross section $(\rho v \tau)^{-1}$ for atomic transitions [from F. Bell and H. D. Betz, *J. Phys. B* **10**, 483 (1977)].

shown that this effect results from a direct competition between electron capture and x-ray transitions as indicated in Fig. 36 and reflects the varying cross sections $\sigma_c(Z_T)$ for capture of any target electron into a particular ion vacancy,[236-238] which are shown in Fig. 37 for the case of 120-MeV chlorine passing through solid targets. Quantitative analysis of the cross sections for production and destruction of chlorine K-shell vacancies is depicted in Fig. 38; it turns out that σ_c represents a dominant

[237] F. Bell and H. D. Betz, *J. Phys. B* **10**, 483(1977).
[238] F. Hopkins, J. Sokolov, and A. Little, *Phys. Rev. A* **15**, 588(1977).

process in a wide range of collision velocities for all solid targets except the light ones. Incidentally, Fig. 38 demonstrates the approximate validity of the LB criterion (Section 3.7.1.): for the heavy targets the curves for σ_l and σ_c intersect at almost the same velocity v_1, which is just the one needed to produce charge states close to $\bar{q} = Z - 1$. In case of the light targets, σ_l is relatively small and the K-shell vacancy distribution results almost entirely from the balance between electron capture σ_c and atomic transitions $(\rho v \tau)^{-1}$, which assume equal magnitude near $v = v_1$.

3.9.2. Equilibrium Excitation of Very Fast Ions in Solids

The buildup and decay of a particular excited-state fraction of projectiles moving inside and outside a solid is demonstrated in Fig. 39, where it is assumed that the state considered is not populated prior to striking the target. The level of excitation N will increase due to collisions until an equilibrium value N_∞ is reached for sufficiently thick targets. When the ions exit from the foil, the excited fraction decays and gives rise to Auger- and x-ray transitions. When all path lengths are expressed in flight times t, the yield of x rays that deexcite the considered states is given by[239,240]

$$Y(t_0) = N_\infty(t_0 + \langle \tau_a \rangle - \langle \tau_i \rangle)/\tau_R, \qquad (3.9.1)$$

where t_0 should correspond to an equilibrium target thickness, $\langle \tau_a \rangle$ and $\langle \tau_i \rangle$ characterize typical times for buildup and decay, respectively, and τ_R signifies the radiative lifetime of the state. When cascading effects can be neglected, $\langle \tau_a \rangle$ equals the total lifetime τ of the state with respect to Auger and X-ray transitions in isolated atoms. The time $\langle \tau_i \rangle$ can be related to a total cross section involving all production and decay modes,

$$\langle \tau_i \rangle \simeq (\rho v \sigma_T)^{-1}, \qquad \sigma_T = \sigma_l + \sigma_c + (\rho v \tau)^{-1}. \qquad (3.9.2)$$

We consider two different cases depending on the collision time $\Delta t_c = [\rho v (\sigma_l + \sigma_c)]^{-1}$, and obtain

$$Y(t_0) \simeq \begin{cases} N_\infty t_0/\tau_R = \omega \sigma_l x_0, & (\Delta t_c \gg \tau), \\ N_\infty(t_0 + \tau)/\tau_R, & (\Delta t_c \ll \tau), \end{cases} \qquad (3.9.3)$$

where ω denotes the fluorescent yield and the corresponding equilibrium fractions are given by

$$N_\infty = \sigma_l/\sigma_T = \begin{cases} \sigma_l v \tau, & (\Delta t_c \gg \tau), \\ \sigma_l(\sigma_l + \sigma_c)^{-1}, & (\Delta t_c \ll \tau). \end{cases} \qquad (3.9.4)$$

[239] H. D. Betz, F. Bell, H. Panke, G. Kalkoffen, M. Welz, and D. Evers, *Phys. Rev. Lett.* **33**, 807(1974).

[240] F. Bell, H. D. Betz, H. Panke, W. Stehling, and E. Spindler, *J. Phys. B* **9**, 3017(1976).

3.9. ATOMIC TRANSITIONS AND CHARGE EQUILIBRATION

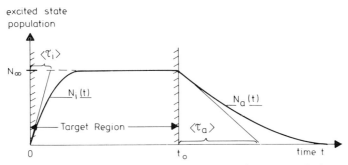

FIG. 39. Schematic illustration of the fraction of ions excited to a particular state inside and outside a solid target foil, as a function of the flight time of the ion.

It is evident that for a dominating capture cross section $\sigma_T \simeq \sigma_c$, the unfamiliar relation $N_\infty \simeq \sigma_I/\sigma_c$ results and becomes effective, for example, in some of the collision systems discussed above (Fig. 38). The production of x rays is then strongly influenced by the magnitude of electron capture directly into the vacancy: for example, when σ_c is large, more vacancies are destroyed and fewer atomic transitions can take place. Results along these lines have been confirmed in several experiments with fluorine, sulfur, and chlorine ions,[55,238,241,242] and are visible, for example, in intensity ratios of satellite and hypersatellite x-ray lines (Fig. 40).

It is worthwhile to point out several other consequences of large electron capture cross sections: x-ray yields do not increase in proportion

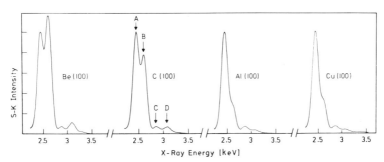

FIG. 40. K x-ray spectrum of 110-MeV sulfur ions emitted during and after passage through various foil targets of thickness 100 μg/cm² [from H. D. Betz, F. Bell, H. Panke, G. Kalkoffen, M. Welz, and D. Evers, *Phys. Rev. Lett.* **33**, 807 (1974)].

[241] F. Hopkins, *Phys. Rev. Lett.* **35**, 270(1975).
[242] T. J. Gray, P. Richard, K. A. Jamison, J. M. Hall, and R. K. Gardner, *Phys. Rev. A* **14**, 1333(1976).

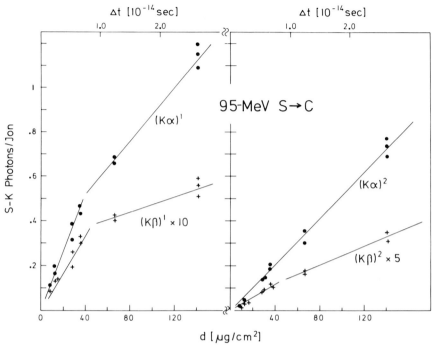

FIG. 41. Absolute yield of four K x-ray lines (identified in Fig. 40) of 95-MeV sulfur ions passing through carbon, as a function of target thickness (lower scale) and foil transit time (upper scale) [from H. D. Betz, F. Bell, H. Panke, G. Kalkoffen, M. Welz, and D. Evers, *Phys. Rev. Lett.* **33**, 807 (1974)].

with target thickness[239,241–243] (Fig. 41), lifetimes τ of well-defined multiplet states in the order of foil transit times ($\sim 10^{-14}$ sec) can be determined from nonproportional target thickness dependence of x-ray yields,[239,243–245] and equilibrium fractions of excited ions inside solids become measurable.[238,240]

3.10. Radiative Electron Capture

The possibility of radiative electron capture (REC) by ions was realized very early[20,246–248]; interest in REC arose in connection with nonradiative Coulomb capture to which REC is a competing phenomenon, and from

[243] H. D. Betz, F. Bell, H. Panke, G. Kalkoffen, M. Welz, and D. Evers, *Phys. Lett.* **49A**, 133(1974).
[244] H. Panke, F. Bell, H. D. Betz, W. Stehling, E. Spindler, and R. Laubert, *Phys. Lett.* **53A**, 457(1975).
[245] H. Panke, F. Bell, H. D. Betz, and W. Stehling, *Nuclear Instr. Meth.* **132**, 25(1976).

3.10. RADIATIVE ELECTRON CAPTURE

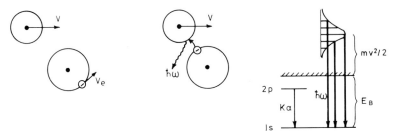

FIG. 42. Schematic illustration for radiative capture of a bound target electron in a collision with a moving ion.

work in astro- and plasmaphysics where the process is called radiative recombination. In contrast to Coulomb capture REC can occur with a completely free electron and is then the inverse of the photoeffect. Bound target electrons can also be captured radiatively: the schematics are illustrated in Fig. 42 and reveal that REC is accompanied by a continuous, though peaked emission spectrum. The first observations of REC have been reported for sulfur, chlorine, and bromine ions with velocities $v \leq 12v_0$ ($E_K \leq 2$ keV) incident on a variety of targets[24,249–251] and prompted many further investigations.[236,252–260] Typical results are

---

[246] B. M. Stobbe, *Ann. Physik* **7**, 682(1930).

[247] E. C. Stückelberg and P. M. Morse, *Phys. Rev.* **35**, 116(1930).

[248] W. Wessel, *Ann. Physik* **5**, 611(1930).

[249] H. D. Betz and H. W. Schnopper, *Proc. Europ. Conf. Nucl. Phys.* (M. Porneuf, ed.), *Compt. Rend. II*, p. 116. (1972).

[250] H. D. Betz, in "Proceedings of the Heavy-Ion Summer Study" (S. T. Thornton, ed.) p. 545. USAEC CONF-720669(1972).

[251] H. W. Schnopper, H. D. Betz, J. P. Delvaille, K. Kalata, A. R. Sohval, K. W. Jones, and H. E. Wegner, *Phys. Rev. Lett.* **29**, 898(1972).

[252] P. Kienle, M. Kleber, B. Povh, R. M. Diamond, F. S. Stephens, E. Grosse, M. R. Maier, and D. Proetel, *Phys. Rev. Lett.* **31**, 1099(1973).

[253] B. R. Appleton, T. S. Noggle, C. D. Moak, J. A. Biggerstaff, S. Datz, H. F. Krause, and M. D. Brown, *in* "Atomic Collisions in Solids" (S. Datz, B. R. Appleton, and C. D. Moak, eds.), Vol. 2, p. 499. Plenum Press, New York, 1975.

[254] H. W. Schnopper and J. P. Delvaille, *in* "Atomic Collisions in Solids" (S. Datz, B. R. Appleton, and C. D. Moak, eds.), Vol. 2, p. 481. Plenum Press, New York, 1975.

[255] F. Bell and H. D. Betz, *in* "Atomic Collisions in Solids" (S. Datz, B. R. Appleton, and C. D. Moak, eds.), Vol. 1, p. 397. Plenum Press, New York, 1975.

[256] H. W. Schnopper, J. P. Delvaille, K. Kalata, A. R. Sohval, M. Abdulwahab, K. W. Jones, and H. E. Wegner, *Phys. Lett.* **47A**, 61(1974).

[257] H. D. Betz, F. Bell, and H. Panke, *J. Phys. B* **7**, L418(1974).

[258] H. D. Betz, M. Kleber, E. Spindler, F. Bell, H. Panke, and W. Stehling, *in* "The Physics of Electronic and Atomic Collisions" (J. S. Risley and R. Geballe, eds.), p. 520. Univ. Washington Press, Seattle, 1976.

[259] J. Lindskog, J. Phil, R. Sjödin, A. Marelius, K. Sharma, R. Hallin, and P. Lindner, *Phys. Scripta.* **14**, 100(1976).

146   3. CHARGE EQUILIBRATION OF HIGH-VELOCITY IONS

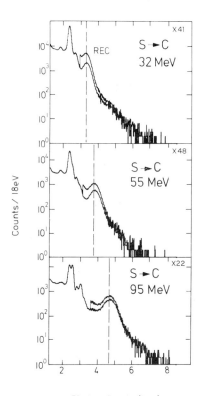

FIG. 43. Experimental x-ray spectra from collisions between sulfur ions and carbon with thickness 150 µg/cm², observed with a 150-µm carbon absorber. The REC distribution has been corrected for absorption effects.

shown in Fig. 43 for sulfur on carbon: the REC emission spectrum is peaked near an energy $\hat{E}_x = E_f + E_K$ and exhibits a width that reflects the initial momentum distributions $\phi_i(p)$ of bound target electrons. The shift of $\hat{E}_x$ with varying E_K is a distinct feature that provides an easy and unambiguous identification of the REC process.

The differential cross section for capture of a free electron moving with initial velocity v by a bare nucleus can be expressed, in analogy with the photoeffect, by

$$d\sigma_R/d\Omega = E_x^2(4\pi^2\hbar^4 c^3 v)^{-1}|\langle\psi_f| - e\mathbf{A}\mathbf{v}/c|\psi_i\rangle|^2, \quad (3.10.1)$$

where E_x is the (sharp) energy of the emitted photon, ψ_i represents a Cou-

---

[260] A. R. Sohval, J. P. Delvaille, K. Kalata, K. Kirby-Docken, and H. W. Schnopper, *J. Phys. B* **9**, L25(1976).

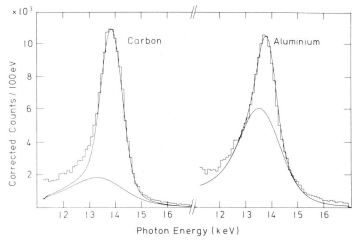

FIG. 44. Comparison between experimental (histogram) and calculated (solid lines) REC profiles for 400-MeV copper ions in carbon and aluminum foils. Contributions from target core electrons are also shown [E. Spindler, H. D. Betz, and F. Bell, *J. Phys. B* **10**, L561 (1977)].

lomb wave that takes into account the deflection of the electron during the collision, and ψ_f is the bound state after capture. Evaluation yields the fairly rigorous expression for the total capture cross section[213]

$$\sigma_R = \sigma_0 \left(\frac{\eta^3}{1+\eta^2}\right)^2 \frac{\exp(-4\eta \arctan \eta^{-1})}{1 - \exp(-2\pi\eta)}, \quad (3.10.2)$$

where $\sigma_0 = 2^8 \alpha^3 (\pi/3) \pi a_0^2 = 9.1 \times 10^{-21}$ cm^2 and $\eta^2 = E_f/E_K$. Extensions to capture of bound electrons have been worked out for simple collision systems[261,262] and yield essentially the following result for $E_i \ll E_f$ and $E_i \ll E_K$:

$$\frac{d^2\sigma_{REC}}{dE_x \, d\Omega} = \frac{3}{8\pi} \int \sigma_R(p) \sin^2 \Theta |\phi_i(\mathbf{p} - \mathbf{p}_0)|^2 \, \delta(E) \, d^3\mathbf{p} \quad (3.10.3)$$

where the δ function ensures energy conservation, Θ is the angle between **p** and the direction of the emitted photon, and $p_0 = mv$ is the momentum of a target electron at rest, relative to the projectile. Proper evaluation of Eq. (3.10.3) allows close reproduction of experimental results obtained with fast heavy ions,[236] provided that (1) correct wavefunctions are used and (2) several corrections are taken into account. As regards (1), it is important to realize that $\phi_i(p)$ of weakly bound target electrons, which

---

[261] J. S. Briggs and K. Dettmann, *Phys. Rev. Lett.* **33**, 1123(1974).
[262] M. Kleber and D. H. Jakubassa, *Nuclear Phys.* **A252**, 152(1975).

contribute most to REC differs in gaseous and solid targets. Point (2) implies effects due to absorbers, instrumental line broadening, Doppler shift, and composition of charge and excitation states in the ion beam. The latter is particularly difficult to handle when the ions are not fully stripped. In almost all experiments where REC into the K shell of ions has been observed, one must disentangle contributions for single and double K-shell vacancies and varying numbers of electrons in higher shells. Incidentally, we note that the presence of other (screening) electrons on the ion does not inhibit REC but merely modifies the binding energies E_f. Figure 44 proves that satisfactory agreement can be obtained between experiment and a theoretical description based on Eq. (3.10.3) and does not contain any adjustable parameter.[263] This result raises hopes that REC may be utilized to determine experimental Compton profiles for complex solids.

REC cross sections are usually below 10^{-21} cm^2 and are thus small compared to the ones for nonradiative Coulomb capture. Only at very large v is it possible, due to the different v dependences, that REC exceeds Coulomb capture.[87] In the high-velocity region, the two cross sections Eqs. (3.4.2) and (3.10.2) are simply given by

$$\sigma_c \simeq 2^{18} Z_1^5 Z_2^5 (v_0/v)^{12} \pi a_0^2 / 5, \quad (1s \longrightarrow 1s),$$
$$\sigma_R \simeq 2^7 \alpha^3 (Z_1 v_0/v)^5 \pi a_0^2 / 3, \quad (\text{``free''} \longrightarrow 1s), \quad (3.10.4)$$

it is seen that approximate equality is not obtained for velocities below $\sim 20 v_0$ and REC dominates at most in the relativistic region.

---

[263] E. Spindler, H. D. Betz, and F. Bell, *J. Phys.* B **10**, L516(1977).

4. INELASTIC ENERGY-LOSS MEASUREMENTS IN SINGLE COLLISIONS

By Bent Fastrup

4.1. Introduction

4.1.1. Introductory Remarks

In an inelastic collision between two atoms, energy from the relative motion of the nuclei is transferred to the surrounding electrons as internal motion. This energy, which is here denoted *the inelastic energy loss,* amounts to only a few percent of the primary kinetic energy of the nuclei. Nonetheless, it is an important quantity, which, being shared between the two collision partners, determines their respective degrees of excitation. The subsequent relaxation of the postcollision particles leads to emission of their surplus excitation energy in the form of photons and/or electrons. For gentle collision events, where only the outer electronic shells interpenetrate, excitation energies are of the order of first-ionization potentials. Violent collisions, on the other hand, causing deep interpenetration of inner electronic shells, may produce significant disturbance of these shells. Excitation energies of hundreds of electron volts have been observed in such collisions. The first type of collision, which primarily concerns only outer electronic shells, is referred to as *outer-shell collisions,* whereas the latter type of collision, concerning also inner electronic shells, is referred to as *inner-shell collisions.* In the past, both kinds of collisions have been extensively studied experimentally as well as theoretically.

Essentially two experimental methods lend themselves to a detailed study of inelastic collisions. The first method takes its starting point in the deexcitation process. By studying intensity, energy, and perhaps angular distributions of emitted photons and electrons, a detailed mapping of the deexcitation process is achieved. Although this method also furnishes some information about the primary-excitation process, it suffers generally from being a total cross-sectional measurement, i.e., inherently integrated over all impact parameters of the collision. The second method, which eliminates the latter drawback, is based on a determination of the inelastic energy-loss spectrum as a function of scattering angle, and provides detailed differential information about the collision process. As the two methods have their individual strengths and domains of applicability, a complete mapping of the collision process generally requires information obtainable only if both of them are used. It is the aim of this overview to describe in detail the latter method, i.e., the inelastic energy-loss measurement, and to illustrate its application to the determination of characteristics of the primary excitation mechanism.

4.1.2. Definition of Inelastic Energy Loss

In a collision process, an atom A, ionized i times, encounters a target atom B, which is normally neutral and considered to be stationary, apart from thermal motion discussed later. As a result of their mutual interaction, the two particles are scattered angles θ and φ away from the direction of beam incidence. The inelastic energy loss absorbed by the electronic shells may subsequently lead to the ejection of electrons, i.e.,

$$A^{i+} + B \longrightarrow A^{m+} + B^{n+} + (m + n - i)e. \tag{4.1.1}$$

In fast, violent collisions, a substantial fraction of the initially bound electrons may be transferred to the continuum, thus leaving the collision partners with partially stripped outer shells.

The inelastic energy loss is defined as

$$Q = E_0 - (E_1 + E_2), \tag{4.1.2}$$

where E_0, E_1, and E_2 are kinetic energies of the incident, the scattered, and the recoil particle, respectively. Q also equals the total excitation energy of the collision system; thus,

$$Q = \sum_i U_i + \sum_i T_{e,i} + \sum_j h\nu_j, \tag{4.1.3}$$

where U_i is the ionization potential of the ith electron, $T_{e,i}$ its kinetic energy, and $h\nu_j$ the energy of the jth photon.

An inelastic energy-loss measurement is based on simultaneous determination of any three out of five collision parameters E_0, E_i, E_2, θ, and φ. It is differential in scattering angle and thus associates the inelastic energy-loss spectrum with the specific impact parameter b of the collision (see Fig. 1). This feature makes inelastic energy-loss measurements an important tool in the detailed study of the collision mechanism. Endeavors to lend the possibility also to spectroscopic techniques of obtaining differential data (e.g., as a function of scattering angle) has led to sophisticated coincidence techniques, whereby the spectroscopic quantity is determined in coincidence with the scattered projectile.

4.1. INTRODUCTION

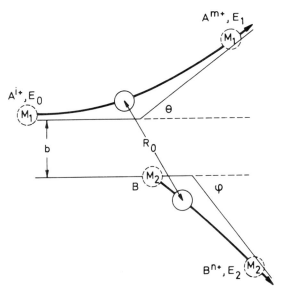

FIG. 1. Schematic diagram of the collision $A^{i+} + B \rightarrow A^{m+} + B^{n+} + (m + n - i)e$ in the laboratory system.

4.1.3. Historical Remarks

It is interesting to note that even though electrostatic and magnetic particle spectrometers had been used in nuclear-reaction studies for a considerable period of time, it was not until the mid-1950s that Everhart and co-workers[1] and Fedorenko and co-workers[2] subjected heavy-ion–atom collisions to a detailed experimental analysis, including differential techniques and the use of electrostatic (magnetic) spectrometers. The first such studies by the two groups primarily aimed at a determination of the charge states of the scattered particles for various scattering angles under single-collision conditions as exemplified in Fig. 2. The unusually high charge states observed in these (essentially quasi-adiabatic) collisions were a surprise and far beyond what could be explained by single-electron excitation processes in atoms such as, for example, formulated in the Born approximation. The molecular approach to an understanding of the collision mechanism, which has become so familiar to us, was already

---

[1] E. Everhart, R. J. Carbone, and G. Stone, *Phys. Rev.* **98**, 1045 (1955); R. J. Carbone, E. N. Fuls, and E. Everhart, *Phys. Rev.* **102**, 1524 (1956); E. N. Fuls, P. R. Jones, F. P. Ziemba, and E. Everhart, *Phys. Rev.* **107**, 704 (1957).

[2] D. M. Kaminker and N. V. Fedorenko, *Zh. Tekhn. Fiz. (USSR)* **24**, 2239 (1955); V. V. Afrosimov and N. V. Fedorenko, *Soviet Phys.—Tech. Phys.* **2**, 2378 (1957).

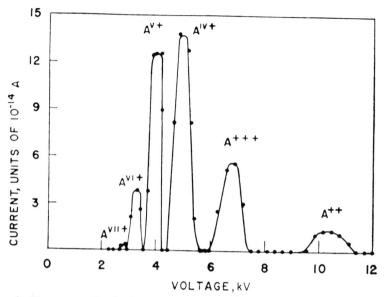

FIG. 2. Charge-state distribution of scattered particles in 138 keV Ar$^+$ + Ar collisions measured at 8° with a Faraday cage and plotted vs. the potential of an electrostatic analyzer. (R. J. Carbone et al.[1]).

suggested by Weizel and Beeck in 1932[3] but had apparently been forgotten ever since. Not until the invention of the molecular-orbital (MO) promotion model by Fano and Lichten in 1965[4] did a satisfactory understanding of heavy-ion–atom collisions become available. The MO model, diagrammed in Fig. 3 for Ar + Ar, which emphasizes the distance of closest approach in the collision (R_0) as the determining quantity in the excitation process, is in contrast to the high-energy models, e.g., the Born approximation, where the velocity of the collision is the important parameter. This feature of quasi-adiabatic collisions makes them especially well suited for differential studies such as inelastic energy-loss measurements.

The first indication of inner-shell excitation in heavy-ion–atom collisions, although initially not identified as such, was provided by Morgan and Everhart in 1962.[5] By analyzing the energy spectrum of the recoil particles in 3–100-keV Ar$^+$ + Ar collisions, (Fig. 4), they observed Q spectra composed of up to three peaks (Fig. 5). These peaks, later labeled Q_I, Q_{II}, and Q_{III}, are associated with the creation of zero, one, and

[3] W. Weizel and O. Beeck, Z. Phys. **76**, 250 (1932).
[4] U. Fano and W. Lichten, Phys. Rev. Lett. **14**, 627 (1965).
[5] G. H. Morgan and E. Everhart, Phys. Rev. **128**, 667 (1962).

4.1. INTRODUCTION

two 2p vacancies, respectively. When plotting the mean inelastic energy loss $\overline{Q}$ as a function of R_0, Morgan and Everhart obtained a single-valued curve, almost independent of the collision energy (Fig. 6). A drastic rise of $\overline{Q}(R_0)$ around $R_0 = 0.25$ Å was later identified as the distance of closest approach, where the diabatic $4f\sigma$ MO experiences a sudden rise, thereby producing one or two 2p vacancies in argon.

The next important improvement in the inelastic energy-loss technique came in the early 1960s, when Fedorenko and co-workers and Everhart and co-workers constructed their new coincidence apparatuses. Based on a simultaneous determination of the emission angles of the scattered particles, the new technique made it possible to measure the charge states

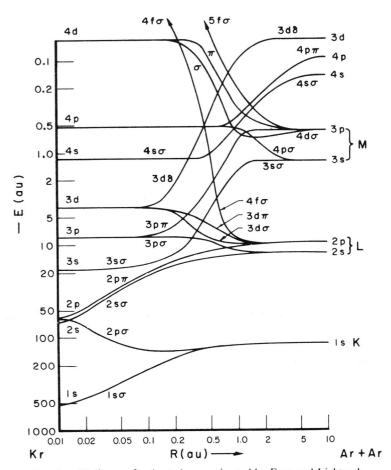

FIG. 3. MO diagram for Ar + Ar as estimated by Fano and Lichten.[4]

154   4. INELASTIC ENERGY-LOSS MEASUREMENTS

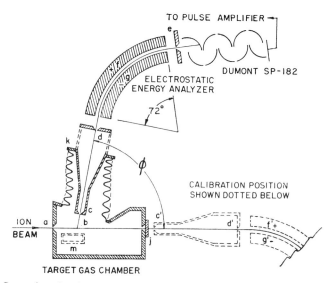

FIG. 4. Scattering chamber used to measure recoil energy spectra vs. recoil angle (G. H. Morgan and E. Everhart[5]).

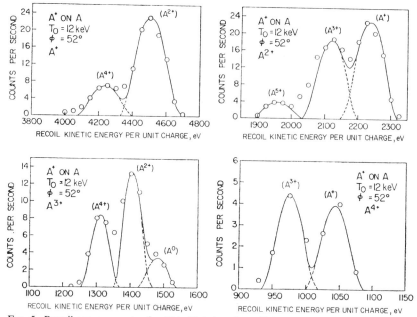

FIG. 5. Recoil energy spectra for 12-keV Ar$^+$ + Ar collisions using the apparatus shown in Fig. 4. Scattering angle and charge state of recoiling ions are indicated in the figure (G. H. Morgan and E. Everhart[5]).

4.1. INTRODUCTION

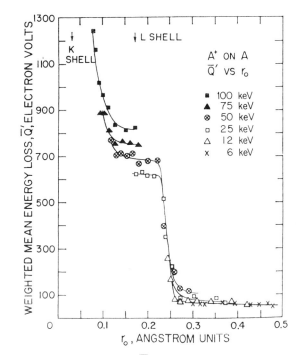

FIG. 6. The mean inelastic energy loss $\overline{Q}$ for Ar^+ + Ar collisions plotted vs. distance of closest approach r_0 (G. H. Morgan and E. Everhart[5]).

of the two scattered particles in the same collision event (Fig. 7). This allowed a more complete mapping of the collision event than did the earlier techniques. The first system to be studied was Ar^+ + Ar. The observations published in 1964/65[6,7] marked a turning point in the study of heavy-ion–atom collisions. Not only did the data reveal a narrow range of R_0 values around 0.25 Å, where the inelastic energy-loss spectrum was triple-valued, but they also gave very detailed information about the charge states of the two particles. This new information gave the inspiration to the MO model (see above) and was actually the starting signal to a series of inner-shell excitation studies in various heavy-ion–atom collision systems.

The first coincidence studies were restricted to rare-gas ion–atom collisions and detected inner-shell excitations only in symmetric combinations

---

[6] V. V. Afrosimov, Yu. S. Gordeev, M. N. Panov, and N. V. Fedorenko, *Soviet Phys.—Tech. Phys.* **9**, 1248, 1256, 1265 (1965); **11**, 89 (1966).

[7] E. Everhart and Q. C. Kessel, *Phys. Rev. Lett.* **14**, 247 (1965); Q. C. Kessel and E. Everhart, *Phys. Rev.* **146**, 16 (1966).

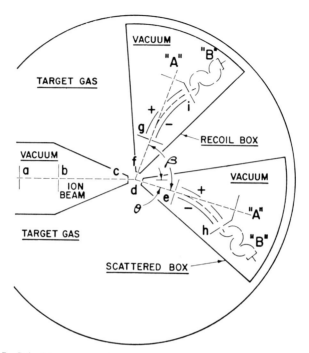

FIG. 7. Coincidence scattering apparatus (Q. C. Kessel and E. Everhart[7]).

such as $Ar^+ + Ar$ $(L_{2,3})$[6,7] and $Ne^+ + Ne$ (K).[8] For reasons to be discussed later, the coincidence method suffers from a large, inherent broadening effect due to thermal target motion. Despite the fact that thermal-motion energies are only a fraction of an electron volt, this motion can be shown to have a severe impact on the width of the measured Q spectra (Fig. 8), actually so severe that the coincidence method in its first version was incapable of tracing M-shell excitations in $Kr^+ + Kr$ collisions.[9]

To improve the coincidence method, Afrosimov and co-workers[10] installed two ion spectrometers in their coincidence apparatus, thus allowing a simultaneous determination of all three kinetic energies entering

---

[8] Q. C. Kessel, M. P. McCaughey, and E. Everhart, *Phys. Rev. Lett.* **16**, 1189 (1966); Q. C. Kessel, M. P. McCaughey, and E. Everhart, *Phys. Rev.* **153**, 57 (1967).

[9] M. P. McCaughey, E. Knystantas, and E. Everhart, *Phys. Rev.* **175**, 14 (1968).

[10] V. V. Afrosimov, Yu. S. Gordeev, A. M. Polyanskii, and A. P. Shergin, *Soviet Phys.-JETP* **30**, 441 (1970); idem, *Soviet Phys.—Tech. Phys.* **17**, 96 (1972); V. V. Afrosimov, *Proc. Int. Conf. Inner-Shell Ioniz. Phenomena Future Appl.*, Atlanta, Ga. 1972. p. 1297. USAEC Report No CONF-720404, Oak Ridge, Tenn. (1973).

4.1. INTRODUCTION

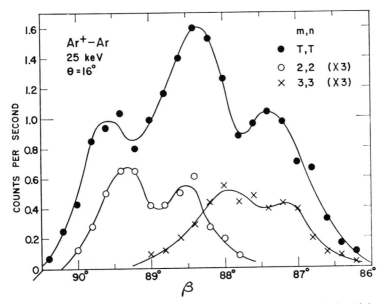

FIG. 8. Coincidence counts vs. sum of scattering angles, $\beta = \theta + \varphi$, in the triple-peak region (E. Everhart and Q. C. Kessel[7]).

in Eq. (4.1.2). The improved coincidence method gave well-resolved Q spectra, an example of which is shown in Fig. 9, but for obvious reasons, the method is time-consuming and only practicable for selected cases.

Subsequent application of a noncoincidence technique by Fastrup and co-workers[11] to the study of asymmetric collision systems revealed a good resolution of the Q spectra obtained. The method, which is based on a determination of the energy spectrum of the scattered particle, was shown to be less influenced by thermal target-broadening effects than the original coincidence method.

Inelastic energy-loss measurements have not only played a significant role in the detailed exploration of inner-shell excitation processes, but have also been used extensively in the study of outer-shell excitation processes. The first such measurements were carried out by Lorents and Aberth in 1965.[12] Utilizing the measured energy-loss distributions to discriminate against inelastic processes, they were able to derive the differential elastic cross section in low-energy He^+-He collisions.

---

[11] B. Fastrup and G. Hermann, *Phys. Rev. Lett.* **23**, 157 (1969); B. Fastrup, G. Hermann, and K. J. Smith, *Phys. Rev. A* **3**, 1591 (1971).

[12] D. C. Lorents and W. Aberth, *Phys. Rev.* **139** A 1017 (1965); F. T. Smith, P. R. Marchi, and K. G. Dedrick, *Phys. Rev.* **150**, 79 (1966).

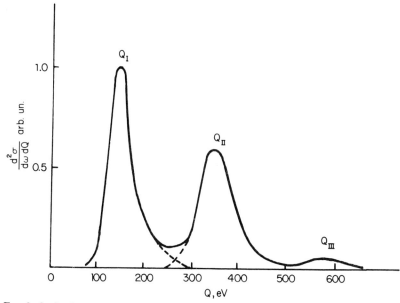

FIG. 9. Inelastic energy-loss spectrum Q for 25-keV Ar$^+$ + Ar, $\vartheta = 16°$, obtained by the improved coincidence method (V. V. Afrosimov[10]).

The use of electrostatic spectrometers is restricted to ionic species. In the case of scattered neutrals, a time-of-flight (TOF) technique has been used successfully.[13] The method, which is limited to low energies, to the author's best knowledge has not been used in the study of inner-shell excitation processes.

Since this overview mainly concerns the experimental aspects of inelastic energy-loss measurements, no systematic discussion is given of the data in terms of current theories. For readers, who might be interested in those important aspects, several review papers are now available.[14–16]

[13] R. Morgenstern, M. Barat, and D. C. Lorents, *J. Phys. B* **6**, L330 (1973); J. C. Brenot, J. Pommier, D. Dhuicq, and M. Barat, *J. Phys. B.* **8**, 448 (1975).

[14] Q. C. Kessel and B. Fastrup, *Case Stud. At. Phys.* **3**, No 3 (1973); B. Fastrup, *in* "Physics of Electronic and Atomic Collisions," IX ICPEAC 1975 (J. S. Risley and R. Geballe, eds.), p. 361. Univ. Wash. Press, Seattle, 1976.

[15] J. D. Garcia, R. J. Fortner, and T. M. Kavanagh, *Rev. Mod. Phys.* **45**, 111 (1973).

[16] J. S. Briggs, *in* "Physics of Electronic and Atomic Collisions," IX ICPEAC 1975 (J. S. Risley and R. Geballe, eds.), p. 384 Univ. Wash. Press, Seattle, 1976; J. S. Briggs, *Rep. Prog. Phys.* **39**, No 3 (1976).

4.2 Binary Encounters

This chapter deals first with the kinematics of a binary encounter, then follows a discussion of interatomic potentials, and finally gives an account of various, inherent broadening effects, which may influence the resolution of the experimental data.

4.2.1. Derivation of Q

As stated previously, an experimental determination of the inelastic energy loss Q may be based upon a simultaneous measurement of any three out of the five collision parameters E_0, E_1, E_2, θ, and φ, characterizing the kinematics of the collision events (see also Fig. 1).

By applying conservation of linear momentum in the collision after some trivial algebra, Eq. (4.1.2) yields the following important formulas:

Scattered-particle method $Q(E_0, E_1, \theta)$:

$$Q = 2\gamma(E_0 E_1)^{1/2} \cos\theta + (1 - \gamma)E_0 - (1 + \gamma)E_1. \quad (4.2.1)$$

Recoil-particle method $Q(E_0, E_2, \varphi)$:

$$Q = 2\left(\frac{E_0 E_2}{\gamma}\right) \cos\varphi - \left(1 + \frac{1}{\gamma}\right) E_2. \quad (4.2.2)$$

Coincidence method $Q(E_0, \theta, \varphi)$:

$$Q = \left(1 - \frac{\sin^2\varphi + \gamma \sin^2\theta}{\sin^2\beta}\right) E_0, \quad (4.2.3)$$

where $\gamma = M_1/M_2$ and $\beta = \theta + \varphi$.

As we shall see later, all these formulas have been used for the derivation of the inelastic energy loss.

4.2.2. Kinetic Energy of Scattered Particles

(i) The kinetic energy E_1 of the scattered incident particle is obtained from Eq. (4.2.1). Two solutions may occur, of which the higher value E_1^+ is called the "soft" component and corresponds to small scattering angle in the center-of-mass (CM) system, whereas the lower value E_1^- is called the "hard" component and corresponds to large scattering angle in the CM system and is therefore a less probable event. For the situations where two solutions may occur, a maximum scattering angle $\theta_{\max}$ in the laboratory system does exist and is given by

$$\sin^2\theta_{\max} = \frac{1}{\gamma^2} - \frac{1 + \gamma}{\gamma^2} \frac{Q}{E_0}. \quad (4.2.4)$$

Only when $\gamma > [1 - (1 + \gamma)Q/E_0]^{1/2}$ and $\theta < \theta_{max}$ will there be two solutions, E_1^+ and E_1^-.

Except for scattering angles in the neighborhood of θ_{max} we may ignore the hard component.

Inverting Eq. (4.2.1), we have for the energy lost by the incident particle

$$\Delta E_1^+ = E_0 - E_1^+$$
$$= \frac{2\gamma}{(1 + \gamma)^2} E_0 \left\{ 1 + \gamma \sin^2\theta + \left(1 + \frac{1}{\gamma}\right) \frac{Q}{2E_0} \right.$$
$$\left. - \cos\theta \left[1 - \gamma^2 \sin^2\theta - (1 + \gamma) \frac{Q}{E_0} \right]^{1/2} \right\}. \quad (4.2.5)$$

Since in most cases we shall have $\theta \ll 1$ and $Q \ll E_0$, Eq. (4.2.5) may be simplified to yield the approximation

$$\Delta E_1^+ \simeq \gamma E_0 \theta^2 + Q. \quad (4.2.6)$$

Utilizing the fact that $E_0 = E_1 + E_2 + Q$, we further have

$$E_2 \simeq \gamma E_0 \theta^2. \quad (4.2.7)$$

Thus ΔE_1^+ is approximately the sum of elastic energy loss $\gamma E_0 \theta^2$ (recoil energy) and inelastic energy loss Q.

Equation (4.2.6) offers an interesting application in connection with molecular targets. At low collision energies, two extreme cases are possible. In one case, the incident particle experiences a binary encounter with one of the atoms constituting the molecular target. In the other case, the incident particle "feels" the molecular target as one massive particle. By plotting ΔE_1^+ as a function of $E_0 \theta^2$ and assuming Q not to vary much, the slope $\gamma = M_1/M_2$ may distinguish between the two cases.

(ii) The kinetic energy E_2 of the recoil particle M_2 is obtained from Eq. (4.2.2). Only for inelastic collisions will there be two solutions, viz., a soft component and a hard (normal) component. Inverting Eq. (4.2.2), we have for the normal component

$$E_2^- = \frac{\gamma E_0}{(1 + \gamma)^2} \left\{ 2 \cos^2\varphi - (1 + \gamma) \frac{Q}{E_0} \right.$$
$$\left. + 2 \cos^2\varphi \left[1 - (1 + \gamma) \frac{Q}{E_0 \cos^2\varphi} \right]^{1/2} \right\}. \quad (4.2.8)$$

A maximum recoil emission angle φ_{max} is given by

$$\cos \varphi_{max} = [(1 + \gamma)Q/E_0]^{1/2}. \quad (4.2.9)$$

Inspecting Eq. (4.2.8) and noting that recoil emission angles φ will normally be close to 90°, we see that recoil energy is very sensitive to the actual value of the inelastic energy loss Q. In the early days of inelastic energy-loss measurements, this led to the misconception that the recoil particle method was superior to the scattered-particle method. As we shall see later, the latter method is by far the better of the two.

4.2.3. Transformation from Laboratory System to CM System

To fully appreciate the implications of inelastic kinematics, we now consider the transformation from the laboratory system to the CM system and vice versa. Figure 10 shows a velocity diagram for $\gamma > 1$, but the results we derive from this diagram will apply for all γ. As compared to the elastic case ($Q = 0$), we note that inelasticity in the collision causes a reduction of CM velocities of the particles, i.e., $v'_{10} \to v'_1$ and $v'_{20} \to v'_2$ for particles M_1 and M_2, respectively. Assuming that $Q \ll E_0$, which is normally the case for quasi-adiabatic collisions, it is easy to show that the

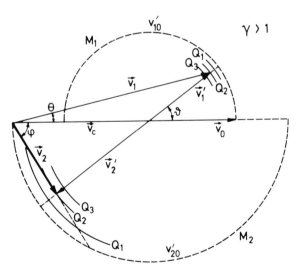

FIG. 10. Velocity diagram for the scattered and recoil particles in the case of $\gamma = M_1/M_2 > 1$. Laboratory velocities of scattered and recoil particles are v_1 and v_2, respectively. Corresponding CM velocities are v'_1 and v'_2. Elastic kinematics correspond to dashed semicircles, while inelastic kinematics are indicated by the arcs labeled Q_1, Q_2, and Q_3. The contraction of CM velocities due to inelasticity Q is readily seen to have a vast influence on kinematics of the recoil particle, while the scattered particle is hardly affected for small ϑ.

relative reduction to first order in Q/E_0 is

$$\frac{v'_{10} - v'_1}{v'_{10}} = \frac{v'_{20} - v'_2}{v'_{20}} = \frac{1+\gamma}{2} \frac{Q}{E_0}, \qquad (4.2.10)$$

where $v'_{10} = v_0/(1 + \gamma)$ and $v'_{20} = v_c = \gamma v_0/(1 + \gamma)$.

Transformation from laboratory scattering angles θ and φ to CM angle ϑ and vice versa is accomplished by the following formulas ($Q \ll E_0$):

$$M_1: \quad \sin(\vartheta - \theta) = \gamma \left(1 + \frac{1+\gamma}{2} \frac{Q}{E_0}\right) \sin \theta$$

or
$\qquad\qquad\qquad\qquad\qquad\qquad\qquad\qquad\qquad\qquad\qquad$ (4.2.11)

$$\tan \theta = \frac{\sin \vartheta}{\gamma\{1 + [(1+\gamma)/2]Q/E_0\} + \cos \vartheta}.$$

$$M_2: \quad \sin(\vartheta + \varphi) = \left(1 + \frac{1+\gamma}{2} \frac{Q}{E_0}\right) \sin \varphi$$

or
$\qquad\qquad\qquad\qquad\qquad\qquad\qquad\qquad\qquad\qquad\qquad$ (4.2.12)

$$\cot \varphi = \tan \frac{\vartheta}{2} + \frac{1+\gamma}{2} \frac{Q}{E_0} \frac{1}{\sin \vartheta}.$$

We note from Eq. (4.2.12) that elastic kinematics ($\varphi \approx \frac{1}{2}\pi - \frac{1}{2}\vartheta$) prevails when $\sin^2 \frac{1}{2}\vartheta \gg (1 + \gamma)Q/(4E_0)$, which for small ϑ and with $\theta \approx \vartheta/(1 + \gamma)$ gives the simpler condition for elastic kinematics,

$$Q \ll (1 + \gamma)E_0 \theta^2.$$

In the coincidence method, transformation from laboratory scattering angles θ and φ to CM angle ϑ is effected via

$$\cot \vartheta = \cot \theta - [\gamma/(1 + \gamma)](\cot \theta + \cot \varphi). \qquad (4.2.13)$$

In practice, one of the laboratory scattering angles is fixed, while the energy spectrum of the scattered particles is being recorded. This may cause entangling implications in the data analysis. We now consider one such implication and defer the discussion of other entangling implications to Sections 4.2.5 and 4.3.1. Keeping laboratory angles fixed in Eqs. (4.2.11) and (4.2.12), differentiation with respect to Q yields for small ϑ,

$$M_1: \quad \Delta \vartheta \simeq \frac{\gamma(1+\gamma)}{2E_0} \theta \, \Delta Q$$

$$M_2: \quad \Delta \vartheta \simeq \frac{1+\gamma}{2E_0} \frac{1}{\cos(\varphi + \vartheta)} \Delta Q. \qquad (4.2.14)$$

Let us first consider the recoil particle method where φ is kept fixed. Since φ is close to 90°, $\cos(\varphi + \vartheta)$ is small, and hence inelasticity affects ϑ appreciably. As seen from, e.g., Fig. 10, inelasticity will also cause a drastic change in recoil velocity v_2. This means that a recoil energy spectrum consisting of, e.g., three peaks, may actually be composed of three different collision events in the CM system, one for each peak or inelastic energy loss. As noted from the formulas given above, such complexities, however, do not occur when the energy spectrum of the scattered, incident particle is studied. Because $\theta \ll 1$, inelasticity will here barely affect the corresponding CM angle ϑ.

4.2.4. Interatomic Potentials and Distance of Closest Approach

It has been well established that the nuclear motion during a collision at not too low velocities to a good approximation may be described by classical mechanics. Thus E_0 and R_0 (or b; see Fig. 1) constitute a convenient set of collision parameters for the discussion of the collision process. In the laboratory, however, Q is measured as a function of E_0 and θ, and by assuming a suitable potential for the atom–atom interaction during the collision, we may convert Q into a function of E_0 and R_0 (or b), i.e., $Q[E_0, R_0(b)]$. As inelastic energies amount to only a few percent of the kinetic energy of the incident particle, it generally is justified to approximate the potential with the elastic potential. In doing so, however, one should exercise some caution when the potential is significantly distorted by the occurrence of inelastic processes, e.g., at crossings (see, for example, Fig. 11).

A reliable calculation of R_0 depends on the choice of a fair approximation to the real potential. This is particularly so for R values close to R_0, where a major part of the angular deflection takes place. For inner-shell processes to occur, these shells must interpenetrate deeply during the collision, and therefore R_0 will be comparable to the radii of the inner shells in question. Hence, the potential function must be well known at small values of R, where electronic screening of the nuclear charges is small. Considering, on the other hand, outer-shell processes, related R_0 values correspond to those regions of the potential function where electronic screening is excessive.

The most commonly used potential function $V(R)$ may be written as a product of a nuclear–nuclear Coulomb potential and a screening function accounting for the screening action of the electrons, i.e.,

$$V(R) = (Z_1 Z_2 e^2/R) F(R/a), \qquad (4.2.15)$$

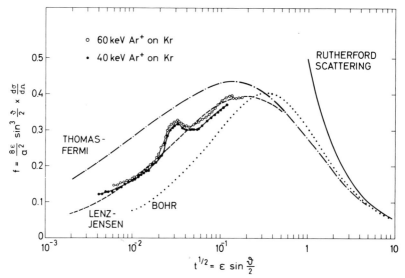

FIG. 11. Reduced scattering cross section $f(t^{1/2})$ in the CM system, see Eq. (4.2.17) for Rutherford, Bohr, Thomas–Fermi, and Lenz–Jensen potentials. Also shown are experimental data [P. Loftager[22]].

where a is a screening distance that may take different forms, and $F(0) = 1$ and $F(\infty) = 0$.

The simplest potential function of that type is the Bohr potential[17] or the exponentially screened Coulomb potential,

$$V_B(R) = (Z_1 Z_2 e^2/R)\, e^{-R/a}. \qquad (4.2.16)$$

In the past, several forms of the screening distance have been proposed. Bohr originally gave the form $a_B = a_0(Z_1^{2/3} + Z_2^{2/3})^{-1/2}$, where $a_0 = 0.53$ Å. On the basis of a comparison with his numerically calculated potential of Thomas–Fermi type, Firsov[18] proposed $a_F = 0.8853 a_0 (Z_1^{1/2} + Z_2^{1/2})^{-2/3}$. Applying Thomas–Fermi arguments, Lindhard et al.[19] suggested a screening distance $a_{TF} = 0.8853 a_0 (Z_1^{2/3} + Z_2^{2/3})^{-1/2}$.

The lack of formal justification for the exponentially screened Coulomb potential is more or less compensated for by its simplicity. In this context, however, it should be pointed out that Eq. (4.2.16) overestimates the screening of the Coulomb potential, particularly at larger R. Some cau-

---

[17] N. Bohr, *K. Dan. Vidensk. Selsk. Mat. Fys. Medd.* **18**, No 8 (1948).

[18] O. B. Firsov, *Soviet Phys.—JETP* **6**, 534 (1958); **7**, 308 (1959).

[19] J. Lindhard, V. Nielsen, and M. Scharff, *K. Dan. Vidensk. Selsk. Mat. Fys. Medd.* **36**, 10 (1968).

tion therefore should be exercised when using it to estimate R_0 for gentle collisions, i.e., outer-shell processes. For inner-shell processes, where $R_0 \ll a$ and the screening function F is close to unity, it is well justified to use Eq. (4.2.16) to reduce scattering data, i.e., to derive R_0 (or impact parameter b) from laboratory data E_0 and θ.

Except at very asymmetric atom–atom combinations, the different screening distances given above differ only slightly. Using a_B for screening distance, Everhart et al.[20] and later Bingham[21] solved the scattering equation numerically for the particle trajectory and gave R_0 and b in a tabular form for a few selected cases.

Another screening function, not expressible analytically, is developed from the Thomas–Fermi description of the atom. Using a_{TF} for the screening distance, Lindhard et al.[19] derived a universal curve for the differential-scattering cross section,

$$d\sigma = \pi a_{TF}^2 \frac{dt}{2t^{3/2}} f(t^{1/2}),$$

where $t^{1/2} = \epsilon \sin(\vartheta/2)$, $\epsilon = a_{TF}/\rho$, and ρ is the collision diameter (distance of closest approach in a head-on collision). The universal function $f(t^{1/2})$ is shown in Fig. 11. It is noteworthy that the distance of closest approach R_0 is approximately a function depending solely on $t^{1/2}$ (or, at small θ, a function of the product $E_0\theta$). The differential-scattering cross section per unit solid angle in the CM system is given by

$$\left(\frac{d\sigma}{d\Omega}\right)_{CM} = \frac{a_{TF}^2}{8} \frac{\epsilon^2}{t^{3/2}} f(t^{1/2}). \qquad (4.2.17)$$

The corresponding cross sections in the laboratory system are (i) scattered particles,

$$\left(\frac{d\sigma}{d\Omega}\right)_L^1 = \left|\frac{\sin \vartheta}{\sin \theta} \frac{d\vartheta}{d\theta}\right| \left(\frac{d\sigma}{d\Omega}\right)_{CM},$$

and (ii) recoil particles, $\qquad (4.2.18)$

$$\left(\frac{d\sigma}{d\Omega}\right)_L^2 = \left|\frac{\sin \vartheta}{\sin \varphi} \frac{d\vartheta}{d\varphi}\right| \left(\frac{d\sigma}{d\Omega}\right)_{CM}.$$

A comparison between differential-scattering cross sections calculated from Eqs. (4.2.16) and (4.2.17) and experimental-scattering data of Loftager[22] is shown in Fig. 11. One notes that the experimental data lie

---

[20] E. Everhart, G. Stone, and R. J. Carbone, *Phys. Rev.* **99**, 1287 (1955).
[21] F. W. Bingham, *J. Chem. Phys.* **46**, 2003 (1967).
[22] P. Loftager and G. Hermann, *Phys. Rev. Lett.* **21**, 1623 (1968); P. Loftager, Univ. of Aarhus (unpublished data).

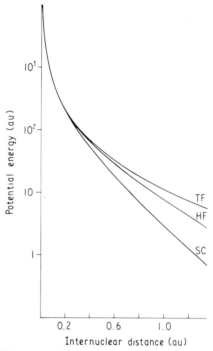

FIG. 12. The potential energy of the Ne–O molecule vs. internuclear distance in atomic units HF, Hartree–Fock; TF, Thomas–Fermi; SC, Bohr (K. Taulbjerg and J. S. Briggs[24]).

approximately midway between the two calculated curves with a tendency of being closer to the Thomas–Fermi cross section. Also shown is a calculated curve using a Lenz–Jensen potential. This curve seems to reproduce reasonably well the experimental data at even smaller values of $t^{1/2}$, where screening is appreciable.

Extensive tabulations of the CM scattering angle, distance of closest approach, and impact parameter for Bohr, Born–Mayer, and Thomas–Fermi potentials have been worked out by Robinson.[23]

Ab initio potentials using, e.g., Hartree–Fock methods, are comparatively scarce. Thus, for example, Briggs and Taulbjerg[24] have investigated the Ne–O system by such methods and obtained a potential that lies approximately halfway between the Bohr and the Thomas–Fermi potential (Fig. 12). This result complies nicely with the experimental-scattering results.[22,25]

[23] M. T. Robinson, Oak Ridge National Laboratory (ORNL-3493, UC-34-Physics, TID-4500, 21st ed.); [ORNL-4556, UC-34-Physics (1970)].
[24] K. Taulbjerg and J. S. Briggs, *J. Phys. B* **8**, 1895 (1975).

At small scattering angles θ, the differential-scattering cross section in the case of pure elastic scattering can be shown to possess certain scaling properties.[12] Thus, by plotting the reduced cross section $\theta \sin \theta (d\sigma/d\Omega)_L^1$ as a function of $\tau = E_0 \theta$, a unique curve, almost independent of the specific values of E_0 and θ, results. Such plots, where reduced units are used, may be convenient even when inelastic processes do indeed contribute to the scattering.

4.2.5. Kinematical Effects in Cross Sections

In Section 4.2.4, we saw that the CM scattering angle ϑ may depend on inelasticity for a fixed laboratory scattering angle. This dependence was most pronounced for recoils where each recoil energy (or Q) corresponded to a specific value of ϑ. That the effect is not purely academic has been shown by Fastrup et al.[11] for the case of 20-keV $P^+ - Ar$ set at a recoil scattering angle of $77°$. Here the Q spectrum revealed three Q peaks, Q_I (100 eV), Q_{II} (250 eV), and Q_{III} (400 eV), corresponding to CM scattering angles 23.83, 22.82, and 20.18°. As the differential-scattering cross section $(d\sigma/d\Omega)_{CM}$ depends strongly upon ϑ (for the case of medium screening, $d\sigma/d\Omega$ goes approximately as ϑ^{-3}), the recoil energy spectrum may be significantly distorted from what it would have been if no such effect was present.

Transformation of solid angle $d\Omega$ from CM system to laboratory system and vice versa is also affected by the kinematics of the collision process. To see this, $d\vartheta/d\theta$ and $d\vartheta/d\varphi$, entering into Eqs. (4.2.18), are evaluated by differentiating Eqs. (4.2.11) and (4.2.12) with respect to ϑ. Allowing Q to be a function of ϑ, after some trivial algebra we obtain for (i) scattered particles,

$$\left(\frac{d\vartheta}{d\theta}\right) \simeq \left(\frac{\sin \vartheta}{\sin \theta}\right)\left[\cos(\vartheta - \theta) - \sin(\vartheta - \theta)\frac{1+\gamma}{2E_0}\left(\frac{dQ}{d\vartheta}\right)\right]^{-1},$$

and for (ii) recoil particles, (4.2.19)

$$\left(\frac{d\vartheta}{d\varphi}\right) \simeq \left(\frac{\sin \vartheta}{\sin \varphi}\right)\left[\cos(\vartheta + \varphi) - \sin(\vartheta + \varphi)\frac{1+\gamma}{2E_0}\left(\frac{dQ}{d\vartheta}\right)\right]^{-1}.$$

We note that a sudden change of Q with ϑ as, for example, occurs at a crossing between MOs, may have a significant impact on the transformation formulas, Eqs. (4.2.18). If this is so, the laboratory differential-scattering cross section may be distorted from what it would have been if

---

[25] Yu. S. Gordeev, Proc. Int. Conf. Inner-Shell Ioniz. Phenomena Future Appl., Atlanta, Ga. (1972) USAEC Report No. CONF-720404, Oak Ridge, Tenn. (1973), p. 1232; I. M. Torrens, "Interatomic Potentials," Academic Press, New York and London, 1972.

only elastic scattering had prevailed. Outside those "active" regions, $dQ/d\vartheta$ is generally small and may thus be neglected.

An interesting feature is seen when either $\vartheta - \theta$ or $\vartheta + \varphi$ becomes $\pi/2$. Then, as noted from Eqs. (4.2.19), $d\vartheta/d\theta$ or $d\vartheta/d\varphi$ experiences a maximum, respectively. For $\vartheta \ll 1$, such pileup effects are essential only for the recoils.[22]

A distinctive feature of inelastic processes in slow ion–atom collisions is the significance of localized couplings at MO crossings. Experimentally one finds excitation probabilities that are approximately unity inside some crossing radius R_x and zero elsewhere. Such situations lead to excitation cross sections of the form[26]

$$\sigma_I = \pi b_x^2 = \pi R_x^2 \left[1 - \frac{V(R_x)}{E_{\text{rel}}}\right], \quad (4.2.20)$$

where $E_{\text{rel}} = E_0/(1 + \gamma)$ and $V(R_x)$ is the interatomic potential at $R = R_x$. The term in parentheses describing the threshold behavior of the cross section is of kinematic origin. Even though other excitation mechanisms such as rotational and long-range couplings cannot be described by a simple form such as Eq. (4.2.20), they also lead to a threshold behavior that is mainly governed by the kinematics of the collision.

4.2.6. Inherent Broadening Effects

Inelastic energy-loss measurements are primarily hampered by the following three broadening effects: (i) thermal target motion; (ii) isotopic composition of target gas, and (iii) momenta carried away by energetic electrons being ejected after the collision. We now consider these effects in more detail and defer a discussion of so-called instrumental broadening effects to the following section, where experimental methods are described.

(i) The average kinetic energy of the thermal target motion E_T amounts to only a fraction of an electron volt and is thus extremely small as compared to the kinetic energies of the particles. Despite this, its effect on the measured Q spectra may be appreciable.[11]

The initial momenta $\mathbf{P}_0$ and $\mathbf{P}_T$ of the particles and their corresponding post collision momenta $\mathbf{P}_1$ and $\mathbf{P}_2$ form a quadrangle (see Fig. 13). Thermal target–particle momentum $\mathbf{P}_T$ has been decomposed into $\alpha_x P_0$ and $\alpha_y P_0$ in the "scattering plane." The third component $\alpha_z P_0$, out of the plane, may be ignored because, to a good approximation, it does not affect the energy-loss distribution.[11]

[26] Q. C. Kessel, *Bull. Am. Phys. Soc.* **14**, 946 (1969); R. K. Cacak, Q. C. Kessel, and M. E. Rudd, *Phys. Rev. A* **2**, 1327 (1970).

4.2. BINARY ENCOUNTERS

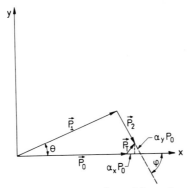

FIG. 13. Momentum quadrangle describing effect of thermal-target momentum $\mathbf{P}_T$ on momenta of scattered and recoil particles. For details, see text.

We first consider the effect of thermal target motion on the scattered particles (M_1).

Conservation of linear momentum in the collision yields two equations. Eliminating the angle φ from these equations, we obtain to first order in α_x and α_y

$$E_2 = -2\gamma(E_0 E_1)^{1/2}[(1 + \alpha_x)\cos\theta + \alpha_y \sin\theta] + \gamma E_0 + \gamma E_1 + 2\gamma \alpha_x E_0.$$

Utilizing Eq. (4.1.2), we have

$$Q = 2\gamma(E_0 E_1)^{1/2}[(1 + \alpha_x)\cos\theta + \alpha_y \sin\theta] \\ + (1 - \gamma)E_0 - 2\gamma\alpha_x E_0 - (1 + \gamma)E_1. \quad (4.2.21)$$

Assuming $Q(E_1, \alpha_x, \alpha_y)$ to be constant with variations of α_x and α_y and utilizing the fact that measured energy spectra are transformed into Q spectra by means of Eq. (4.2.1), we obtain

$$\Delta Q_T = \alpha_x(\partial Q/\partial \alpha_x) + \alpha_y(\partial Q/\partial \alpha_y),$$

and by squaring and averaging

$$\overline{\Delta^2 Q_T} = \overline{\alpha_x^2}(\partial Q/\partial \alpha_x)^2 + \overline{\alpha_y^2}(\partial Q/\partial \alpha_y)^2,$$

where we have used the fact that α_x and α_y are independent stochastic variables. The x and y components of the momentum of the thermal motion obey a Gaussian-distribution law, i.e.,

$$n_{x,y} \propto \exp(-M_2 v_{x,y}^2/2kT) \propto \exp[-\alpha_{x,y}^2/(kT/\gamma E_0)].$$

For the mean values $\overline{\alpha_x^2}$ and $\overline{\alpha_y^2}$, we have accordingly

$$\overline{\alpha_x^2} = \overline{\alpha_y^2} = KT/(2\gamma E_0).$$

Evaluating $\partial Q/\partial \alpha_x$ and $\partial Q/\partial \alpha_y$ from Eq. (4.2.21), we finally have $(\delta Q_T^2 = \overline{\Delta^2 Q_T})$,

$$\delta Q_T = (2E_2 kT)^{1/2} = 7(E_2)^{1/2} \quad \text{eV} \tag{4.2.22}$$

(E_2 in keV). For small θ, we may make use of Eq. (4.2.7) and hence,

$$\delta Q_T = (2\gamma E_0 kT)^{1/2}\theta \simeq 7(\gamma/E_0)^{1/2}(E_0\theta) \quad \text{eV} \tag{4.2.23}$$

(E_0 in keV). Since for small θ, the distance of closest approach R_0 depends on the product $E_0\theta$ only and not on the individual parameters, Eq. (4.2.23) shows that for a fixed value of R_0, the thermal broadening is inversely proportional to the square root of the kinetic energy E_0 of the incident particle.

A similar calculation for the recoil particle yields the result

$$\delta Q_T = (2E_1 kT/\gamma)^{1/2} \simeq 7(E_1/\gamma)^{1/2} \quad \text{eV} \tag{4.2.24}$$

(E_1 in keV).

For the coincidence method, one may show[11] that if $\varphi \simeq 90°$, then the thermal-target broadening is approximately given by

$$\delta Q_T = (2E_0 kT/\gamma)^{1/2} = 7(E_0/\gamma)^{1/2} \quad \text{eV} \tag{4.2.25}$$

(E_0 in keV).

A comparison between the three formulas shows that the recoil particle method and the coincidence method yield approximately the same broadening, whereas the scattered-particle method for small θ is least affected by thermal-target motion. Let us illustrate this by an example and consider a 20-keV Kr$^+$–Kr collision with a scattering angle of 11.9°. For this case, by inserting into Eqs. (4.2.23)–(4.2.25) we have the following values for the broadening effect:

Scattered-particle method: $\delta Q_T \simeq 6.5$ eV.
Recoil particle method: $\delta Q_T \simeq 30.5$ eV.
Coincidence method: $\delta Q_T \simeq 30.5$ eV.

(ii) As noted from, e.g., Eq. (4.2.6), the energy lost by the incident particle depends on γ, the mass ratio M_1/M_2. If the target gas consists of several isotopes, each isotope will give rise to a specific energy loss. If the relative mass shift from one isotope to another is appreciable, as is the case for a neon [^{20}Ne (90.92%) and ^{22}Ne (8.82%)] gas, the isotope effect may give rise to the occurrence of additional and separated peaks in the measured energy spectra.[27] An example of that is shown in Fig. 14. If, on the other hand, the relative shift from one isotope to another is small

---

[27] B. Fastrup, G. Hermann, Q. C. Kessel, and A. Crone, *Phys. Rev. A* **9**, 2518 (1974).

4.2. BINARY ENCOUNTERS

(such as for, e.g., krypton), the additional peaks will coalesce and form one broad peak.[9] We now estimate this broadening effect. The relation $dQ/d\gamma$ may be obtained from Eqs. (4.2.1)–(4.2.3) and (4.2.7). Thus

$$\delta Q_\gamma = E_2(\delta\gamma/\gamma) \quad \text{(scattered-particle method)}, \quad (4.2.26)$$

$$\delta Q_\gamma = |\{Q + [1 - (1/\gamma)]E_2\}(\delta\gamma/2\gamma)| \quad \text{(recoil particle method)}, \quad (4.2.27)$$

$$\delta Q_\gamma = E_2(\delta\gamma/\gamma) \quad \text{for} \quad \theta \ll 1 \quad \text{(coincidence method)}. \quad (4.2.28)$$

Let us consider as an example a 20-keV Kr–Kr collision. The krypton gas contains four significant isotopes: ^{80}Kr (2.27%), ^{82}Kr (11.56%), ^{83}Kr (11.55%), and ^{84}Kr (56.9%). Thus $\delta\gamma/\gamma$ is estimated to be 0.015. Assuming again a scattering angle of 11.9°, we have

$\delta Q_\gamma = 13$ eV [scattered-particle method, Eq. (4.2.26)],
$\delta Q_\gamma = 1$ eV [recoil particle method, Eq. (4.2.27)],
$\delta Q_\gamma = 13$ eV [coincidence method, Eq. (4.2.28)].

(iii) The third broadening effect stems from the momenta carried away by the electrons emitted after the collision, but before the scattered particle is energy-analyzed. The effect causes the source particle to recoil and thus changes its kinetic energy and apparent emission angle. Although at first sight the effect may seem analogous to the thermal broadening effect, it differs from this by involving only that particle from which the electron(s) is (are) ejected, whereas the thermal target motion affects both collision partners. In the following we make the plau-

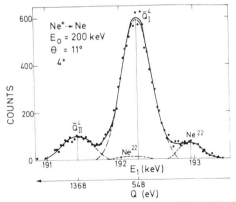

FIG. 14. Energy spectrum of scattered Ne^{4+} ions from 200-keV Ne$^+$–Ne collisions with scattering angle 11°. Contributions from ^{22}Ne target atoms are indicated (B. Fastrup et al.[27]).

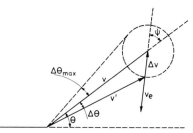

FIG. 15. Velocity diagram describing the effect of electron emission on source particle. Emission angle relative to the direction of motion of source particle is denoted by ψ. Recoil velocity of source particle is Δv.

sible assumption that the electron distribution is isotropic in the source frame. For the sake of simplicity, we consider only the case where one fast electron is emitted. The effect of emittance of more electrons will probably diminish the broadening effect since the individual momenta of the electrons will add incoherently.

As seen from Eqs. (4.2.1)–(4.2.3), the calculated Q depends upon energy and emission angle of the scattered particle. Making the simplifying assumption that the energy of the scattered (recoil) particle changes insignificantly with scattering angle within a small range $\Delta\theta$ around θ (Fig. 15), and letting the emitted electron have kinetic energy E_e, the maximum change of energy E of the scattered particle is $\Delta E_{max} = 2(m/M)^{1/2}(E_e E)^{1/2}$, where m is the electron mass and M the particle mass. Let the reduced energy be given by $t = E/E_0$, where E_0 as usual is the primary energy of the projectile; then $\Delta t_{max} = 2(m/M)^{1/2}(E_e/E_0)^{1/2} t^{1/2}$. As seen from Fig. 15, the maximum change of scattering (recoil) angle due to emission of an electron is $\Delta\theta_{max} \approx \Delta v/v = (m/M)^{1/2}(E_e/E)^{1/2}$ or by using the reduced energy t, $\Delta\theta_{max} \approx (m/M)^{1/2} t^{-1/2}$. One notes that $\Delta t_{max}/t = 2\Delta\theta_{max}$.

When inspecting the formulas Eqs. (4.2.1) and (4.2.2), used to derive Q, one thus finds that the scattered-particle and the recoil particle methods are equally affected by the emitted electron when $\gamma = 1$. This may also be seen by inserting the above expressions for Δt_{max} and $\Delta\theta_{max}$ in the formulas Eqs. (4.3.10) and (4.3.11), to be discussed in Section 4.3.3. One notes that Q derived from Eq. (4.2.1) is predominantly sensitive to a change of energy E_1, i.e., Δt_{max}, while Q derived from Eq. (4.2.2) is predominantly sensitive to a change of recoil angle φ, i.e., $\Delta\varphi_{max}$. That the effect on the measured Q is not negligible may be illustrated by the following example. A scattered argon ion with kinetic energy 50 keV emits an Auger electron with energy 200 eV. Using the above formulas for Δt_{max}

and $\Delta\theta_{max}$, we find by inserting in Eq. (4.3.10) that the maximum change of Q is $\Delta Q_{max} = 22$ eV. If we had used the recoil particle method, an equally large change of Q would have been found.

4.3. Experimental Method of Determining Q

As described in Chapter 4.2, the determination of the inelastic energy loss Q requires a knowledge of any three of the five collision parameters E_0, E_1, E_2, θ, and φ. The methods that have been used so far to obtain Q fall naturally into two categories:

(1) Noncoincidence methods, where either the scattered or the recoil particle is energy-analyzed for a given scattering angle [see, e.g., Eqs. (4.2.1) and (4.2.2)]. Depending upon the particle being charged or not, energy analysis may be performed by either deflection in an electric (magnetic) field or by use of a TOF technique.

(2) The coincidence methods, where both of the scattered particles are analyzed simultaneously by either measuring their scattering angles, Eq. (4.2.3), or their kinetic energies, Eq. (4.1.2). These methods, which were used extensively in earlier studies of inner-shell excitations, provide the most complete information of the collision process, including the charge states of both of the scattered particles.

As regards the primary excitation process, the interesting quantity, however, is the inelastic energy loss. Here, as we saw in Chapter 4.2, it may even be advantageous to employ a noncoincidence method, for example the scattered-particle method, which not only is a fast and simple method, but which is also insignificantly affected by thermal target broadening.

4.3.1. Noncoincidence Methods. General

These methods, which have been widely used for experimental studies of inelastic processes in inner and outer shells of the collision partners, are the scattered-particle[11,27-29] and the recoil particle[1] methods. In the former method, the inelastic energy loss is obtained by measuring the energy of the scattered particle, whereas in the latter method the energy of the recoil particle is measured. However, before describing the two

---

[28] B. Fastrup and G. Hermann, *Phys. Rev. A* **3**, 1955 (1971).
[29] D. C. Lorents, W. Aberth, and V. W. Hesterman, *Phys. Rev. Lett.* **17**, 849 (1966); A. Pernot, M. Abignoli, M. Barat, J. Baudon, and A. Septier, *Rev. Phys. Appl.* **2**, 203 (1967); D. J. Biermann and W. C. Turkenburg, *Physica* **60**, 357 (1972).

methods in more detail, we first summarize a few of their characteristic features, partially discussed in Chapter 4.2.

(i) For a given collision in the CM system (fixed E_0 and ϑ), the intensities per unit solid angle of the scattered particles I_1 and the recoil particles I_2 are

$$I_{1,2} = I_0 N_t l_{\text{eff}} (d\sigma/d\Omega)_L^{1,2}, \qquad (4.3.1)$$

where I_0 is the number of incident particles per second and N_t the number of target atoms per cubic centimeters; l_{eff}, the effective target length, is proportional to $1/\sin\theta$ (scattered particles) or $1/\sin\varphi$ (recoil particles), and $(d\sigma/d\Omega)_L$ is the differential-scattering cross section in the laboratory system.

By using Eqs. (4.2.18), (4.2.19), and (4.3.1), the ratio I_1/I_2 between intensities of scattered and recoil particles corresponding to a given solid angle in the CM system is

$$\frac{I_1}{I_2} = \left(\frac{\sin\varphi}{\sin\theta}\right)^2 \frac{(d\vartheta/d\theta)}{(d\vartheta/d\varphi)} \approx \frac{(1+\gamma)^3}{\vartheta^3} \left| \cos(\vartheta+\varphi) \right.$$
$$\left. - \sin(\vartheta+\varphi) \frac{1+\gamma}{2E_0} \left(\frac{dQ}{d\vartheta}\right) \right|. \qquad (4.3.2)$$

Except when $\cos(\vartheta + \varphi) = 0$ or $\vartheta + \varphi = \pi/2$, we note that the scattered yield I_1 is much higher than the corresponding recoil yield I_2. The problem at $\vartheta + \varphi = \pi/2$ is partially removed when we make allowance for the analyzer acceptance angle not being infinitesimally small but finite, i.e., by replacing $d\vartheta/d\theta$ and $d\vartheta/d\varphi$ by $\Delta\vartheta/\Delta\theta$ and $\Delta\vartheta/\Delta\varphi$, respectively.

(ii) Transformation of the measured energy spectrum of the scattered particles $d^2\sigma/dE_1\, d\Omega$ into a double-differential cross section $d^2\sigma/dQ\, d\Omega$ is straightforward at small ϑ since from Eq. (4.2.6) we have $dE_1/dQ = -1$. For the recoil spectrum, however, the transformation, which is facilitated by Eq. (4.2.8), is a much more complicated operation.

(iii) As was shown in Chapter 4.2, the transformation of the recoil angle φ into corresponding CM angle ϑ depends significantly upon Q. Thus, an energy spectrum obtained at fixed φ is actually a weighted sum of contributions from different collisions with deviating cross sections in the CM system. Further, where for small ϑ, $\cos(\vartheta + \varphi)$ is likely to be much smaller than unity, the weights are functions of ϑ, Q, and its derivative $dQ/d\vartheta$ [see also Eqs. (4.2.18) and (4.2.19)]. It is therefore no surprise that the data analysis of the recoils is an intricate problem.

For the scattered particles, no such problems occur for small ϑ. As transformation formulas, Eqs. (4.2.11) and (4.2.19), are essentially insensitive to Q, the measured spectral intensities are readily converted into corresponding CM intensities.

(iv) Thermal target motion, which is the most prominent among the different inherent broadening contributions, was shown in Chapter 4.2 to affect recoil spectra most severely.

4.3.1.1. Setting of Experimental Conditions. The optimal conditions for measuring inelastic energy loss will always be a compromise between conflicting requirements such as energy resolution and scattered intensity. However, before examining these requirements in any detail, we describe a scattering chamber for measuring the energy of the scattered particles as a function of their scattering angle (θ or φ). For this purpose, and without any loss of generality, we consider a tilted, sliding-seal scattering chamber (Fig. 16).

The chamber consists of two parts, of which the lower is stationary and the upper is rotatable from 0 to 132°. Inside the chamber is placed a differentially pumped gas cell furnished with a bell-shaped Faraday cup for measuring the instantaneous beam current. Admittance to the gas cell is provided through beam collimators (a) and (b). Attached to the upper half of the scattering chamber is a cylindrical, electrostatic analyzer with radius 12 cm and sector angle $\Phi = 63.65°$, aimed so as to analyze the charged particles that are scattered through exit collimators (c) and (d). The detector, which is an open multiplier, has an intrinsic efficiency of ~ 100% for keV ions. In front of the detector is situated a removable Faraday cup used at zero-angle setting for analysis of the incident-beam energy. As the axis of rotation A–A is tilted 24° off vertical, the angle of rotation is not exactly equal to the scattering angle, but a relation may easily be established.

An inelastic energy-loss measurement does not require any specific knowledge of absolute gas pressure, but in order to achieve good counting statistics, the gas pressure must be kept high, however, without violating the single-collision conditions, which are assumed to prevail when the relative charge-state distribution of scattered particles does not change with gas pressure.

The experimental conditions further comprise setting of entrance and exit collimators, thus defining the target volume and the angular dispersion of the scattered particles to be analyzed. This, combined with the choice of energy analyzer, determines the energy resolution of the apparatus.[30]

The optimal experimental conditions being a compromise between sev-

---

[30] G. Hermann, Thesis, Univ. of Aarhus, unpublished (1968).

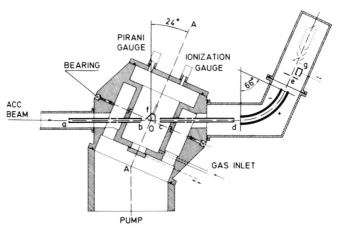

FIG. 16. Noncoincidence scattering chamber of the tilted sliding-seal type. Upper half of the chamber may rotate about an axis A–A, which is tilted off vertically (B. Fastrup et al.[11]).

eral factors are intimately related to the collision system under investigation. To achieve a good understanding of the significance of the various factors (effective target length, angular dispersion, etc.) of importance for the experimental conditions, we now consider those in more detail.

4.3.1.2. Target Length and Solid Angle. For the sake of simplicity, we assume that the incident particles form a parallel and homogeneous beam and that also $\Delta\theta \ll \theta$. Diaphragm (c) normally has a slit aperture with width s_c defining the target length, whereas diaphragm (d) may be of any reasonable shape. However, also here we let this be a slit with width s_d and height h_d ($A_d = s_d h_d$). The effective value of (target length × solid angle) is (see Fig. 17):

$$(l\, d\Omega)_{\text{eff}} = \int_{\text{col-volume}} dz\, d\Omega \approx \frac{A_d}{L_1 + L_2} \frac{s_c}{L_1 \sin \theta}.$$

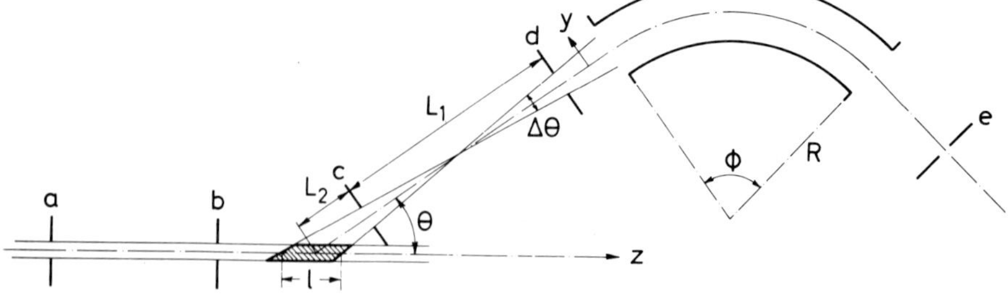

FIG. 17. Scattering geometry.

4.3. EXPERIMENTAL METHOD OF DETERMINING Q

Variation of s_c and s_d with $\Delta\theta = (s_c + s_d)/L_1$ kept constant yields a maximum value of

$$(I\,d\Omega)_{\text{eff}} = \frac{h_d(\Delta\theta)^2}{4(L_1 + L_2)\sin\theta} L_1 \qquad (4.3.3)$$

for $s_c = s_d = \tfrac{1}{2}L_1\,\Delta\theta$.

As the energy of the scattered particle depends upon angle θ, the finite acceptance angle of the energy analyzer will inevitably cause an energy spread of the particles.

4.3.1.3. Energy Spread Due to Finite Acceptance Angle.
For the scattered, incident particles, the relative energy spread $\Delta E/E$ due to $\Delta\theta$ may be found by differentiating Eq. (4.2.1) with respect to θ. After some trivial algebra, we have

$$\lambda_1 = \frac{\Delta E_1}{E_1} = \frac{-2\gamma \sin\theta}{(1 - \gamma^2 \sin^2\theta)^{1/2}} \Delta\theta. \qquad (4.3.4)$$

In the derivation, we have assumed that $Q = 0$, which is justified for the scattered particles.

For the recoils, we have correspondingly

$$\lambda_2 = \frac{\Delta E_2}{E_2} = -\left[2\tan\varphi \bigg/ \left(1 - \frac{(1+\gamma)}{\cos^2\varphi}\frac{Q}{E_0}\right)^{1/2}\right]\Delta\varphi. \qquad (4.3.5)$$

It is immediately recognized that $\lambda_2 \gg \lambda_1$ when $\theta \ll 1$ or $\varphi \sim 90°$. At first glance, this might seem to be a severe drawback in the recoil particle method as compared to the scattered-particle method (see also Section 4.3.1.4). However, when we consider the Q spectrum of the collision, the corresponding spreads in Q due to the finite analyzer acceptance angle are approximately equal. This is shown below. For a given collision, i.e., fixed values of E_0 and ϑ, one can show that

$$\lambda_1 = -2\gamma \tan(\vartheta - \theta)\,\Delta\theta \quad \text{and} \quad \lambda_2 = 2\tan(\vartheta + \varphi)\,\Delta\varphi.$$

Here, once again starting from Eqs. (4.2.1) and (4.2.2) and doing a little algebra, we get

$$(\Delta Q)_{\lambda_1} = -\lambda[2\gamma/(1+\gamma)]E_0[1 - (1+\gamma)Q/E_0]^{1/2} \sin\vartheta\,\Delta\theta$$
$$\text{(scattered particles)}, \qquad (4.3.6)$$

or, approximately, $(\Delta Q)_{\lambda_1} \approx -[2\gamma/(1+\gamma)]E_0 \sin\vartheta\,\Delta\theta$ and

$$(\Delta Q)_{\lambda_2} = [2/(1+\gamma)]E_0[1 - (1+\gamma)Q/E_0]^{1/2} \sin\vartheta\,\Delta\varphi$$
$$\text{(recoils)}, \qquad (4.3.7)$$

or, approximately, $(\Delta Q)_{\lambda_2} \approx [2/(1+\gamma)]E_0 \sin\vartheta\,\Delta\varphi$.

From the ratio $(\Delta Q)_{\lambda_1}/(\Delta Q)_{\lambda_2} = \gamma\,\Delta\theta/\Delta\varphi$, we note that the two methods yield identical spreads in Q for symmetric collision systems and constant acceptance angle ($\Delta\theta = \Delta\varphi$). For asymmetric collisions, however, significant differences may result. For small values of ϑ, $(\Delta Q)_{\lambda_1}$ and $(\Delta Q)_{\lambda_2}$ to a very good approximation depend only on the product $E_0\theta = E_0\vartheta/(1 + \gamma)$, i.e., distance of closest approach R_0.

4.3.1.4. Electrostatic Energy Analyzer. Electrostatic energy analyzers, which are well suited for determination of the energy of the scattered particles as a function of their scattering angle, are cylindrical-sector field and parallel-plate analyzers. Since both of these have been described elsewhere,[31] we restrict our discussion here to those aspects of particular importance for setting the experimental conditions.

The position y of a particle is measured as its distance from a central ray (Fig. 17). Let y_s and y_i be the positions at source (first focal plane) and image (second focal plane), respectively; then, to first order,

$$y_i = My_s + D_E\lambda, \qquad (4.3.8)$$

where M is the magnification and D_E the energy dispersion ($\lambda = \Delta E/E$). Due to the energy spread of the particles accepted by the analyzer, a point source will be imaged into a line of length $y_{i,\lambda} = D_E\lambda$. Considering also that the source has a definite size $2y_s$, an additional contribution to the image formation, $2y_{i,s} = M2y_s$, will result. If slit (c) is situated at the first focal plane (source), then particles with the higher energies (and smaller scattering angles) will preferentially pass through the higher part of slit (c). By the same argument, particles with the lower energies will preferentially pass through the lower part of slit (c). We can thus arrange it so that the two contributions appearing in Eq. (4.3.8) tend to cancel each other. If substantial, the effect will allow us to improve analyzer resolving power R correspondingly ($R^{-1} = y_i/D_E$ and is equal to the relative

TABLE I. Magnification M and Energy Dispersion D_E for Various Electrostatic Analyzers[a]

	Cylindrical-sector field		Parallel plate (45° inc.)
	$\Phi = \pi/2\sqrt{2}$ (63.65°)	$\Phi = \pi/\sqrt{2}$ (127.3°)	
M	$-l_2\sqrt{2}/r$	-1	-1
D_E	$r/2 + l_2/\sqrt{2}$	r	L

[a] For the cylindrical-sector field analyzer, r is the radius of the central ray and l_2 the distance from analyzer boundary (exit) to image. For the parallel-plate analyzer, L is the distance between entrance and exit slits.

[31] P. Dahl, "Introduction of Electron and Ion Optics," Academic Press, New York and London, 1973.

4.3. EXPERIMENTAL METHOD OF DETERMINING Q

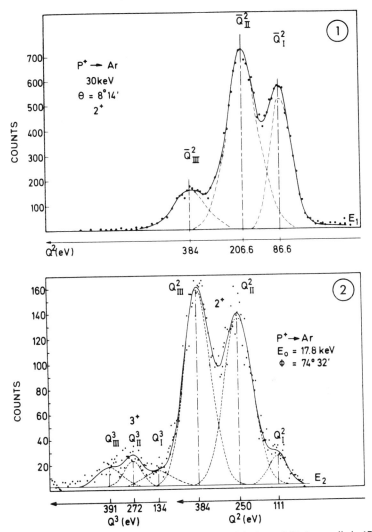

FIG. 18. Energy spectra of (1) scattered P in 30-keV P^+–Ar and (2) Ar recoils in 17.8-keV P^+–Ar collisions. Scattering angles and charge states are indicated (B. Fastrup et al.[11]).

energy change necessary to sweep the beam at the image a distance $y_i = My_s + D_E\lambda$). For the analyzer shown in Fig. 16, such conditions occur when the analyzer bends the scattered particles back toward the direction of the incident beam (see also Fig. 17). As $\lambda_2 \gg \lambda_1$ for $\theta \ll 1$, this partial cancellation is most pronounced for the scattered particles.

Table I gives the magnification and energy dispersion for

cylindrical-sector field analyzers with sector angles $\Phi = \pi/2\sqrt{2}$ (63.65°) and $\pi/\sqrt{2}$ (127.3°) and parallel-plate analyzer with 45° incidence. Only for the cylindrical-sector field analyzer with $\Phi = \pi/2\sqrt{2}$ are the focal planes situated outside the boundaries of the analyzer. The distances l_1 and l_2 from the boundaries to the focal plane can be shown to obey the relationship $l_1 l_2 = r^2/2$, where r is the radius of the central ray.

As demonstrated in Section 4.3.1.3 for small values of ϑ, the broadening of Q due to a finite analyzer acceptance angle depends only upon the product $E_0 \vartheta$ or upon the distance of closest approach R_0. The other broadening (due to the finite size of the source), however, can be shown to increase with increasing energy E_0. For the scattered-particle method, where the latter contribution may be comparable to the first, it thus seems preferable to choose low values of E_0 when $E_0 \vartheta$ or R_0 is to be kept fixed.

In Fig. 18 are shown energy spectra for (1) scattered particles and (2) recoils in P$^+$–Ar collisions at 30 and 17.8 keV. One notes that Ar^{2+} and Ar^{3+} recoil spectra are not completely separated, thus making a deconvolution of the spectral lines nontrivial. This complicating feature of the recoil spectrum is a consequence of its dramatic dependence on Q for fixed scattering angle φ (see, e.g., Fig. 10). For the scattered phosphorous particles, where different charge-state spectra are well separated, such effects do not occur.

4.3.1.5. Time-of-Flight Energy Analysis. The methods considered so far for energy analysis apply, of course, only for charged particles. If neutrals are present in the scattered beam, their energy analysis constitutes a special problem. For violent collisions likely to produce inner-shell excitation, only a minor or negligible fraction of the scattered particles will be neutrals. Thus, it is permissible to exclude those particles from the energy analysis. For slow and soft collisions, on the other hand, where a substantial fraction of neutrals may be present in the scattered beam, one can no longer disregard their presence.

As has been demonstrated during recent years, the TOF technique is very powerful for the study of neutrals in low-energy collisions of up to a few thousand electron volts of primary energy. A TOF apparatus constructed by Brenot et al. is shown in Fig. 19. Here, an accelerated beam of particles is pulsed by passing it between two parallel capacitor plates, over which a time-varying voltage, for example, a step function with fast rising time (a few nanoseconds), is applied. If desired, the beam may be subsequently neutralized by passing it through a gas-filled charge exchange cell. The remaining charged particles are, in turn, swept away in an electric field between two capacitor plates situated after the charge exchange cell. Placed at a fairly large distance from the target cell is a particle detector set to register the pulsed particles being scattered at angle θ.

4.3. EXPERIMENTAL METHOD OF DETERMINING Q

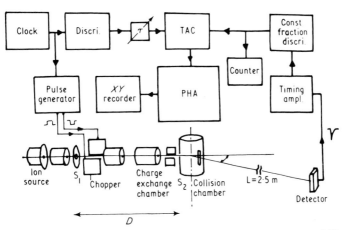

FIG. 19. Schematic diagram of a TOF apparatus (J. C. Brenot et al.[13]).

The starting signal to a time-to-amplitude converter (TAC) is supplied by the detector pulse. The preceding chopper pulse, which has been delayed, sets the stop signal. The output of the TAC is directed to a pulse height analyzer (PHA) and stored. The pulse height spectrum representing the time spectrum of the particles received by the detector may be

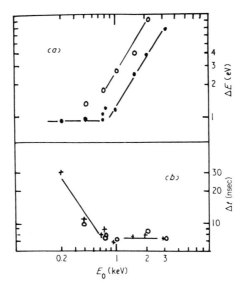

FIG. 20. The effect of apparatus length L on measured energy resolution of TOF spectra as a function of beam energy in He^+–He collisions. (a) Energy resolution ΔE; (b) time resolution Δt. ○, $L = 2.5$ m; +, ●, $L = 1.25$ m (J. C. Brenot et al.[13]).

readily converted into an energy spectrum of the scattered particles. Expressing the flight time t in nanoseconds, the flight length L in meters, the particle mass M in atomic mass units, and the particle energy E in kiloelectron volts, we have $E = 5.3 \times 10^6 L^2 M/t^2$. A time resolution dt will cause a corresponding energy spread $dE/E = (0.87 \times 10^{-3}/L)(E/M)^{1/2} dt$.

Figure 20 shows time and energy resolutions of the above TOF apparatus for various energies and for two different flight lengths. One notes that the time resolution is independent of the flight length and also of energy above 1 keV for He$^+$–He. Let us assume this constant value, which is $dt = 7.5$ nsec, and further let $L = 2.5$ m. Then $dE/E = 2.6 \times 10^{-3}(E/M)^{1/2}$. At ratios of E/M approximately equal to unity, this energy resolution compares qualitatively well with a corresponding value for electrostatic energy analysis for charged particles, where dE/E is determined by the settings of the analyzer, such as width of entrance and exit slits.

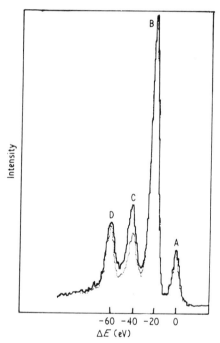

FIG. 21. Energy-loss spectra (He$^+$–He) obtained using the TOF technique for 3-keV incident-beam energy. (A) Capture into the ground state, (B) capture into excited state, (C) 40-eV energy loss due to multiple collisions, (D) electron capture and simultaneous excitation of both particles (J. C. Brenot et al.[13]).

In Fig. 21 is shown a TOF energy spectrum of scattered He in 3-keV He⁺–He obtained by Brenot et al.[13] One notes that the method is capable of resolving four peaks, each one, except the elastic peak (zero energy loss), consisting of a group of closely spaced lines.

4.3.1.6. High-Resolution Energy Spectrometry. When a high-energy resolution is needed, for example, in outer-shell excitations by fast, light particles, retardation of the scattered particles has been successfully applied. Such a system, devised by Park et al.,[32] is shown in Fig. 22. Its main feature is the decelerating column in front of the energy analyzer. The analyzer and its accessories are encased in a metal housing connected electrically to the positive terminal of a high-voltage supply (V). Another metal housing encasing the ion source is electrically connected to the former housing through an adjustable voltage ΔV. By means of an offset voltage, the potential of the ion source is raised relative to its metal housing by a value V_0.

When the target chamber is set at ground potential, incident particles are accelerated through a potential drop $V + V_0 + \Delta V$, before entering the target region. By a successive passage of a switching magnet, unde-

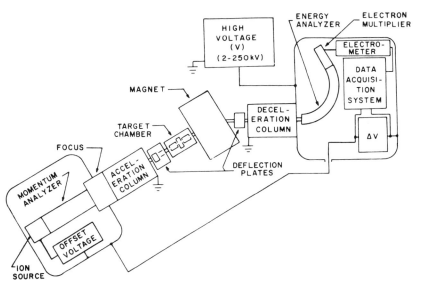

FIG. 22. Schematic diagram of an accelerating–decelerating system for obtaining high-resolution energy spectra. For details, see text (V. Pol et al.[32]).

---

[32] J. T. Park and F. D. Schowengerdt, *Rev. Sci. Instr.* **40,** 753 (1969); idem, *Phys. Rev.* **185,** 152 (1969); V. Pol, W. Kauppila, and J. T. Park, *Phys. Rev. A* **8,** 2990 (1973); J. T. Park, J. E. Aldag, J. M. George, and J. L. Peacher, *Phys. Rev. A* **15,** 508 (1977).

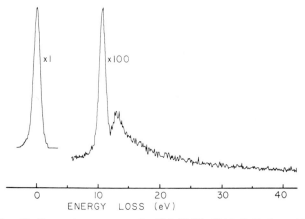

FIG. 23. Energy-loss spectrum for 50-keV H^+–H (J. T. Park et al.[32]).

sired ions are removed from the beam before the main beam enters a decelerating column with voltage V. The particles with charge state q reach the entrance of the energy analyzer with kinetic energy $qV_0 + E_p - \Delta E$, where E_p is the most probable energy of an ion in the ion source and ΔE the energy loss of particles in the target gas. With no gas in the target chamber, the analyzer is set so that for $\Delta V = 0$, a maximum detector response is achieved. In order to keep transmission of decelerating column and energy analyzer constant, these settings (including V_0) are not altered during measurement.

With gas in the target chamber, ΔV is varied until maximum detector response is reached. Thus, the energy loss of the particles just equals $\Delta E = q \, \Delta V$. As it can be shown that the transmitted ion current at the entrance slits of the analyzer does not change significantly when ΔV is varied over a small range (up to a few percent of V), the method immediately yields the energy-loss spectrum of the incident particles. An example of such a spectrum is shown in Fig. 23. One notes that a few electron volts of energy resolution are obtained even for 50-keV incident protons on hydrogen.

To make possible differential-scattering measurements, the entire accelerator can be rotated about the collision point.

4.3.2. Coincidence Methods. General

The coincidence method aims at a determination of emission angles (energies) and charge states of both partners. The coincidence rate is determined mainly by the channel with the lowest count rate, i.e., the recoil channel, and thus some of the arguments leading to Eq. (4.3.2) may also

4.3. EXPERIMENTAL METHOD OF DETERMINING Q

be applied here. It is interesting to note that although the coincidence method has frequently been used in inner-shell excitation studies, it has hardly been used in outer-shell excitation studies.

4.3.2.1. Setting of Experimental Conditions. In a coincidence experiment, either angle of emission θ or φ is fixed. By varying the other angle and recording the corresponding coincidence rate, the Q spectrum may be derived either from Eq. (4.2.3) or Eq. (4.1.2). It is favorable to choose θ as the fixed angle, the main reasons being, first, that the CM angle ϑ (or R_0) stays fixed as φ is varied, and, second, as seen from Eq. (4.2.7), the kinetic energy of the recoil particle is approximately independent of recoil angle φ (or Q) for θ fixed. Thus, the analyzer in the recoil channel, which determines the charge state, may have an approximately fixed setting even as φ is varied.

To achieve a high rate of coincidence, the two channels must ideally view the same target volume. Since target volumes for scattered and recoil particles, respectively, are proportional to $1/\sin \theta$ and $1/\sin \varphi$, this requirement can be fulfilled only partially.

The angular dispersions of the two channels must correspond to the same angular dispersion $\delta\vartheta$ in the CM system. Only then will the coincidence rate be a maximum. Let us look at this requirement in more detail and assume that Q is constant within the angular ranges accepted by the two channels.

The angular dispersions $\delta\theta$ and $\delta\varphi$ corresponding to angular dispersions $\delta\vartheta'$ and $\delta\vartheta''$ of CM angle ϑ, respectively, are derived from Eqs. (4.2.11) and (4.2.12) by differentiating these with respect to ϑ. For $Q \ll E_0$ and $\vartheta \ll 1$, the following estimates result:

$$\delta\theta \simeq (1 + \gamma)^{-1} \delta\vartheta', \qquad \delta\varphi \simeq [-\tfrac{1}{2} + \tfrac{1}{2}(1 + \gamma)(Q/E_0\vartheta^2)] \delta\vartheta''.$$

By equating $\delta\vartheta'$ and $\delta\vartheta''$,

$$(\delta\gamma)_\theta \simeq \{-\tfrac{1}{2}(1 + \gamma) + \tfrac{1}{2}[(1 + \gamma)^2](Q/E_0\vartheta^2)\} \delta\theta. \qquad (4.3.9)$$

We note the dependence of $(\delta\varphi)_\theta$ on Q, which is especially significant at small values of ϑ. In practice, Eq. (4.3.9) cannot be fulfilled during a complete Q scan. Thus, an average Q may be inserted into Eq. (4.3.9) to obtain a relation between the two angular dispersions.

A Q spectrum is then measured by sweeping $\beta(\varphi)$ over a range corresponding to the appropriate Q distribution.

The finite acceptance angles $\Delta\theta$ and $\Delta\varphi$ cause a spread in the measured Q that is estimated by differentiating Eq. (4.2.3) with respect to θ and φ.

A coincidence apparatus constructed by Kessel and Everhart[7] is shown in Fig. 7. Immersed in a big cylindrical container filled with target gas are

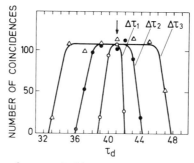

FIG. 24. The number of true coincidences as a function of delay time τ_d (μsec) (scattered-particle signal) for various values of the resolving time $\Delta\tau$ in the coincidence circuit. Redrawn from Fig. 4 in V. V. Afrosimov, Yu. S. Gordeev, M. N. Panov, and N. V. Fedorenko, *Sov. Phys.-Techn. Phys.* **9**, 1248 (1965).

three boxes, of which one is stationary and carries entrance collimators (a) and (b) defining the incident beam, and the other two are rotatable and select the particles that have been scattered to angles θ and ϑ.

Besides the collimators, which define the acceptance angles of scattered particles $\Delta\theta$ and $\Delta\varphi$, the two boxes are equipped with charge-state analyzers such that a measurement with prescribed charge states of the scattered particles can be accomplished. During a Q scan, recoil angle φ is varied, while θ is kept fixed.

A coincidence circuit selects events where both of the scattered particles from a collision event have been detected. Corresponding scattering angles θ and φ are thus measured, and by means of Eq. (4.2.3), Q is derived. To ensure that the pulses from the detected recoil particle and the scattered particle arrive at the coincidence circuit at the same time, the fast, scattered particle has to be delayed. Furthermore, the coincidence resolving time $\Delta\tau$ must be large enough for account to be made of the energy spread of recoils but still small enough to discriminate against accidental coincidences. Figure 24 shows the coincidence rate as a function of delay time of scattered particles in a 50-keV Ar^+ + Ar collision with scattering angle $\theta = 7.5°$ for different values of $\Delta\tau$ (Ref. 6).

4.3.3. Error Analysis

Due to fluctuations and long-term drifts in beam energy and analyzer voltage and due to errors in the setting of scattering angles, the experimental determination of the inelastic energy loss Q is encumbered by an uncertainty that, as we shall see, depends upon the method employed.

Let us introduce reduced units in Eqs. (4.1.2) and (4.2.1)–(4.2.3), i.e.,

4.3. EXPERIMENTAL METHOD OF DETERMINING Q

$t_1 = E_1/E_0$ and $t_2 = E_2/E_0$, and further assume that $\theta \ll 1$ ($\varphi \sim 90°$). Then we have by differentiating with respect to $t_1, t_2, \theta, \varphi$, and $\beta(\varphi)$

$$\frac{\Delta Q}{E_0} \approx -\frac{\Delta t_1}{t_1} - 2\gamma\theta\,\Delta\theta \qquad \text{(scattered-particle method),} \quad (4.3.10)$$

$$\frac{\Delta Q}{E_0} \approx -\left(\frac{1+\gamma}{2\gamma}t_2 - \frac{Q}{2E_0}\right)\frac{\Delta t_2}{t_2} - 2\left(\frac{t_2}{\gamma}\right)^{1/2}\Delta\varphi$$
$$\text{(recoil particle method),} \quad (4.3.11)$$

or, by using Eq. (4.2.7)

$$\frac{\Delta Q}{E_0} \approx -\left(\frac{1+\gamma}{2}\theta^2 - \frac{Q}{2E_0}\right)\frac{\Delta t_2}{t_2} - 2\theta\,\Delta\varphi,$$

$$\frac{\Delta Q}{E_0} \approx -2\gamma\theta\,\Delta\theta - 2\theta\,\Delta\beta \qquad \text{(coincidence method),} \quad (4.3.12)$$

$$\frac{\Delta Q}{E_0} \approx -\Delta t_1 - \Delta t_2 \qquad \text{(improved coincidence method).} \quad (4.3.13)$$

In the derivations, we made use of the approximation $\Delta(Q/E_0) \approx \Delta Q/E_0$, which is valid when $Q \ll E_0$.

If energy analysis of scattered and recoil particles is performed with the same particle spectrometer, then $\Delta t_1/t_1 = \Delta t_2/t_2$. Of the two sources contributing to $\Delta t_1/t_1 (\Delta t_2/t_2)$, fluctuations in beam energy are most likely to be dominant. Although long-term drifts may be compensated for by periodically checking of the beam energy with the spectrometer set at zero scattering angle, fast fluctuations may still constitute a significant problem. For example, when using the scattered-particle method, a variation of the beam energy by 0.1% may bring about a change of Q by 10% [see also Eq. (4.2.6)]. However, as we shall see now, not all four methods for determining Q are equally sensitive to variations in E_0. Inspection of Eqs. (4.3.10)–(4.3.13) clearly shows that for $\theta \ll 1$, the recoil particle and coincidence methods are those least affected by fluctuations in E_0.

The precision with which the scattering angles can be set is not only a mechanical problem. In order to achieve good beam intensities, entrance apertures defining direction of the incident beam should not be made too small. Thus, a corresponding uncertainty of the scattering angles is found. In their experimental setup, Afrosimov et al.[6] reported that the incident beam had a divergence of ±4' and that the analyzer angle of rotation was known to better than ±1'. This corresponds to a maximum uncertainty in scattering angle of ±0.08°.

Instrumental errors and fluctuations give rise to two distinctive effects.

The former causes a shift in absolute Q values, whereas the latter causes a broadening of Q peaks. Since in many contexts one is often more interested in Q differences than in their absolute values, the first effect, although appreciable, may not be a serious problem. Instrumental broadening, the second effect, should be compared to the inherent broadenings discussed in Chapter 4.2, of which thermal target motion gave rise to the largest contribution. A third contribution is the natural linewidth, which is normally the smallest of the three contributions.

For a specific inner-shell excitation, e.g., creation of a K vacancy in Ne + Ne, the many close-lying states corresponding to various excited outer-shell configurations often produce a broad, unresolved Q distribution.

Of the four methods for determination of Q, it may be concluded that the scattered-particle and the improved-coincidence methods result in the best resolved Q spectra. This is also confirmed by experiments.

4.4. Data Reduction and Comparison with Auger Electron/X-Ray Yields

4.4.1. Data Reduction

From measured Q spectra such as those shown in Fig. 18, one may derive information about excitation energy, excitation probability, and charge-state distributions of scattered (recoil) particles for a given excitation mode.

For the noncoincidence methods, only the charge state of the particle being analyzed is specified. This may not be a serious problem since, as was originally shown by Kessel et al.[33] in their coincidence study of, e.g., the $Ar^+ - Ar$ collision, the charge states of the collision partners are almost uncorrelated except for the obvious case, where the experiment was set to select collisions with only one inner-shell excitation in either partner.

Since our primary concern is the excitation process, i.e., excitation energy and probability, without any loss of generality we may base our discussion on data reduction on noncoincidence Q spectra such as those displayed in Fig. 18. From such spectra consisting of more Q peaks (here up to three), the individual peaks may be unfolded by assuming suitable line profiles, e.g., Gaussians. Thus, if the area of peak j, representing all scattered particles with charge state m and energy loss $\overline{Q_j^m}$, is denoted by N_j^m, we may derive the following quantities[11]

---

[33] Q. C. Kessel, A. Russek, and E. Everhart, *Phys. Rev. Lett.* **14**, 484 (1965).

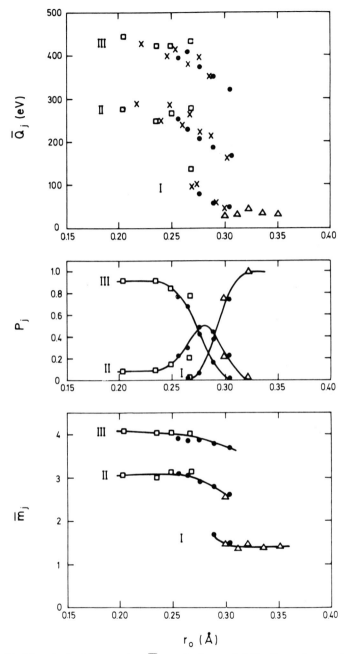

FIG. 25. Mean inelastic energy loss $\overline{Q}_j$, excitation probability P_j, and mean charge state $\overline{m}_j$ of the scattered P particles at 20, 30, and 50 keV P^+ + Ar, represented by Δ, ●, and □, respectively. Also shown are $\overline{Q}_j$ values obtained by energy analysis of recoil particles at 17.8 and 19.3 keV incident energy (×) (B. Fastrup et al.[11]).

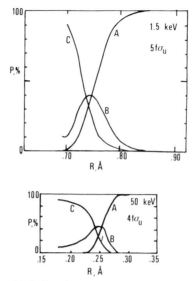

FIG. 26. Upper curve, M-shell excitation; lower curve, L-shell excitation in Ar + Ar collisions. (A) No excitation; (B) one-electron excitation; (C) two-electron excitation (J. C. Brenot et al.[34]).

(1) Charge-state distribution P_j^m of particles having been excited in mode j, $P_j^m = N_j^m / \Sigma_{m,j} N_j^m$.
(2) Mean inelastic energy loss $\overline{Q}_j$ averaged over the charge state distribution, $\overline{Q}_j = \Sigma_m P_j^m \overline{Q}_j^m / \Sigma_m P_j^m$.
(3) Probability P_j of excitation to mode j, $P_j = \Sigma_m P_j^m$.
(4) Mean charge state of particles being excited to mode j, $\overline{m}_j = \Sigma_m m P_j^m / \Sigma_m P_j^m$.

Figures 25 and 26 show such reduced data for inner-shell excitation in P^+ + Ar (Ref. 11) and for outer-shell excitation [$P(A) = P_I$, $P(B) = P_{II}$, and $P(C) = P_{III}$] in Ar^+ + Ar (Ref. 34), respectively. One notes the sudden changes of P_j (j = I, II, III) in a narrow region of R_0 values. This is a typical feature of a MO excitation induced by a steeply rising MO.[14]

From the above data, the total excitation cross section integrated over all impact parameters is given by

$$\sigma_j = 2\pi \int_0^\infty P_j b \, db. \qquad (4.4.1)$$

The study of inner/outer-shell excitations by the inelastic energy-loss

---

[34] J. C. Brenot, D. Dhuicq, J. P. Gauyacq, J. Pommier, V. Sidis, M. Barat, and E. Pollack, *Phys. Rev. A* **11**, 1245 (1975).

method is, of course, restricted to gaseous targets. Moreover, individual inner-shell excitations can only be evaluated quantitatively if they occur with a probability of at least 5%. Only then will it be possible to resolve successfully a given Q spectrum into separate peaks, each corresponding to a specific mode of excitation. If this is not so, one may turn to spectroscopic methods.[14,35]

4.4.2. Spectroscopic Methods

Relaxation of an inner-shell vacancy is accompanied by emission of either an Auger electron or an x-ray quantum, the energies of which are determined by the properties of the electronic shells participating in the relaxation. For excitation among the outer shells, the corresponding emission products are known as an autoionizing electron and a photon.

The fluorescence yield, i.e., the probability of emitting an x-ray quantum, is an increasing function of the atomic number of the particle to undergo a transition. Thus, to obtain information of the excitation cross section, Auger spectroscopy is more adequate for the lighter elements, while x-ray spectroscopy is more adequate for the heavier elements.

For medium-range elements, where the fluorescence yield is neither close to zero nor to unity, a further complication arises because here the fluorescence yield may depend strongly upon the state of excitation/ionization of the collision partner before decay. Thus, the lack of precise knowledge of charge state of the emitting collision partner, combined with the lack of reliable knowledge of fluorescence yields for multiply ionized atoms, may make the interpretation of the measured emission yields in terms of the primary excitation mechanism an intricate problem.

When performing emission-yield studies, one must also bear in mind that the angular distribution in the source frame may not be completely isotropic.[35,36] Although strong anisotropy has been observed for specific transitions in light systems, it is generally assumed that when summing over all transitions associated with the decay of an inner-shell vacancy, the anisotropy becomes less important.

When using solid targets, special complications arise. So-called solid effects,[37] due to multiple-collision processes and to recoiling target atoms, must be carefully considered before the measured emission yield can be reduced to a well-defined excitation cross section.

Despite these serious problems, the spectroscopic methods possess several important features, of which we shall now consider three.

---

[35] M. E. Rudd and J. H. Macek, *Case Stud. Atom. Phys.* **3**, No 2 (1972).
[36] P. Dahl, M. Rødbro, B. Fastrup, and M. E. Rudd, *J. Phys. B* **9**, 1567 (1976).
[37] K. Taulbjerg and P. Sigmund, *Phys. Rev. A* **5**, 1285 (1972); K. Taulbjerg, B. Fastrup, and E. Laegsgaard, *Phys. Rev. A* **8**, 1814 (1974).

First, spectroscopic methods are fast. This is particularly so when we consider x-ray methods, where solid-state devices may be used to detect the characteristic radiation. Second, and more important, spectroscopic methods in either version may be applied to almost any projectile–target combination, whereas inelastic energy-loss measurements are restricted to a few selected cases with gaseous targets.[14] It is also noteworthy that spectroscopic methods, in particular x-ray methods, are capable of tracing specific decays over several decades of intensity; an example is the K-vacancy-sharing process in ion–atom collisions.[38] Third, the spectroscopic methods are able to resolve finer details in the excitation process such as probabilities of populating specific terms and multiplets.

Most emission-yield data are total cross sections, i.e., inherently integrated over all impact parameters, although in a few cases differential data have also been obtained by measuring the radiation in coincidence with the scattered projectile.[39]

Acknowledgments

The author is grateful to Dr. Jørgen Østgaard Olsen for many valuable comments to the manuscript, and to Mrs. Alice Grandjean for skillful preparation of the manuscript for the publishers. Last, but not least, the author wishes to thank Dr. Georg Hermann, with whom he collaborated for several years (1967–1974). Many of the ideas discussed in this part originated from this fruitful collaboration.

---

[38] W. E. Meyerhof, *Phys. Rev. Lett.* **31**, 1341 (1973).

[39] H. O. Lutz, in "Physics of Electronic and Atomic Collisions," IX ICPEAC (1975) (J. S. Risley and R. Geballe, eds.), p. 432. Univ of Wash. Press, Seattle, 1976.

5. TARGET IONIZATION AND X-RAY PRODUCTION FOR IONS INCIDENT UPON SOLID TARGETS

By Tom J. Gray

5.1. Introduction

With the discovery of x rays by Roentgen[1] a new area of study was initiated. A vast amount of information on the generation and properties of x rays has been accumulated since that time. Chadwick[2] first observed the production of characteristic x rays that resulted from the exposure of thick solid targets of several elements to alpha particles from a radium source. Franz and Bothe[3] observed x-ray emission from a wide range of targets using polonium alpha particles to bombard solid targets. In their studies they observed K-, L- and M-shell x rays for targets from $Z = 32-79$ and $Z = 83$, respectively. In 1930, Barton[4] unsuccessfully attempted to produce target x rays by the bombardment of solid targets with low-energy protons. Grethsen and Reusse[5] successfully performed the first experiment in which target x rays were observed using 30–150-keV protons on solid targets. The first absolute K-shell yield measurement was reported by Peter[6] in 1936 for 132-keV protons on thick aluminum targets. In 1937, Livingston et al.[7] studied characteristic x rays from solid targets in the range $Z = 12-82$, which were produced by proton bombardment at incident energies up to 1.76 MeV.

These early works served to establish a basis from which present experimental studies in ion–atom collisions leading to excitation of the inner atomic shells have evolved. The evolution of the studies in inner-shell

---

[1] W. C. Roentgen, *Sitzrengsberichte der Würzburger Physik—Medig. Gesellschaft,* (1895); *Ann. Phys. Chem.* **64**, 1 (1898).
[2] J. Chadwick, *Phil. Mag.* **24**, 594 (1912).
[3] H. Franz and W. Bothe, *J. Franklin Inst.* **52**, 466 (1928).
[4] H. A. Barton, *J. Franklin Inst.* **209**, 1 (1930).
[5] C. Gerthsen and W. Reusse, *Phys. Z.* **34**, 478 (1933).
[6] O. Peter, *Ann. Phys., Leipzig* **27**, 299 (1936).
[7] M. S. Livingston, F. Genevese, and E. J. Konopinski, *Phys. Rev.* **51**, 835 (1937).

ionization by high-velocity ions has been closely tied to advances in particle accelerators, x-ray detection technology, data accumulation and handling hardware, and computational facilities. The adaptation of experimental facilities, particularly particle accelerators of the cyclotron, Van de Graaff, or linear type, which were originally established for nuclear physics investigations, has lead to the present advances in ion–atom collision studies. The availability of highly stripped ions ranging from protons to uranium with energies up to the giga-electron volt/nucleon range (for protons) and the mega-electron volt/nucleon range for heavier ions has provided a means of studying the processes of ionization, charge transfer, and radiation from atomic systems that can be in a high degree of excitation. The information derived from such studies is not only important to the understanding of the physics of the atom but forms an important basis for advances in the areas of astrophysics, plasma physics, and radiation damage.

The intent of Part 5 is to survey the field of study that encompasses the production of target x rays arising from the bombardment of solid targets by high-velocity ions. In the limited space available not all of the many experiments that have been reported can be discussed. Hence the emphasis here is on recent results. In addition, there are several excellent review articles and data tabulations[8] resident in the literature that serve to provide as complete as overview as is currently possible.

Part 5 begins with a discussion of the relevant parameters and the theoretical descriptions of the inner-shell ionization process as used to describe the interactions between the high-velocity ion and the target atom. It is not intended to give a detailed treatise on theory but to provide a basis for comparison between theory and experiment. General experimental considerations involving methods of x-ray detection, detector efficiencies, and solid target configurations are discussed. Specific results from K-, L- and M-shell x-ray studies are presented for both light and heavy ions incident upon solid targets. Other considerations are presented, including the effects of changes in fluorescence yields and residual inner-shell vacancies in the projectile (for incident heavy ions). The role that inner-shell vacancies play in determining charge state fractions for heavy ions penetrating thin solids is discussed as it relates to observations made of target x-ray production.

[8] J. D. Garcia, R. J. Fortner, and T. M. Kavanagh, *Rev. Mod. Phys.* **45**, 111 (1973); R. L. Kauffman and P. Richard, in "Methods of Experimental Physics" (Dudley Williams, ed.), Vol. 13A, pp. 148–203. Academic Press, New York, 1976; P. Richard, in "Atomic Inner-Shell Processes" (Bernd Crasemann, ed.), pp. 74–159. Academic Press, New York, 1975; C. H. Rutledge and R. L. Watson, *At. Data. and Nucl. Data Tables* **12**, 195 (1973); T. L. Hardt and R. L. Watson, *At. Data and Nucl. Data Tables* **17**, 107 (1976); R. K. Gardner and T. J. Gray, *At. Data and Nucl. Data Tables* **21**, 515 (1978).

5.2. Definitions of Parameters

5.2.1. Relationship of Ionization and X-Ray Cross Sections

The study of inner-shell ionization using solid targets typically involves the measurement of x-ray production cross sections for the target atoms and/or projectiles. The quantity with which such data is compared is the ionization cross section, which is calculated from one or more theoretical models. The comparison of the experimental results and theoretical calculations must necessarily involve relationships in which the radiative probabilities are involved. The process of comparing the decay of an excited atomic system via observation of the emitted x ray is thus linked to the formation of that state as represented by σ_I, the ionization cross section, and the decay of that state with some probability ω. For studies of K-shell ionization, the relationship is

$$\sigma_{KX}(Z_1, Z_2, E_1, \alpha) = \omega_K(Z_2, \alpha)\sigma_{KI}(Z_1, Z_2, E_1), \quad (5.2.1)$$

where $\omega_K(Z_2, \alpha)$ is the fluorescence yield and $\sigma_{KI}(Z_1, Z_2, E_1)$ the K-shell ionization cross section. The quantities Z_i ($i = 1, 2$) represent the projectile and target atom species, respectively, and E_1 is the incident projectile energy. The fluorescence yield $\omega_K(Z_2, \alpha)$ includes the parameter α, which allows for the dependence of this quantity upon the state of the atom initially after the ionizing collision. For light ions incident upon heavy targets, i.e., $Z_1/Z_2 \ll 1$, the fluorescence yield has generally been considered to be independent of the ionization of the atom. The process of x-ray generation has thus been considered as a two-channel process: the ionization of the atom in the entrance channel and the decay of the excitation as given by ω_K in the exit channel. For a system such as hydrogen incident upon copper this assumption seems valid to within the typical errors of the experimental measurements. Under conditions where ω_K is independent of σ_{KI}, neutral atom values of ω_K as given by Bambynek et al.[9] have been generally utilized to provide the fluorescence yields required for comparisons between experiment and theory.

In studies of L-shell ionization the relationships between the observed x rays and theory become more complex because of the existence of additional decay mechanisms and the three L subshells. A qualitative representation of the decay of L-shell vacancies is given in Fig. 1. If consideration of the $L\alpha_{1,2}$ doublet, $L\gamma_1$ and $L\gamma_4$ x-ray lines is undertaken, the expressions for the x-ray production cross sections in terms of the

---

[9] W. Bambynek, B. Crasemann, R. W. Fink, H. U. Freund, H. Mark, C. D. Swift, R. E. Price, and P. Vanugopala Rao, Rev. Mod. Phys. **44**, 716 (1972).

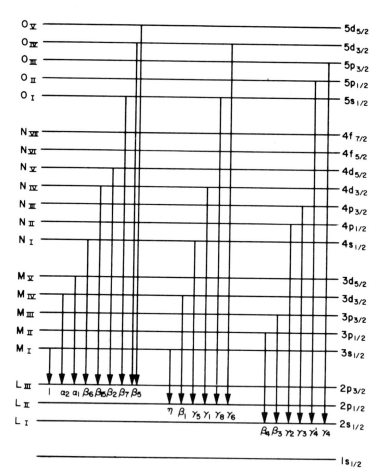

FIG. 1. A qualitative diagram of the x-ray spectral lines associated with the transitions to the L shell.

L-subshell ionization cross sections and radiative and nonradiative parameters are given as follows:

$$\sigma_{L\gamma_4} = \sigma_{LI}\, \omega_1 \frac{\Gamma_{1\gamma_4}}{\Gamma_1},$$

$$\sigma_{L\gamma_1} = (\sigma_{LI} f_{12} + \sigma_{LII})\omega_2 \frac{\Gamma_{2\gamma_1}}{\Gamma_2}, \qquad (5.2.2)$$

$$\sigma_{l\alpha_{1,2}} = [\sigma_{LI}(f_{12}f_{23} + f_{13}) + \sigma_{LII}f_{23} + \sigma_{LIII}]\omega_3 \frac{\Gamma_{1\alpha_1} + \Gamma_{1\alpha_2}}{\Gamma_3},$$

where σ_{Li} is the ionization cross section for the ith L subshell, ω_i the fluorescence yield for the ith L subshell, f_{ij} the Coster–Kronig probability for transitions between the ith and jth subshells, and Γ_i the partial and total radiative widths of the transitions under study. As in the case for the K shell when $Z_1/Z_2 \ll 1$, the values of ω_i, f_{ij}, and Γ_i are assumed independent of the interaction that gives rise to the formation of the states of L-shell ionization. Values for the radiative and nonradiative parameters are available from Bambynek et al.,[9] McGuire,[10] Scofield,[11] and Crasemann et al.[12] The expressions for L-shell x-ray production can be inverted to give experimentally deduced subshell ionization cross sections. Such a procedure, however, results in a representation of the experimental data that is not directly comparable to the data taken in the laboratory.

5.2.2. Parameters Associated with Theoretical Considerations

Resident in the literature are a number of works that report and describe various theoretical models for inner-shell ionization by fast ions. Madison and Merzbacher[13] have reviewed the development of the various theoretical models. The models are constructed around a number of parameters that are set forth to provide convenient scaling procedures, appropriate dimensionless quantities, or physical quantities characteristic of the ionization process. Before discussing the different theories, it is appropriate to define the salient parameters used in theory in order to draw attention and understanding to the language used in calculations.

Consider a collision between a bare point projectile (particle 1) and a target atom in its ground state (particle 2) in which an amount of energy ΔE is transferred from the projectile to the atom. The transferred energy can go into the excitation of an electron from state n to n' and thus $\Delta E = \hbar\omega_{nn'}$, where $\omega_{nn'}$ is the characteristic frequency associated with the excitation. For the case where the atom is ionized,

$$\Delta E = \hbar\omega_{2s} = U_{2s} + E_e, \quad (5.2.3)$$

where U_{2s} is the electron binding energy in its initial state and E_e the free

---

[10] E. J. McGuire, *Phys. Rev.* **185**, 1 (1969); *Phys. Rev. A* **2**, 273 (1970); *Phys. Rev. A* **3**, 587 (1971).

[11] J. H. Scofield, *Phys. Rev.* **179**, 9 (1969); *Phys. Rev. A* **9**, 1041 (1974).

[12] B. Crasemann, M. H. Chen, and V. O. Kostroun, *Phys. Rev. A* **4**, 2161 (1971); M. H. Chen, B. Crasemann, and V. O. Kostroun, *Phys. Rev. A* **4**, 1 (1971).

[13] D. H. Madison and E. Merzbacher, in "Atomic Inner-Shell Process" (Bernd Crasemann, ed.), pp. 1–72. Academic Press, New York, 1975.

electron kinetic energy. Experimentally most of the ejected electrons have $E_e \ll U_{2s}$.

The change in momentum of the projectile viewed in the reduced mass system is

$$h\mathbf{q} = \mu(\mathbf{v}_{1i} - \mathbf{v}_{1f}) = h(\mathbf{K}_i - \mathbf{K}_f) \qquad (5.2.4)$$

or

$$\mathbf{q} = \mathbf{K}_i - \mathbf{K}_f. \qquad (5.2.5)$$

Upon squaring the momentum transfer

$$q^2 = K_i^2 + K_f^2 - 2\mathbf{K}_i \cdot \mathbf{k}_f, \qquad (5.2.6)$$

it is seen that the minimum momentum transfer occurs for forward scattering, i.e., $\mathbf{K}_i \cdot \mathbf{K}_f = K_i K_f$. The minimum momentum that can be given to an electron in some initial state s is then

$$hq_s = hK_i - hK_f = \mu(v_{1i} - v_{1f}). \qquad (5.2.7)$$

This is expressable in terms of the particle velocities and the initial and final energies of the projectile or the energy transfer during the collision. Thus one has

$$hq_s = (2\mu E_1)^{1/2}[1 - (1 - \Delta E/E_1)^{-1/2}]. \qquad (5.2.8)$$

If one assumes that $\Delta E \ll E_1$, this reduces to

$$hq_s = \Delta E/v_1, \qquad (5.2.9)$$

which becomes

$$q_s = \omega_{nn'}/v_1, \qquad (5.2.10)$$

in the case of excitation of a target electron between states n and n'. For ionization the minimum momentum transfer is

$$q_s = \omega_{2s}/v_1 = U_{2s}/hv_1. \qquad (5.2.11)$$

The maximum momentum transfer in the reduced mass system is

$$hq_{max} = (2\mu E_1)^{1/2}[1 + (1 - \Delta E/E_1)^{-1/2}] \qquad (5.2.12)$$

or

$$q_{max} = 2\mu v_1/h, \quad \text{for} \quad \Delta E \ll E_1, \qquad (5.2.13)$$

where $\mu = m_1 m_2/(m_1 + m_2)$. For typical values of μv_1 encountered in ion–atom collisions $q_{max} \simeq \infty$.

The theoretical developments of the ionization cross section in ion–atom collisions employ screened hydrogenic wavefunctions in order to

5.2. DEFINITIONS OF PARAMETERS

describe the electron involved in the collision with the projectile. The use of this simplification in lieu of the Hartree–Fock description for the electrons introduces general scaling laws that have some applicability for the cases of light ions. The average electron shell radius a_{2s} and the average electron velocity v_{2s} are given as

$$a_{2s} = (n^2/Z_{2s})\, a_0, \quad (5.2.14)$$

$$v_{2s} = (Z_{2s}/n)\, v_0, \quad (5.2.15)$$

where $a_0 = 0.529$ Å $= 1$ a.u. and $v_0 = 2.19 \times 10^8$ cm/sec $= 1$ a.u. The quantity Z_{2s} is the screened charge, i.e.,

$$Z_{2s} = Z_2 - s, \quad (5.2.16)$$

where s is the Slater screening coefficient and n the principal quantum number of the shell of interest.

The characteristic times involved in a collision between the ion and an electron are discussed as establishing the range of validity for a given theoretical treatment of the collision process. The collision time is approximated by

$$t_c = a_{2s}/v_1. \quad (5.2.17)$$

This quantity is compared to the characteristic orbital time for the electron,

$$\tau = nh/2U_{2s}, \quad (5.2.18)$$

where U_{2s} is the observed binding energy of the electron shell in question. The ratio of τ and t_c is an important consideration in several of the theories of inner-shell ionization.

With the use of the screened hydrogenic description of the target electrons it is convenient to define a set of dimensionless parameters that follow scaling laws. In the plane-wave Born approximation (PWBA) there are two such parameters: the reduced velocity variable η and the reduced binding energy parameter θ. These two quantities are defined as follows:

$$\eta_{2s} = \frac{1}{n^2} \frac{m_e}{m_1} \frac{E_1}{E_{2s}} \quad (5.2.19)$$

or

$$\eta_{2s} = 40.3 \frac{E_1(\text{MeV})}{m_1(\text{amu})} \frac{1}{Z_{2s}^2}, \quad (5.2.20)$$

and

$$\theta_{2s} = U_{2s}/E_{2s}. \quad (5.2.21)$$

The quantity E_{2s} is the screened hydrogenic binding energy of an electron in the s shell, i.e.,

$$E_{2s} = (Z_{2s}^2/n^2) R_y, \tag{5.2.22}$$

where $R_y = 13.6 \text{ eV} = 1/2$ a.u., and U_{2s} is the observed electron binding energy.

The inclusion of velocity-dependent effects leading to modifications of the electron binding energy during the collision as developed by Brandt *et al.*[14] introduces the parameter ξ_{2s}, defined as

$$\xi_{2s} = \tau/t_c = (v_1/v_{2s})(2/\theta_{2s}), \tag{5.2.23}$$

or in terms of PWBA variables

$$\xi_{2s} = 2n(\eta_{2s})^{1/2}/\theta_{2s}. \tag{5.2.24}$$

Bang *et al.*[15] introduce the variable ξ in their treatment of the effects of the ion–nucleus interaction on the ionization process. Their parameter is related to ξ_{2s} by

$$\xi = \xi_{2s}^{-1}. \tag{5.2.25}$$

In the description of the ionization process by the binary encounter approximation (BEA) the scaled-velocity parameter is given by

$$E_1/\lambda U_{2s} = (v_1/\bar{v})^2, \tag{5.2.26}$$

where $\lambda = m_1/m_e$ and $\bar{v}$ is the average velocity of the s-shell electrons obtained by use of the virial theorem, i.e.,

$$\bar{v} = (2U_{2s}/m_e)^{1/2}. \tag{5.2.27}$$

In terms of the parameters used in the other theories,

$$E_1/\lambda U_{2s} = n^2(\eta_{2s}/\theta_{2s}) = \tfrac{1}{4}(\theta_{2s}\xi_{2s}^2). \tag{5.2.28}$$

5.3. Theoretical Models of Inner-Shell Ionization

5.3.1. Direct Coulomb Ionization

5.3.1.1. Binary Encounter Approximation. Several basic approaches to the theoretical description of the process of inner-shell ionization have evolved. In all cases the theories are intended to be applicable for $Z_1 \ll$

---

[14] W. Brandt, R. Laubert, and I. A. Sellin, *Phys. Rev.* **151**, 56 (1966).
[15] J. Bang and J. M. Hansteen, *KGL. Danske Videnskab. Selskab, Mat-Fys. Medd.* **31**, No. 13 (1959); J. M. Hansteen and O. P. Mosebekk, *Z. Phys.* **234**, 281 (1970); *Nucl. Phys. A* **201**, 541 (1973).

5.3. THEORETICAL MODELS OF INNER-SHELL IONIZATION

Z_2. The classical binary encounter approximation (BEA) is based upon the classical energy transfer process wherein a projectile interacts with an inner-shell electron having a velocity distribution representative of its binding energy. In the case of K-shell ionization, the ionization cross section is represented by a universal function

$$\sigma_1(E_1) = (N_S Z_1^2 / U_K^2)\, \sigma_0 G(V), \qquad (5.3.1)$$

where E_1 is the laboratory energy of the projectile, Z_1 the projectile atomic number, U_K the K-shell electron binding energy, $\sigma_0 = 6.56 \times 10^{-14}\, \text{cm}^2\, \text{eV}^2$, and $V = v_1/v_{2S}$. The function $G(V)$ carries all of the dynamics of the interaction between the projectile and the electron, and it is only a function of the scaled velocity parameter V. Rearrangement of Eq. (5.3.1) for $\sigma_1(E_1)$ gives

$$U_K^2 \sigma_1(E_1)/Z_1^2 \propto G(V). \qquad (5.3.2)$$

This expression states the universal character of the ionization process that is the major feature of the BEA. General features of the BEA and specific functions for $G(V)$ have been reported by Gerjuoy,[16] Vriens,[17] Garcia,[18] Garcia et al.,[19] and McGuire and Richard.[20] The utility of the BEA calculations are their simple character and the universal scaling laws that are the result of using appropriate hydrogenic wavefunctions to obtain the K-shell electron velocity distributions. In general, for light ions, i.e., protons, incident on a wide range of Z_2 the BEA description provides an adequate representation of the measured K-shell ionization for $v_1/v_K \lesssim 1$. However, it is to be noted that v_K is related to the 1s velocity distributions and hence the BEA calculations of Refs. 16–20 are appropriate only to K-shell ionization processes. Specific examples are discussed in the section on data.

In the case of L-subshell ionization it is not appropriate to simply scale the BEA results for K-shell ionization. Scaling may provide approximate results for ionization of the 2p levels by light projectiles. However, the nodal character of the 2s electron state is not contained in the scaled cross section, which uses a $G(V)$ function based upon 1s electron properties. Hansen[21] reported constrained BEA calculations (CBEA) that were constructed using screened hydrogenic wavefunctions for the 1s, 2s, and 2p states and hence the features of the nodal structure inherent in the 2s

---

[16] E. Gerjuoy, *Phys. Rev.* **148**, 54 (1966).
[17] L. Vriens, *Proc. R. Soc. London* **90**, 935 (1966).
[18] J. D. Garcia, *Phys. Rev. A* **4**, 955 (1971).
[19] J. D. Garcia, E. Gerjuoy, and J. Welker, *Phys. Rev. A* **1**, 280 (1970).
[20] J. H. McGuire and P. Richard, *Phys. Rev. A* **8**, 1374 (1973).
[21] J. S. Hansen, *Phys. Rev. A* **8**, 822 (1973).

states is incorporated in the calculations. When data for L_I-subshell ionization by protons are compared to these CBEA predictions it is observed that there are definite problems with this theory. Specifically, while the CBEA does indicate structure in the energy dependence of the cross section for L_I-subshell ionization, it is concluded that the calculations are not in agreement with measurements of x-ray production associated with ionization of this 2s level. Examples of this feature of L-subshell ionization are given in the data section.

5.3.1.2. Semiclassical Approximation. Bang and Hansteen[15] and Hansteen and Mosebekk[15] have developed the semiclassical approximation (SCA) theory for inner-shell ionization by incident light ions ($Z_1 \ll Z_2$). The necessary condition for the ionization process to be treated classically is that the wave packet of the projectile be small compared to the closest distance of approach to the target nucleus. The trajectory of the projectile is taken to be the path of the classical particle and ionization cross section is calculated in terms of an impact-parameter dependent function I_b:

$$\sigma_I = 2\pi \int_0^\infty I_b b \, db, \qquad (5.3.3)$$

where b is the classical impact parameter. The function I_b is related to the differential cross section per unit final electron energy for the scattering of the electron by the projectile. The SCA ionization cross sections obey the same scaling as the BEA predictions with respect to the Z_1 dependences, i.e.,

$$\sigma_{I_1}(Z_1) = Z_1^2 \sigma_I \qquad (5.3.4)$$

Hansteen and Mosebekk[15] present tabulated ionization cross sections for the K shell over a range of energies and targets for incident protons.

5.3.1.3. Plane-Wave Born Approximation. Madison and Merzbacher[13] review the state of theoretical development for inner-shell ionization by fast light ions. Of the various models proposed, the plane-wave Born approximation (PWBA) represents the purely quantum-mechanical approach to the problem of inner-shell ionization. For straight line particle trajectories, the PWBA and SCA have been shown to give equivalent results.[15] In the PWBA the initial and final states of the projectile are assumed to be plane waves and the initial and final electron states are taken to be describable by screened hydrogenic wavefunctions. The scattering amplitude is given by

$$f = \langle n' e^{i\mathbf{K}' \cdot \mathbf{r}} | V | n e_i^{\mathbf{K} \cdot \mathbf{r}} \rangle, \qquad (5.3.5)$$

where V is taken to be the Coulombic interaction between the projectile

5.3. THEORETICAL MODELS OF INNER-SHELL IONIZATION

and the electron initially in state $|n\rangle$. The cross section is

$$\sigma_1 \sim |f|^2 \sim Z_1^2 F(\eta_K, \theta_K), \qquad (5.3.6)$$

where $F(\eta_K, \theta_K)$ have been tabulated for the K shell by Khandelwal et al.[22] and for the L shell by Choi et al.[23] As in the case of the classical theories, the PWBA predicts a Z_1^2 scaling of the ionization cross section at a fixed projectile velocity.

5.3.2. Corrections to the Theory of Direct Coulomb Ionization

As heavier ions are employed as projectiles, the data for both K- and L-shell ionization are no longer in agreement with the predictions of any of these theories for direct Coulomb ionization. The deviations between the theories and the data can be associated with one or more of the following considerations:

(1) Corrections for modifications of the binding energy of the electrons in the shell from which the ionization may occur.

(2) Electron transfer processes may become comparable to or exceed the direct Coulomb excitation channel.

(3) Ion–solid effects may appear depending on the processes that govern inner-shell vacancy populations in the projectile and the degree to which such vacancies are coupled to the target electron shell of interest.

(4) Modification of the fluorescence yields from atomic values may be required if multiple ionization is dominant.

The consideration of such corrections to the theories of inner-shell ionization is not unreasonable in that the original theories were formulated for structureless point charges as the projectiles. It is appropriate then that a heavy ion moving in a solid target be observed as not having all of the simplifications associated with a simple point charge.

In evaluating the validity of the PWBA, Basbas et al.[24] have enumerated a hierarchy of conditions under which the PWBA should be valid. These are listed below and are subject to the overriding stipulation that $Z_1 \ll Z_2$:

(1) a. The initial states of the target electrons are unperturbed by the projectile.
 b. The electron response time $n/\omega_{2S} \gg a_{2S/v_1}$. This condition is met for $\tfrac{1}{2} n\, \theta_{2S}(v_{2S}/v_1) \ll 1$.

---

[22] G. S. Khandelwal, B. H. Choi, and E. Merzbacher, *Atom. Data* **1**, 103 (1969).
[23] B. H. Choi, E. Merzbacher, and G. S. Khandelwal, *Atom. Data* **5**, 291 (1973).
[24] G. Basbas, W. Brandt, and R. Laubert, *Phys. Rev. A* **7**, 983 (1973).

(2) a. The projectile acts as a point charge.
   b. The projectile K-shell radius is much larger than the target s-shell radius, where s denotes the electron involved in the interaction.
   c. The amplitude of the scattered wave is negligible compared to the incident wave. This condition is fulfilled for $Z_1/Z_{2s} \ll v_1/v_{2s}$.
(3) a. The initial and final states are approximated as plane waves. This condition is fulfilled for

$$\tfrac{1}{2} n \, \theta_{2s}(v_{2s}/v_1)^3 \, Z_1(m_e/m_1) \ll 1.$$

Some of the conditions are not met for low-velocity projectiles and can be accounted for, in part, by modification of the electron binding energy. In the limit as $v_1 \to 0$,

$$\frac{U_{2s}}{R_y} \to \frac{\theta_{2s}}{n^2}(Z_{2s} + Z_1)^2 = \frac{\theta_{2s}}{n^2} Z_{2s}^2 \left(1 + \frac{2Z_1}{Z_{2s}}\right), \qquad (5.3.7)$$

i.e., the binding energy increases by $2Z_1/Z_{2s}$, where $Z_1 \ll Z_{2s}$.

In the case of light ions Basbas et al.[24] have reported modifications of the binding energy for the target electron K shell, which result from the presence of the projectile. Brandt and Lapicki[25] have extended the same ideas to allow for corrections of the L-subshell binding energies. Such calculations for K-shell ionization result in replacing θ_K by $\epsilon_K \theta_K$, where

$$\epsilon_K \theta_K = [1 + 2(Z_1/Z_{2K} \, \theta_K) g(\xi_K)] \, \theta_K, \qquad (5.3.8)$$

$$g(\xi_K) = (1 + 5\xi_K + 7.14\xi_K^2 + 4.27\xi_K^3 + 0.947\xi_K^4)(1 + \xi_K)^{-5}. \qquad (5.3.9)$$

The limiting values for $g(\xi_K)$ are given as

$$g(\xi_K) = \begin{cases} 1 - O(\xi_K^2), & \text{for } \xi_K \ll 1, \\ 0.95\xi_K^{-1}, & \text{for } \xi_K \gg 1. \end{cases} \qquad (5.3.10)$$

For the L subshells, Brandt and Lapicki[25] give the correction factors as

$$g_{L_1}(\xi) = (1 + 9\xi + 30.2\xi^2 + 66.8\xi^3 + 100\xi^4 + 94.1\xi^5 \\ + 51.3\xi^6 + 15.2\xi^7 + 1.891\xi^8)/(1 + \xi)^9, \qquad (5.3.11)$$

$$g_{L_{2,3}}(\xi) = (1 + 9\xi + 36\xi^2 + 84\xi^3 + 112\xi^4 + 89.8\xi^5 \\ + 43.7\xi^6 + 12\xi^7 + 1.42\xi^8)/(1 + \xi)^9. \qquad (5.3.12)$$

The expression for $g_{L_{2,3}}(\xi)$ is taken from the work of Lapicki and Losonsky[26] and represents a correction of the expressions for g_{L_2} and g_{L_3} as originally given by Brandt and Lapicki.[25]

---

[25] W. Brandt and G. Lapicki, *Phys. Rev. A* **10**, 474 (1974).
[26] G. Lapicki and W. Losonsky, *Phys. Rev. A* **15**, 896 (1977).

5.3. THEORETICAL MODELS OF INNER-SHELL IONIZATION

The binding energy correction is a low-velocity effect and as $\xi_K \propto E_1^{1/2}$ it is seen that $\lim g_K(\xi_1) \to \infty$; hence at high velocities $E_1 \to \infty$, $\xi_K \theta_K \to \theta_K$, and the results of the "uncorrected" PWBA are obtained.

Another low-velocity effect that arises as first developed by Bang et al.[15] in the SCA is the Coulomb deflection correction for the ion–nucleus interaction. Instead of a straight line trajectory for the projectile, a hyperbolic trajectory results. The correction for Coulomb deflection is incorporated into the PWBA as

$$\sigma_s = (n_s - 1) E_{n_s}(\pi \, dq_{os}) \sigma_s^{PWBA}, \quad (5.3.13)$$

where $n_s = 10$ for $s = K$ or L_I and $n_s = 12$ for $s = L_{II}$ or L_{III}. The function $E_{n_s}(x) = \int_0^\infty t^{-n_s} e^{-xt} \, dt$ is approximated as

$$E_{n_s}(x) = [(n_s - 1)/(n_s - 1 + x)] e^{-x}, \quad (5.3.14)$$

as suggested by Brandt and Lapicki.[25]

The inclusion of both the binding energy effect and the Coulomb deflection effect leads to the cross section σ_s^{PWBABC}, which becomes

$$\sigma_s^{PWBABC} = 8\pi a_0^2 \left(\frac{Z_1}{Z_{2s}}\right) \frac{1}{\eta_{Ls}} \frac{n_1^{-1}}{n_s - 1 + \pi \, d\epsilon q_{os}} \exp(-\pi \, d\epsilon q_{os}) f_{2s}(\eta_{2s}, \epsilon \sigma_{2s}). \quad (5.3.15)$$

The effects of the corrections are shown in Fig. 2 for protons on copper and nitrogen ions on the L_I ionization for ytterbium. In the case of protons as the projectiles, the binding energy and Coulomb deflection correction are approximately equal in their effect on the calculated ionization cross section for $E_1 \simeq 0.5$ MeV in comparison to the PWBA results. However, for heavier ions the binding energy effect (B) is much larger in its effect on the calculated ionization cross section than in the Coulomb deflection effect (C). A comparison of the PWBA calculations for each correction is given in Table I for ^1H, ^{12}C, ^{14}N, and ^{16}O ions on nickel at an incident energy of 1.0 MeV/amu. It is noted that C should be the same when m_1/Z_1 is a constant. Data from Gray et al.[27] for ^2H and ^4He on dysprosium L-shell ionization illustrate this feature of the C calculations for the PWBA.

When heavy ions are employed in the study of inner-shell ionization several complications arise that have only recently been investigated. These effects are associated with the added effects associated with the competing charge exchange interaction, the role that the target medium

---

[27] T. J. Gray, C. Shank, G. M. Light, and R. K. Gardner, *Bull. Am. Phys. Soc.* **20**, 674 (1975).

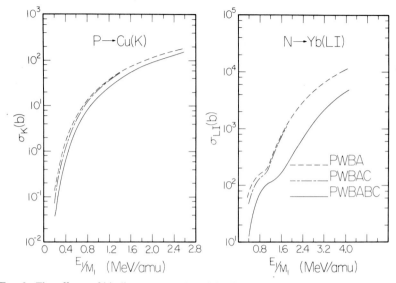

FIG. 2. The effects of binding energy (B) and Coulomb deflection corrections (C) on the PWBA predictions for p + Cu(K) and N + Yb(L$_I$) systems are illustrated. The B and C calculations are taken from the perturbed–stationary-state approach as given in Ref. 24 and Ref. 25 for the K and L shells, respectively.

plays in modifying the projectile states, and the coupling of those states to the ionization process involved in the ionization of the target shell of interest. Halpern and Law[28] analyzed data by Macdonald et al.[29] for the collision of fully stripped projectiles of H^+, C^{6+}, N^{7+}, O^{8+}, and F^{9+} with argon atoms under single-collision conditions. Their analysis included

TABLE I. Theoretical PWBA K-Shell X-Ray Production[a]

Ion	^{1}H	^{12}C	^{14}N	^{16}O
PWBA	14.2	517	704	919
PWBA(C)	13.4	494	671	875
PWBA(B)	9.8	54	53	50
PWBABC	9.3	52	50	47
PWBA:PWBABC	1.5:1	10:1	14:1	20:1

[a] With individual values for perturbations arising from consideration of the Coulomb deflection (C) and binding energy (B) corrections for 1.0-MeV/amu ^{12}C, ^{14}N, and ^{16}O on nickel with ω_K = 0.432. Cross-section units are 10^{-24} cm^2.

[28] A. M. Halpern and J. Law, *Phys. Rev. Lett.* **31**, 4 (1973).
[29] J. R. Macdonald, L. M. Winters, M. D. Brown, L. D. Ellsworth, T. Chiao, and E. W. Pettus, *Phys. Rev. Lett.* **30**, 251 (1973).

5.3. THEORETICAL MODELS OF INNER-SHELL IONIZATION

the competing process of K-shell charge exchange from the target to the bound states of the projectile using the Brinkman–Kramers formulation of the charge exchange cross section as given by Nikolaev.[30] The formula for charge exchange for K transfer to a given projectile state of principal quantum number n is

$$\sigma_{BK} = \sum_n \frac{2^{19}}{5} \pi a_0^2 \frac{Z_1^5}{n^3} Z_2^5 \left(\frac{v_1}{v_0}\right)^8$$

$$\times \left\{ \left[\left(\frac{v_1}{v_0}\right)^2 + Z_2^2 - \frac{Z_1^2}{n^2} \right]^2 + 4 \left(\frac{v_1}{v_0}\right)^2 \frac{Z_1^2}{n^2} \right\}^{-5}, \quad (5.3.16)$$

where a_0 is the Bohr radius, v_1 the projectile velocity, $v_0 = U/Z_2$, and Z_2 the target atomic number. The σ_{BK} calculations overestimate the experimental charge exchange cross section by factors of ~ 3–10. Recent studies by Lapicki and Losonsky[26] for electron capture from inner shells by fully stripped ions have been reported in which the effects of binding energy and Coulomb deflection have been incorporated into the Nikolaev formulation. Their development follows the earlier work of Basbas et al.[24] and Brandt and Lapicki.[25] The inclusion of these two corrections lowers the calculated charge exchange cross sections for bare projectiles; hence the need for a semiempirical scaling factor is eliminated.

5.3.3. Additional Considerations

5.3.3.1. Effects of Multiple Ionization on Fluorescence Yields.

Multiple ionization of the inner atomic shells has been studied extensively as discussed by Hopkins and Matthews in Parts 8 and 9, respectively. For the purpose of this section, one may use such information as derived from the high-resolution x-ray spectra for K-shell spectroscopy to estimate the changes in the K shell fluorescence yield. The majority of cross-section measurements reported use single-hole values of ω_K to relate the data to predictions of the theory. For light ions (^{1}H, ^{4}He, . . . , $Z_1 \ll Z_2$) this is an adaquate procedure. However, for heavier ions several manifestations of multiple ionization appear in Si(Li) data. There are (1) energy shifts in the K_α and K_β lines, (2) peak broadening, and (3) changes in the relative intensities of the K_α and K_β transitions. These features are well understood when data from high-resolution studies are considered.

One may obtain an estimate of the influence of multiple ionization on ω_K from Si(Li) data by determining the shifts in energy for the observed K_α and K_β transitions from the accepted energies of these x rays for a given target species. Hartree–Fock calculations for the transition energies using initial configurations $(1s)^{-1}(2s)^2(2p)^{-n}(3s)^2(3p)^{-m}$, where n and m are

---

[30] V. S. Nikolaev, *Sov. Phys.—JETP* **24**, 847 (1967).

the number of vacancies in the shells indicated, give the appropriate transition energies. The measured energy shifts ΔK_α and ΔK_β then give an average initial configuration, say $n = 2.5$ and $m = 0$. In comparison to a high-resolution measurement, $n = 2.5$ would mean that the $(2p)^{-2}$ and $(2p)^{-3}$ components would be equal in intensity, along with equal intensities for the $(2p)^{-1}$ and $(2p)^{-4}$, etc. Once the average configuration is determined, the values of ω_K can be estimated using a scaling technique by Larkins.[31] In the approximation

$$\omega_K = \Gamma^{x'}/(\Gamma^{x'} + \Gamma^{A'}), \qquad (5.3.17)$$

where $\Gamma^{x'}$ and $\Gamma^{A'}$ are given by

$$\Gamma^{x'} = (n/n_0)\Gamma^x, \qquad \Gamma^{A'} = \Gamma'(K-LL) + \Gamma'(K-LM), \qquad (5.3.18)$$

with

$$\Gamma'(K-LL) = \Gamma(K-L_1L_1) + \left(\frac{6-n}{6}\right)\Gamma(K-L_1L_{23})$$
$$+ \left(\frac{6-n}{6}\right)\left(\frac{5-n}{5}\right)\Gamma(K-L_{23}L_{23}), \qquad (5.3.19)$$

$$\Gamma'(K-LM) = \Gamma(K-L_1M_1) + \left(\frac{6-m}{6}\right)\Gamma(K-L_1M_{23})$$
$$+ \left(\frac{6-n}{6}\right)\Gamma(K-L_{23}M_1)$$
$$+ \left(\frac{6-n}{6}\right)\left(\frac{6-m}{6}\right)\Gamma(K-L_{23}M_{23}). \qquad (5.3.20)$$

For $Z = 10$ to 30, the K–LL rates are nearly constant at K–LL $= 0.73(K-L_{23}L_{23}) + 0.23(K-L_1L_{23}) + 0.4(KL_1L_1)$. Using the calculated branching ratios of Kostrum et al.[32] for the K–LM widths gives

$$K-LM = 0.23(K-L_1M_1) + 0.12(K-L_1M_{23})$$
$$+ 0.39(K-L_{23}M_1) + 0.26(L_{23}M_{23}).$$

Using these ratios gives

$$\Gamma^{A'} \simeq \left[0.04 + 0.23\left(\frac{6-n}{n}\right) + 0.73\left(\frac{6-n}{6}\right)\left(\frac{5-n}{5}\right)\right]\Gamma(K-LL)$$
$$+ \left[0.23 + 0.12\left(\frac{6-m}{6}\right) + 0.39\left(\frac{6-n}{6}\right)\right.$$
$$\left. + 0.26\left(\frac{6-n}{6}\right)\left(\frac{6-m}{6}\right)\right]\Gamma(K-LM). \qquad (5.3.21)$$

---

[31] F. P. Larkins, *J. Phys. B* **4**, L 29 (1971).
[32] V. O. Kostroun, M. H. Chen, and B. Crasemann, *Phys. Rev. A* **3**, 533 (1971).

5.3. THEORETICAL MODELS OF INNER-SHELL IONIZATION

This expression thus allows a first-order estimate to be obtained for $\Gamma^{A'}$. The quantity n_0 is the number of electrons in the shell being considered; e.g., for the 2p shell $n_0 = 6$. As a sample calculation the values of ω_K calculated using values of $\Gamma(K-LL)$ and $\Gamma(K-LM)$ from E. J. McGuire[10] for titanium with $n = 0-3$, $m = 0$, are as follows:

n	ω_K
0	0.233
1	0.256
2	0.277
3	0.295

When high-resolution spectra are available for the projectile–target combination a more complex procedure can be employed utilizing the tabulated multiplet rates and relative line strengths determined from the spectra.

5.3.3.2. Corrections for Relativistic Electron Wavefunctions. In the PWBA, BEA, and SCA calculations for inner-shell ionization, nonrelativistic screened hydrogenic wavefunctions are utilized. For $Z_2 \geq 50$

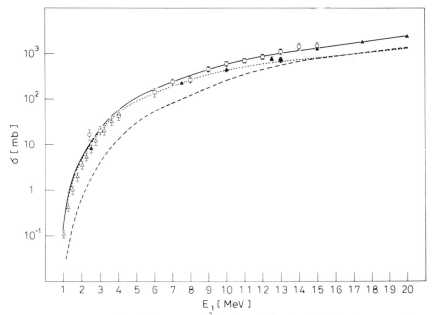

FIG. 3. Total K-shell ionization cross section for gold bombarded by hydrogen as a function of E_1. Solid line, relativistic wavefunctions, all final states; dotted line, relativistic wavefunctions, $l = 0$, final states only. Experimental points are defined in Ref. 36.

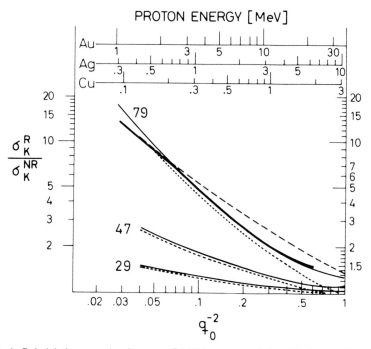

FIG. 4. Relativistic correction factors σ_K^R/σ_K^{NR} for targets of $Z_2 = 79$, 47, and 29, as functions of projectile energy E_1 (upper scale) and energy scaling parameter q_0^{-2} as defined in the text (lower scale). The bold curve is from complete SCA calculations for protons on gold. The broken curve is from the equation given in the text and the full and dotted curves are as defined in Ref. 37.

these calculations underestimate the K-shell cross section considerably. Jamnik and Zupančič[33] calculated the K-shell cross sections using Dirac wavefunctions in the PWBA approximation, obtaining ionization cross sections larger than the corresponding nonrelativistic calculations. Choi[34] carried out the corresponding L-shell calculations using relativistic wavefunctions with similar results. Hansen[21] incorporated relativistic effects into the BEA calculations by taking into consideration the relativistic mass–energy relationship for the electrons. McDaniel et al.[35] have used Hansen's approach as an ad hoc correction for PWBA calculations.

Amundsen[36] recently reported calculations for electronic relativistic ef-

[33] D. Jamnik and C. Zupančič, K. Danske Videnskab. Selskab, Mat-Fys. Medd. **31**, 1 (1957).
[34] B.-H. Choi, Phys. Rev. A **4**, 1001 (1971).
[35] F. D. McDaniel, J. L. Duggan, P. D. Miller, and G. D. Alton, Phys. Rev. A **15**, 846 (1977).
[36] P. A. Amundsen, J. Phys. B **9**, 971 (1976).

5.3. THEORETICAL MODELS OF INNER-SHELL IONIZATION

fects in K-shell Coulomb ionization for heavy charged particles. His calculations were performed using the straight-line semiclassical approximation with relativistic Coulomb wavefunctions for the electrons. Using this approach good agreement is achieved for the K-shell ionization data for protons on gold in the energy range 1–20 MeV as shown in Fig. 3. Amundsen et al.[37] recently presented an approximate correction formula for Coulomb K-shell ionization that agrees quite well with the more complex full calculations and is easier to use. Their relationship for the relativistic cross section is

$$\sigma_K^R \simeq q_0^{4(1-\gamma)} \left(1 + (1-\gamma)\frac{\pi}{2}\frac{3+2\gamma}{5+4\gamma} q_0\right)^2 \sigma_K^{NR}, \quad (5.3.22)$$

where $q_0 = E_B/\hbar V_1 a_K$, $a_K = a_0/Z_2$, and $\gamma = (1 - \alpha^2 Z_2^2)^{1/2}$. The quantities required in these calculations are E_B, K-shell binding energy; V_1, projectile laboratory velocity; Z_2, projectile atomic number; α, fine-structure constant; a_0, Bohr radius 0.542 Å; and σ_K^{NR}, nonrelativistic K-shell ionization cross section. The results of the approximate calculations for protons on copper, silver, and gold are compared to the full calculations in Fig. 4.

5.3.3.3. Relativistic Projectiles. Studies of inner-shell ionization by highly relativistic heavy particles have been reported for only one experiment. Jarvis et al.[38] reported the results for K-shell ionization by 160 MeV protons where no dramatic deviations from the predictions of the nonrelativistic (projectiles) PWBA were observed. A comparison was presented to relativistically calculated cross sections for incident electrons with the result that these authors suggested that at higher proton energies the cross section should begin to increase with increasing projectile energy. Anholt et al.[39] reported measured cross sections for K-vacancy production by 4.88 GeV protons on elements between nickel and uranium. In the course of their work they found a target thickness effect that was related to (1) protons creating high-energy secondary electrons in the target, which then contribute to the excitation of K vacancies, and (2) photoelectric excitation of the K-shell from secondary electron bremsstrahlung. Effects from heavy-particle bremsstrahlung were not considered. Their data were compared to a theoretical model based upon the PWBA. Two contributions arise that we associated with a longitudinal contribution or normal nonrelativistic PWBA contribution and a transverse contribution. Thus $\sigma_K = \sigma_K^L + \sigma_K^t$. The σ_K^L contri-

---

[37] P. A. Amundsen, L. Kocbach, and J. M. Hansteen, *J. Phys.* B **9**, L203 (1976).
[38] O. N. Jarvis and C. Whitehead, *Phys. Rev.* A **5**, 1198 (1972).
[39] R. Anholt, S. Nagamiya, J. O. Rasmussen, H. Bowman, J. G. Ioannou-Yannou, and E. Rauscher, *Phys. Rev.* A **14**, 2103 (1976).

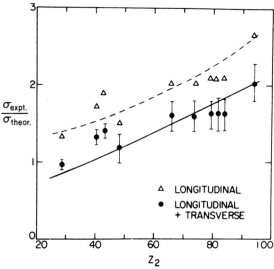

FIG. 5. Ratios of experimental and theoretical cross sections for 4.88-GeV protons. Solid lines are guides to the eye only.

butions were calculated using standard tabulated results. Using a dipole approximation the transverse part is given as

$$\sigma_K^t = 1.056 \times 10^4 \frac{Z_1^2}{Z_2^2} \frac{\ln(\gamma^2) - \beta^2}{\beta^2} \text{ (barns)}, \quad (5.3.23)$$

where $\beta = V_1/c$ and $\gamma^2 = 1 - \beta^2$. The authors give a numerical form for σ_K^t that does not employ the dipole approximation. The difference between the two methods of calculation is 10% for protons on uranium, where the deviation is the largest. The data of their work at 4.88 GeV proton energy are compared to the theoretical estimates in Fig. 5.

5.3.4. Molecular Orbital Excitation

An area of particular interest is the study of slow collisions between heavy-collision partners. In such collisions the relative velocity of the two colliding atoms is small in comparison to the orbital velocities of the inner-shell electrons. In such cases the collision is described in terms of the formation of a temporary molecule by the ion and target atom. Fano and Lichten[40] first suggested that during such collisions a discrete number of inner-shell vacancies were formed owing to the idea that the inner-shell electrons move independently during the collision. The inner-shell elec-

---

[40] U. Fano and W. Lichten, *Phys. Rev. Lett.* **14**, 627 (1965).

trons during the collision occupy one-electron molecular orbitals that belong to the transient molecule formed during the collision. In principle, as two atoms collide an inner-shell vacancy on one of the collision partners may be transferred to the other collision partner, leading to a final inner-shell ionization of that partner when the atoms have separated following the collision.

The subject of molecular orbital (MO) excitation has been discussed at length by Briggs,[41] Kessel,[42] Kessel and Fastrup,[43] Garcia et al.,[8] Lichten,[44] and Barat and Lichten.[45] Recent works by Briggs and Macek,[46] Taulbjerg and Briggs,[47] Briggs and Taulbjerg,[48] Piacentini and Salin,[49] and Taulbjerg et al.[50] give the theoretical developments of the total cross section and impact parameter dependences for symmetric collisions (homonuclear collisions) and asymmetric collisions (heteronuclear collisions). The majority of comparisons of the MO model predictions have been made to data from heavy-ion collisions on gaseous targets where single-collision conditions prevailed. Attempts to compare inner-shell x-ray yield data derived from heavy ions incident upon solid targets have not proved successful because of the complexity encountered in ion–solid experiments. These complications are discussed later in this chapter.

5.4. General Experimental Consideration

5.4.1. Experimental Arrangement

The typical experimental arrangement employed for the measurement of inner-shell ionization cross sections for the target is shown in Fig. 6. The x-ray detector can be a scintillation detector, gas proportional counter, or Si(Li) detector. Dispersive devices such as the bent crystal spectrometer have not been used for x-ray cross-section measurements be-

---

[41] J. S. Briggs, *Rep. Prog. Phys.* **39**, 217 (1976).
[42] Q. C. Kessel, "Case Studies in Atomic Collision Physics" (E. W. McDaniel and M. R. C. McDowell, eds.), Vol. 1, pp. 401–462. North-Holland, Amsterdam, 1969.
[43] Q. C. Kessel and B. Fastrup, "Case Studies in Atomic Collision Physics" (E. W. McDaniel and M. R. C. McDowell, eds.), Vol. 3, pp. 137–213. North-Holland, Amsterdam, 1973.
[44] W. Lichten, *Phys. Rev.* **164**, 131 (1967).
[45] M. Barat and W. Lichten, *Phys. Rev. A* **6**, 211 (1972).
[46] J. S. Briggs and J. Macek, *J. Phys. B* **5**, 579 (1972).
[47] K. Taulbjerg and J. S. Briggs, *J. Phys. B* **8**, 1895 (1975).
[48] J. S. Briggs and K. Taulbjerg, *J. Phys. B* **8**, 1909 (1975).
[49] R. D. Piacentini and A. Salin, *J. Phys. B* **7**, L 311 (1974).
[50] K. Taulbjerg, J. S. Briggs, and J. Vaaben, *J. Phys. B* **9**, 1351 (1976).

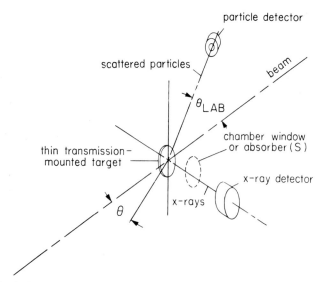

FIG. 6. Typical experimental configuration for thin target geometries. The angle θ is the angle between the beam direction and the target normal.

cause of inherent low efficiency and the effects of crystal reflectivities and continuously variable solid angles. The systems used employ a monitor of the incident particle flux. Several types are generally used. Either the target current or the scattered beam is used to provide the measure of the incident particle flux. The method of monitoring the number of ions incident upon the target is generally related to the types of solid targets being used in the experiment and/or the incident ion energy.

5.4.2. X-Ray Detector Efficiency

Central to the techniques of measuring the x-ray production cross section is the determination of the absolute overall system counting efficiency ϵ_X for the x-ray detector. This efficiency must necessarily include the effects of absorbers between the target and detector regions as well as the solid angle subtended by the x-ray detector. In the case of a Si(Li) detector there are two general methods used to give ϵ_X. The first of these methods relies on the general efficiency characteristics of the Si(Li) detector itself. At a photon energy of 14.4 keV the efficiency of a Si(Li) detector is 1.0. Thus by the use of tabulated absorption coefficients, detector efficiency can be constructed relative to the efficiency at 14.4 keV. Data on the size and placement of the silicon crystal within the cryostat are generally available from the detector manufacturer and hence the geo-

5.4. GENERAL EXPERIMENTAL CONSIDERATION

metrical solid angle is determined for a given experimental configuration by the dimension of the detector and the system. Hence the total absolute system efficiency is

$$\epsilon_X(h\nu) = \epsilon_X(14.4 \text{ keV}) \exp\left(-\sum_j U_{lj} X_j\right) d\Omega, \quad (5.4.1)$$

where U_{lj} is the linear absorption coefficient for the jth absorber in the system, and $d\Omega$ the detector solid angle. For a point source $d\Omega$ is $d\Omega = A/R^2$, where A is the detector area taken as its projection onto the line joining the source and detector and R the source to detector distance.

The method of determing ϵ_X by using the above procedure may be complicated when thin absorbers and low photon energies are involved. Uncertainties in $X_j U_{lj}$ are the largest sources of error in determining ϵ_X. Tabulated values of U_{lj} are given in Table II.

An alternative method for determing ϵ_X is to use calibrated sources whose γ-ray activities have been determined. Work by several groups has established the number of x rays emitted per γ decay (for a particular γ transition) and hence this method allows for a direct determination of ϵ_X that includes the effects of geometry and absorption automatically. The efficiency ϵ_X is thus

$$\epsilon_X(h\nu) = I_m(h\nu)/I(h\nu) = [I_m(h\nu)/I_0(\gamma)]\alpha_{\gamma X}, \quad (5.4.2)$$

TABLE II. Mass Absorption Coefficients for $K_{\alpha_{1,2}}$ Radiation in Several Materials[a]

	Material (cm/g)		
Element	Be	Al	Fe
Ne	985.473	1982.659	—
Al	186.256	425.919	3726.282
Si	117.618	3595.554	2437.469
Cl	35.492	1144.648	806.243
K	17.908	595.493	428.706
Ti	7.281	252.034	186.736
Cr	4.270	151.382	114.090
Fe	2.615	94.752	72.538
Ni	1.663	61.485	374.185
Zn	1.091	41.093	254.969
Ge	0.736	28.212	178.238
Se	0.608	19.825	127.391
Rb	0.303	12.107	79.663
Y	0.220	8.902	59.443
Mo	0.140	5.769	39.331

[a] R. Theisen and D. Vollath, "Tables of X-Ray Mass Attenuation Coefficients," Verlag Stahleisen M. B. H. Dusseldorf, 1967.

TABLE III. Relative Photon Intensities for Calibration Sources

Isotope	Half-life	Transition type	Energy (keV)	Relative photon intensities			
				Ref.[a]	Ref.[b]	Ref.[c]	Ref.[d]
^{54}Mn	312.16 ± 0.11 days	$K_\alpha + K_\beta$	5.47	0.2514 ± 0.0013	0.250 ± 0.002	0.251	—
		γ	834.8	1.000	1.000	1.000	—
^{57}Co	271.41 ± 1.54 days	$K_\alpha + K_\beta$	6.46	0.660 ± 0.013	0.646 ± 0.015	0.57	
		γ	14.41	0.112 ± 0.002	0.111 ± 0.002	0.102	0.1116 ± 0.0023
		γ	122.1	1.000	1.000	1.000	1.000
		γ	136.5			0.122	
^{65}Zn	244.0 ± 0.2 days	$K_\alpha + K_\beta$	8.15	0.7507 ± 0.006	0.126 ± 0.003	0.779	0.7505 ± 0.010
		γ	1115.5	—			
^{85}Sr	64.5 ± 0.5 days	K_α	13.38	0.5020 ± 0.0033	—	—	0.5034 ± 0.0040
		K_β	14.97	0.0880 ± 0.0011	—	—	0.0866 ± 0.0014
		γ	514.0	1.000	—	—	1.000
^{88}Y	106.6 ± 0.1 days	K_α	14.12	0.5491 ± 0.0066	$K_\alpha + K_\beta$	—	0.5503 ± 0.0165
		K_β	15.85	0.0989 ± 0.0019	0.694 ± 0.003	—	0.0974 ± 0.0037
		γ	898.0	1.000	1.000	—	1.000
^{109}Cd	450 ± 5 days	K_α	22.10	22.02 ± 1.08	—	21.9	—
		K_β	25.0	4.68 ± 0.23	—	4.7	—
		γ	88.0	1.000	—	1.000	—
^{137}Cs	30.5 ± 0.3 years	K_α	32.1	0.0666 ± 0.0018	0.0666 ± 0.0018	0.0643	—
		K_β	36.6	0.0159 ± 0.00048	0.01580 ± 0.00048	0.0151	—
		γ	661.6	1.000	1.000	1.000	—
^{139}Ce	137.63 ± 0.10 days	K_α	33.29	0.808 ± 0.081	—	—	—
		K_β	38.0	0.195 ± 0.020	—	—	—
		γ	165.9	1.000	—	—	—

Isotope	Half-life	Transition	Energy (keV)	Intensity	Intensity	Intensity
^{145}Pm	18 years	L x	5.7	—	—	0.208 ± 0.026
		K_α	37.2	—	—	1.000
		K_β	42.5	—	—	0.258 ± 0.020
		γ	67.0	—	—	0.013 ± 0.002
		γ	72.2	—	—	0.041 ± 0.004
^{155}Eu	1.8 years	L x	6.38	—	—	0.259 ± 0.028
		γ	18.776	—	—	0.0017 ± 0.0003
		γ	26.513	—	—	0.010 ± 0.001
		K_α	42.8	—	—	0.58 ± 0.03
		γ	45.2972	—	—	0.041 ± 0.003
		K_β	49.0	—	—	0.154 ± 0.008
		γ	60.010	—	—	0.039 ± 0.003
		γ	86.5452	—	—	1.000
		γ	105.308	—	—	0.68 ± 0.04
^{241}Am	432.9 ± 0.8 years	M x	3.30	—	0.470 ± 0.006	—[a]
		L_α x	11.9	0.0650 ± 0.0020	1.000	0.0225 ± 0.0020
		L_β x	13.9	1.000	1.556 ± 0.04	0.351 ± 0.025
		L_γ x	17.75	1.458 ± 0.0248	0.133 ± 0.010	0.532 ± 0.038
			20.08	0.367 ± 0.012	0.185 ± 0.019	0.062 ± 0.005
		γ	26.345	0.181 ± 0.007	—	0.0029 ± 0.003
		γ	33.119	—	—	0.0016 ± 0.005
		γ	43.463	—	—	1.000
		γ	59.537	—	2.659 ± 0.053	

[a] J. L. Campbell and L. A. McNelles, *Nucl. Instr. Meth.* **125**, 205 (1975).
[b] J. S. Hansen, J. C. McGeorge, D. Nix, W. D. Schmidt-Ott, I. Unus, and R. W. Fink, *Nucl. Instr. Meth.* **106**, 365 (1973).
[c] R. J. Gehrke and R. A. Lokken, *Nucl. Instr. Meth.* **97**, 219 (1971).
[d] J. L. Campbell and L. A. McNelles, *Nucl. Instr. Meth.* **117**, 519 (1974).

where I_m is the measured intensity at photon energy $h\nu$, $I_0(\gamma)$ the absolute γ-decay rate for the calibrated source, and $\alpha_{\gamma x}$ the number of x rays per γ transition. Table III (pp. 216 and 217) gives the published values of $\alpha_{\gamma x}$ for a number of standard radioactive sources.

Inherent in the typical measurement of x-ray production cross sections for solid targets is the assumption that the radiation is emitted isotropically. Experimental results by Lewis et al.[51] at low resolution have verified that the yield of target x rays is isotropic to less than 3.6% for ^{4}He-induced tin K_β radiation. Recent results by Jamison and Richard[52] show however that in high resolution, polarization of +52.5% exists for 1.9-MeV He$^+$ ions on thick aluminum targets. These authors observe large polarizations for the main multiplets of the K_α L^1 satellite of aluminum. They also report that the net polarization of the K_α L^1 satellite line is $\leq 1\%$, thus indicating that large polarization effects may be hidden by inappropriate averaging. If their result is indicative of the general case to be expected for multiplet structures, then the lower resolution Si(Li) detectors would be expected to exhibit minimal sensitivity to polarization and hence yield isotropic distributions for target K x rays.

5.4.3. Target Considerations for Light Ions

Generally two types of solid targets have been utilized in the study of inner-shell ionization by fast ions. These two systems are classified as thick targets wherein the projectiles lose all of their energy in traversing the target and thin targets wherein only a small fraction of the incident energy of the projectile is lost upon passage through the target. At low projectile energies ($E_1 \leq 200$ keV/amu) the thick-target situation is generally necessary in view of considerations of energy loss and count rates for the x-ray channel. At higher energies the thin-target geometry offers a definite advantage because of the simplicity of the equations that govern the yield of target x rays in terms of the parameters describing the system.

5.4.3.1. Thick Targets. Consider the case of low-energy ions incident upon a thick target. The x-ray production cross section at energy E_1 is $\sigma_X(E_1)$ and is related to the experimental measurement by the Merzbacher–Lewis (ML) relationship

$$\sigma_X(E_1) = \frac{4\pi}{\eta \Omega \gamma} \left[\left(\frac{dY_X(E_1')}{dE_1'} \frac{dE_1}{dR_p(E_1)} \right)_{E_1'=E_1} + \frac{\mu \cos \theta}{\cos \phi} Y(E_1) \right], \quad (5.4.3)$$

where n is the target number density, Ω and γ the solid angle and efficiency of the x-ray detector, respectively, and $Y(E_1)$ the average number

---

[51] C. W. Lewis, R. L. Watson, and J. B. Natowitz, *Phys. Rev A* **5**, 1773 (1972).
[52] K. A. Jamison and P. Richard, *Phys. Rev. Lett.* **38**, 484 (1977).

5.4. GENERAL EXPERIMENTAL CONSIDERATION

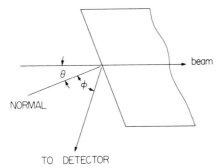

FIG. 7. Definition of the angles used in the Merzbacher–Lewis (ML) formulation of the x-ray production cross section for ions incident upon infinitely thick solid targets.

of target characteristic x rays per incident particle. A best fit is made through the points $Y_1(E_1)$ over a range of E_1 and the slope $dY_1/dE_1'|_{E_1'=E_1}$ is determined for each energy E_1 of interest. The angles θ and ϕ are given in Fig. 7. The quantity μ is the absorption coefficient of the target for its own characteristic x rays and the quantity $dE_1'/dR_p(E_1')$ is related to the stopping power for the incident ions in the medium. For most systems

$$dE_1'/dR_p(E_1') = S(E_1), \tag{5.4.4}$$

where $S(E_1)$ is the stopping power as given in compilations such as Northcliffe and Schilling.[53] (See Part 2 of this work for a detailed discussion of energy loss and stopping power.)

Generally the experimental geometry is chosen so that $\theta = \phi = 45°$. Hence

$$\sigma_X(E_1) = \frac{4\pi}{n\Omega\gamma}\left(S(E_1)\frac{dY(E_1)}{dE_1} + \mu Y(E_1)\right). \tag{5.4.5}$$

It is found that the determination of the slope $dY_1(E_1)/dE_1$ is the largest source of error in thick-target measurements of $\sigma_X(E_1)$. Second, uncertainties in $S(E_1)$ also contribute on the order of $\geq 10\%$ to the error in such measurements. Basbas et al.[24] conclude that thick-target measurements of σ_{KX} are uncertain by $\pm 20–30\%$ because of these factors.

5.4.3.2. Thin Targets. In the case of thin-target geometry, particularly when the monitor of the beam is a detector that records nuclear scattering events, much greater accuracy can be achieved in measuring $\sigma_X(E_1)$. The nuclear scattering detector acts like a dynamic monitor in that fluctuations in its count rate associated with beam and/or target instabilities are directly correlated to fluctuations in the x-ray detection channel. The

---

[53] L. C. Northcliffe and R. F. Schilling, *Nucl. Data Tables* **7**, 2331 (1970).

yield expression for characteristic x rays from a thin target is

$$Y_X(E_1) = N_2 \sigma_X(E_1) \epsilon_X(h\nu) \phi_1, \quad (5.4.6)$$

where N_2 is the number of target atoms/cm^2, ϕ_1 the incident particle flux, and $\epsilon_X(h\nu)$ the total x-ray detection efficiency including the detector solid angle. The scattered particle yield Y_p is given by

$$Y_p(E_1, \theta) = N_2 \phi_1 [d\sigma(E_1, \theta)/d\Omega] \, d\Omega, \quad (5.4.7)$$

where $d\sigma(E_1)/d\Omega$ is the nuclear differential scattering cross section in laboratory detection angle θ and $d\Omega$ the solid angle for the nuclear scattering detector. The ratio $Y_X(E_1)/Y_p(E_1, \theta)$ is thus

$$\frac{Y_X(E_1)}{Y_p(E_1, \theta)} = \frac{\epsilon_X(h\nu)}{d\Omega} \sigma_X(E_1) \frac{d\sigma(E_1, \theta)}{d\Omega}. \quad (5.4.8)$$

Hence the result for $\sigma_X(E_1)$ is given as

$$\sigma_X(E_1) = \frac{d\Omega}{\epsilon_X(h\nu)} \frac{Y_X(E_1)}{Y_p(E_1, \theta)} \frac{d\sigma(E_1, \theta)}{d\Omega}. \quad (5.4.9)$$

For a large range of systems $d\sigma(E_1, \theta)/d\Omega$ is given by the Rutherford scattering result

$$\frac{d\sigma}{d\Omega}(E_1, \theta) = 1.2965$$

$$\times 10^{-3} \left(\frac{Z_1 Z_2}{E}\right)^2 \frac{1}{\sin^4 \theta/2} \left[1 - 2\left(\frac{M_1}{M_2}\right)^2 \sin^4\left(\frac{\theta}{2}\right)\right] \quad (5.4.10)$$

in laboratory coordinates. As a general rule of thumb, the nuclear scattering cross section will be given by the Rutherford scattering expression for

$$E_1 \leq 0.7 \frac{m_1 + m_2}{m_2} \frac{1.44 \text{ MeV } f}{1.4 f} \frac{Z_1 Z_2}{(A_1^{1/3} + A_2^{1/3})} \quad (5.4.11)$$

or

$$E_1 \leq 0.72 \frac{m_1 + m_2}{m_2} \left(\frac{Z_1 Z_2}{A_1^{1/3} + A_2^{1/3}}\right), \quad (5.4.12)$$

where A_1 and A_2 are the atomic number of projectile and target, respectively, m_1 and m_2 the nuclear masses, and Z_1 and Z_2 the atomic numbers.

The advantage to using the thin-target geometry with a scattering detector is seen in the expression for $\sigma_X(E_1)$. The particle flux ϕ_1 and the target density N_2 cancel out of this expression. This result holds so long

5.4. GENERAL EXPERIMENTAL CONSIDERATION

as there are *no* target thickness effects in the x-ray channel. Later in this part the case of target thickness effects is discussed in detail.

5.4.4. Target Considerations for Heavy Incident Ions

5.4.4.1. General Considerations.
In the study of inner-shell ionization involving heavy ions incident upon solid targets with projectile energies in the keV/amu range, all measurements of target and/or projectile x-ray yields must be considered as arising from thick targets. The purpose of such investigations has been to study the production of inner-shell vacancies with a view toward testing the concepts associated with the MO excitation model. At first glance it is tempting to apply the Merzbacher–Lewis (ML) formulation for thick-target x-ray yields to the analysis of the observed x-ray emission for heavy ions incident upon thick solid targets. Such a procedure has been used both for low and high incident projectile energies by Kavanagh *et al.*[54] in the study of copper L-shell excitation for energies in the keV/amu range and more recently by Meyerhof *et al.*[55] in the study of K-shell ionization at energies in the range of MeV/amu for a wide range of projectile–target combinations. In order for the ML relation to be valid the following assumptions must be fulfilled:

(1) The ions slow down along straight trajectories.

(2) Energy straggling of the beam is small compared to the energy loss along the path of the ion in the solid.

(3) Recoil atoms do not contribute to the production of x rays in subsequent collisions.

(4) Inner-shell vacancy configurations are not a major consideration in the production of target ionization.

The first three assumptions are not valid for heavy ions incident upon solid targets at low energies. The fourth assumption will fail in varying degrees pending the relative Z_1/Z_2 ratio. The neglect of the effects associated with these assumptions can introduce relatively large errors into the conversion of the measured target x-ray yields into target x-ray production cross sections for a particular ion species incident upon a particular target. The errors in such analyses (based upon the ML approach) can approach orders of magnitude.

---

[54] T. M. Kavanagh, M. E. Cunningham, R. C. Der, R. J. Fortner, J. M. Khan, E. J. Zaharis, and J. D. Garcia, *Phys. Rev. Lett.* **25**, 1473 (1970).

[55] W. E. Meyerhof, T. K. Saylor, S. M. Lazarus, A. Little, R. Anholt, and L. F. Chase, Jr., *Phys. Rev. A* **14**, 1653 (1976).

5.4.4.2. Recoil and Straggling Effects. Taulbjerg and Sigmund[56] and Taulbjerg et al.[57] discuss the contributions to the target x-ray production of recoil and straggling effects for heavy ions on solid targets. Brandt and Laubert[58] report analytical relationships that can be used to estimate such contributions to the target x-ray yield. Following the development given by Brandt and Laubert, one can write $\sigma_X(E_1) = L(E_1) - K(E_1) - R(E_1)$, where $L(E_1)$ is the ML relation, i.e.,

$$L(E_1) = S_{12}(E_1)\frac{dY_2(E_1)}{dE_1} + \frac{\mu}{N} Y_2(E_1). \quad (5.4.13)$$

In the limit of negligible absorption of x rays by the target,

$$L(E_1) = S_{12}(E_1)\frac{d^2Y_2(E_1)}{dE_1^2}, \quad (5.4.14)$$

The term $K(E_1)$ accounts for the energy straggling of the projectile in the target and is given by

$$K(E_1) = 1.31 \times 10^5 \left(\frac{Z_1Z_2M_1}{M_1 + M_2}\right)^2 \frac{d^2Y_2(E_1)}{dE_1^2}, \quad (5.4.15)$$

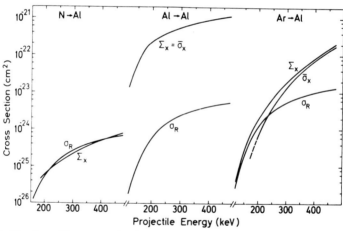

FIG. 8. Aluminum K x-ray production cross sections vs. E_1 for nitrogen, aluminum, and argon ions. The *observed* target x-ray cross section is $\Sigma_X = \overline{\sigma_X} + \sigma_R$, where $\overline{\sigma_X}$ is the x-ray production cross section from the ions incident on the target species and σ_R the contribution to the observed cross section associated with the collisions between recoiling target atoms and the target species. Within uncertainties of the experimental data, $\Sigma_X = \sigma_R$ in the case of N → Al such that only an upper limit to $\overline{\sigma_X}$ may be obtained.

[56] K. Taulbjerg and P. Sigmund, *Phys. Rev. A* **5**, 1285 (1972).
[57] K. Taulbjerg, B. Fastrup, and E. Laegsgaard, *Phys. Rev. A* **8**, 1814 (1973).
[58] W. Brandt and R. Laubert, *Phys. Rev. A* **11**, 1233 (1975).

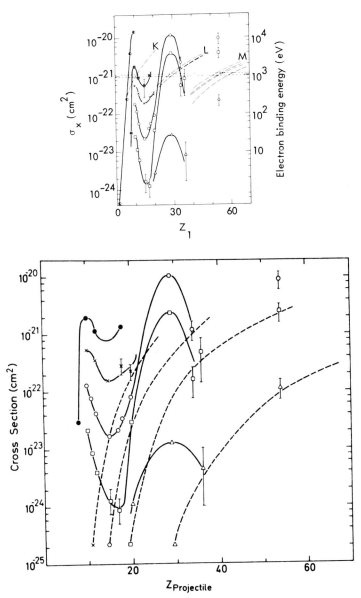

FIG. 9. (a) Cross sections for copper L x-ray production vs. Z_1, for fixed incident velocities from Ref. 54 using the ML analysis for thick targets: △, 1.0 keV/amu; □, 2.0 keV/amu; ○, 3.0 keV/amu, ×, 5.0 keV/amu; ⊗, 10 keV/amu; and ◐, 50 keV/amu. (b) Cross sections for copper L x-ray production vs. Z_1 for fixed incident ion velocities, symbols the same as in (a) with one exception; ●, 10 MeV/amu. The cross sections are reevaluated on the bases of yield data from Ref. 54 by including the effects contained in $\Sigma_X = \overline{\sigma_X} + \sigma_R$ (see Fig. 8). The solid lines represent Σ_X. Each dashed line corresponds to one of the solid lines and represents the recoil contribution σ_R. For $Z_1 < 30$, $\overline{\sigma_X} \approx \Sigma_X$, while for $Z_1 \geq 30$, $\overline{\sigma_X}$ is the small difference between σ_X and σ_R and cannot be accurately determined (see Ref. 57).

where $K(E_1)$ is in units of 10^{-24} cm² for the second derivative of the yield expressed in (x rays/projectile) keV⁻². The propagation of errors limits the accuracy in the x-ray production cross section of $\sim \pm 20\%$ and hence $K(E_1)$ need not be evaluated to higher accuracy than inherent in this expression.

The recoil contribution $R(E_1)$ can be approximated as

$$R(E_1) = \frac{2.4 \times 10^{-2}}{(l^2 - 1/4)} \frac{Z_1^2 Z_2^{5/6}(M_1 + M_2)}{M_1 M_2 E_1^{3/2}} \sigma_X(Z_2, Z_2, T_M), \quad (5.4.16)$$

where $l \geq 1$, E_1 is in keV/amu and $\sigma_X(Z_2, Z_2, T_M)$ is expressed in barns. The cross section $\sigma_X(Z_2, Z_2, T_M)$ accounts for the recoil–target contribution and can be approximated by

$$\sigma_X(Z_2, Z_2, E) = A(1 - B/E)\theta(B - E), \quad (5.4.17)$$

where

$$A = \omega_K P \pi r_c^2, \quad B = 2Z_2^2 e^2 r_c^{-1} \exp(-2Z_2^{1/3} r_c) \geq T_M.$$

The quantity $T_M = [4M_1 M_2/(M_1 + M_2)^2] E_1$ and $\theta(B - E)$ is the step function, with $P \simeq 0.1$ and $r_c \simeq a_0 Z_2^{-1}$, and ω_K is the target fluorescence yield.

Evaluations of the relative contributions of the various terms to the analysis of data for heavy ions on aluminum are given in Fig. 8. It is seen that at the lower projectile energies the considerations of recoil and straggling contributions are a most important feature. In the case of the analysis of copper L x-ray production data for a number of different incident ion species, a comparison of the analysis by Kavanagh et al.[54] using the ML formula and Taulbjerg et al.[56] using corrections for recoil and straggling effects is given in Fig. 9.

5.4.4.3. Multiple-Collision Effects. Macek et al.[59] treat the case of inner-shell vacancies in the projectile that arises from the passage of the beam through the solid target. Specifically these authors analyze the thick-target yields of silicon K x rays and projectile argon L x rays in terms of a double-collision scattering mechanism whereby silicon K vacancies are produced by MO promotion to an argon $L_{2,3}$ vacancy formed in a prior collision within the solid. Their treatment is for thick targets. The derivation of the vacancy cross section for the target (labeled 2) is found in the work of Macek et al.[59] The expression for the vacancy cross section as related to the experimentally related quantities is

$$\sigma_2(E) = \left(\frac{S(E) \, dY_2(E)}{dE} + \mu_2 Y_2(E)\right)\left(\frac{S(E) \, dY_1(E)}{dE} + \mu_1 Y_1(E)\right)^{-1}$$
$$\times \omega_1(E)[\omega_2(E)\tau_1(E)v_1(E)n]^{-1}, \quad (5.4.18)$$

[59] J. Macek, J. A. Cairns, and J. S. Briggs, *Phys. Rev. Lett.* **28**, 1298 (1972).

5.4. GENERAL EXPERIMENTAL CONSIDERATION

where μ_j are the absorption coefficients for the x rays from the target and projectile, $\omega_i(E)$ the fluorescence yields of target and projectile, τ_1 the lifetime of the excitation in the 2p level of the projectile, and v_1 the incident projectile velocity. It is noted that the target x-ray production cross section is

$$\sigma_{2X}(E) = \frac{\omega_2(E)}{\omega_1(E)} \sigma_2(E). \tag{5.4.19}$$

This formulation was used to analyze silicon K x-ray data for argon ions in the energy range 130–240 keV. The measurements involve the yields of both the silicon K x rays and argon L x rays. The results for these measurements are shown in Fig. 10. A comparison of these data to results for argon on aluminum as given in Fig. 8 shows a serious discrepancy between energy dependences that are derived from the two different models for analysis of the target x-ray yields.

Garcia[60] has developed an analysis of thick-target x-ray yield data that includes the effects of inner-shell vacancies in the projectile on the observed target x-ray yield. His analysis is of a more complex nature than that presented by Macek *et al.*[59] in that it requires first and second derivatives of the target x-ray yield taken with respect to the total range R_0 of the ion in the target. Additionally, the excitation cross sections for the inner shells of the projectile must be known. Specifically, the target x-ray production cross section is given as

$$\omega_t \sigma^t(E_0) = \frac{1}{N\sigma^P(E_0)} \left[T_X''(E_0) \right.$$
$$\left. + \left(\frac{1}{N\sigma^P(E_0)} \frac{\partial \sigma^P(E_0)}{\partial R_0} \frac{\partial E_0}{\partial R_0} - \frac{\lambda(E_0)}{N} \right) T_X'(E_0) \right]. \tag{5.4.20}$$

where

$$T_X''(E_0) = \frac{\partial^2 T_X}{\partial E_0^2} \left(\frac{\partial E_0}{\partial R_0} \right)^2 - M_t^2 T_X(E_0) + 2\mu_t T_X^1(E_0),$$

$$T_X' = \frac{\partial T_X}{\partial r_0} + M_t T_X.$$

The quantity $\partial R_0/\partial E_0$ is the stopping power, μ_t the x-ray absorption coefficient for the target characteristic radiation, and $\lambda(E_0) = N\sigma^P(E_0) + 1/v(E_0)\tau$, with τ the lifetime of the excited inner vacancy of the projectile. The target x-ray production cross section is thus related to the properties of the projectile and its excitations. This is not surprising since from an MO point of view the projectile is required to

[60] J. D. Garcia, Phys. Rev. A **11**, 87 (1975).

bring a vacancy into the interaction in order for target inner-shell excitation to occur. Garcia also presents a similar analysis of the projectile x-ray production cross section. Recoil effects are not included in the analyses of target x-ray yield data. As of 1977, there were no published analysis of thick-target x-ray data using Garcia's analysis. It is appropriate that consideration be given to the application of this complex analysis

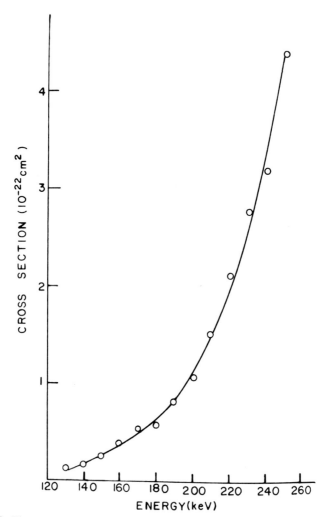

FIG. 10. The cross section for the production of silicon K vacancies by collision with argon ions having an $L_{2,3}$ vacancy. A formulation based upon a double-scattering mechanism is used to analyze target x-ray yield data for a thick target (see Ref. 59).

5.4. GENERAL EXPERIMENTAL CONSIDERATION

to systems like those reported by workers such as Meyerhof *et al.*[55] It would be interesting to compare such analyses to target x-ray production cross sections for thin targets bombarded by heavy ions using the analysis outlined in Section 5.5.5.4.

5.4.4.4. Dynamic Fluorescence Yield Considerations. The analysis of target x-ray data is further complicated by the question of what fluorescence yield is appropriate for the conversion of the measured target x-ray yields for a thick target. In addition, work by Matthews and Fortner[61] suggests that the dynamic fluorescence yield for the projectile moving in the thick target is given by

$$\bar{\omega}_j = \omega_j[1 + N(\sigma_{Qj} + \sigma_{Vj})v_1\tau_j]^{-1}, \quad (5.4.21)$$

where σ_{Qj} is the total cross section for quenching the jth state, σ_{Vj} the vacancy production cross section for the jth state with lifetime τ_j, ω_j the fluorescence yield for the jth state, and N the target density (cm^{-3}). This result is an important consideration for beam foil measurements. It is also important to the question of target x-ray production in the formulations of Macek *et al.*[59] and Garcia,[60] where one requires not only target vacancy production but also projectile vacancy production.

5.4.4.5. Summary Comments. In summary, the technique of measuring and interpreting target and/or projectile x-ray yields for the purpose of extracting x-ray production cross sections or vacancy production cross sections for heavy ions on thick targets is a complicated and unresolved question. Clearly the ML formula is applicable for systems like protons on copper. The application of the recoil correction is appropriate to those systems wherein vacancy configuration effects are not an important consideration. There are regions where vacancy configuration effects are an overriding consideration and hence cannot be neglected. It is very important that each projectile–target system be viewed with the complexities outlined above in consideration. It is common for published papers to give only the reduced results, i.e., a graph or table of "measured" x-ray production cross sections. As has been shown for heavy ions on thick targets there is the potential for such data to be analyzed by differing and contrasting methods of analysis, thereby leading to reported ionization cross sections or x-ray production cross sections that are strictly model dependent. It is most important to give the data for $Y_1(E)$ [and $Y_2(E)$, if available] and thus afford others the opportunity to arrive at the cross sections relevant to their choice of model analysis. It is

---

[61] D. Matthews and R. Fortner, *in* "Program and Extended Abstracts International Conference on the Physics of X-Ray Spectra" (R. D. Deslattes, ed.), p. 183. Nat. Bur. Stand., Gaithersburg, Md., 1976.

suggested that a model that includes recoil effects, effects for residual vacancies, and dynamic fluorescence yields would be useful in placing the analysis of thick-target x-ray measurements on a uniform basis so that comparison of the relevant cross sections would have a universal meaning.

At higher energies, care needs to be exercised in the design and interpretation of x-ray yield experiments for heavy ions incident on thick targets. In particular, the measurement of x-ray yields for thick targets and thin gas targets show differences of factors of 2 (see Ref. 55). Care needs to be exercised in making the gas target measurements. It is *not* appropriate to pass a heavy-ion beam through an entrance window on a gas cell for the purpose of measuring the target x-ray production rate under "single-collision" conditions. Consider a 202-MeV krypton beam incident upon a gas cell with an entrance window of 2.5 μm of nickel. Upon passage through this window all projectile vacancy configurations will have reached their equilibration values. It is true that a majority of these vacancies will decay before entering the interaction region, viewed by the x-ray detector because of lifetime considerations. However, a few percent will survive in metastable components and hence the state of the incident beam is undetermined. Further, the peak in the charge state distribution will be around Kr^{28+}, i.e., krypton with eight electrons. Since the x-ray production in the target can be very sensitive to the state of ionization of the projectile, the existence of n vacancies in the projectile a priori may be strongly reflected in the measured x-ray yield. Clearly, the measurement of x-ray production in gas cells with entrance windows already includes the very effect that one is hoping to understand when such data are compared to results from adjacent solid targets with incident projectiles at the same energy. One may not see a clear, solid-state effect under such conditions and one would not expect to see such effects as they are potentially resident in both sets of data.

5.5. X-Ray Production Cross-Section Measurements

5.5.1. Introduction

In this section a discussion of results for inner-shell ionization and/or x-ray production cross-section measurements is presented. The emphasis is placed initially on the results for light ions on solid targets where $Z_1/Z_2 \ll 1$. This procedure is adopted because of the clearer relation between theory and experiment for such systems. Later in this chapter the condition $Z_1/Z_2 \ll 1$ is relaxed with the discussion of inner-shell ion-

5.5 X-RAY PRODUCTION CROSS-SECTION MEASUREMENTS

ization by heavy ions moving in solid targets. However, in this phase of the development of the topic of this chapter it will become clear that critically important and complex factors have been neglected by many workers, thus leading to a situation in which each experiment where $Z_1/Z_2 \geq 0.3$ needs to be carefully reviewed to assess the potential for problems resident in the reported results. In the case of heavy ions incident upon solid targets, the quantity measured has not been a cross section that is directly comparable to the single-collision theories of inner-shell ionization because of the effects of the interactions of the solid on the ion and the changes in the observed x-ray production associated with these multiple-collision effects.

5.5.2. K-Shell X-Ray Measurements ($Z_1/Z_2 \ll 1$)

The main emphasis of K x-ray production cross-section measurements in solids has been to establish a systematic picture of inner-shell ionization for comparison to the present theories. Historically the PWBA, SCA, and BEA theories were first advanced to explain the phenomenon of inner-shell ionization by fast bare point projectiles. The extensive experimental studies have provided information that has led to modifications of these original ideas with the introduction of relativistic corrections for the description of the inner-shell electrons in heavy-target species, modifications of the electron binding energy and particle trajectories, and attempts to include screening effects on the projectile charge as it moves through the solid. The PWBA, SCA, and BEA theories all predict that for a fixed scale velocity v_1/v_e the cross section should scale as Z_1^2. Further, the dependence on the projectile energy E_1 and target atomic number Z_2 is given. A study of the published reviews, data compilations, and references in Table IV shows that most of the work to date has been done for $v_1/v_e < 1$, with only a few results reported for $v_1/v_e > 1$. Further, the majority of recent works on inner-shell ionization use thin solid targets where the energy loss of the projectile is small and self-absorption of the target x rays is minimized. There are exceptions to this generalization, which are discussed in the course of this chapter.

Langenberg and van Eck[62] have compared the measured cross section for K-shell ionization of carbon by hydrogen ions from gases and thin solid targets. They find agreement between the measured cross sections to within the quoted errors. They present the summary of results for carbon K-shell ionization by several groups ranging in proton energies from 0.015 to 16.0 MeV as shown in Fig. 11. Comparisons are given to calculations based upon the BEA, PWBA, and PWBABC theories. This

---

[62] A. Langenberg and J. van Eck, *J. Phys. B* **9**, 2421 (1976).

TABLE IV. Recent Results for K and L Excitation Using Ion Beams on Solid Targets since 1973[a]

Target(s)	Projectile	Projectile energy	Target configuration	Remarks	Ref.[a]
Cr, Cu, In,	^{1}H	0.9–2.5 MeV	Thin	K shell	118
Cr, Cu	^{4}He	0.9–2.5 MeV			
Al, V, Fe, Cu	^{4}He	1–5 MeV	Thick and thin	Illustrates that unperturbed theories give σ's larger than data for K shell	119
Y, Mo, Ag, Sn S, K, Ti, V, Cr, Mn, Fe, Co, Ni, Cu, Ge, Br	^{16}O	7–24 MeV	Thin	Target thickness effects not included in experiment or analysis; K shell	93
K–Ag	^{2}H	2.88–12.50 MeV/amu	Thin	Measured K_α/K_β ratios, no cross sections reported	120
	^{4}He				
	^{12}C	2.35–8.33 MeV/amu	Thin	Measured K_α/K_β ratios, no cross sections reported	120
Zn	^{1}H	1.4–4.4 MeV	Thin	K shell	121
Sn	^{1}H	1.4–4.4 MeV	Thin	K shell compared to PWBA with relativistic corrections	122
Ag	^{16}O	12–50 MeV	Thin	K shell	72
S, Sc, Fe, Cu	^{1}H, ^{16}O, ^{19}F	0.83–3.28 MeV/amu	Thin	K shell, tests Z_1^2 dependence of σ, raises question of contributions from charge exchange	75
Ti–Y	^{4}He	0.5–2.5 MeV	Thin	K shell	68
Ti–Sb	^{7}Li	1.0–5.0 MeV/amu	Thin	K shell	70
C	^{1}H	1–18 MeV	Thin	K shell, shows PWBA fits energy dependence in high energy region	123
Si, Al	^{1}H	0.75–4 MeV/amu	Thin	K shell	63
	^{3}He	0.75–4 MeV/amu			
Ti–Se	^{4}He	1–4.4 MeV	Thin	K shell	67
Cu–Pd	^{12}C	8–36 MeV	Thin	K shell	124
Cr–Pb	^{4}He	5 MeV/amu	Thin	K shell, measures K_α/K_β and energy shifts in E_X for K_α/K_β	125
	^{14}N	5 MeV/amu			

Ni, Rb, Ag, Sb	^{12}C, ^{14}N	0.4–2.4 MeV/amu	Thin	K shell, illustrates effects of binding energy correction	73
C	^{16}O	0.29–16 MeV	Thin	K shell	126
	^{1}H	18–26 MeV	Thin	K shell	127
Ca–Zn	^{14}N	0.5–2.5 MeV/amu	Thin	K shell, includes multiple corrections to PWBA	128
Ge–Ag	^{14}N	0.5–2.5 MeV/amu	Thin	K shell, same analysis as Ref. 130	129
Al	^{1}H^{+}, ^{1}H$_2^{+}$, ^{1}H$_3^{+}$	90–150 MeV/amu	Thick	K shell, shows effects of clusters at low energies	130
Ti–Sb	^{1}H	1–3 MeV	Thin	K shell, measures K_α production cross section only	131
Ni, Ge, Ag		1–3.7 MeV			
Fe–Au	^{4}He	30–100 MeV	Thin	K shell	132
Mn–Cd	^{1}H	2.5–12 MeV	Thin	K shell	66
Fe–As	^{1}H	0.5–2.0 MeV	Thin	K shell	64
Ti–Cu	^{16}O, ^{18}O	0.556, 0.741, 1.111 MeV/amu	Thin	K shell, test for Coulomb deflection effects	133
S–Br	^{19}F	4–28 MeV	Thin	K shell, does not account for target thickness effects for $Z_2 \leq 29$	134
Ti–Ag	^{1}H	1–3 MeV	Thick	K shell	135
Ca–Cd	^{4}He, ^{10}B, ^{12}C, ^{14}N, ^{16}O, ^{19}F, ^{20}Ne	7.1 MeV/amu	Thin	K shell, tests of Z_1^2 dependence, measurements made for targets having equilibration thickness for K shell of projectile	
Mn–Ge	S	5–48 MeV	Thin	K shell, measurement of x-ray production with no considerations of target thickness effects	94
Al	^{16}O	0.15–91 MeV	Thin	K shell, broad energy range measurements with no consideration of target thickness effects	95
	Ne	10 MeV	Thin	K shell, consideration of target thickness effects on target x-ray production using two-component model: x-ray yield versus target thickness	107

(continued)

TABLE IV. (Continued)

Target(s)	Projectile	Projectile energy	Target configuration	Remarks	Ref.[a]
Cu	Cl	40–60 MeV	Thin	K shell, consideration of target thickness effects on target x-ray production using two-component model with data for hydrogen-like ions; x-ray yield versus target thickness	109
	F, Al, Si, S	1.71 MeV/amu	Thin	K shell, consideration of target thickness effects on target x-ray production using three-component model with data for base and hydrogen-like ions; x-ray yield versus target thickness	113
	S	110 MeV	Thin	K shell, x-ray yield versus target thickness, no analysis.	101
Al, Ni	Al	0.5–40 MeV	Thin	K shell, broad-range energy dependence study of K shell ionization for symmetric collisions; target thickness effects not included	97
	Ni	10–63 MeV			
C + Cu	S	70, 140 MeV	Thin Sandwich	K shell, layered-foil technique to measure projectile K-vacancy processed in carbon foils	105
	F	14–62 MeV	Thin Sandwich	K shell, layered-foil technique to measure projectile K-vacancy processes in carbon foils	106
	S	64 MeV			
	Cl	35–140 MeV			
Sc, Ti, Cu, Ge	Si	52 MeV	Thin	K shell, measurements of K-shell charge exchange contributions to target K x-ray production	112
Au	^{1}H	0.5–30 MeV	Thin	L shell, total x-ray production cross section	76
Pb	^{1}H	0.5–14 MeV	Thin	L shell, total x-ray production cross section	92

Target	Projectile	Energy	Thickness	Description	Ref.
Ta, Lu, Pb, U, Th	^{4}He	26 MeV	Thin	L shell, measured relative x-ray yields, no cross sections	136
Ta, Au, Bi, Pt	^{4}He	0.3–5.5 MeV	Thin	L shell, measured L_{III}/L_{II} ratio of ionization cross sections	137
Au	^{1}H	0.25–5.2 MeV	Thin	L shell, measured L_{III}, L_{II}, and L_{I} ionization cross sections show 2s structure in L_I x-ray production	78
	He	1–12 MeV			
Pb, Bi	^{1}H	0.5–4 MeV	Thin	L shell, measurement of L_I and L_{II} ionization cross sections, shows 2s structure in L_I cross section	79
Sm	^{1}H	0.3–2.0 MeV	Thin	L shell, measurement of L_α, L_γ, and $L_{\gamma_{2,3}}$ x-ray production, shows 2s structure associated with L_I cross section	80
Pb	^{1}H	1.4–4.4 MeV	Thin	L shell, total x-ray production, and individual L x-ray production cross sections	121
	^{3}He	1.4–4.4 MeV			
Sn–Bi	^{4}He	1–5 MeV	Thin	L shell, measured total L-shell ionization production	138
Sn, Ta	^{1}H	1.4–4.4 MeV	Thin	L shell, measured total x-ray production	122
Pr, Eu, Gd, Dy	^{1}H	0.3–2 MeV	Thin	L shell, measured L x-ray production cross section and ratios	139
Ag, Au	^{16}O	12–50 MeV	Thin	L shell, measured L x-ray production cross section	72
Ta, Au, Bi	^{1}H	1–5.5 MeV	Thin	L shell, measured L_I, L_{II}, and L_{III} cross sections	81
	^{4}He	1–11 MeV			
Au, Bi, U	^{1}H	1–4.5 MeV	Thin	L shell, measured total x-ray production and individual L x-ray production cross sections	140
	^{3}He	3–9 MeV			
La, Ce, Pr, Sm, Eu, Dy, Ho	^{1}H	3 MeV	Thin	L shell, measured individual L x-ray production cross sections for ^{16}O ions, high-resolution spectra for 3-MeV ^{1}H and 25-MeV ^{16}O ions on lanthanum	84
	^{16}O	8–36 MeV			

(continued)

TABLE IV. (Continued)

Target(s)	Projectile	Projectile energy	Target configuration	Remarks	Ref.[a]
Ta–Pt	^{4}He	1–4 MeV	Thin	L shell, measured ratio of L_{III} to L_{II} cross sections	141
Sm, Yb, Pb	^{1}H ^{4}He ^{7}Li	0.3–2.4 MeV/amu 0.15–4.4 MeV/amu 0.9–3 MeV/amu	Thin	L shell, measured individual L x-ray production cross sections	82
Rb–Gd	^{1}H	950 keV	Thin	L shell, measured L x-ray production cross sections	142
Ag, Sb, Au	^{1}H	200–900 keV	Thin	L shell, measured L_I and L_{II} cross sections	83
Ta–Pb	^{1}H	0.4–2 MeV	Thin	L shell, measured L x-ray production cross section and individual L x-ray production cross sections	143
Pd, Ag, Sn	^{1}H ^{16}O	3–12 MeV 15–40 MeV	Thin	L shell, measured total L-shell ionization production	144
Sm–Pb	^{4}He	30–80 MeV	Thin	L shell, L_I, L_{II}, and L_{III} cross sections measured	145
Pb	^{1}H	2, 3, 6 MeV	Thin	L shell, measured $P(b)$; report total L-shell production cross section	146
U	^{1}H ^{4}He	2–18 MeV 3–16 MeV	Thin	L shell, measured L_I, L_{II}, and L_{III} cross section	147
U	^{1}H ^{4}He ^{12}C ^{16}O	0.5–6 MeV/amu 0.5–6 MeV/amu 0.5–6 MeV/amu 0.5–6 MeV/amu	Thin	L shell; measured ratio of L_{III} to L_{II} and L_I to L_{II} cross sections	148
Pt, Au, Hg	^{1}H	0.4–2.0 MeV	Thin	L shell, measured L_I, L_{II}, and L_{III} cross sections	149
Au–U	^{1}H	0.5–3.5 MeV	Thin	L shell, measured ratios of L-subshell cross sections	85

| Ag | ¹⁹F | Thin | L shell, measurement of L x-ray production for base, hydrogen-like, and multi-electron fluorine ions; study of target thickness effects, illustrate effects of charge exchange for F^{8+} and F^{9+} ions | 88 |
| Ce–U | ¹H | 1–3 MeV | Thin | L shell, measured L_α x-ray production cross section only | 131 |

[a] Complete references for Table IV are listed on p. 278.

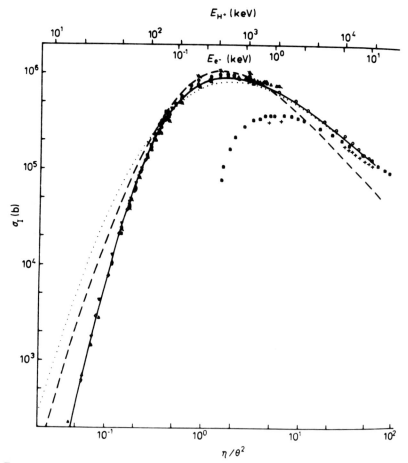

FIG. 11. Carbon K-shell ionization cross section σ_I as a function of η/θ^2 for protons and electrons. For x-ray measurements $\omega_K = 0.00236$ is applied to obtain σ_I. Full curve, PWBABC; dotted curve, PWBA; broken curve, BEA. See Ref. 62 for data definitions.

case supports the PWBABC approach to the description of K-shell ionization by protons. In the higher-energy limit, the binding energy and Coulomb deflection effects become minimal, thus allowing PWBABC → PWBA. Clearly the $E_1 \ln E_1$ dependence of the PWBA fits the data for $v_1/v_e > 2$. In the low-velocity range, the importance of the binding energy and Coulomb deflection effects is illustrated.

Other studies involving protons on heavier elements have been reported. In the intermediate range of Z_2, i.e., $13 \leq Z_2 \leq 32$, the works of

5.5 X-RAY PRODUCTION CROSS-SECTION MEASUREMENTS

Tawara et al.[63] for hydrogen on aluminum and silicon and Lear and Gray[64] for hydrogen on Fe–As in the range of $E_1 \leq 4$ and 2 MeV, respectively, have shown the need to include corrections in the PWBA calculations. However, for protons on iron, for example, the Coulomb deflection correction and binding energy correction are effects of $\sim 25\%$ each on the calculated cross sections at 1 MeV for protons. For heavier target elements $Z_2 > 40$, the Coulomb deflection factor dominates over binding energy modifications as has been shown by Khelil and Gray[65] for protons on elements up to lanthanum. At highly incident energies, Liebert et al.[66] have shown that the predictions of the BEA and PWBA for K x-ray production are satisfactory for 2.5–12 MeV protons on elements from manganese to cadmium. In all of these cases the use of neutral-atom fluorescence yields is assumed to be valid because of the close agreement between the K_α/K_β ratios measured in proton and electron or photon bombardment. With the exception of the data for protons on carbon the data for inner-shell ionization by protons have not provided a test that clearly establishes the role of the perturbation effects for the ionization of the heavier elements. These modifications are, however, relatively strong functions of the projectile energy E_1 and the projectile charge Z_1 (assuming a bare nucleus).

Work for heavier ions has proceeded to establish the role the binding energy modificating plays in giving better agreement between the data and theory for K-shell ionization. An overview of data for K-shell ionization for $Z_1 > 1$ reveals several features common to all reported experiments. First, the PWBA, SCA, and BEA theories become progressively worse in their predictions of K-shell ionization as Z_1 is increased for a given target Z_2. These unmodified theories all overpredict the magnitude of the K x-ray production cross sections by increasingly larger factors ~ 2 (He), ~ 3 (Li), ~ 5 (Be), etc., for incident energies less than about a few MeV/amu. In the case of ^{4}He ionization of midrange elements, this feature of K-shell ionization is illustrated in Fig. 12 from the work of Soares et al.[67] The works of McDaniel et al.[68], and McKnight et al.[69] show similar characteristics. In the case of beryllium ions, Fig. 13 for low energy beryllium on vanadium shows the comparison of the PWBA, BEA, and PWBABC

---

[63] H. Tawara, Y. Hachiya, I. Ishii, and S. Morita, *Phys. Rev. A* **13**, 372 (1976).
[64] R. D. Lear and T. J. Gray, *Phys. Rev. A* **8**, 2469 (1973).
[65] N. Khelil and T. J. Gray, *Phys. Rev. A* **11**, 893 (1975).
[66] R. B. Liebert, T. Zabel, D. Miljanic, H. Larson, V. Valkovic, and G. C. Phillips, *Phys. Rev. A* **8**, 2336 (1973).
[67] C. G. Soares, R. D. Lear, J. T. Sanders, and H. A. Van Rinsvelt, *PRA* **13**, 953 (19̄ ͻ).
[68] F. D. McDaniel, T. J. Gray, and R. K. Gardner, *Phys. Rev. A* **11**, 1607 (1975).
[69] R. H. McKnight, S. T. Thornton, and R. R. Karlowicz, *Phys. Rev. A* **9**, 267 (1974).

238   5. TARGET IONIZATION AND X-RAY PRODUCTION

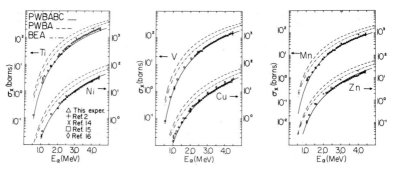

FIG. 12. K-shell x-ray production cross sections for ^{4}He ions incident on thin targets from titanium to zinc.

theories to the measured K x-ray production cross sections. For beryllium ions the PWBABC underestimates the observed cross sections by ≤40% for K x-ray production. A possible source of such a discrepancy is the use of neutral atomic fluorescence yields. However, measurements of K_α/K_β by McDaniel et al.[70] for incident lithium ions in the range 1–5 MeV/amu and results by Li and Watson[71] for ^{2}H, ^{4}He, and ^{12}C ions on elements from potassium to silver show that the changes in fluorescence

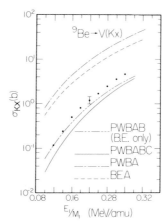

FIG. 13. K-shell x-ray production cross section for ^{9}Be ions incident on a thin vanadium ($Z_2 = 23$) target.

[70] F. D. McDaniel, T. J. Gray, R. K. Gardner, G. M. Light, J. L. Duggan, H. A. Van Rinsvelt, R. D. Lear, G. H. Pepper, J. W. Nelson, and A. R. Zander, *Phys. Rev. A* **12**, 1271 (1975).
[71] T. K. Li and R. L. Watson, *Phys. Rev. A* **9**, 1574 (1974).

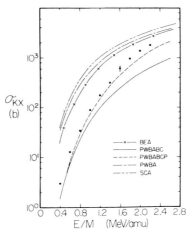

Fig. 14. K-shell x-ray production cross section for ^{12}C ions incident on a thin nickel target.

yields due to multiple ionization should be $\leq 10-15\%$ for incident ions up to carbon on the midrange Z_2 targets.

The results of the PSS theory when used to adjust the PWBA predictions (PWBABC) for K-shell ionization lead to an improved description of observed K x-ray production cross sections for the lighter ions hydrogen to beryllium in the scaled velocity range of $v_1/v_e \ll 1$. It is observed that as v_1/v_e approaches 1 the measured K-shell cross sections begin to deviate from the PWBABC predictions and approach the PWBA values of the cross section. This behavior has been associated with the high-energy polarization effect wherein the cross section for ionization is increased through an additive Z_1^3 term. Results for ^{16}O on silver by Bissinger et al.[72] show that in the energy range 12–50 MeV the measured K-shell cross section is in better agreement with the PWBABC with the added polarization correction. Work by Gray et al.[73] for carbon, nickel, and oxygen ions on the nickel K shell in the energy range 0.4–2.4 MeV/amu shows that the added polarization effect improves the agreement between theory and experiment. Their results for ^{12}C ions on nickel are shown in Fig. 14.

The PWBABC with the polarization effect predicts that the ratio $R = Z_1^2 \sigma(Z_1)/Z_1^2 \sigma(Z_1)$ should be unity at $\eta^{1/2} = \frac{1}{2}\theta_K$. Results by Cue et

---

[72] G. Bissinger, P. H. Nettles, S. M. Shafroth, and A. N. Waltner, *Phys. Rev. A* **10**, 1932 (1974).

[73] T. J. Gray, P. Richard, R. L. Kauffman, T. C. Holloway, R. K. Gardner, G. M. Light, and J. Guertin, *Phys. Rev. A* **13**, 1344 (1976).

al.[74] for bare ions of He, B, C, N, O, F, and Ne on their solid targets of Ca, Fe, Y, and Cd at the fixed incident energy of 7.1 MeV/amu show that the measured ratios $R = 4\sigma(Z_1)/Z_1^2\sigma(Z_1' = 2)$ agree with the predictions of the PSS calculations, which include binding energy and polarization effects. These workers further conclude that charge exchange contributions to the K shell is not an important process. However, their target thicknesses were sufficiently thick so that any inner-shell vacancy fractions would have come to equilibrium. Later in this section we see what implications arise in association with target thickness considerations for heavy ions moving through solids.

In contrast to the work of Cue et al.,[74] results reported by Hopkins et al.[75] for tests of the Z_1^2 dependence of K-shell ionization by ^{1}H, ^{16}O, and ^{19}F ions on thin targets of S, Sc, Fe, and Cu in the energy range 0.83–3.28 MeV/amu show that $R = \sigma(Z_1)/Z_1^2\sigma(Z_1' = 1)$ does not agree with the predictions of the PSS approach including polarization and binding effects. These authors suggest that charge exchange may play an important role in the process of K-shell ionization for heavy ions moving through solid targets.

5.5.3. L-Shell X-Ray Measurements ($Z_1/Z_2 \ll 1$)

Work on the ionization of the L shell by incident ions has served to provide further tests of the theories of inner-shell ionization by high-velocity ions. The results of the studies reported to date have raised questions about the applicability of the classical BEA theory to the description of L-shell ionization, particularly ionization of the L_1 subshell. The majority of the results to date are related to ionization of the L shell by ^{1}H, ^{3}He, and ^{4}He ions. Some work for lithium, carbon, and oxygen ions has been reported but is limited in its scope. The general method used by most workers has been to measure the total L x-ray production cross section (similar to K shell measurements) and/or to measure specific x-ray production cross sections for the x-ray transitions resolvable with a Si(Li) detector ($L_{\alpha 1,2}$, L_β, $L_{\gamma 1}$, $L_{\gamma 2,3\ (6)}$, $L_{\gamma 5}$, and/or $L_{\gamma 4}$). While the deduction of the total x-ray production cross section does provide a test of the theoretical predictions for L-shell ionization, such measurements are not as specific as the data derived from knowledge of the individual x-ray yields for the resolvable x-ray transitions. Given that the theories predict the probabilities of ionization for the individual subshells, the $L_{\alpha 1,2}$, $L_{\gamma 1}$, and $L_{\gamma 2,3,(6)}$ x-ray transitions provide direct information about the L_{III}, L_{II}, and

---

[74] N. Cue, V. Dutkiewicz, P. Sen, and H. Bakhru, *Phys. Rev. Lett.* **32**, 1155 (1974).
[75] F. Hopkins, R. Brenn, A. R. Whittemore, N. Cue, and V. Dutkiewicz, *Phys. Rev. A* **11**, 1482 (1975).

5.5 X-RAY PRODUCTION CROSS-SECTION MEASUREMENTS

L_I subshell ionization cross sections. This aspect of L-subshell ionization has provided a critical test of the theories in that the structure of the 2s wavefunction is reflected in both the theoretical and experimental cross sections, which are related to the $L_{\alpha 2,3,(6)}$ transition. An alternative method of reporting information related to L-subshell ionization has been the use of ratios of the yields of the resolved transitions. This method provides partial cancellation of detector efficiency effects and disagreements between theory and data. Two basic types of measurements have evolved. Some works reported "measured" ionization cross sections inverting the equations given in Section 5.2.1. Other works have reported the measured x-ray production cross sections for these same transitions with the x-ray production cross sections calculated using the results of Section 5.2.1. While it may be a matter of choice, the latter procedure has the utility of reporting the measured experimental data, while in the former procedure the measurements have been filtered through a set of radiative and nonradiative parameters. This procedure renders the data of little use for direct comparison to what one actually measures in the laboratory and hence should be avoided as a sole mechanism for the reporting of data. Comparisons to theory are certainly important but little, if anything, is lost by comparing calculated x-ray production cross sections directly to laboratory measurements.

Shafroth et al.[76] reported the measurement of the total L x-ray production cross section σ_{LX} for protons in the energy range 0.5–30 MeV incident on thin gold targets. Comparisons were made to the predictions of the PWBA, BEA(K), SCA, and PWBA with relativistic corrections by Choi.[77] As in all such cases there is not a clear differentiation between the various theoretical predictions for σ_{LX} and hence no definitive conclusions are available by comparing σ_{LX} data to the predictions. Measurements of $\sigma_{L\alpha}/\sigma_{L\beta}$ and $\sigma_{L\alpha}/\sigma_{L\gamma}$ did suggest that the BEA theory using scaled K-shell wavefunctions was not appropriate. Similar measurements of L x-ray production for 0.5–14-MeV protons on lead showed essentially the same characteristics as the measurements for protons on gold.

Datz et al.[78] reported the subshell ionization cross sections, σ_{LI}, σ_{LII}, and σ_{LIII} are shown in Fig. 15. The structure in the σ_{LI} cross section is related to the momentum distribution for the 2s state. Later works have used this feature of the ionization of the 2s state to distinguish between

---

[76] S. M. Shafroth, G. A. Bissinger, and A. M. Waltner, *Phys. Rev. A* **7**, 556 (1973).
[77] B.-H. Choi, *Phys. Rev. A* **4**, 1002 (1971).
[78] S. Datz, J. L. Duggan, L. C. Feldman, E. Laegsgaard, and J. U. Anderson, *Phys. Rev. A* **9**, 192 (1974).

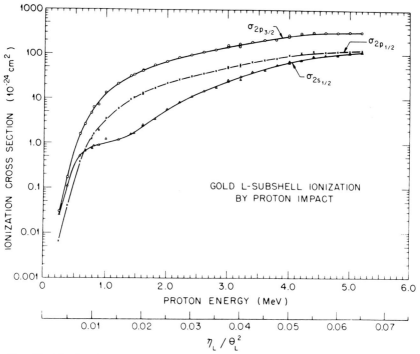

FIG. 15. Absolute ionization cross sections for the L_I, L_{II}, and L_{III} subshells of gold as a function of incident proton energy and the reduced units η/θ^2. Lines are guides to the eye only.

the various theoretical descriptions of inner-shell ionization by incident ions having atomic numbers up to $Z_1 = 8$.

Madison et al.[79] studied the L subshell ionization of lead and bismuth by incident protons in the energy range 0.5–4 MeV. Comparisons of their data for the L_I and L_{II} ionization cross sections were made to the predictions of the PWBA, BEA (using scaled K-shell wavefunctions), and the BEA (using L-shell wavefunctions). Similarly, Abrath and Gray[80] reported the x-ray production cross sections for the L_α, $L_{\gamma 1}$, and $L_{\gamma 2,3}$ transitions in samarium arising from 0.3–2.0 MeV proton bombardment of thin SmF_3 targets. Comparisons of their data were made to the predictions of the PWBA and CBEA theories. Both these cases showed that the various theories predicted the energy dependence and magnitudes of the

[79] D. H. Madison, A. B. Baskin, C. E. Busch, and S. M. Shafroth, *Phys. Rev. A* **9**, 675 (1974).
[80] F. Abrath and T. J. Gray, *Phys. Rev. A* **9**, 682 (1974).

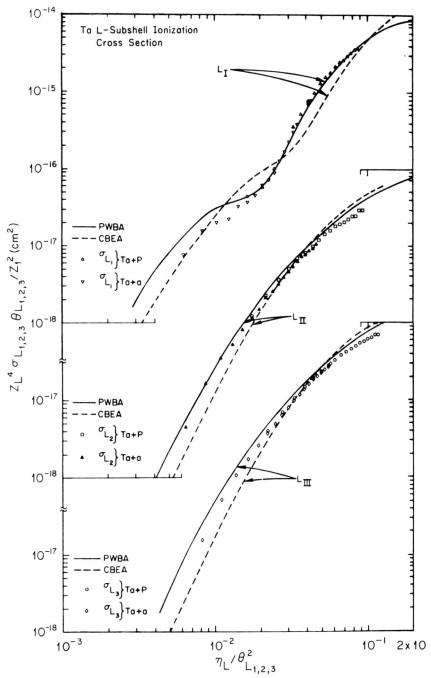

FIG. 16. L-subshell ionization cross sections of tantalum for hydrogen and helium bombardment in reduced units η/θ^2.

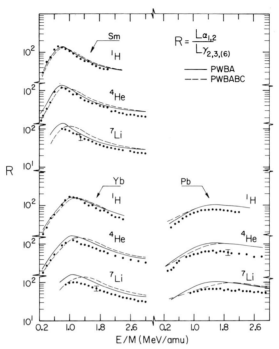

FIG. 17. Ratios of the $L_{\alpha 1,2}$ and $L_{\gamma 2,3(6)}$ x-ray production cross sections compared to PWBA and PWBABC predictions.

cross sections related to the L_{II} and L_{III} cross sections. However, they also clearly established that the classical theories [BEA(K), BEA(L), and CBEA] did not properly describe the L_I-subshell cross section. The structure associated with the 2s electron wavefunction as observed in these measurements was predicted by the PWBA. The classical theories failed to reproduce the observed structure. This was not surprising for the BEA(K) as no structure is predicted. However, in the case of the two BEA calculations using L-shell wavefunctions this result cast serious doubt on the validity of the classical approach to calculating inner-shell ionization cross sections using velocity distributions for screened hydrogenic wavefunctions. The results of Chang et al.[81] for the L-subshell cross sections of tantalum by 1.0–5.5-MeV ^{1}H and 1.0–11.0-MeV ^{4}He ions are shown in Fig. 16. These measurements are typical of these types of data in comparison to the BEA and PWBA theories.

As the incident ion energy is decreased, the measured cross sections begin to fall below the theoretical predictions. This feature of the L-shell data for ionization is similar to trends discussed for K-shell ionization.

[81] C. N. Chang, J. F. Morgan, and S. L. Blatt, *Phys. Rev. A* **11**, 607 (1975).

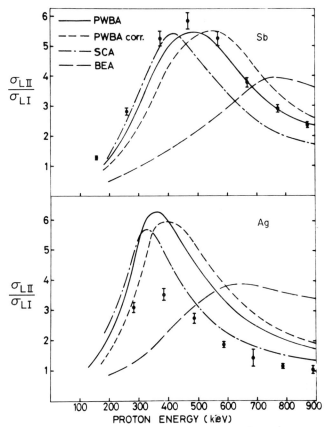

FIG. 18. Ratios of the σ_{LII} and σ_{LI} ionization cross sections for protons on antimony and silver.

Attempts to account for this discrepancy between observation and calculation led to the introduction of binding energy and Coulomb deflection corrections incorporated in the PWBA calculations based upon the PSS approach of Brandt and Lapicki.[25]

Measurements of the x-ray production cross sections for the $L_{\alpha 1,2}$, $L_{\gamma 1}$, and $L_{\gamma 2,3,(6)}$ transitions provide a distinct test of the PWBA and PWBABC predictions. Gray et al.[82] measured these x-ray production cross sections for ^{1}H, ^{4}He, and ^{7}Li ions on thin targets of samarium, ytterbium, and lead. The ratio $R = \sigma L_{\alpha 1,2}/\sigma L_{\gamma 2,3,(6)}$ shows a distinct peak, which reflects the structure in the 2s ionization cross section of the L_I subshell. The results of Gray et al.[82] for R are given in Fig. 17. Their data are compared

[82] T. J. Gray, G. M. Light, R. K. Gardner, and F. D. McDaniel, *Phys. Rev. A* **12**, 2393 (1975).

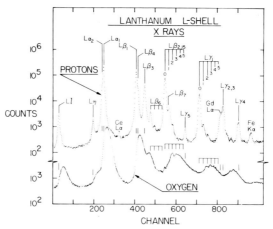

FIG. 19. Curved crystal high resolution x-ray energy spectra for 3-MeV protons and 25-MeV oxygen ions on a thick lanthanum target. The vertical lines on $L_{\beta 2,15}$ mark the calculated energies of the major satellite lines as predicted from Hartree–Fock calculations.

to the PWBA and PWABC predictions. Similar measurements for the ratio $\sigma_{L_{II}}/\sigma_{L_I}$ from Benka et al.[83] are given in Fig. 18 for protons on thin targets of antimony and silver. Both of these works show that the PWBA calculations predict the general properties of the ratios. Further, the work of Gray et al.[82] shows that the PWBA predicts the peak position in the values of R for increasing projectile atomic number. The peak in R occurs at a fixed velocity. The PWBABC calculations shift the peak position to higher bombarding energies for heavier projectiles because of the increase in the binding energy correction with increasing Z_1. This is not in agreement with the measurements.

In terms of predicting the ionization cross section for the L_{III} subshell, the PWBABC calculations improve the agreement between theory and experiment. However, the calculated PWBABC L_{II} and L_I cross sections are decreased in magnitude to the extent that they fall below the data. Furthermore, the shift in the structure in the L_I cross section is not supported by the data.

Measurements of L-shell ionization by ^{16}O ions have been reported by Bissinger et al.[72] for thin silver and gold targets and Pepper et al.[84] for thin targets of Ce, Pr, Sm, Eu, Dy, and Ho. Work in this area has been lacking because of several factors: (1) the loss of spectral resolution be-

[83] O. Benka, M. Geretschläger, A. Korpf, H. Paul, and D. Semrad, J. Phys. B **9**, 779 (1976).

[84] G. H. Pepper, R. D. Lear, T. J. Gray, R. P. Chaturvedi, and C. F. Moore, Phys. Rev. A **12**, 1237 (1975).

5.5 X-RAY PRODUCTION CROSS-SECTION MEASUREMENTS

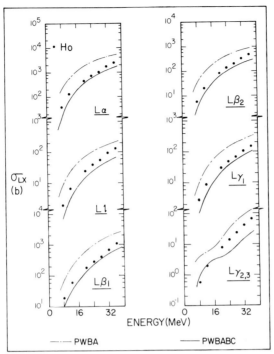

FIG. 20. L x-ray production cross sections for ^{16}O ions incident upon thin holmium targets. Neutral atomic radiative and nonradiative parameters are used to compare theory to experiment.

cause of peak broadening and energy shifts associated with multiple ionization of the inner shells of the target, and (2) the questions of the effects of multiple ionization of the radiative and nonradiative parameters associated with the conversion of ionization cross sections to x-ray production cross sections and vice versa. Pepper et al.[84] reported high-resolution x-ray spectra for ionization of the lanthanum L shell by 3-MeV ^{1}H and 25-MeV ^{16}O ions as shown in Fig. 19. The results of multiple ionization by ^{16}O are in evidence through the general broadening and shifting of the various L-shell x-ray transitions. Having the high-resolution data allows for the construction of a systematic procedure for fitting the lower resolution Si(Li) data in order to extract the intensity information needed for the cross section measurements. By assuming that neutral atom radiative and nonradiative parameters are a first approximation to the values required for the measured x-ray production cross sections, the results for ^{16}O ions on holmium are as shown in Fig. 20. With these approximations the PWBABC gives an improved description of the measured x-ray production cross sections in comparison to the PWBA predictions.

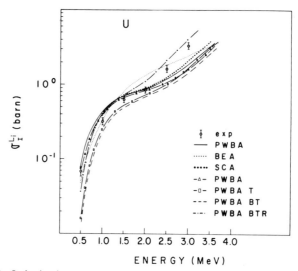

FIG. 21. The L_I-ionization cross section for protons on uranium includes analysis for relativistic effects. The symbol T in PWBAT is equivalent to the Coulomb deflection correction (C) in PWBAC.

Work by Leite et al.[85] illustrates the general state of studies of L-shell ionization for the light ions on heavier elements. The influence of each correction to the PWBA including a relativistic correction for the bound electron in the target along with the CBEA and SCA theories is given in Fig. 21 for 0.5–3.5-MeV protons on a thin target of uranium.

In summary, the ionization of the L shell by ions up to ^{16}O at high velocities is in general agreement with the predictions of the PWBABC. Questions involving the shift in the cross sections to higher projectile energy correction are unanswered. Further, the $\sigma_{\text{LIII}}/\sigma_{\text{LII}}$ ratio for ^{4}He ions on gold (e.g., Ref. 73) is not understood, i.e., the decrease in this ratio with decreasing projectile energy is not predicted by present calculations. Efforts to relate such behavior to multiple ionization or outer-shell electron effects[86,87] have not resolved the questions. Prospects for high-velocity heavy-ion studies are not promising even with high-resolution spectrometry because of the many components associated with multiple ionization effects. Target thickness effects have yet to be fully investigated but initial work by Schiebel et al.[88] for ^{19}F ions on silver shows that

---

[85] C. V. Barros Leite, N. V. de Castro Faria, and A. G. de Pinho, *Phys. Rev. A* **15**, 943 (1977).
[86] C. N. Chang, J. F. Morgan, and S. L. Blatt, *Phys. Lett. A* **49**, 365 (1974).
[87] T. K. Li, D. L. Clark, and G. W. Greenlers, *Phys. Rev. Lett.* **37**, 1209 (1976).
[88] U. Schiebel, T. J. Gray, R. K. Gardner, and P. Richard, *J. Phys. B* **10**, 2189 (1977).

5.5 X-RAY PRODUCTION CROSS-SECTION MEASUREMENTS

consideration of these effects is minimal when the incident ion species does not contain a K-shell vacancy initially.

5.5.4. M-Shell X-Ray Measurements ($Z_1 \ll Z_2$)

Work on ionization of the M shell by light ions has been very restricted because of spectral resolution considerations and the lack of sufficient information on the radiative and nonradiative parameters required for comparisons between measurements and theory. The measurements reported to date thus offer qualitative tests of total M-shell x-ray production theories on a very limited basis. Where comparisons have been made to theory, average values of $\bar{\omega}_M$ have been employed to get an estimate of σ_{MX}, i.e., $\sigma_{MX} = \bar{\omega}_M \sigma_{MI}$. Thornton et al.[89] report the measurement of σ_{XM} for 1–5-MeV ^{4}He ions on thin targets of terbium, gold, and bismuth. They observe a value of $\sigma_{MX} \sim 10^{-21}$ cm^2 for bismuth at an incident energy of 3 MeV. Ishii et al.[90] measured σ_{MX} for 1–4.5-MeV ^{1}H and 3–9 MeV-^{3}He ions on thin targets of gold, bismuth, and uranium. Their results for protons on gold are shown in Fig. 22. Comparisons to the

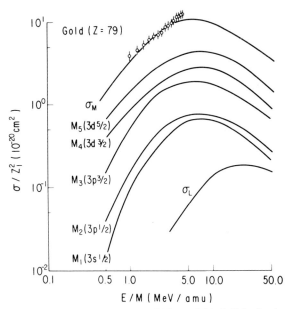

FIG. 22. Comparison of the experimental gold L- and M-shell ionization cross sections with PWBA prediction.

[89] S. T. Thornton, R. H. McKnight, and R. R. Karlowicz, Phys. Rev. A **10**, 219 (1974).
[90] K. Ishii, S. Morita, H. Tawara, H. Kaji, and T. Schiokawa, Phys. Rev. A **11**, 119 (1975).

PWBA calculations of Choi[91] show reasonable agreement with their results considering the uncertainty in $\bar{\omega}_M$. Busch et al.[92] reported measured values of σ_{MX} for 0.5–14-MeV protons on thin lead targets. Comparisons to the predictions of scaled BEA calculations showed the theory to have a systematic 20% deviation from the data, with the theory being lower than the measurements. Further, the peak in the predicted cross section occurs at an incident energy of 6 MeV while the data showed a maximum of 4.4×10^{-22} cm² at 8 MeV. Considering the approximations contained in the calculations, the agreement between theory and experiment is acceptable.

It would be appropriate to pursue a series of measurements for M-subshell ionization similar to those reported for the L subshells. The elucidation of the structures in the subshell cross sections for the 3s and 3p levels for comparison to the PWBA predictions would allow further quantitative tests of the use of screened hydrogenic wavefunctions as a method of calculating M-shell ionization processes. However, high-resolution techniques employing crystal spectrometers are required from a resolution point of view, and the problems with efficiency calibration and crystal reflectivities hinder the speed with which data can be accumulated. The study of M-shell ionization is further dependent upon adequate information about the necessary radiative and nonradiative processes linking theory with measured quantities. The elucidation of M-shell ionization processes awaits further work, both theoretical and experimental in character.

5.5.5. Target K-Shell Cross Sections for Heavier Ions

5.5.5.1. Introduction. Further results for a variety of heavy ions, typically over the energy range of a few MeV/amu, have been reported for thin solid targets. The target thicknesses employed range from ~10 to ~100 μg/cm² in most cases. Data from Knaf et al.[93] for 7–24-MeV ¹⁶O ions on targets from sulfur to bromine are shown in Fig. 23. Similarly, data from sulfur ions on thin targets[94] are shown in Fig. 24. Both of these typical cases show similar characteristics, in that for $Z_1/Z_2 \gtrsim 0.3$ the measured cross sections begin to deviate from the PWBABC calculations with the data becoming larger than the theory as $\eta_K/(\epsilon \theta_K)^2$ increases. Re-

---

[91] B.-H. Choi, Phys. Rev. A **7**, 2056 (1973).
[92] C. E. Busch, A. B. Baskin, P. H. Nettles, and S. M. Shafroth, Phys. Rev. A **7**, 1601 (1973).
[93] B. Knaf, G. Presser, and J. Stähler, Z. Phys. A **282**, 25 (1977).
[94] I. Tserruya, R. Schulé, H. Schmidt-Böcking, and K. Bethge, Z. Phys. A **277**, 233 (1976).

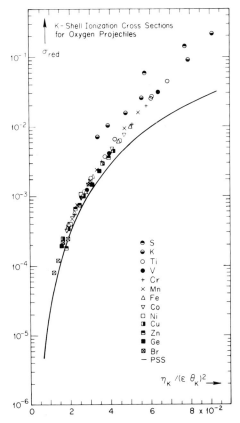

FIG. 23. Experimental K-shell ionization cross sections measured for ^{16}O ions incident upon thin (10–100 μg/cm^2) targets of elements from sulfur to bromine. The PSS calculations are referred to as PWBABC by others.

cent results by Laubert and Losonsky[95] for ^{16}O ions on aluminum over the energy range 0.15–91 MeV show that the inclusion of the polarization effect in the PWBABC can account for the observed increase in the target x-ray production cross section as shown in Fig. 25.

It would seem that the representation of the higher energy range of K-shell ionization by heavy ions can be fairly well understood in terms of the PSS theory with its corrections for binding energy, Coulomb deflection, and polarization effects. This is deceiving, however, because a major complexity in the measurements and interpretations of heavy-ion-induced x-ray production for thin solid targets had been previously

[95] R. Laubert and W. Losonsky, *Phys. Rev. A* **14**, 2043 (1976).

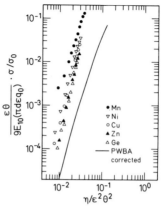

FIG. 24. Universal plot of the ionization cross sections for sulfur ions on thin targets of elements from manganese to germanium. The solid curve is the PSS (or PWBABC) prediction (see Ref. 94). The "nonuniversality" observed here is due to target thickness effects discussed in Section 5.5.5.4.

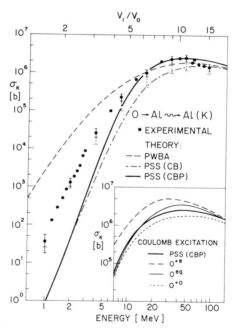

FIG. 25. Ionization cross section for ^{16}O ions incident on a thin aluminum target ($\rho_X \simeq 20$ μg/cm²) as a function of E_1 and V_1/V_0 ($V_0 = 1$ a.u.). Comparison between theory and experiment involves the use of $\omega_K = 1.12\omega_{K0} = 0.0426$ as the fluorescence yield. Insert shows effects of electron capture for O^{8+} ions; equilibrium distributions of projectile charge states, O^{eq}; and for a neutral projectile, O^{+0}.

overlooked. The measured target K x-ray yields for many of the data reported in the literature are in error by factors of 2 or more in many cases because the reported x-ray yields are strong functions of the target thickness. This fact was not included in most works reported prior to 1976 or in some works published in 1977.

5.5.5.2. Dynamic Screening. Brandt et al.[96] suggested that a heavy ion moving in a solid would have an effective charge that depended upon the plasma frequency of the medium in which the ion was moving. A basic property of all unperturbed direct Coulomb ionization theories is that the ionization cross section is proportional to Z_1^2. Hence if a heavy ion had an effective charge different from Z_1, then variations in the target x-ray yield with target thickness, target material, etc., might be expected. Brandt and co-workers reported results for oxygen ions incident on aluminum targets of varying thickness. They measured the aluminum K x-ray yield at the different thicknesses. Their data normalized to the yield at $D = \infty$ are shown in Fig. 26. They presented the dynamic screening model in which the charge on the ion was a function of the target thickness. Hence the idea that the charge state dominated the ionization process for heavy ions incident upon solid targets was advanced. It is noted that the K-shell x-ray yield per steradian in their model is

$$Y(E_1) \simeq \frac{\rho}{4\pi} \frac{\sigma_{X(Z_1,E_1)}}{Z_1^2} \int_0^D q_1^2(n, X, E_1) \, dX, \quad (5.5.1)$$

where $q_1(X) = q_1(X) \exp(-\lambda X/v_1) + q_1(\infty)[1 - \exp(-\lambda X/v_1)]$. The quantities Z_1 and D are the projectile atomic number and target thickness, respectively. The parameter λ is the dynamic screening rate for the ion in the solid and is associated with the plasma frequency for the material, a measure of the electron response time to a perturbation such as expected by the passage of a heavy ion. The velocity of the ion is given by v_1. The model as proposed gave reasonable agreement with the data for oxygen ions on thin aluminum targets. It is noted that the model is based upon

$$\sigma_I \propto \frac{1}{D} \int_0^D q_1^2(X) \, dX. \quad (5.5.2)$$

With the form of $q_1(X)$ as given, taking the limit $D \to 0$ gives

$$\sigma_I \propto q_1^2(0), \quad (5.5.3)$$

i.e., the ionization cross section is proportional to the square of the inci-

---

[96] W. Brandt, R. Laubert, M. Mourino, and A. Schwarzschild, *Phys. Rev. Lett.* **30**, 358 (1973).

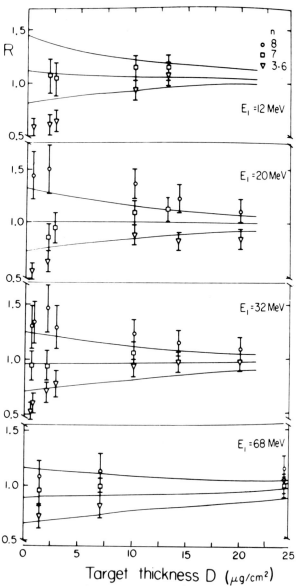

FIG. 26. Variations of the aluminum K x-ray yield for ^{16}O ions expressed in terms of $R(n, D, E_1)$ as a function of target thickness 0. The ratio $R(n, D, E_1) = (N_X/N_p)_D/(N_X/N_p)_\infty$, where N_x is the x-ray yield and N_p the scattered ion yield. The curves are calculations using the dynamic screening model with $\lambda = 5.5 \times 10^{-14}$ sec^{-1}. The upper curve for each energy is for $n = q_1(x = 0) = 8+$, the intermediate curve for $n = 7$, and the lower curve for $n = 6$.

dent charge state in the limit of vanishing target thickness. It is predicted then that for any two incident charge states of a heavy ion the ratio of the ionization cross sections will be

$$\sigma_{\rm I}(q)/\sigma_{\rm I}(q') = [q(0)/q'(0)]^2 \qquad (5.5.4)$$

in the limit of a vanishingly thin target. It will be shown that this prediction fails badly for heavier ions and hence the applicability of the dynamic screening model as a general description of target ionization for heavy ions moving in solids is inappropriate at high velocity ($E/m \geqslant 1$ MeV/amu). At lower velocities the dynamic screening model may have regions of validity but this is an area in need of appropriate study.

5.5.5.3. Target Thickness Effects on Target X-Ray Production. The basic theories of inner-shell ionization depend upon the target and projectile in specific ways, and the perturbations such as the binding energy correction are relatively strong functions of Z_1/Z_2. We have seen a definite interest in extending the studies of inner-shell ionization to include heavy ions incident upon solid targets. There has been a basic question of what type of interaction picture, i.e., MO or direct Coulomb ionization, best describes the observed ionization process. A study of the tabulated data resources at the end of this part shows many experiments reported wherein the goal was to measure the target x-ray production cross sections for heavy ions in the MeV/amu energy range incident upon single-thickness solid targets. More extensive studies[97] such as aluminum on aluminum and nickel on nickel, where K-shell ionization processes, investigated over a wide energy range of 0.5–40-MeV (Al–Al) and 10–63-MeV (Ni–Ni) have been reported as an attempt to span the regions from predominantly MO excitation to predominantly direct Coulomb excitation. Another work of a similar nature is the recent report[95] for ^{16}O ions on aluminum. Other types of studies have evolved wherein the systematic variations in target K x-ray production have been reported as a function of E_1 and Z_2 for a particular species of incident projectile. Work by McDaniel and Duggan[98] for incident heavy ions fluorine, silicon, and chlorine on a variety of targets with $Z_2 \geqslant 22$ and results by Tserruya et al.[94] for sulfur ions on elements from manganese to germanium are representative of these types of studies. All such measurements were reported for single target thicknesses and the analysis was performed as though the quantity measured was the target K x-ray production cross section independent of the physical state or thickness of the target.

[97] R. Laubert, H. Haselton, J. R. Mowat, R. S. Peterson, and I. A. Sellin, *Phys. Rev. A* **11**, 135 (1975).

[98] F. D. McDaniel and J. L. Duggan, in "Beam Foil Spectroscopy," (Ivan A. Sellin and David J. Pegg, eds.) p. 519. Plenum Press, New York, 1976.

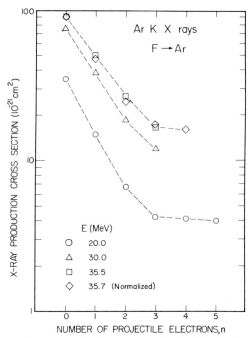

FIG. 27. Experimental x-ray production cross section for fluorine ions incident upon an argon gas target under single collision conditions as a function of the incident ion-charge state.

The study and understanding of the process of inner-shell ionization of an atom in a solid environment by incident high velocities is a complex undertaking. Information on charge-state effects observed in x-ray production for a colliding heavy-ion system under single-collision conditions suggested in the early 1970s that the situation in a dense target medium should somehow depend upon the charge state of the ion incident upon the target. Macdonald et al.[99] studied the dependence of argon K x-ray production resulting from incident fluorine ions in charge states 5+ to 9+. Their results are shown in Fig. 27. These measurements were made under single-collision conditions in a thin argon gas target. The data exhibit a dominant characteristic, namely, there is a large increase in the target K x-ray production as the incident ion charge state approaches that of the bare ion. Halpern and Law[28] analyzed the data of Macdonald et al.[99] by including a charge exchange contribution in addition to the direct Coulomb ionization of the argon K shell. Thus the enhancements ob-

---

[99] J. R. Macdonald, L. Winters, M. D. Brown, T. Chiao, and L. D. Ellsworth, *Phys. Rev. Lett.* **29**, 1291 (1972).

served in target x-ray production were associated with the charge state of the incident ion and hence they were called "charge-state effects." In the case of solid targets, the term "solid-state effects" was used to describe differences in x-ray production between collisions in solids and gases of the same atomic species (e.g., 86-MeV argon ions on solid silicon and gaseous SiH_4 targets[100]). For heavy ions incident on solids it will be seen that both of these previous terminologies may be replaced by the description of what one observes in terms of the inner-shell vacancies. Hence it is suggested that "vacancy configuration effect" is more appropriate as a description of the physical processes evolved for heavy ions moving in solids.

Betz et al.[101] reported the observation of large target thickness variations in target x-ray production for 110-MeV sulfur ions incident upon copper targets of 50–250 $\mu g/cm^2$ thickness. A factor of 4 increase in the target x-ray yield per incident ion in going from 50 to 250 $\mu g/cm^2$ was observed. They suggested either charge exchange involving K vacancies in the projectile or dynamic screening as a reason for their observations. However, no quantitative analyses were pursued and hence the question was not answered.

5.5.5.4. The Two- and Three-Component Models for Target X-Ray Production. The differential equations that describe the change in the K-shell vacancy fractions are related to the vacancy production and quenching processes that establish those fractions. Using the earlier work of Allison,[102] which was developed to describe the charge states of hydrogen and helium moving in gases, one may picture the relationships between the vacancies for the two-component and three-component systems as shown in Fig. 28. The cross sections σ_{ij} are descriptive of the probabilities for the vacancy state to change from i to j. For $i > j$, quenching, i.e., filling, of the vacancy occurs, and for $j > i$, the process of vacancy production occurs.

The differential equation for the two-component system is

$$dY_1/dX = \sigma_{01}Y_0 - \sigma_{10}Y_1, \quad (5.5.5)$$

where X is measured in appropriate units, e.g., if σ_{01} is in $cm^2/atom$, then X is in $atoms/cm^2$ and Y_1 and Y_0 are the fractions with 0 or 1 K vacancies, respectively. This differential equation is subject to the boundary conditions on the incident ion species, i.e., does the incident ion contain a K-shell vacancy? Furthermore, the conservation of probability re-

---

[100] S. Datz, B. R. Appleton, J. R. Mowat, R. Laubert, R. S. Peterson, R. S. Thoe, and I. A. Sellin, *Phys. Rev. Lett.* **33**, 733 (1974).
[101] H. D. Betz, F. Bell, and H. Panke, *J. Phys. B* **7**, L418 (1974).
[102] S. K. Allison, *Rev. Mod. Phys.* **30**, 1137 (1958).

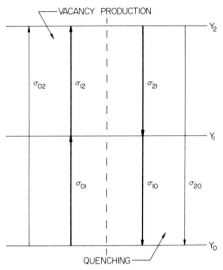

FIG. 28. Qualitative diagram for two- and three-component model. For two-component model, neglect the Y_2 state and its associated processes.

quires that $Y_0(X) = 1 - Y_1(X)$. The general solution of the differential equation for the two-component system is

$$Y_1(X) = \frac{\sigma_{01}}{\sigma_{10} + \sigma_{01}} \{1 - \exp[-(\sigma_{10} + \sigma_{01})X]\} + A \exp[-(\sigma_{10} + \sigma_{01})X]. \quad (5.5.6)$$

The quantity A is the boundary condition, with $A = 1$ for $Y_1 = 1$ incident and $A = 0$ for $Y_0 = 1$ incident. An alternative notation found in the literature is

$$Y_1(X) = \frac{\sigma_V}{\sigma} (1 - e^{-\sigma X}) + A e^{-\sigma X}, \quad (5.5.7)$$

where $\sigma_V \equiv \sigma_{01}$, $\sigma = \sigma_V + \sigma_Q$, where $\sigma_Q \equiv \sigma_{10}$. It is stressed that σ_V and σ_Q are the atomic cross sections for the production and quenching of vacancies in the *projectile* as it moves through the target.

In the case of the three-component model the description becomes more complex because of the coupled differential equations describing the K-shell states with 0, 1, or 2 vacancies. Using Fig. 28 the differential equations can be constructed as

$$dY_2(X)/dX = \sigma_{12} Y_1(X) + \sigma_{02} Y_0(X) - (\sigma_{21} + \sigma_{20}) Y_2(X), \quad (5.5.8)$$

$$dY_1(X)/dX = \sigma_{01} Y_0(X) + \sigma_{21} Y_2(X) - (\sigma_{12} + \sigma_{10}) Y_1(X), \quad (5.5.9)$$

with
$$Y_1(X) + Y_2(X) + Y_3(X) = 1. \quad (5.5.10)$$

The solutions for the fraction $Y_i(X)$ are of the form

$$Y_i(X) = Y_i(\infty) + [P(Z, i)e^{qX} + N(Z, i)e^{-qX}]\exp[-\tfrac{1}{2}f(\sigma ij)X], \quad (5.5.11)$$

where $q = \tfrac{1}{2}[(g + a)^2 + 4bf]^{1/2}$ and $f(\sigma ij) = -(a + g)$. The quantities a, b, f, and g are defined as

$$a \equiv -(\sigma_{10} + \sigma_{12} + \sigma_{21}), \quad (5.5.12)$$

$$b \equiv (\sigma_{01} - \sigma_{21}), \quad (5.5.13)$$

$$f \equiv (\sigma_{10} - \sigma_{20}), \quad (5.5.14)$$

$$g \equiv -(\sigma_{01} + \sigma_{02} + \sigma_{20}). \quad (5.5.15)$$

The equilibration values $F_i(\infty)$ are given as

$$F_0(\infty) = (f\sigma_{21} - a\sigma_{20})/D, \quad (5.5.16)$$

$$F_1(\infty) = (b\sigma_{20} - g\sigma_{21})/D, \quad (5.5.17)$$

$$F_2(\infty) = [\sigma_{20}(a - b) + g(a + \sigma_{21}) - f(b + \sigma_{21})]/D, \quad (5.5.18)$$

where

$$D \equiv ag - bf = \sigma_{12}(\sigma_{01} + \sigma_{02} + \sigma_{20}) + \sigma_{10}(\sigma_{20} + \sigma_{21} + \sigma_{02})$$
$$+ (\sigma_{01} + \sigma_{02})\sigma_{21} + \sigma_{01}\sigma_{20}. \quad (5.5.19)$$

The boundary conditions represented by $P(Z, i)$ and $N(Z, i)$ are given in Table V.

The use of the two-component description of the K-shell vacancy populations as applied to heavy ions moving in thin solids was first advanced

TABLE V. Values of $P(Z, i)$ and $N(Z, i)$ for Three-Component System[a]

$P(0, 1) = \dfrac{1}{2q}[bF_2(\infty) + (b + s - q)F_1(\infty)]$	$N(0, 1) = -\dfrac{1}{2q}[bF_2(\infty) + (b + s + q)F_1(\infty)]$
$P(1, 1) = -\dfrac{1}{2q}[(s - q)(1 - F_1(\infty)) + bF_0(\infty)]$	$N(1, 1) = \dfrac{1}{2q}[(s + q)(1 - F_1(\infty)) + bF_0(\infty)]$
$P(2, 1) = \dfrac{1}{2q}[F_1(\infty)(s - q) - bF_0(\infty)]$	$N(2, 1) = -\dfrac{1}{2q}[F_1(\infty)(s + q) - bF_0(\infty)]$
$P(Z, 0) = P(Z, 1)(s + q)/b$	$N(Z, 0) = N(Z, 1)(s - q)/b$
$P(Z, 2) = -P(Z, 1)(b + s + q)/b$	$N(Z, 2) = -N(Z, 1)(b + s - q)/b$
$Z = 0, 1, 2$	

[a] Z is the incident K-vacancy state and i the K-vacancy state in the solid.

by Betz et al.[103] in their study of a technique for determing the lifetimes of inner-vacancy states for sulfur ions moving in thin foils of beryllium, copper, aluminum, copper, and germanium. Their technique is related to beam-foil measurements and is discussed elsewhere in this volume. It suffices to note that the K-shell quenching cross section is composed of two contributions, i.e.,

$$\sigma_Q = \sigma_c + (1/nv_1\tau), \qquad (5.5.20)$$

where σ_c is the direct capture cross section of an electron, which leads to the quenching of the vacancy, and $1/nv_1\tau$ is the contribution arising from the decay of the vacancy characterized by a lifetime τ of the state in question. Cocke et al.[104] have used a similar analysis in the study of helium-like $(1s2p)^3P_2$ and lithiumlike $(1s\ 2s\ 2p)^4P_{5/2}$ states for chlorine, silicon, and sulfur ions moving in thin foils of copper, aluminum, titanium, and nickel. Such beam-foil measurements depend upon the existence of strong excitation of the projectile K shell.

Hopkins[105] reported the results of a study involving the production of copper x rays from a thin copper probe layer placed upon carbon foils of varying thickness. He employed chlorine beams at energies of 70 and 140 MeV. The experimental technique used in the probe-layer technique is to compare the copper x-ray yield for differing target-beam geometries. The yield of copper K x rays is measured for the beam incident on the copper probe layer (σ_X^E) and then the sandwich foil assembly is rotated such that the incident beam passes through the carbon backing material before striking the copper probe layer (σ_X^S). The ratio of the copper K x-ray yields for the (σ_X^S) and (σ_X^F) geometries gives a measure of $Y_1(X)$ for the sulfur ions moving in the carbon foil. The copper probe layer is then the detector of the residual K-vacancy populations resident in the beam after it has traversed a given thickness ρX of carbon foil. In order for this experimental technique to be feasible, one essential condition has to be met. The probe layer material has to satisfy the condition that sufficiently strong coupling exists between the K shell of the probe material and the residual K vacancy of the projectile to allow for the detection of the enhanced x-ray production in the probe material that may be associated with the charge transfer channel for K shell (target) to bound-state (projectile) electron transfer. The results of Hopkins' initial work using the probe-layer technique are shown in Fig. 29. A strong thickness

---

[103] H. D. Betz, F. Bell, H. Panke, G. Kalkoffen, M. Solz, and D. Evers, *Phys. Rev. Lett.* **33**, 807 (1974).

[104] C. L. Cocke, S. L. Varghese, and B. Curnutte, *Phys. Rev. A* **15**, 874 (1977).

[105] F. Hopkins, *Phys. Rev. Lett.* **35**, 270 (1975).

5.5 X-RAY PRODUCTION CROSS-SECTION MEASUREMENTS

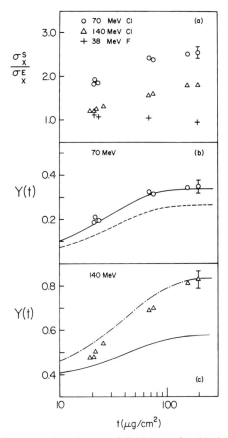

FIG. 29. (a) σ_X^S/σ_X^E as a function of carbon-foil thickness for chlorine and fluorine beams. (b) Experimental and theoretical values for chlorine K-vacancy fraction at 70 MeV. (c) Experimental and theoretical values for chlorine K-vacancy fraction at 140 MeV.

dependence is observed for variations in the carbon foil thickness. The vacancy fractions for the sulfur ions are shown in Fig. 29 together with calculations for $Y_1(X)$ based upon values of σ_v and σ.

Hopkins et al.[106] report additional measurements using the probe-layer technique. These results are for fluorine, sulfur, and chlorine ions on carbon foils with a 1-μg/cm² copper probe layer. The probe-layer technique provides an alternative means of determining the K-shell vacancy configuration for an ion moving in a material in contrast to the beam-foil spectroscopic technique, which views the projectile radiation as a mea-

---

[106] F. Hopkins, J. Sokolov, and A. Little, *Phys. Rev. A* **15**, 588 (1977).

sure of the projectile vacancy states. The probe-layer technique is a valuable experimental technique but it does not provide a direct measure of the target x-ray production rates.

In consideration of the facts that (1) the target species may be sensitive to the K-shell vacancy configuration of the projectile and (2) the projectile K-vacancy population is dependent upon the target thickness, it is appropriate to measure the radiation from the target under heavy-ion bombardment.

Groeneveld et al.[107] have studied both the target x-ray yield and projectile x-ray yield for 10-MeV Ne^{2+} ions on aluminum targets of varying thicknesses. They observe a target thickness dependence for the target x-ray yield that is described using a two-component model. Typical results for the aluminum K_α x-ray yield as a function of aluminum target thickness are shown in Fig. 30. Similar studies have been reported by

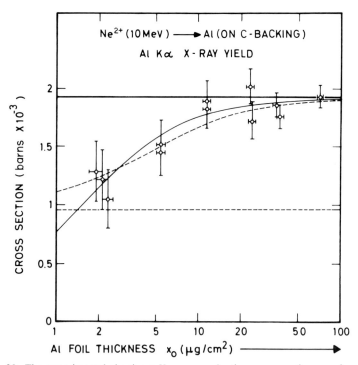

FIG. 30. The experimental aluminum K x-ray production cross section as a function of aluminum target thickness for incident 10-MeV Ne^{2+} ions. The solid curves are predictions of a two-component analysis given in Ref. 107.

[107] K. O. Groeneveld, B. Kolb, J. Schader, and K. D. Sevier, Z. Phys. A **277**, 13 (1976).

5.5 X-RAY PRODUCTION CROSS-SECTION MEASUREMENTS 263

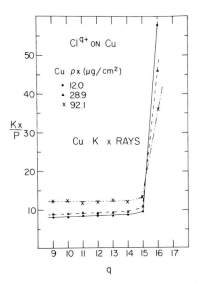

FIG. 31. The variation in the ratio of detected x rays per scattered ion for chlorine ions incident on thin copper targets as a function of the charge state of the incident ion.

Feldman et al.[108] for argon on aluminum from 0.4 to 2.0 MeV. Once again, a two-component model was utilized to analyze the data for target thickness effects.

Gray et al.[109] have reported the study of 60-MeV (1.71 MeV/amu) chlorine ions incident on copper targets over a range of copper target thicknesses. This was the first experimental study in which one-electron heavy ions were used as projectiles in order to further test the full ramifications of the two-component model as applied to the description of target x-ray production. The question of the effects of projectile L-shell electrons on the production of target K x rays was answered in this work. Shown in Fig. 31 is the copper K x-ray yield per scattered particle for three different target thicknesses and incident charge states of 9+ through 16+ for 60-MeV chlorine ions. For the thinnest target (12.1 $\mu g/cm^2$) the copper x-ray yield is essentially constant until the one-electron system (Cl^{16+}) is incident upon the target. For the incident ion with $Y_1(0) = 1$ there is an enhancement in the target x-ray yield of ~8. Hence the existence of a nonnegligible fraction of Y_1' in the copper target as a function of

---

[108] L. C. Feldman, P. J. Silverman, and R. J. Fortner, *Nucl. Instr. Meth.* **132**, 29 (1976).
[109] T. J. Gray, P. Richard, K. A. Jamison, J. M. Hall, and R. K. Gardner, *Phys. Rev. A* **14**, 1333 (1976).

the target thickness will result in a target thickness dependence of the copper K x-ray yield.

With this in mind a two-component model was used by Gray et al.[109] to describe the target K x-ray yield as a function of target thickness. It is assumed that the Y_1 fraction is

$$Y_1(X) = \frac{\sigma_v}{\sigma}(1 - e^{-\sigma X}) + Ae^{-\sigma X}, \quad (5.5.21)$$

as given previously by Allison.[102] The boundary conditions are $A = 1$ for Cl^{16+} ions incident and $A = 0$ for chlorine ions incident that do not contain a K-shell vacancy. The two-electron systems are not used in the study of target thickness effects because they contain a small percentage of $(1s2s^3S)$ metastables, which carry K-shell vacancies into the interaction. It is noted that Fig. 31 shows an ~10% increase in the target x-ray yield for Cl^{15+} ions, which is indicative of the small metastable fraction in the incident 15+ charge state.

The target x-ray production is then considered as represented by an averaged cross section $\overline{\sigma_{KX}}$. This quantity is really not a cross section but rather an integrated target x-ray yield that is averaged over the thickness of the target. In the two-component model, the fraction of the beam without K vacancies is associated with a target x-ray production cross section σ_{K0}, while the cross section for target x-ray production for the Y_1 fraction is given by σ_{K1}. Hence $\overline{\sigma_{KX}}$ is given by

$$\overline{\sigma_{KX}} = \frac{1}{T}\int_0^T (Y_0\sigma_{K0} + Y_1\sigma_{K1})\,dX, \quad (5.5.22)$$

where X is in the appropriate "thickness" units. By the use of $Y_0 = 1 - Y_1$, the expression for $Y_1(X)$, and $\sigma_{K1}(E_1) = \alpha(E_1)\sigma_{K0}(E_1)$, where $\alpha(E_1)$ is the enhancement factor in the target x-ray yield for a vanishingly thin target, the expression for $\overline{\sigma_{KX}}(X)$ is given as

$$\overline{\sigma_{KX}}(X) = \sigma_{K0}\left[1 + (\alpha - 1)\frac{\sigma_v}{\sigma} + \frac{\alpha - 1}{\sigma X}\left(\frac{\sigma_v}{\sigma}A\right)(1 - e^{-\sigma X})\right]. \quad (5.5.23)$$

The quantities σ_v and σ are the projectile cross sections as previously defined. There are several very important features of this result for the two-component model description of target x-ray production:

(1) The target x-ray production depends critically upon the K-vacancy state of the incident ion.

(2) For a finite thickness solid target the quantity measured is $\overline{\sigma_{KX}}$ at that thickness and it is *not* the cross section for target x-ray production unless $\alpha \to 1$.

(3) The equilibration values of the target x-ray yield are given by

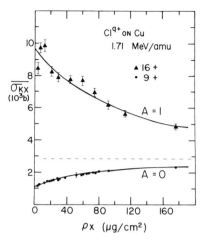

FIG. 32. The averaged target K x-ray yields in units of cross section as a function of the copper target thickness for incident Cl^{9+} ($A = 0$) and C^{16+} ($A = 1$) ions at a bombarding energy of 1.71 MeV/amu (60 MeV). The curves are predictions of the two-component model discussed in Section 5.5.5.4.

$\overline{\sigma_{KX}}(\infty) = \sigma_{K0}[1 + (\alpha - 1)\sigma_v/\sigma]$, which in general is what has been reported for the value of the target x-ray production cross section for measurements made on targets of the order of 60 to 100 $\mu g/cm^2$ by works such as McDaniel and Duggan.[98] This quantity is *not* the target x-ray production cross section and hence comparisons to the predictions of the theories such as the PWBA with or without corrections such as the binding energy correction are fortuitous.

Shown in Fig. 32 are the results of Gray et al.[109] for the measurement of $\overline{\sigma_{KX}}$ for 60-MeV chlorine ions (with and without an incident K-shell vacancy) on copper targets ranging in thickness up to 180 $\mu g/cm^2$. The curves shown in the figure are model calculations for $\overline{\sigma_{KX}}$ using the two-component model. The values of σ_v and σ obtained from the model are in good agreement with measurements of Hopkins[106] for σ_v and estimated from a scaled Brinkman–Kramers capture cross section using the results of Nikolaev.[30] The measured value of α (taken as $\alpha = \sigma_{K1}/\sigma_{K0}$, where σ_{K1} and σ_{K0} are determined for a vanishingly thin target) are compared to an estimate that assumes that $\alpha = (\sigma_{K0} + \pi R^2 \omega_K W)/\sigma_{K0}$. This assumption states that σ_{K1} is enhanced over σ_{K0} through the addition of an electron transfer channel that is present for the single K-vacancy system. The quantity W is taken as the Meyerhof[110] single-pass electron transfer

[110] W. E. Meyerhof, *Phys. Rev. Lett.* **31**, 1341 (1973).

probability given by

$$W = 1/(1 + e^{2|\chi|}), \qquad (5.5.24)$$

where

$$\chi = \frac{13.6}{v_1 \text{ (a.u.)}} \frac{I_2 - I_1}{I_2^{1/2} + I_1^{1/2}}.$$

The quantities I_1 and I_2 are the K-shell binding energies for the target and one-electron ion, respectively. The quantity R is taken as the radius at which the dynamic coupling peak elements as given by Taulbjerg et al.[111] In the case of 60-MeV chlorine on copper the measured value of α is 8.1. This is to be compared with the calculated value of 8.3. Recent results by McDaniel et al.[112] employ the charge exchange calculation of Lapicki and Losonsky[26] in order to obtain the cross sections associated with ion incident on the target with K-shell vacancies.

A comparison of the results for the copper K x-ray production for vanishingly thin targets gives

$$\sigma(16+)/\sigma(14+) = 8.1. \qquad (5.5.25)$$

This is to be compared to the predictions of the dynamic screening model,[93] where

$$\sigma(16+)/\sigma(14+) = 1.3. \qquad (5.5.26)$$

Further, the dynamic screening model predicts that each ratio $\sigma(16+)/\sigma(q')$, $q' = 14, 13, 12, \ldots$, is different. The results of Gray et al.[109] do not support such predictions. Furthermore, in work to be discussed in connection with data for aluminum, silicon, and sulfur ions on copper similar disagreements between the data for $\overline{\sigma_{KX}}$ for vanishingly thin targets and the predictions of the dynamic screening model persist. Hopkins et al.[106] have noted similar disagreements with the dynamic screening model.

The two-component model is applicable to those incident ion systems where the fraction of ions with two K-shell vacancies is small in comparison to Y_1. However, for lighter ions or experiments that employ bare incident projectiles, considerations of the three-component model in which vacancy fractions for 0, 1, and 2 K-shell vacancies may be important. Gardner et al.[113] have extended the ideas upon which the two-component

[111] K. Taulbjerg, J. Vaaben, and B. Fastrup, Phys. Rev. A **12**, 2325 (1975).

[112] F. D. McDaniel, J. L. Duggan, G. Barbas, P. D. Miller, and G. Lapicki, Phys. Rev. A **16**, 1375 (1977).

[113] R. K. Gardner, T. J. Gray, P. Richard, C. Schmiedekamp, K. A. Jamison, and J. M. Hall, Phys. Rev. A **15**, 2202 (1977).

5.5 X-RAY PRODUCTION CROSS-SECTION MEASUREMENTS

model for target x-ray production have been based. Their work is a natural extension of the development of the two-component model given by Gray et al.[109]

In the three-component model, the averaged target x-ray yield is given by

$$\overline{\sigma_{KX}} = \frac{1}{T} \int_0^T (Y_0 \sigma_{K0} + Y_1 \sigma_{K1} + Y_2 \sigma_{K2}) \, dX. \quad (5.5.27)$$

The factor $Y_2 \sigma_{K2}$ is added to account for the contributions to the target x-ray yield associated with the double K-vacancy state of the projectile moving in the target. Using the results of Allison[102] for the fractions $Y_i(X)$ as given for the three-component system, the expression for $\overline{\sigma_{KX}}$ is found to be

$$\overline{\sigma_{KX}}(T) = \sigma_{K0} \left\{ 1 + (\alpha - 1) \left[F_1(\infty) + \frac{P(Z,1)}{Tf(\sigma_t)} \{\exp[Tf(\sigma_t)] - 1\} \right. \right.$$
$$\left. - \frac{N(Z,1)}{Tf'(\sigma_t)} \{\exp[-Tf'(\sigma_t)] - 1\} \right] + (\beta - 1) \left[F_2(\infty) \right. \quad (5.5.28)$$
$$\left. \left. + \frac{P(Z,2)}{Tf(\sigma_t)} \{\exp[Tf(\sigma_t)] - 1\} - \frac{N(Z,2)}{Tf'(\sigma_t)} \{\exp[-Tf'(\sigma_t)] - 1\} \right] \right\},$$

where T is the target thickness in appropriate units. The quantities α and β are the enhancements in $\overline{\sigma_{KX}}$ for the one and two K-vacancy systems measured for vanishingly thin targets, i.e., $\alpha \equiv \sigma_{K1}/\sigma_{K0}$ and $\beta \equiv \sigma_{K2}/\sigma_{K0}$. The functions $f(\sigma_t)$ and $f'(\sigma_t)$ are defined as

$$f(\sigma_t) \equiv \{[(d - a)^2 + 4bc]^{1/2} + \sigma_T\}/2,$$
$$f'(\sigma_t) \equiv \{[(d - a)^2 + 4bc]^{1/2} - \sigma_T\}/2, \quad (5.5.29)$$

where $\sigma_T \equiv -(a + d)$ and the quantities a, b, d, and $c \equiv f$ are as defined previously. Further, $F_i(\infty)$ are as previously defined and the boundary conditions for Z as the incident vacancy state and i as the final vacancy state are given in Table V, p. 259.

The three-component model is much more complicated than the two-component model in that it involves six projectile cross sections, one target x-ray production cross section, and the two enhancement factors α and β. Of the nine parameters in the three-component model, three of them are relatively fixed by the target x-ray production data in the limit of a vanishingly thin target. These are σ_{K0}, α, and β, i.e., the parameters associated with the target. Shown in Fig. 33 are the general features of the model predictions for a typical system, e.g., aluminum on copper at 1.71 MeV/amu. The shapes of the curves in Fig. 33 are determined by the six projectile cross sections as are the equilibration values for $\overline{\sigma_{KX}}$. In an application of the three-component model to data for $\overline{\sigma_{KX}}$, where data

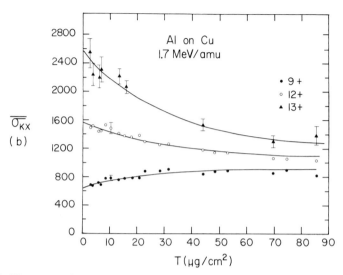

FIG. 33. The averaged target K x-ray yield in units of cross section for 1.71-MeV/amu aluminum ions incident upon thin copper targets as a function of target thickness. The aluminum ions were incident upon the targets with initial charge states, 9+, 12+, and 13+, i.e., $Y_0 = 1$, $Y_1 = 1$, and $Y_2 = 1$, respectively. The curves are calculations based upon the three-component model.

for the Z_1, $Z_1 - 1$, and $\leq Z_1 - 3$ charge states have been employed as projectiles, the procedure is to fix σ_{K0}, α, and β within bounds dictated by the measurement of σ_{K0}, σ_{K1}, and σ_{K2} for vanishingly thin targets. The data are taken over a target thickness range of sufficient breadth so as to clearly establish the thickness dependences of $\overline{\sigma_{KX}}$ for each incident ion species. The data for $\overline{\sigma_{KX}}$ is then fitted using an appropriate mathematical fitting routine such as the least-squares method employing the expression for $\overline{\sigma_{KX}}$. Values of the projectile parameters are varied in order to minimize χ^2. There is the question of what initial values of σ_{ij} should be employed in the calculations at the start of the fitting procedure. The results of Gardner et al.[113] suggest that the initial values of σ_{21} and σ_{10} can be taken as the Nikolaev OBK capture cross sections if scaled by a factor of 0.1. The double-quenching cross section σ_{20} should be set to $< \sigma_{10} \times 0.1$. The initial estimates of σ_{01} have been taken from the work of Winters et al.[114] for sulfur ions on krypton as being appropriate for sulfur ions on copper. The value of σ_{12} is taken initially as $\sigma_{12} \simeq \sigma_{01}/2$ and σ_{02} is taken as $\sigma_{02} < \sigma_{12}$.

[114] L. Winters, M. D. Brown, L. D. Ellsworth, T. Chiao, E. W. Pettus, and J. R. Macdonald, Phys. Rev. A 11, 174 (1975).

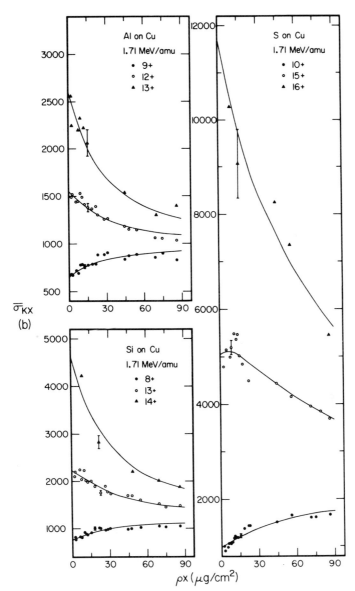

FIG. 34. The averaged target K x-ray yield for thin copper targets bombarded by 1.71-MeV/amu aluminum, silicon, and sulfur ions as functions of target thickness and incident ion K-vacancy configuration. The solid lines are calculations based upon the three-component model.

Shown in Fig. 34 are the results of Gardner *et al.*[113] for aluminum, silicon, and sulfur ions incident on thin solid targets of copper at 1.71 MeV/amu. In each case the values of $\overline{\sigma_{KX}}$ were measured for projectiles with 0, 1, and 2 K-shell vacancies. The curves given in Fig. 34 are the results of fitting the data for $\overline{\sigma_{KX}}$ with the three-component model. The val-

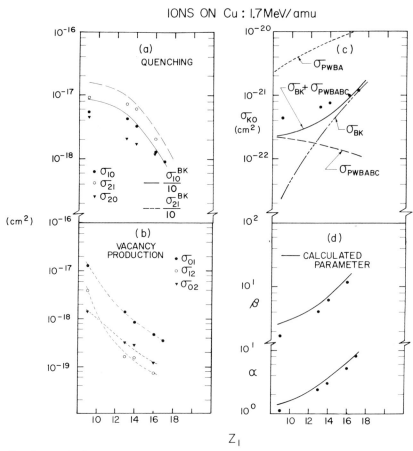

FIG. 35. Projectile and target parameters obtained from the three-component model calculations for the fits to the experimental data of Fig. 34. (a) Projectile K-quenching cross sections are compared to scaled Brinkmann–Kramers calculations for charge exchange (i.e., electron transfer) to the projectile K shell. Radiative decay of projectile K vacancies is small in comparison to electron capture in these cases. (b) Projectile K-vacancy cross sections. The dashed lines are guides to the eye only. (c) Values of σ_{K0} measured in the limit of vanishing target thickness, i.e., $\overline{\sigma_{KX}}(x \to 0)$ for incident ions without a K-shell vacancy. (d) Calculated enhancement factors α and β compared to the observed experimental values. (Ions incident on Cu at 1.71 MeV/amu.)

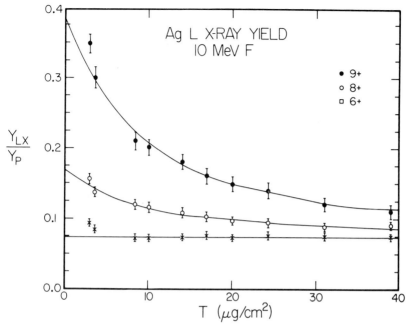

FIG. 36. The ratio of detected L-shell x rays and scattered ions for silver bombarded by fluorine ions as functions of the silver target thickness and incident fluorine charge state. For incident ions with $q \leq 6+$, there is no observed target thickness dependence. The values of Y_{LX}/Y_p in the limit of a vanishingly thin target show the effects of Ag(L) → F(K) charge exchange for the one-electron (8+) and bare (9+) incident fluorine ions.

ues of the projectile parameters derived from the fits to $\overline{\sigma_{KX}}$ are given in Fig. 35. The single-quenching cross sections σ_{10} and σ_{21} are in good agreement with the scaled OBK predictions (given as a function of Z_1 for fixed Z_2). No reliable estimates of σ_{20} are currently available. The values of the single-vacancy production cross sections for the projectiles follow the general trends that one may expect from estimates based upon the PWBA in that $\sigma_{01} > 2\sigma_{12}$ because of the number of electrons available in the ion with a full K shell. The magnitude of these two cross sections is, however, approximately 0.1 of the PWBA predictions. The vacancy cross sections as determined from the work of Gardner et al.[113] do not agree with what one would expect from the measurement of electron loss cross sections as inferred from cross-section measurements for heavy ions incident on thin gas targets under single-collision conditions. The fact that the relative values of σ_{01} and σ_{12} do not agree with what is expected from the single-collision gas studies may be a ramification of the effects of differences in collision frequency encountered in dilute gases in comparison to solid targets. Further, the three-component model is sufficiently complex that the double loss and capture cross sections cause little effect on the calculations. Experiments that are more sensitive to the Y_2 fraction directly would be appropriate to a better understanding of the processes affecting the Y_2 vacancy fractions.

An application of the three-component model for target x-ray production for L x rays has been reported by Schiebel et al.[88] Their results for fluorine ions incident upon thin silver targets are given in Fig. 36. For fluorine ions incident without a K-shell vacancy no target thickness effects in $\overline{\sigma_{LX}}$ are observed, whereas both the single-vacancy (F^{8+}) and double-vacancy (F^{9+}) data for $\overline{\sigma_{LX}}$ show strong target thickness effects. This is to be expected because the quenching processes leading to the filling of the K-shell vacancies of the projectile dominate the process.

Certainly the study of heavy ions incident on thin solid targets is a most complex and possibly fruitful area of investigation. Clearly, those measurements in which the effects of the enhancements in target x-ray production for ions with one or two K-shell vacancies have been either overlooked or ignored may not provide an adequate test of the process of target ionization by heavy ions. Further, the role of the projectile cross sections is intimately associated with all facets of the physical properties of heavy ions moving in solids.

5.5.6. Charge-State Fractions for Heavy Ions in Solids as Related to the Two-Component Model

The results presented so far build a strong case for the dominant influence of projectile K-vacancy configurations as related to beam-foil measurements, target x-ray production, and projectile x-ray yields for heavy ions penetrating solid targets. This dominance rests on the simple idea that the collision frequency is sufficiently high that the projectile excitations are not allowed to relax. This situation is distinctly different from that encountered in the study of heavy ions passing through gases under single-collision conditions. Indeed, the very basic equations as given by Betz,[115] which are the foundation of charge-changing collision studies for single-collision conditions involving heavy ions moving in dilute gas targets, do not allow excitations. Hence cascading effects and elastic and inelastic collisions are removed from consideration at the outset. When a target x-ray yield is measured for an incident heavy ion on a solid target the results are sensitive to the vacancy configuration of the projectile, not

---

[115] H. D. Betz, *Rev. Mod. Phys.* **44**, 465 (1972).

5.5 X-RAY PRODUCTION CROSS-SECTION MEASUREMENTS

its charge state. It is the role of the inner-shell vacancies in addition to other interactions such as direct Coulomb ionization and charge exchange that is important. A heavy ion may reside at any given instance in a specific charge state within a solid. In the case of target x-ray production, what its particular charge state happens to be is of minimal importance in comparison to what its inner-shell vacancy configuration is at that point.

In order to further illustrate the dominant role of K vacancies in determining the measurable quantities that have been regimes of intense study for heavy ions moving in solids consider now the question of charge-state fractions for heavy ions moving in such media. For simplicity consider that the charge state fractions of a heavy ion can be written as

$$\begin{aligned}
\phi_1 &= Y_1 g_0, \\
\phi_2 &= Y_0 g_0 + Y_1 g_1, \\
& \vdots \\
\phi_j &= Y_0 g_{j-2} + Y_1 g_{j-1},
\end{aligned} \qquad (5.5.30)$$

where ϕ_j is the charge state fraction for an ion having j electrons, g_j the probability of having j electrons outside the K shell, and Y_i the fraction of ions with i K-shell vacancies. By writing the ϕ_j in this manner a two-component system is being considered. Such a system is appropriate when the Y_2 fraction is small, i.e., $Y_2 \leq 0.1 Y_1$.

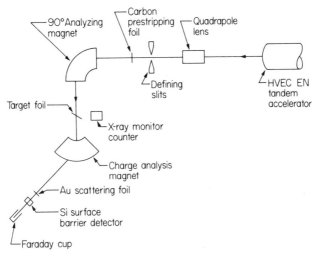

FIG. 37. An experimental arrangement for measuring charge-state fractions resulting from the passage of heavy ions through thin self-supporting targets.

The experimental arrangement that is typically employed in determining charge state fractions for ion beams is as shown in Fig. 37. The lifetimes of K-shell excitations for heavy ions (exclusive of metastable states) is typically of the order to 10^{-14}–10^{-15} sec. Hence after leaving the interaction region the excited systems will undergo electron rearrangement and/or loss prior to detection. With this in mind the charge-state fractions ϕ_j as previously written are not the charge state fractions that are detected. If it is assumed that decay of the K vacancies is dominated by nearest-neighbor cascades, then the ϕ_j can be rewritten as

$$\phi_1 = Y_1 g_0,$$
$$\phi_2 = Y_0 g_0 + Y_1 g_1 \omega_2 + Y_1 g_2 (1 - \omega_3),$$
$$\vdots \qquad (5.5.31)$$
$$\phi_j = Y_0 g_{j-2} + Y_1 [g_{j-1} \omega_j + g_j(1 - \omega_{j+1})].$$

As a two-component model is being used, the expressions for Y_0 and Y_1 are well known:

$$Y_0 = 1 - Y_1, \qquad (5.5.32)$$

$$Y_1 = \frac{\sigma_v}{\sigma}(1 - e^{-\sigma X}) + A e^{-\sigma X}. \qquad (5.5.33)$$

Hence

$$\phi_j(X) = g_{j-2} + \left[\frac{\sigma_v}{\sigma}(1 - e^{-\sigma X}) + A e^{-\sigma X}\right][\beta_j - g_{j-2}], \qquad (5.5.34)$$

where $\beta_j \equiv g_{j-1}\omega_j + g_j(1 - \omega_{j+1})$, the term that accounts for cascading in the next-nearest-neighbor approximation. The limiting cases for $\phi_j(X)$ are

Case I. $A = 0$: $\phi_j(0) = g_{j-2}$.
Case II. $A = 1$: $\phi_j(0) = \beta_j$.

Case III. $A = 0,1$: $\phi_j(\infty) = g_{j-2} + \frac{\sigma_v}{\sigma}(\beta_j - g_{j-2})$.

This two-component model for the charge-state fractions of a heavy-ion beam moving in a solid target thus predicts that:

(1) All charge-state fractions will have a target thickness target dependence governed by σ.

(2) The charge-state fractions will depend upon the initial conditions i.e., $A = 1$ or $A = 0$.

(3) The equilibration values are dependent upon σ_v/σ and hence are Z_2 dependent.

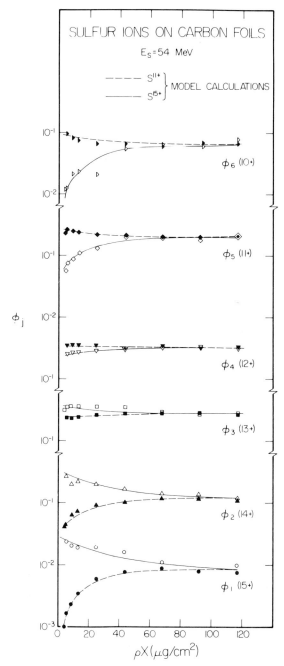

FIG. 38. The measured charge-state fractions for sulfur ions after the ions have passed through carbon foils as a function of carbon target thickness and incident ion K-vacancy configuration. The solid and broken curves are calculated ion charge-state fractions using the two-component model for $\phi_j(x)$.

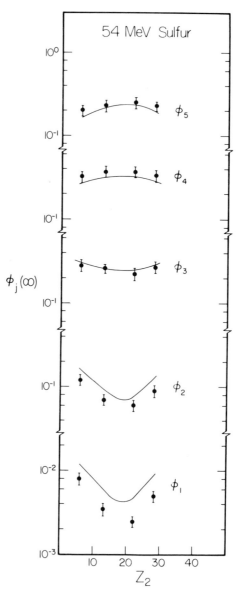

FIG. 39. Measured charge-state fractions for sulfur ions at equilibration target foil thicknesses as a function of foil atomic species. The curves are calculated equilibrium charge-state fractions using the two-component model.

5.5 X-RAY PRODUCTION CROSS-SECTION MEASUREMENTS

(4) The intercept values at $X = 0$ are dependent on the initial conditions of the incident ion beam.

Shown in Fig. 38 are the results of Gray et al.[116] for the charge state fractions $\phi_1-\phi_6$ for 54-MeV sulfur ions incident on carbon foils of thicknesses from ~ 4 to 120 $\mu g/cm^2$ with incident charge states of 11+ and 16+. The curves are model calculations using the development given above. Additional experiments on the equilibration charge states for 54-MeV sulfur ions incident on carbon, aluminum, titanium, and nickel foils are shown in Fig. 39. The two-component model calculations predict the salient features of these data. The calculations for Fig. 39 assume that the g_j are independent of Z_2. The g_j are further assumed to be equilibrated and relaxed at the time of detection. These two assumptions are in need of additional investigation.

While the question of charge-state fractions for heavy ions moving in solids is seemingly unrelated to the question of target x-ray production for heavy ions at first glance, it is now critically clear that the role of inner-shell vacancies is a dominant consideration in such studies at high energies. Had the development of the target dependence of charge-state fractions occurred earlier, the incorporation of the K-vacancy concepts for the projectile would have been naturally included into the development of studies in target inner-shell ionization studies. The properties of charge-state fractions, beam-foil spectroscopic measurements of inner-shell excitations, and target x-ray production are all related in the same manner through the quantities σ_v and σ for the projectile. As stated previously the description of such processes as have been developed in this section as being "charge-state" effects is a misnomer because the inner-shell vacancy configuration and not the charge state is responsible for what is observed.

Indeed, serious questions can be raised concerning the design and interpretation of past experiments designed to determine the charge state of an ion as it penetrates a solid by comparing the solid target x-ray yield to that of a thin gas target of the same element. Discussions of experiments such as argon ions moving in SiH_4 gas in comparison to 52-$\mu g/cm^2$ solid silicon targets given by Datz[100] must necessarily be influenced by the role of inner-shell vacancies in the x-ray generation process. Such considerations are in addition to the problems of the relative fluorescence yields in the same experiment as previously pointed out by Kauffman et al.[117] Extreme care must be exercised when one approaches problems

---

[116] T. J. Gray, C. L. Cocke and R. K. Gardner, *Phys. Rev.* A **16**, 1907 (1977).

[117] R. L. Kauffman, K. A. Jamison, T. J. Gray, and P. Richard, *Phys. Rev. Lett.* **36**, 1074 (1976).

that may be strongly dependent upon the medium in which the ions move at high velocities.

References Cited in Table IV (p. 230)

[118] E. Koltay, D. Berényi, I. Kiss, S. Ricz, G. Hock, and J. Bacsó, *Z. Phys. A.* **278**, 299 (1976).

[119] R. H. McKnight, S. T. Thornton, and R. R. Karlowicz, *Phys. Rev. A* **9**, 267 (1974).

[120] T. K. Li and R. L. Watson, *Phys. Rev. A* **9**, 1574 (1974).

[121] H. Tawara, K. Ishii, S. Morita, H. Kaji, C. N. Hsu, and T. Schiokawa, *Phys. Rev. A* **9**, 1617 (1974).

[122] K. Ishii, S. Morita, H. Tawara, H. Kaji and T. Schiokawa, *Phys. Rev. A* **10**, 774 (1974).

[123] D. Burch, *Phys. Rev. A* **12**, 2225 (1975).

[124] R. M. Wheeler, R. P. Chaturvedi, J. L. Duggan, J. Tricomi, and P. D. Miller, *Phys. Rev A* **13**, 958 (1976).

[125] Y. Awaya, K. Iyumo, T. Hamada, M. Okano, T. Takahuhi, A. Hashiyume, Y. Tendow, and T. Katou, *Phys. Rev. A* **13**, 992 (1976).

[126] G. Bissinger, J. M. Joyce, and H. W. Kugel, *Phys. Rev. A* **14**, 1375 (1976).

[127] G. Bissinger, J. M. Joyce, B. L. Doyle, W. W. Jacobs, and S. M. Shafroth, *Phys. Rev. A* **16**, 443 (1977).

[128] F. D. McDaniel, J. L. Duggan, P. D. Miller, and G. D. Alton, *Phys. Rev. A* **15**, 846 (1977).

[129] J. Tricomi, J. L. Duggan, F. D. McDaniel, P. D. Miller, R. P. Chaturvedi, R. M. Wheeler, J. Lin, K. A. Kuenhold, L. A. Rayburn, and S. J. Cipolla, *Phys. Rev. A* **15**, 2269 (1977).

[130] F. K. Chen, R. Laubert, and W. Brandt, *Phys. Rev. A* **15**, 2227 (1977).

[131] R. C. Bearse, D. A. Close, J. J. Malanify, and C. J. Umbarger, *Phys. Rev. A* **7**, 1269 (1973).

[132] T. L. Hardt and R. L. Watson, *Phys. Rev. A* **7**, 1917 (1973).

[133] B. Knaf, G. Presser, and J. Stähler, *Phys. Lett.* **56A**, 165 (1976).

[134] B. Knaf and G. Presser, *Phys. Lett.* **56A**, 167 (1976).

[135] M. D. Rashiduzzaman Khan, D. Crumpton, and P. E. Francois, *J. Phys. B* **9**, 455 (1976).

[136] R. K. Wyrick, and T. A. Cahill, *Phys. Rev. A* **8**, 2288 (1973).

[137] C. N. Chang, J. F. Morgan, and S. L. Blatt, *Phys. Lett.* **49A**, 365 (1974).

[138] S. T. Thornton, R. H. McKnight, and R. R. Karlowicz, *Phys. Rev. A* **10**, 219 (1974).

[139] F. Abrath, and T. J. Gray, *Phys. Rev. A* **10**, 1157 (1974).

[140] H. Tawara, K. Ishii, S. Morita, H. Kaji, and T. Schiokawa, *Phys. Rev. A* **11**, 1560 (1975).

[141] J. F. Morgan, C. N. Chang, and S. L. Blatt, *Phys. Rev. A* **12**, 1731 (1975).

[142] M. Milazzo and C. Ricobono, *Phys. Rev. A* **13**, 578 (1976).

[143] J. R. Chen, J. D. Reber, D. J. Ellis, and T. E. Miller, *Phys. Rev. A* **13**, 941 (1976).

[144] R. P. Chaturvedi, R. M. Wheeler, R. B. Liebert, D. J. Miljanic, T. Zabel, and G. C. Phillips, *Phys. Rev. A* **12**, 52 (1975).

[145] T. L. Hardt and R. L. Watson, *Phys. Rev. A* **14**, 137 (1976).

[146] K. E. Stiebing, H. Schmidt-Böcking, R. Schulé, and K. Bethge, *Phys. Rev. A* **14**, 146 (1976).

[147] T. K. Li, D. L. Clark, and G. W. Greenless, *Phys. Rev. A* **14**, 2106 (1976).

[148] T. K. Li, D. L. Clark, and G. W. Greenless, *Phys. Rev. Lett.* **37**, 1209 (1976).

[149] J. R. Chen, *Phys. Rev. A* **15**, 487 (1977).

6. CHARGE DEPENDENCE OF ATOMIC INNER-SHELL CROSS SECTIONS

By James R. Macdonald

6.1. Introduction

In heavy-ion–atom collisions, various cross sections for producing excited states have been observed to show a pronounced dependence on the charge state of the incident ions. This effect results from the complexity of multielectron systems, which can be excited to a myriad of possible final states with probabilities strongly influenced by the occupation number of some of the levels of the initial projectile–target complex. The effect is indicated schematically in Fig. 1, where the dashed arrows indicate different channels that open for the change of occupation of K levels of atom, Z_1.

Further complicating our understanding of the cross sections for such inelastic atomic collisions is the coupling of the charge state effects to the decay modes of the projectile and target excited states. These states relax rapidly after the collision by the emission of photons and electrons at decay rates that reflect the precise structure of the multielectron state at the time of emission, and hence depend strongly on the excitation channels available for all electrons at the start of the collision.

A general study of multielectron excitation for each initial state is not currently accessible either theoretically or experimentally. In the former case, calculations of the simultaneous excitation of several electrons from different atomic shells within the strong field of a heavy-ion collision have not been formulated for a large number of interacting particles. Experimentally, detector resolution is rarely adequate to separate all the states formed in a collision. When we define atomic cross sections for particular excitation processes, it is our practical goal to isolate a transition between relevant atomic levels and treat the multielectron effects of other levels in an average sense. Traditionally, with excitation by swift point charges (electrons, protons, and alpha particles) this isolation was readily accomplished for inner-shell excitations. As research has progressed to excitation with structured heavy ions (Z_1^{q+}) it has been tacitly assumed that the required isolation of interaction channels for the inner electrons

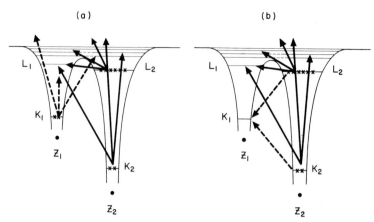

FIG. 1. Schematic representation of inner-shell transitions that occur between the levels of two interacting atoms Z_1 and Z_2. (a) Initial state with inner shells full. (b) Initial state with an empty inner shell that opens up the additional excitation channels indicated by the broken arrows, and closes channels shown in this way in (a).

still can be accomplished by careful control of the initial (final) charge state $q(q')$ of the projectile and reasonable detector resolution of excited-state decay. In this chapter, the charge state effects of atomic cross sections are surveyed to establish the techniques used and the assumptions made in isolating primary excitation mechanisms from analysis of these effects.

6.2. Conditions for Measuring Charge-Dependent Collision Processes

In an experiment to determine a charge-dependent cross section using a heavy-ion projectile, the first requirement is to gain separate access to a range of initial-charge-state ions at the same velocity. This has been accomplished both using sources of multiply charged ions prior to acceleration to the final velocity, and producing multiple-charge distributions after acceleration for separate selection to the experimental apparatus. For the former technique, high-power Penning ionization sources[1,2] have made charge states with ionization potentials $\leqslant 120$ eV available, while the prospect of extending the range to charge states bound to ~ 3–10 keV

---

[1] M. L. Mallory and D. H. Crandall, *IEEE Trans. Nucl. Sci.* **NS-23**, 1069 (1976).
[2] R. Keller and M. Müller, *IEEE Trans. Nucl. Sci.* **NS-23**, 1049 (1976).

6.2. MEASURING CHARGE-DEPENDENT COLLISION PROCESSES

has been tantalizingly close for a decade.[3,4] With such sources, charge state effects can be studied easily even at relatively low energy ($\gtrsim 10$ keV).

At higher energy, a range of charge state ions is readily available for experimental study by the traditional post acceleration stripping technique illustrated in Fig. 2a. Using self-supporting carbon stripping foils ($\sim 5-20$ μg/cm² in thickness) reasonable intensity can be obtained for several high-charge states near the equilibrium value at the beam energy.[5] The intensity of lower charge states can be increased using a windowless gaseous poststripper to maximize a particular charge state before equilibrium is reached. With this technique the requirement for a variable target thickness in a range from $\sim 1/10$ to 100 Torr-cm (1/3 to 300 $\times$ 10¹⁶ molecules/cm²) produces the technical complication of a large gas flow in the vacuum system. Solutions to the problem will be particular to each experimental system and are potentially expensive. Charge states lower than the initially accelerated beam can never be produced in abundance by this technique.

Intensity considerations for the postacceleration stripping technique preclude that studies be restricted for a particular charge state (q) to energies above $E_{min}(q)$, where

$$E_{min}(q) \sim \frac{M}{m} I(q), \qquad (6.2.1)$$

with M, m, and $I(q)$ denoting the ion mass, electron mass, and ionization potential of the ion, respectively. Pertinent to work with incident ions containing K-shell vacancies, this condition severely limits the velocity range available for study with a particular accelerator. The high-velocity limit is extended readily by multiple-stripping stages within an accelerating system. In principle, sufficient intensity for experimentation at lower velocities can be obtained by a combination accelerator–decelerator system with a charge-stripping stage prior to the deceleration. Such techniques are being investigated[6] using double tandem accelerators shown in Fig. 2b, and the success of the development will depend on the ability to focus high-charge, low-energy ions in the decelerating column.

After an ion beam of separable charge states at a given velocity has been developed, usual transport techniques can be used to deliver each

---

[3] E. D. Donets, V. I. Ilyushchenko, and V. A. Alpert, JINR P7-4124, Dubna, (1968).
[4] J. Arianer and Ch. Goldstein, *IEEE Trans. Nucl. Sci.* **NS-23**, 979 (1976).
[5] H. D. Betz, *Rev. Mod. Phys.* **44**, 465 (1972); H. D. Betz, in "Methods of Experimental Physics" (P. Richard, ed.), Vol. 17, Chap. 3. Academic, New York (1980).
[6] T. K. Saylor, L. D. Gardner, R. Stephenson, and J. E. Bayfield, *BAPS* **23**, 1107 (1978).

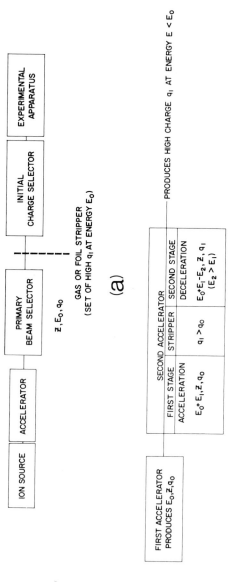

FIG. 2. (a) Traditional postacceleration stripping configuration to produce a set of high-charge states for experimental study at the same relatively high projectile energy. (b) Acceleration–deceleration stripping configuration to produce high-charge-state ions at low projectile energy.

beam to an experimental arrangement. Because maximum charge exchange cross sections rarely exceed πa_0^2, residual gas in the vacuum system generates $\sim 1/10\%$ charge impurity for a long (~ 4 m) beam path at normal high vacuum ($\sim 10^{-6}$ Torr). Ultra-high-vacuum techniques are not necessary except at very low velocity, where the cross sections can reach $10^2 \pi a_0^2$. A more serious contribution to initial charge state impurity results from small-angle scattering from all collimators used to define the beam. This scattering produces a charge equilibrated beam without significant energy loss. Reasonable design and focusing care are required to keep the slit-scattered charge impurity below a few percent, and extreme care is required to keep charge impurity to 1/10% with a heavy-ion beam.

Within the apparatus for measurement of a charge-dependent atomic cross section a thin target of variable thickness is essential. The particular target optimum for an experiment may be a gas or vapor jet, a differentially pumped cell, or even a thin solid foil. The choice should not be generalized but depends on the magnitude of cross sections for processes competing with that to be observed. Usually these range over many orders of magnitude for excitation, ionization, and charge transfer of the electrons of different binding energy in an atomic collision with a particular charge state ion. If the particular observation to be made in the experiment has a cross section comparable to the largest of all competing processes available to the incident state, then it is straightforward to select a target of thickness π atom/cm^2 that both preserves the incident charge state purity and also provides sufficient intensity of the detected event to evaluate charge-dependent cross sections from the observed yield per incident ion

$$Y(q) = \sigma(q)\pi. \qquad (6.2.2)$$

However, in the more usual case, when the initial state can be changed with a cross section large compared to the process of interest, the selection of the target is crucial to successful interpretation of experimental data. Often one cannot choose a thin enough target to preserve the initial state and from Equation (6.2.2) one deduces at best a cross section $\bar{\sigma}(q)$ averaged over all intermediate states populated in the collision. In these cases a proper evaluation of $\sigma(q)$ requires a detailed study of the target thickness and density dependence of the observed yield $Y(q, x)$ and analysis in terms of rate equations coupling the important intermediate states. If processes coupling three states can be isolated, the analysis developed for charge exchange measurements[7] provides a technique to evaluate

---

[7] S. K. Allison, *Rev. Mod. Phys.* **30**, 1137 (1958).

cross sections for competing processes. If more than three states are coupled strongly, direct analysis will not yield all cross sections uniquely and a model must be fabricated to identify which average cross sections can be interpreted from experimental data.

6.3. Projectile Charge Dependence of Target Inner-Shell Cross Sections

Dramatic charge-dependent effects were first observed in target x-ray production cross sections for highly charged ions incident on thin gaseous targets ($10^{15} - 10^{16}$ atoms/cm^2). In all cases studied there is a monotonic increase in x-ray cross section (seen in Fig. 3) with increasing charge state q. In these experiments the x-ray yield grows linearly with target thickness and the q-dependent cross sections were obtained from the slope of growth [Eq. (6.2.2)], in a range where charge impurity was kept below 10%. However, the magnitude of some outer-shell excitation cross sections may approach πa_0^2, several orders of magnitude larger than those observed for the inner shells. Hence, the cross sections represent averages over initial excited states populated from the incident

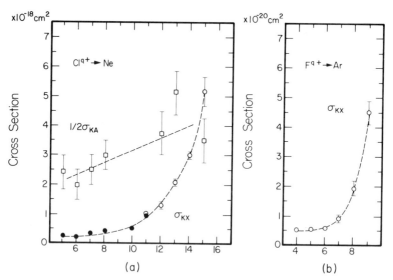

FIG. 3. Projectile charge dependence of target inner-shell cross sections (a) $Z_1 > Z_2$. Neon K cross sections excited by Cl^{q+} ions. X-ray data[8,10] taken at 1.4 (solid points) and 1.2 MeV/amu (open circles); Auger data[10] at 1.4 MeV/amu. For graphical purposes $\tfrac{1}{2}\sigma_A$ has been plotted. (b) $Z_1 < Z_2$. Argon K x-ray cross sections[11] excited by F^{q+} at 1.9 MeV/amu.

6.3. CHARGE DEPENDENCE OF TARGET CROSS SECTIONS

charge state. Experimentally, the defined x-ray cross section is the sum of a product of a collision cross section times a radiative fluorescence yield, where the summation is over all those excited states * that include a vacancy in a target inner shell, that is,

$$\sigma_x(q) = \sum_* \sigma^*(q)\omega_x^*. \qquad (6.3.1)$$

Already the implicit approximation has been made that the state formation in the collision and the decay process are independent, a good approximation for low-Z atoms since typical collision times are shorter than the lifetimes of most excited states. For higher-Z or particularly fast autoionizing states, such independence may not be realistic and a time integral over the transition probability would replace Eq. (6.3.1) for the interpretation of the experimental results.

Distinctly different effects occur in asymmetric collisions of $Z_1^{q+} \to Z_2$ in the case of $Z_1 > Z_2$ compared to $Z_1 < Z_2$. The former case has been studied extensively for the production cross section of neon K x rays by Sellin and co-workers.[8,9] The observed x-ray cross sections increase exponentially with charge state. Although the collisional effects have never been quantitatively explained, this is a perfect example of an experimentally defined inner-shell cross section varying strongly with q, for which the partial cross sections $\sigma^*(q)$ need vary only slightly with q to produce more highly excited states with rapidly increasing ω_x^*. The same collisions have been studied by Burch et al.[10] through observation of the neon K Auger electron decay channel for which the experimentally defined cross section (shown in Fig. 3a) in terms of a nonradiative Auger yield ω_A^* is

$$\sigma_A(q) = \sum_* \sigma^*(q)\omega_A^*. \qquad (6.3.2)$$

By definition, $\varphi_x^* + \omega_A^* = 1$. However, in the present nomenclature pertinent to heavy-ion collisions the excited state * may include stable plasma states with all excited electrons in the continuum. For these states $\omega_x^* = \omega_A^* = 0$ and the total collision cross section

$$\sigma(q) \equiv \sum_* \sigma^* \geq \sigma_x(q) + \sigma_A(q), \qquad (6.3.3)$$

---

[8] J. R. Mowat, D. J. Pegg, R. S. Peterson, P. M. Griffin, and I. A. Sellin, *Phys. Rev. Lett.* **29**, 1577 (1972); *Phys. Rev. A* **9**, 644 (1974).

[9] J. R. Mowat, I. A. Sellin, D. J. Pegg, R. S. Peterson, M. D. Brown, and J. R. Macdonald, *Phys. Rev. Lett.* **30**, 1289 (1973).

[10] D. Burch, N. Stolterfoht, D. Schneider, H. Wieman, and J. S. Risley, *Phys. Rev. Lett.* **32**, 1151 (1974).

where the inequality will become important for higher exciting charge states q. The neon K-Auger cross sections in Fig. 3a show only a small q dependence and it is clear that the variation in fluorescence yield (discussed in Chapter 6.4) of the collisionally excited states is the predominant cause of the exponential q dependence in the x-ray cross sections. An explanation of the gradual increase with q in the outer-shell excitation awaits study of the outer-shell populations produced in these collisions.

For the case of $Z_1 < Z_2$ studied for the production cross section of argon K x rays by $Z_1 < 10$ by Macdonald et al.,[11] Winters et al.,[12] and Hopkins et al.,[13] the cross section shown in Fig. 3b is relatively constant for low-charge states but increases by significant factors successively as the last three electrons are removed from the projectile. The explanation of this increase was given first by Halpern and Law,[14] who pointed out that direct capture of target electrons to a vacant 1s projectile level would dominate over excitation to the continuum (ionization) for sufficiently large projectile atomic number Z_1. These authors argued on the basis of a Z_1^5 scaling in an Oppenheimer–Brinkman–Kramers (OBK) formulation of the first Born approximation for direct capture compared to a Z_1^2 dependence for Coulomb ionization.

Various modifications (for example, Losonsky and Lapicki[15]) to the OBK approximation have given agreement with observed charge-dependent x-ray cross sections for restricted projectile and target combinations by taking into account binding energy considerations for electron transfer to high-charge projectiles. However, calculated first Born transition probabilities for electron transfer often exceed unity for heavy-ion collisions and this theoretical basis is unsound to quantitatively describe the process. On a stronger foundation, numerical calculations by Lin et al.[16] for 1s to 1s electron capture cross sections in a two-state atomic expansion model formulated by Bates[17] have quantitatively confirmed the magnitude of the observed cross sections with incident bare nuclei at velocities near the peak of the cross section.

With incident one-electron ions compared to bare nuclei, a simple reduction of a factor of two in the capture cross section can be estimated

[11] J. R. Macdonald, L. Winters, M. D. Brown, T. Chiao, and L. D. Ellsworth, *Phys. Rev. Lett.* **29**, 1291 (1972).

[12] L. M. Winters, J. R. Macdonald, M. D. Brown, T. Chiao, L. D. Ellsworth, and E. W. Pettus, *Phys. Rev. A* **8**, 1835 (1973).

[13] F. Hopkins, R. Brenn, A. R. Whittemore, N. Cue, V. Dutkiewicz, and R. P. Chaturvedi, *Phys. Rev. A* **13**, 74, (1976).

[14] A. M. Halpern and J. Law, *Phys. Rev. Lett.* **31**, 4 (1973).

[15] G. Lapicki and W. Losonsky, *Phys. Rev. A* **15**, 896 (1977).

[16] C. D. Lin, S. C. Soong, and L. N. Tunnell, *Phys. Rev. A* **17**, 1646 (1978).

[17] D. R. Bates, *Proc. Roy. Soc. (London) A* **245**, 299 (1958).

6.3. CHARGE DEPENDENCE OF TARGET CROSS SECTIONS

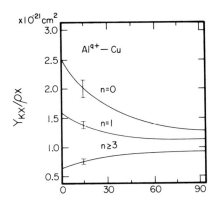

FIG. 4. The general trend in solid targets of the charge state and target thickness dependence of the normalized copper K x-ray yield per projectile per target atom $Y_x/\rho x$, for excitation by silicon ions with $n = 0, 1$, and $n \geq 3$ ($n = Z - q$). The lines are model fits[18] to experimental data points (not shown) taken with many thin solid copper targets of thickness from 2 to 90 $\mu g/cm^2$. The uncertainty of the experimental data fit by the model is indicated by the error bars on the line for each charge state.

from the availability of initial states for the collisionally induced transition. In the absence of theoretical guidance, this estimate has been used by Gardner et al.[18] to fit the charge and target thickness dependence of target x-ray cross sections from thin solid films in which the variation of projectile charge state with depth is inherent to the analysis. These experiments represent the interesting situation in which the large cross sections for outer-shell excitation and transfer provide some unknown equilibrium distribution of outer electrons in the solid medium. Hence in the analysis the implicit approximation is made[19] that the collisionally coupled atomic states are those identified by the presence of 0, 1, or 2 K-shell vacancies in the projectile. Rate equations defining the growth and depletion of these three experimentally defined states have been analyzed using the formulation of Allison[7] to describe the target thickness and incident charge state dependence for observed x-ray yields shown in Fig. 4. Accurate target x-ray production cross sections are obtained from the extrapolation of the experimental data to zero target thickness. The cross section for incident one-electron ions is midway between that for

---

[18] R. K. Gardner, T. J. Gray, P. Richard, C. Schmiedekamp, K. A. Jamison, and J. M. Hall, *Phys. Rev. A* **15**, 2202 (1977).

[19] H.-D. Betz, F. Bell, H. Panke, G. Kalkoffen, M. Welz, and D. Evers, *Phys. Rev. Lett.* **33**, 807 (1974).

bare nuclei and lower-charge-state ions, indicating the factor of two reduction in the capture cross section. However, the final atomic-state binding energy is a sensitive parameter influencing both the velocity dependence and magnitude of direct transfer amplitudes; hence one must not expect the factor of two decrease in 1s capture cross sections by one-electron ions compared to bare nuclei to be universal.

With incident two-electron ions the target x-ray cross section, slightly larger than with three-electron ions, reflects the presence of $(1s2s)^3S_1$ heliumlike metastable ions in any accelerator beam. The precise fraction f, ranging from 10 to 50%, of these ions to the ground-state two-electron ion has been studied for Si^{12+} beams by Schiebel et al.[20] and depends on many factors such as the ion formation conditions and the drift time to the collision region. For target x-ray production the metastable fraction f containing a 1s vacancy behaves like the one-electron hydrogenlike ion, while the ground state, with a full K shell, has an effect similar to three-electron ions. Thus the fraction of metastables can be estimated from

$$f \simeq (\sigma_2 - \sigma_3)/(\sigma_1 - \sigma_3), \qquad (6.3.4)$$

where the subscript represents the target x-ray cross section for n-electron incident ions.

In asymmetric collisions with $Z_1 < Z_2$, the small variation of target cross section for incident projectile states with three or more electrons depends on the final-state fluorescent yield consequent from the outer-shell configurations produced in the collision and possibly on effects from L-shell electron screening of the direct 1s ionization process. Since the probability for the latter process falls rapidly at impact parameters outside the K-shell radius, we must expect that the charge dependence from changes in screening to be a relatively minor effect on x-ray cross sections. Estimates of these effects have been made by Brandt et al.[21] to account for the charge dependence of integrated x-ray yields from solid foils.

In nearly symmetric collisions, the strong charge dependence of target K x-ray cross sections not only includes the effects with 0, 1, and 2 electrons but extends throughout the removal of the projectile L-shell electrons as well. The importance of 2p vacancies in permitting electron promotion of inner-shell electrons along molecular orbitals introduces this dependence to the inner-shell excitation cross section on the initial projectile charge state as the L shell is depleted. For excitation by Cu^{q+} ions of argon ($Z_1 > Z_2$) and krypton ($Z_1 < Z_2$) a strong dependence seen in

[20] U. Schiebel, B. L. Doyle, J. R. Macdonald, and L. D. Ellsworth, *Phys. Rev. A* **16**, 1089 (1977).

[21] W. Brandt, R. Laubert, M. Mourino, and A. Schwarzschild, *Phys. Rev. Lett.* **30**, 358 (1973).

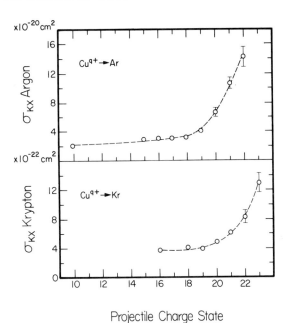

FIG. 5. K x-ray cross sections excited by Cu^{q+} ions. (a) $Z_1 > Z_2$. Argon K x rays at 1 MeV/amu.[22] (b) $Z_1 < Z_2$. Krypton K x rays at 1.4 MeV/amu.[23] The strong charge dependence for $q \geq 19$ occurs as the copper L-shell electrons are removed from the incident ion.

Fig. 5 for $q > 19$ has been observed by Schiebel et al.,[22] Warczak et al.,[23] and Lennard et al.[24] The charge dependence observed in these cases has been used to study the cross section for rotational coupling between $2p\pi$ and $2p\sigma$ molecular orbitals as calculated by Taulbjerg et al.,[25] as well as the radial coupling between $3d\pi$ and $2p\pi$ orbitals as discussed by Lennard et al.[24]

6.4. Fluorescence Yield Effects in High-Charge Collisions

The purpose of an experiment measuring x-ray cross sections is usually to determine atomic collision cross sections. Interpretation of the data always involves knowledge of the radiative fluorescence yield for the

[22] U. Schiebel and B. L. Doyle, Z. Phys. A **285**, 241 (1978).
[23] A. Warczak, D. Liesen, J. R. Macdonald, and P. H. Mokler, Z. Phys. A **285**, 235 (1978).
[24] W. N. Lennard, J. V. Mitchell, J. S. Forster, and D. Phillips, J. Phys. B **10**, 2199 (1977).
[25] K. Taulbjerg, J. S. Briggs, J. Vaaben, J. Phys. B **9**, 1351 (1976).

states produced in the collision. For each individual excited state of an isolated atom the fluorescence yield is defined as $\omega_x = \Gamma_x/\Gamma_{tot}$ in terms of the radiative x-ray and total transition probabilities Γ for the particular state. The transition probabilities, which depend on the angular momentum quantum numbers, the numbers of electrons available for transitions, as well as the excitation energy, can be calculated[26] using a pertinent atomic model for any state of a multiplet formed with a particular electronic configuration. Comprising any x-ray line observed in an atomic collision experiment are contributions from unresolved multiplet states, usually from different configurations. The fluorescence yield needed to reduce x-ray data and extract collision cross sections is the average fluorescence yield of all the states contributing to the x-ray line, where the average must be taken over the population distribution of all the states. This latter distribution is virtually never known and assumptions about the distribution, implicit or explicit, have always been made in using a particular theoretical fluoresence yield for an x-ray line. For example, if N excited states are produced with population distribution n_i each having fluorescence yield $\omega_x{}^i$, then the average fluorescence yield for the distribution is given by

$$\omega_X^{av} = N^{-1} \sum_i n_i \omega_X{}^i. \qquad (6.4.1)$$

A decade ago, the primary use of calculated fluorescence yields was in the interpretation of x rays following nuclear decay, or photoionization in which inner-shell vacancy production occurred with relatively little disruption of outer shells. The assumption of statistical populations of states within each multiplet, each configuration, and all possible configurations for the available electrons often has been made to permit the use of angular momentum addition rules to compute ω_X^{av} without calculating $\omega_X{}^i$ for individual states. In atomic collisions, however, outer-shell excitation concurrent with inner-shell vacancy production is the rule rather than the exception, electrons are unlikely to be distributed statistically to all possible configurations, and the distribution will depend strongly on the charge state of the incident projectile.

From comparison of the measured x-ray and Auger electron cross sections for the same atomic collision, the average fluorescence yield for the excited-state distribution produced in the collision is determined from

$$\omega_X^{av} = \sigma_X/(\sigma_X + \sigma_A). \qquad (6.4.2)$$

The charge dependence of ω_{KX}^{av} for K-shell x rays shown in Figs. 6 and 7

[26] W. Bambynek, B. Crasemann, R. W. Fink, H. U. Freund, H. Mark, C. D. Swift, R. E. Price, and P. V. Rao, *Rev. Mod. Phys.* **44**, 716 (1972).

6.4. FLUORESCENCE YIELD IN HIGH-CHARGE COLLISIONS

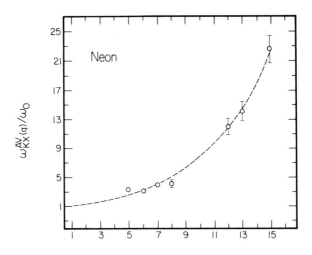

Projectile Charge State

FIG. 6. Projectile charge dependence of the average K x-ray fluorescence yield of neon[10] excited by 1.4 MeV/amu Cl^{q+} ions expressed as a ratio to the neutral atom fluorescence yield $\omega_0 = 1.6 \times 10^{-2}$.

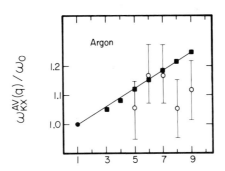

Projectile Charge State

FIG. 7. Projectile charge dependence of the average K x-ray fluorescence yield of argon excited by Z_1^{q+} ions expressed as a ratio to the neutral atom fluorescence yields, $\omega_0 = 0.12$. The data with error bars were determined[27] from σ_A and σ_X measurements for excitation by F^{q+} ions. The points represented by squares are mean values of the values determined with different incident ions by measurements of K x-ray energy shifts and relative intensities[12] and comparison with configuration average calculations[33] of these quantities and ω_K (Table II).

has been measured for neon excited by 1.4 MeV/amu Cl^{q+} ions by Burch et al.[10] and for argon excited by F^{q+} ions by Randall et al.[27] In the latter case for $Z_1 < Z_2$, a small but significant increase in ω_{KX}^{av} above the neutral atom fluorescence yield occurs with increasing charge, but in the former case for $Z_1 > Z_2$ the effect is enormous, with ω_{KX}^{av} more than a factor of 20 greater than the neutral-atom value for the highest charge state.

To examine this effect theoretically, calculations of the K x-ray fluorescence yields ω_{KX}^i for a large number of individual excited states of neon have been made by Chen[28] and Bhalla.[29] A selection of these theoretical results is given in Table I, where average fluorescence yields for various multiplets of particular electron configurations have been computed assuming a statistical distribution of states within the multiplet. In general there is an increase in fluorescence yield from the single 1s vacancy value of 1.6×10^{-2} as outer electrons are removed from the ion, and in particular there is a dramatic increase to unity as the last few L-shell electrons are removed from the atom, making Auger decay impossible. Comparison of the measured average fluorescence yield in Fig. 6 for neon reaching $\sim 2.2\omega_0$ when excited by high-charge-state ions to the calculated values in Table I shows that a significant fraction of the x rays must come from Auger forbidden states with $\omega_{KX} = 1$.

Another technique used to study the charge state dependence of the population of K-vacancy states produced in heavy-ion collisions is through high-resolution spectroscopy of either the x-ray or Auger decay channel. In the x-ray measurements, pioneered by Richard and co-workers,[30] the relative intensities of $K\alpha$ satellite lines χ_i are used together with theoretical fluorescence yields for the lines to calculate a semiempirical average fluorescence yield by

$$1/\omega_K^{av} = \sum_i (\chi_i/\omega_{KX}^i). \quad (6.4.3)$$

For Auger measurements, performed by Stolterfoht et al.,[31] the relative intensities A_i of KLL Auger lines may be determined and then a semiempirical average K x-ray fluorescence yield is given by

$$\omega_{KX}^{av} = \sum_i \omega_{KX}^i A_i. \quad (6.4.4)$$

---

[27] R. R. Randall, J. A. Bednar, B. Curnutte, and C. L. Cocke, Phys. Rev. A **13**, 204 (1976).
[28] M. H. Chen, B. Crasemann, and D. L. Matthews, Phys. Rev. Lett. **34**, 1309 (1975).
[29] C. P. Bhalla, Phys. Rev. A **12**, 122 (1975).
[30] P. Richard, in "Atomic-Inner Shell Processes" (B. Crasemann, ed.), Chap. II. Academic, New York (1974); R. L. Kauffman, F. Hopkins, C. W. Woods, and P. Richard, Phys. Rev. Lett. **31**, 621 (1973).
[31] N. Stolterfoht, D. Schneider, P. Richard, and R. L. Kauffman, Phys. Rev. Lett. **33**, 1418 (1974).

6.4. FLUORESCENCE YIELD IN HIGH-CHARGE COLLISIONS

TABLE I. Theoretical Fluorescence Yield for a $2p \to 1s$ X-Ray Transition for Some Highly Ionized States of Neon[a]

Number of L-shell vacancies	Initial configuration	Initial state	Final state	X-ray energy (eV)	ω_{KX} (%)	ω_{KX}/ω_0
0	$1s2s^22p^6$	^2S	^2P	848.8	1.6	1.0
1	$1s2s^22p^5$	^1P	^1S	850.4	0.5	0.3
			^1D	855.4	2.3	1.4
	$1s2s2p^6$	^1S	^1P	850.9	1.4	0.9
2	$1s2s^22p^4$	^2D	^2P	859.1	0.4	0.3
		^2P	^2P	860.0	1.6	1.0
		^2S	^2P	865.1	1.8	1.1
	$1s2p^6$	^2S	^2P	861.7	2.2	1.4
3	$1s2s^22p^3$	^1P	^1S	869.6	1.9	1.2
			^1D	875.5	2.4	1.5
	$1s2p^5$	^1P	^1S	865.5	0.7	0.4
			^1D	871.1	3.4	2.1
4	$1s2s^22p^2$	^2D	^2P	881.6	1.5	0.9
		^2P	^2P	882.9	10.5	6.7
		^2S	^2P	888.3	2.0	1.3
	$1s2p^4$	^2D	^2P	876.5	0.6	0.4
		^2P	^2P	877.7	2.7	1.7
		^2S	^2P	882.9	3.2	2.0
5	$1s2s^22p$	^1P	^1S	895.6	9.6	6.0
	$1s2s2p^2$	^1D	^1P	888.2	1.6	1.0
		^1P	^1P	895.2	20.8	13.0
		^1S	^1P	895.1	2.3	1.4
	$1s2p^3$	^1P	^1S	889.6	3.7	2.3
			^1D	895.9	4.8	3.0
6	$1s2s2p$	$^3P_+$	^2S	914.0	1.2	0.8
		$^2P_-$	^2S	906.6	46.8	29.3
	$1s2p^2$	^2P	^2P	905.7	100.0	62.5
		^2D	^2P	904.0	3.2	2.0
		^2S	^2P	911.2	7.7	4.8

[a] From Ref. 29.

Using either (6.4.3) or (6.4.4), reliable values of ω_{KX}^{av} can be obtained if the set of states formed in the collision has values of ω_{KX}^i in a rather small range around ω_{KX}^{av}. However, if states with a wide range of ω_{KX}^i are populated in the collision, states with low ω_{KX}^i will be lost in the x-ray spectrum while those with large ω_{KX}^i would be missing in the Auger spectrum. In this case Eq. (6.4.3) would overestimate the average fluorescence yield while Eq. (6.4.4) would underestimate it. Indeed, for the case of neon excited by Cl^{q+} ions, Brown et al.[32] were led to erroneous conclusions on

[32] M. D. Brown, J. R. Macdonald, P. Richard, J. R. Mowat, and I. A. Sellin, Phys. Rev. A **9**, 1470 (1974).

the magnitude of the increase in fluorescence yield with charge state using this x-ray technique. For excitation of neon by lower-charge states, however, this technique has given the correct average fluorescence yield.[30]

When the fluorescence yield of an inner-shell vacancy state varies by an order of magnitude or more with charge state as is the case for decay energies of ~ 1 keV or less, high-resolution spectroscopy is necessary to determine the distribution of states and hence the average fluorescence yield. However, for decay energies in excess of ~ 3 keV, the fluorescence yield variation with charge state is much less. Reliable estimates of ω_X^{av} can be obtained from analysis of x-ray data taken with a solid-state detector of modest resolution that does not separate individual satellite lines. For the case of argon K x rays observed with a resolution of ~ 200 eV, Winters et al.[12] have used the configuration-average fluorescence yields calculated by Bhalla[33] and shown in Table II to deduce the dependence of ω_{KX}^{av} on the charge state of exciting ions for $q < 10$. The basis of the technique is to measure the centroid energy and the relative intensity of the two groups of x-ray lines, nominally a distribution of Kα and Kβ satellites observed after excitation with each incident charge state. Then these three parameters are used to select configurations with comparable calculated values from Table II that are likely to be dominant in the x-ray spectrum. The calculated fluorescence yields for configurations fitting these parameters are found comparable and provide an estimate of the ω_K^{av} that will be accurate as long as the incident charge state is not so large that all but the last few outer-shell electrons are removed from the atom.

Fluorescence yields obtained in this way are in agreement with those determined using Eq. (6.4.2) from a comparison of Auger and x-ray cross sections.[29] In the argon K x-ray data that have been analyzed[12] for incident ions with $1 \leq Z \leq 9$ and $1 \leq q \leq 9$, the value of ω_{KX}^{av} has depended only on q but not on Z. The derived projectile q dependence of ω_{KX}^{av} for argon is shown in Fig. 7 in comparison with the measured values. For more symmetric collisions of the same projectiles exciting neon, values of ω_{KX}^{av} found by applying Eq. (6.4.3) to high-resolution x-ray satellite distributions[30] have confirmed that ω_{KX}^{av} does not depend on Z_1, but only on q. Such a conclusion is not likely to be valid for $Z_1 > Z_2$, where projectile structure can be expected to play an important role in the target outer-shell structure that primarily defines the inner-shell fluorescence yield. For $Z_1 \leq Z_2$ the projectile structure is apparently not as important as the total charge q of the exciting projectile.

All the preceding discussion is pertinent to observations from dilute gas

---

[33] C. P. Bhalla, *Phys. Rev. A* **8**, 2877 (1973).

6.4. FLUORESCENCE YIELD IN HIGH-CHARGE COLLISIONS

TABLE II. Calculated $K\alpha$ and $K\beta$ X-Ray Energy Shifts,[a] Ratio of $K\beta$ and $K\alpha$ Intensities, and K-Shell Fluorescence Yields for Various Vacancy Configurations of Argon[b]

Vacancy configuration $[1s,2p^m,3p^n]$		ΔE ($K\alpha$) (eV)	ΔE ($K\beta$) (eV)	I ($K\beta$)/I ($K\alpha$)	ω_K
m	n				
0	0	—	—	0.084	0.120
1	0	15.7	43.6	0.121	0.126
2	0	32.9	89.8	0.176	0.136
3	0	51.2	138.4	0.266	0.145
4	0	71.3	189.4	0.443	0.151
5	0	91.9	242.8	0.971	0.147
1	1	16.8	48.1	0.110	0.127
2	1	34.2	95.3	0.157	0.136
3	1	53.0	145.4	0.236	0.145
4	1	73.3	196.9	0.390	0.151
5	1	94.7	251.3	0.849	0.145
6	1	—	308.2	—	0.106
1	2	18.0	53.4	0.095	0.128
2	2	36.0	101.7	0.134	0.136
3	2	55.2	152.3	0.200	0.145
4	2	75.9	205.4	0.328	0.151
5	2	97.8	260.8	0.705	0.143
6	2	—	318.7	—	0.096
1	3	19.8	59.8	0.076	0.128
2	3	38.0	108.9	0.106	0.137
3	3	57.8	160.6	0.158	0.146
4	3	78.9	214.7	0.258	0.151
5	3	101.3	271.2	0.556	0.140
6	3	—	330.1	—	0.082
1	4	21.7	67.0	0.054	0.129
2	4	40.5	117.2	0.075	0.137
3	4	60.7	170.0	0.110	0.146
4	4	82.3	225.0	0.180	0.150
5	4	105.3	282.6	0.386	0.135
6	4	—	342.4	—	0.063
1	5	24.2	74.4	0.029	0.129
2	5	43.4	126.6	0.040	0.138
3	5	64.1	180.2	0.058	0.147
4	5	86.2	236.4	0.093	0.149
5	5	109.7	294.8	0.200	0.129
6	5	—	355.8	—	0.036

[a] Energies of the normal $K\alpha$ and $K\beta$ x rays are 2957.01 and 3190.5 eV, respectively.
[b] From Ref. 33.

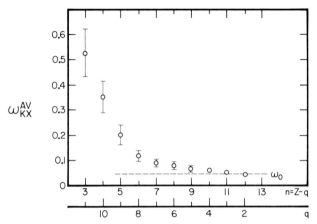

FIG. 8. Average fluorescence yield of Si^{q+} projectile ions carrying n electrons into collisions with helium[35] shown as a ratio to the neutral atom fluorescence yield ω_0. The values of ω_{KX}^{av} were deduced from measurements of the charge dependence of silicon K x-ray cross sections.

targets in which the excited atom decays as a free highly ionized atom without interaction with other target atoms. Even for the most highly ionized recoiling target atoms that interact with outer-shell cross sections comparable to atomic dimensions, free decay will occur for states with lifetimes less than 1 nsec at pressures up to 100 mTorr as long as the recoil energy does not exceed a few kilo-electron volts. However, in solids, the target density is sufficiently high that outer-shell relaxation occurs prior to inner-shell decay and the inner-shell fluorescence yield is independent of the collision parameters including the incident charge state and is dominated by the relaxation process in the solid. This relaxation is not yet understood and may involve intraatomic transitions[34] as well as interatomic decay; however, the effects in solids are so large that they dominate over variations in collision parameters.

Another technique that has been used to determine the average fluorescence yield of the states produced in atomic collisions is to examine the excitation of projectiles in well-defined initial states. Doyle et al.[35] have measured the x-ray cross sections for Si^{q+} ions incident on helium and used the results to deduce ω_{KX}^{av} for the states of silicon excited in the collisions under the assumption that the silicon K-shell ionization cross sec-

[34] R. L. Watson, A. K. Leeper, B. I. Sonobe, T. Chiao, and F. E. Jenson, *Phys. Rev. A.* **15**, 914 (1977).
[35] B. L. Doyle, U. Schiebel, J. R. Macdonald, and L. D. Ellsworth, *Phys. Rev. A* **17**, 523 (1978).

tion is independent of the charge state of the incident ion. In the analysis, 1s–2p excitation was considered to make a significant contribution to the K-vacancy production, and this component was calculated using the formulation of Bates[17] suitably modified for the binding energy of the perturbed silicon atom. The deduced average fluorescence yields for the different charge states of the silicon ions are shown in Fig. 8. Theoretical calculations for these ions are not currently available for comparison, but the charge state dependence is equivalent to that calculated for argon ions[33] normalized by a factor equal to $\omega_0(\mathrm{Si})/\omega_0(\mathrm{Ar})$, the ratio of the neutral atom fluorescence yields for the different atoms.

In interpreting the charge state dependence of measured x-ray cross sections in terms of collision processes, the appropriate fluorescence yield to be used presents the largest uncertainty that must be considered. In virtually all cases, even for rather high Z where ω_0 is large, the variation in outer-shell configurations in a collision that excites an inner-shell electron will depend on the projectile charge state. The consequent variation in fluorescence yield will influence both x-ray and Auger measurements.

6.5. Charge Dependence of Projectile Cross Sections

In an atomic collision the most pronounced charge state effects occur for inelastic cross sections to form particular states of excitation in the projectile that is preselected with a definite electron configuration. As shown in Fig. 1, some excitation channels are open with a particular incident charge state and closed for others. Except at very high velocity, electron transfer of target electrons to high-lying projectile states is always a dominant process that has a maximum cross section of atomic dimensions ($\sim \pi a_0^2$ per electron) at velocities comparable to the mean velocity of the electrons. At low velocities target outer-shell electrons are captured to projectile states having binding energy comparable to the initial state in the target. As the interaction velocity increases, the cross section for capture of a particular target shell decreases rapidly, but that for the next inner shell reaches a maximum, so that the cross section for forming excited states of the projectile is dominated by the electron capture process except at the very highest velocities (compared to the most tightly bound electrons in the target).

The charge dependence of target cross sections discussed in Chapter 6.3 has illustrated the importance only of the capture of target inner shell electrons, while the variation of target fluorescence yield (discussed in Chapter 6.4) has illustrated the importance of L and outer-shell capture in

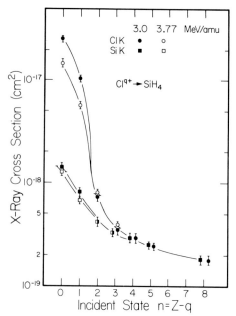

FIG. 9. Projectile charge dependence of projectile and target x-ray cross sections. $n = Z - q$ is the number of electrons on the incident chlorine ions.[36]

the collisions. The capture of these electrons to excited states of the projectile accounts for the charge dependence of projectile x-ray cross sections, even though the capture process likely does not depend strongly on the projectile electron configuration. For lower-charge-state projectile ions the total cross sections for x-ray production from target and projectile are determined by ionization cross sections, but when the incident ions carry K vacancies into the collision, the K x rays from the projectile form the dominant x-ray spectral features, and the projectile x-ray cross section is more strongly dependent on the incident charge state than that for the target.

The overall charge dependence is evident in Fig. 9, in which the K x-ray cross sections for Cl^{q+} incident on SiH_4 are shown for ions carrying $n = Z - q$ electrons into the collision.[36] Although the increase in the silicon target cross section with charge state is approximately a factor of five, similar to that for argon excited by low-Z projectiles (Fig. 3), the increase of the chlorine projectile cross section is approximately two orders

[36] J. R. Macdonald, M. D. Brown, S. J. Czuchlewski, L. M. Winters, R. Laubert, I. A. Sellin, and J. R. Mowat, *Phys. Rev. A* **14**, 1997 (1976).

6.5. CHARGE DEPENDENCE OF PROJECTILE CROSS SECTIONS

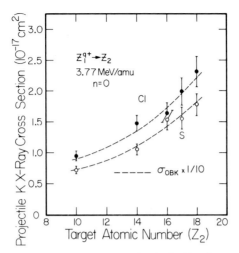

FIG. 10. Variation of projectile x-ray cross section for incident bare nuclei of chlorine and sulfur on targets of varying Z_2. The dashed lines represent the results of an OBK calculation of the electron capture cross section to projectile excited states, reduced by a factor of 10.

of magnitude when the bare nucleus is the incident state. With incident bare nuclei ($n = 0$), projectile x rays signal the decay of the excited states formed by electron capture predominantly from outer shells of the target. For incident one-electron ions ($n = 1$), the capture cross section may be reduced slightly because of screening of the projectile charge, but basically one expects the collision cross section to be comparable to that for $n = 0$. However, the resultant excited states of the two-electron projectile ion formed by the collision include numerous metastable states that do not decay by the emission of an x ray within view of the detector. This may qualitatively account for the lower x-ray cross section for $n = 1$ compared to $n = 0$, but at present there is no definitive model that accounts quantitatively for the factor of 2.4 reduction seen in Fig. 9. Indeed, for incident bare nuclei capturing outer-shell electrons, no reliable theoretical model is currently available in the literature, although a calculation in the OBK approximation to the capture process[37] reproduces the relative trends in cross sections but grossly overestimates the magnitude of the cross section. For example, the variation with target of the x-ray cross section for bare nuclei of sulfur and chlorine at 3.77 MeV/amu shown in Fig. 10 is well reproduced by calculated OBK cross sections to projectile excited states, reduced by a factor of 10.[36]

[37] J. A. Guffey, Ph.D. Thesis, Kansas State University, 1977, unpublished; J. A. Guffey, L. D. Ellsworth, and J. R. Macdonald, *Phys. Rev. A.* **15**, 1863 (1977).

In general, the magnitude of the cross section for capture to excited states (and consequent projectile x-ray decay) decreases rapidly with increasing collision velocity, while excitation and ionization cross sections reach a maximum only when the collision velocity is comparable to the expectation value of the velocity of the inner-shell electrons. Thus the charge dependence of projectile cross sections depends markedly on the projectile energy. Illustrating this pronounced variation, K x-ray cross sections for Si^{q+} ions incident on helium[20] are shown as a function of energy in Fig. 11. For the lower-charge states, the variation in fluorescence yield discussed in the last section accounts for the almost constant charge dependence of the cross section over the entire energy range. However, for the $n = 0$ and $n = 1$ incident states, the rapidly falling cap-

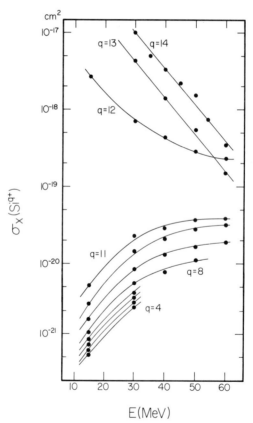

FIG. 11. Charge state and energy dependence of projectile x-ray cross sections for Si^{q+} ions incident on helium.[20,35] For the higher-charge states the q dependence of the cross section changes dramatically over the energy range.

6.5. CHARGE DEPENDENCE OF PROJECTILE CROSS SECTIONS

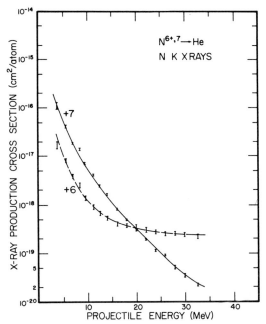

FIG. 12. Projectile x-ray cross sections for $N^{6+,7}$ incident on helium.[37] The crossover at $E_C = 20$ MeV is a general feature as the contribution to the one-electron cross section increases with energy compared to the contribution of capture to excited states.

ture cross section dramatically reduces the overall charge dependence of the x-ray cross section at higher energies. At the highest energy shown in Fig. 11, the $n = 2$ cross section exceeds that for $n = 1$ because the $^3S_1 \rightarrow {}^3P_1$ excitation cross section for the $\sim 20\%$ metastable component of the incident two-electron ions exceeds the total capture cross section.[20]

For incident ions with $n = 1$ direct excitation of the 1s electron is a process in direct competition with electron capture to excited states for the production cross section of projectile K x rays. The energy dependence of the former process reaches a broad maximum at the highest energy shown for Si^{q+} ions in Fig. 11, while the latter process rapidly becomes less important for x-ray production. Indeed, for lower-atomic-number ions, for example, $N_{14}^{6+,7}$, shown in Fig. 12, the projectile x-ray cross section for $n = 1$ becomes larger than that for $n = 0$ at a projectile energy in excess of

$$E_C/A_1 \simeq 0.2 Z_1 \quad \text{MeV/amu.} \qquad (6.5.1)$$

No theoretical significance is attached to the observed value of the crossover energy E_C, because no reliable formulation of the absolute magnitude of the contributing capture cross section is presently available. Further, the systematics of heavy-ion excitation cross sections have received virtually no attention in the literature. To remedy this situation, the first work studying excitation of well-defined states of high-charge projectile ions by application of high-resolution x-ray spectroscopy has recently been begun by Tawara and co-workers.[38] Quantitative understanding of the charge dependence of projectile excitation and capture cross sections near the peak in the excitation cross section awaits further research.

6.6. Summary

At the present time, the charge dependence of atomic cross sections observed primarily for the production of x rays and electrons from atomic collisions is being used to deduce contributions of electron capture, ionization, and excitation to the total cross section as well as the branching ratio for decay of particular excited states. Experimental techniques are readily available to extend such measurements over a wide range of collision velocity with well-defined incident states that can be followed as a function of Z along an isoelectronic sequence. Because the magnitude of competing atomic collision processes varies widely with collision parameters, general statements summarizing charge-dependent effects are inappropriate. However, experiments in which careful selection of incident and final charge state are straightforward will be applied to a wide range of measurements of differential cross sections as well as total cross sections by utilizing coincidence techniques, and can be expected to add significantly to our understanding of particular collision mechanisms.

Acknowledgments

I acknowledge support of the U.S. Department of Energy, Division of Chemical Sciences, during the course of this work. I am particularly grateful to Dea Richard for her assistance in preparation of this manuscript.

---

[38] H. Tawara, P. Richard, K. A. Jamison, and T. J. Gray, *J. Phys. B* **11**, L615 (1978).

7. COINCIDENCE EXPERIMENTS FOR STUDYING IMPACT-PARAMETER-DEPENDENT AND CHARGE-CHANGING PROCESSES*

By C. Lewis Cocke

7.1. Introduction

The last decade has brought us a great deal of experimental information on violent ion–atom collisions involving inner-shell processes. Representing the core of this body of data are measurements of total cross sections for inner-shell ionization, excitation, and charge exchange. These "singles" experiments usually have relatively high count rates and lend themselves to systematic studies in which target and projectile species, bombarding energy, etc., are varied in a rather continuous fashion. A total cross section measurement, however, integrates out a great deal of the detail characterizing each individual ion–atom encounter. Some of this detail may be recovered by detecting two or more of the reaction products in time coincidence, thus ensuring that the simultaneously gained data characterize the same target–projectile encounter. The coincidence experiments have necessarily lower count rates, since they observe at one time only a small fraction of the total encounter cross section. However, in return they allow systematic determination of the dependence of an individual encounter on, for example, the impact parameter characterizing the encounter or the charge state of the reaction products. The differential cross sections thus obtained are complementary to the coarse-grained singles surveys, in most cases allowing more stringent tests of theoretical descriptions of the collisions.

This part describes experimental considerations and surveys results of some coincidence experiments relevant to inner-shell processes involving ion–atom collisions primarily at energies in the MeV/amu range. We restrict our discussion to experiments in which x rays or Auger electrons are detected in coincidence with scattered ions. We exclude discussion of coincident inelastic energy loss (IEL) measurements, which are covered

---

* See also Vol. 7A of this series, Part 4, particularly Section 4.2.5.

in Part 4. The largest coherent set of data available to date are measurements of the impact parameter dependence of inner-shell vacancy production. Part of this information comes to us from IEL experiments, part from the coincident detection of scattered projectiles and x rays or Auger electrons emitted following the encounter. Discussion of the latter type of experiment forms the heart of this part in Chapter 7.2. Related experiments involving coincidences between the same reaction products and charge-analyzed ions are discussed in Chapter 7.3.

7.2. Impact Parameter Dependence of Inner-Shell Vacancy Production

7.2.1. Introduction and History

From the early experiments of Afrosimov et al.[1] and Morgan and Everhart,[2] it was apparent that inner-shell excitation probabilities in violent ion–atom encounters are quite sensitive to the impact parameter (b), or equivalently to the distance of closest approach (r_0), characterizing the collision. For the case of Ar–Ar collisions the probability of vacating the L shells was found to jump abruptly from nearly zero for large r_0 to a large value for collisions in which the L shells interpenetrate, and to depend only weakly on bombarding energy. These experiments and their successors, most of which employ the IEL technique, are described in several excellent review articles [3-4] to which the reader is referred, as well as in Part 4 of this volume. Thomson et al.[5] were the first to use ion-electron coincidences to measure the r_0 dependence of inner-shell vacancy production probabilities. They detected argon L-Auger electrons from Ar–Ar collisions in coincidence with scattered argon ions and found results that confirmed conclusions developed from the IEL data (see Section 7.3.1).

Shortly thereafter began the explosion of experimental information on inner-shell vacancy production that forms the subject of much of this book. Attention was called early to the importance of measuring inner-shell vacancy production probabilities (P), as distinguished from

---

[1] V. V. Afrosimov and N. V. Fedorenko, *J. Tech. Phys. (USSR)* **27**, 2557 (1957) [*Sov. Phys.—Tech. Phys.* **2**, 2591, (1957)]; V. V. Afrosimov, Yu, S. Gordeev, M. N. Pavov, and N. V. Fedorenko, *Zh. Tekn. Fiz.* **34**, 1613, 1624, 1637 (1964); [*Sov. Phys.—Tech. Phys.* **9**, 1248, 1256, 1265 (1965)].

[2] G. H. Morgan and E. Everhart, *Phys. Rev.* **128**, 667 (1962).

[3] Q. C. Kessel, "Coincidence Measurements", in "Case Studies in Atomic Physics I," (E. W. McDaniel and M. R. C. McDowell, eds.), p. 399. N. Holland-Amsterdam, 1969.

[4] Q. C. Kessel and B. Fastrup, *Case Stud. Atom. Phys.* **3**, 137 (1973).

[5] G. M. Thomson, P. C. Laudieri, and E. Everhart, *Phys. Rev. A* **1**, 1439 (1970).

7.2. INNER-SHELL VACANCY PRODUCTION

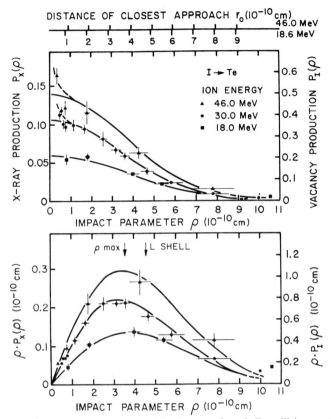

FIG. 1. P_x and bP_x vs. impact parameter for L x rays from I–Te collisions at three bombarding energies (Stein et al., Phys. Rev. A **5**, 2126, 1972). High points at small b may be due to recoil effects (D. Burch and K. Taulbjerg, Phys. Rev. A **12**, 508, 1975).

only total cross sections. For example, the shift to higher energy of characteristic K x rays produced in heavy-ion bombardments of solid targets[6] is due to simultaneous K–multiple-L vacancy production. Such a process is only important because P_L is of order unity for such collisions.

The first x-ray–ion coincidence measurements were made by Stein et al.,[7] who measured L x rays from iodine bombardment of tellurium in coincidence with the scattered iodine ions. They deduced a probability for L-vacancy production as high as 0.5 for b inside the L-shell radius (see Fig. 1). These early measurements were on the L shell and invited inter-

---

[6] For a review, see P. Richard, "Ion–Atom Collisions," in "Atomic Inner-Shell Processes," (B. Crasemann, ed.), Vol. 1, p. 73. Academic Press, New York, 1975.

[7] H. J. Stein, H. O. Lutz, P. H. Mokler, K. Sistemich, and P. Armbruster, Phys. Rev. Lett. **24**, 701 (1970); and Phys. Rev. A **2**, 2575 (1970); and Phys. Rev. A **5**, 2126 (1972).

pretation in terms of the Fano–Lichten[8] molecular-promotion model. Subsequent work has concentrated on K-shell vacancy production in both Coulomb and molecular-promotion regimes.

7.2.2. The Single-Encounter Experiment

7.2.2.1. Schematic.
A schematic of a classical encounter between a projectile of nuclear charge Z_1 and a target of nuclear charge Z_2 is shown in Fig. 2. So long as the deBroglie wavelength λ of the projectile nucleus is small compared to the dimensions of the scattering center to be probed (typically of order of $r_n = n^2 a_0 / Z_2$, where $a_0 = 0.53 \times 10^{-10}$ cm, and n characterizes the target shell of interest), the projectile may be treated as following a classical trajectory through the scattering region. Nonrelativistically $\lambda = 29/(mE)^{1/2}$, where m (amu) and E (MeV) are the projectile mass and laboratory energy, respectively, and λ is in fm. Thus for a 1-MeV proton probing the K shell of a copper atom, $\lambda = 29$ fm, much smaller than $r_K = 1827$ fm. If the entire trajectory of the projectile may be described classically, θ may be calculated uniquely from b, provided the effective internuclear potential is known. Since most experiments have ments have involved x-ray–ion coincidence measurements, a schmatic experimental arrangement for measuring the probability per collision for x-ray production, $P_x(b)$, is shown in Fig. 3. We restrict the following discussion of experimental detail to the x-ray case, with the understanding that most of the conclusions pertain to particle–Auger-electron coincidence experiments as well.

Experimentally $P_x(b)$ is determined from the ratio between the number of detected scattered particles and the number of true coincidences:

$$P_x(b) = (N_t/N_p)(4\pi/\Delta\Omega_x \epsilon_x), \qquad (7.2.1)$$

where $\Delta\Omega_x$ is the solid angle subtended by the x-ray detector, ϵ_x the detection efficiency of this detector, N_t the number of true coincident events, and N_p the total number of scattered particles detected.

Equation (7.2.1) assumes a point collision center and isotropic x-ray emission. If the interaction region is extended, as is the case when a gas target is used, $(\Delta\Omega_x \epsilon_x / 4\pi)$ must be replaced by the average probability that an x ray produced within the scattering volume will be registered by the x-ray detector. For very large angle scattering in symmetric collision, account must be taken of the fact that both projectiles and recoils enter the particle detector.[9,10] For small-angle scattering this correction is quite small.

---

[8] U. Fano and W. Lichten, *Phys. Rev. Lett.* **14**, 627 (1965); W. Lichten, *Phys. Rev.* **164**, 131 (1967).
[9] S. Sackmann, H. O. Lutz, and J. Briggs, *Phys. Rev. Lett.* **32**, 805 (1974).
[10] D. Burch, *Phys. Lett.* **47A**, 437 (1974).

7.2. INNER-SHELL VACANCY PRODUCTION

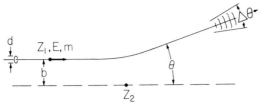

FIG. 2. Collision schematic.

The efficiency of the particle detector ϵ_p does not enter. However, if ϵ_p and the solid angle $\Delta\Omega_p$ subtended by this detector are known, an experimental check may be performed by comparing total cross sections for x-ray production obtained by two semi-independent methods:

Singles: At any θ,

$$\sigma_x = [(d\sigma/d\Omega)_p](4\pi\epsilon_p \Delta\Omega_p/\epsilon_x \Delta\Omega_x)[N_x/N_p], \quad (7.2.2)$$

where N_x is the total number of singles x rays detected and $(d\sigma/d\Omega)_p$ the differential cross section for total particle scattering. If the relationship between b and θ is known, one may use

$$(d\sigma/d\Omega)_p = (b/\sin\theta)(db/d\theta). \quad (7.2.3)$$

Coincidence:

$$\sigma_x = 2\pi \int_0^\infty P_x(b) b \, db. \quad (7.2.4)$$

The equality of σ_x calculated from these equations is insensitive to errors in x-ray detection efficiency. The probability for the production of

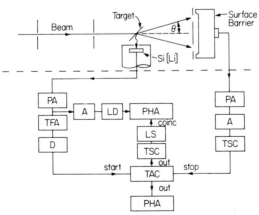

FIG. 3. Schematic of experimental arrangement for solid target and block diagram of typical electronics. PA, Preamplifier; TFA, fast amplifier; A, amplifier; D, fast discriminator; TSA, timing single-channel analyzer; PHA; pulse height analyzer; TAC, time-to-amplitude converter; LD, linear delay; LS, logic shaper.

the vacancy $P(b)$ is equal to $P_x(b)/\omega$, where ω is the fluorescence yield characterizing the appropriate radiative decay of the collisionally excited system. This analysis assumes ω to be b independent, although there is some experimental evidence that this assumption may not always be a safe one.[11]

7.2.2.2. Is the Assumption of Classical Scattering Justified? Bohr[12] has pointed out that one's ability to associate θ with a unique b is not guaranteed by $\lambda \ll r_n$. While the projectile wave packet may be rather tight while passing through the scattering region, it will subsequently spread. Thus, if a wave packet is confined at b to a Gaussian of full width d, a diffraction width $\phi \sim \lambdabar/d$ will result (See Fig. 2). The scattering angle is thus rendered uncertain by an amount $\Delta\theta$, which has contributions from both the diffraction and differential deflection by the scattering potential. Since for classical small-angle Coulomb scattering $\theta = d_0/d$, where d_0 is the collision diameter $Z_1 Z_2 e^2/E$, the latter contribution may be estimated as $(d\theta/db)\,\Delta b \simeq (d_0/b^2)(d/2)$. Combining these two uncertainties quadratically and optimizing the aperture diameter leads to a minimum uncertainty in b given by $\Delta b/b \geqslant [\lambdabar/d_0]^{1/2}$. Thus the parameter $\chi \equiv d_0/\lambdabar = 2Z_1 Z_2 e^2/\hbar v$ must be large compared to unity in order that θ and b be uniquely related. Of course, a $P(\theta)$ can always be deduced. One loses only the physical picture of a collision taking place at a well-defined impact parameter.

7.2.2.3. Classical Relationship between b and θ. For a light projectile penetrating deeply into the K shell of a high-Z target, the relationship between b and the center-of-mass (c.m.) scattering angle θ_c will be nearly that characteristic of Rutherford scattering, $b = d_0'(\tfrac{1}{2})\cot(\theta_c/2)$, where $d_0' = Z_1 Z_2 e^2/E_{cm}$ and E_{cm} is the center of mass energy. The usual laboratory c.m. transformation[13] may be made to go from θ_c to θ. For small θ this reduces to $\theta = d_0/b$. If the scattering is to probe a target shell of binding energy U, an impact parameter near the classical radius of the shell, $\theta \sim 2Z_1 U/E$. Thus, the small-angle expression will often be excellent. For example, a 25-MeV oxygen ion scattering at $b = r_K$ from copper is deflected to $\theta = 7.3$ mrad.

For collisions between higher-Z systems and at larger b, the influence of electronic screening on the trajectory will not be negligible. The reader is referred to the article of Lindhart et al.[14] for an in-depth discus-

---

[11] H. O. Lutz, *Proc. Second Int. Conf. Inner Shell Ioniz. Phenomena*, p. 617. Freidburg, 1967.
[12] N. Bohr, *Mat.-Fys. Medd. Dan. Vidensk. Selsk.* **18**, No. 8 (1948); see also ref. 14, p. 10.
[13] R. D. Evans, "The Atomic Nucleus," p. 828. McGraw-Hill, New York, 1955.
[14] J. Lindhart, V. Nielsen, and M. Scharff, *Mat. Fys. Medd. Dan. Vidensk, Selsk.* **36**, No. 10 (1968).

sion of the problem. For most purposes a screened Coulomb potential of the form suggested by Bohr,[12]

$$V(R) = [(Z_1 Z_2 e^2)/R]e^{-R/a}, \quad \text{with} \quad a = 0.88 a_0/(Z_1^{2/3} + Z_2^{2/3})^{1/2},$$

has been used to represent the scattering potential. Calculation of $b(\theta)$ may be performed numerically following the prescription given by Everhart *et al.*[15] Direct verification of the validity of this procedure may be made by experimentally determining the differential cross section for projectile scattering under the same conditions as those used for the $P(b)$ measurements. If the correct relationship for $b(\theta)$ is being used, the experimental $(d\sigma/d\Omega)_D$ must follow that deduced from Eq. (7.2.3). Although more realistic potentials could be used, the Bohr potential appears quite adequate for $P(b)$ measurements made to date.

7.2.2.4. Multiple Scattering and Finite Angular Resolution. The expressions given above are valid only under single-scattering conditions. Solid-target experiments must be done with finite target thickness, and thus the deflection angle θ may be produced by multiple soft collisions instead of a single hard one. Studies of multiple scattering have been made by many authors[14,16] and go well beyond the scope of this part.

While it is possible to make corrections for multiple scattering effects to the analysis outlined above, it may be difficult to do so correctly. In principle, one must know the shape of $P(b)$ in order to make appropriate corrections. Thus an iteration procedure must be used. The better procedure is to avoid the problem by using sufficiently thin targets. Roughly speaking, it is necessary that the angular half-width of the beam profile, after passage through the target, remain much smaller than the smallest angle to be used in $P(b)$ measurements. Extensive calculations of multiple-scattering angular distributions, based on Thomas–Fermi and Lenz–Jensen potentials, have been made by Sigmund and Winterbon.[17] On the basis of their tabulated distribution functions, Schmidt-Böcking *et al.*[18] have presented, in a particularly useful form, the ratio Λ of the angular distribution for multiple scattering to that for single scattering as a function of reduced scattering angle $\tilde{\alpha}$ and reduced target thickness τ (Fig. 4). Here $\tilde{\alpha} = Ea/2Z_1Z_2e^2$ and $\tau = \pi a^2 Nx$, where Nx is the areal target number density. One should ideally confine one's measurements to a

---

[15] E. Everhart, G. Stone, and R. J. Carbone, *Phys. Rev.* **99**, 1287 (1972).

[16] L. Meyer, *Phys. Stat. Sol.* **44**, 253 (1971); G. Moliere, *Z. Naturforsch.* **3a**, 78 (1948); W. T. Scott, *Rev. Mod. Phys.* **35**, 231 (1963).

[17] P. Sigmund and K. B. Winterbon, *Nucl. Inst. Meth.* **119**, 541 (1974); and H. Knudsen, Private communication.

[18] H. Schmidt-Böcking, A. Gruppe, and W. Lichtenberg, private communication.

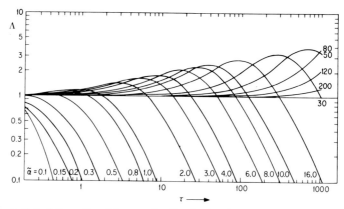

FIG. 4. A plot of the ratio of the angular distribution for multiple scattering to that for single scattering vs. reduced target thickness for fixed values of reduced scattering angle. See text for further detail (H. Schmidt-Böcking, A. Gruppe, and W. Lichtenberg, *Frankfurt Annual Report*, 1975).

region of $\tilde{\alpha}$ and τ such that Λ is near unity. In practice, the best experimental test one can make for the validity of the single collision assumption is to measure $P(b)$ over a range of target thicknesses. Since the multiple-scattering effect increases with τ, the signature of the single-scattering regime is the lack of dependence of $P(b)$ on τ.

In both solid and gas target experiments finite angular resolution effects may be important, since measurements often need be made at very small θ. If all contributions to the experimental resolution function, including beam collimation, finite detector aperture, and multiple scattering, are represented by a single resolution function, the problem may be dealt with rather simply.[19,20] It can be shown that if P varies slowly over the resolution function, the measured P given by Eq. (7.2.1) is equal to the true $P(\langle \theta \rangle)$, where $\langle \theta \rangle$ is the average angle accepted by the system, weighted according to $(d\sigma/d\Omega)_p$. If the resolution function is taken to be of the form $\exp(-\theta^2/\sigma^2)$, numerical evaluation of $\langle \theta \rangle$, using the Rutherford relationship between b and θ, shows that the angle of the geometrical center of the detection system, θ_g, differs from $\langle \theta \rangle$ by less than 10% for $\theta_g > 5\sigma$. For small θ_g, a correction for this effect may be made. Were $(d\sigma/d\Omega)_p$ not so strongly forward peaked, the effect would be extremely small. Of course, if P varies substantially over an angular range σ, the structure will be washed out and cannot be regained in the data analysis.

[19] R. Randall, Ph.D. dissertation, Kansas State University, Manhattan (1975).
[20] C. Annett, M.S. dissertation, Kansas State University, Manhattan (1977).

7.2.3. Techniques

7.2.3.1. General Timing Considerations.
Referring to the schematic of Fig. 3, we label the count rates in particle and x-ray detectors by R_p and R_x, respectively. We designate by p the probability that, given a detected scattering, an associated x ray will be detected ($p = P_x 4\pi/\epsilon_x \Delta\Omega_x$). The decay time of an inner shell vacancy is typically $\leq 10^{-14}$ sec, very much less than the time resolution of the detectors or electronics. If the beam current has no time structure, a spectrum of time delays t between x-ray and particle detection will be composed of a background of random coincidences, whose rate in each time channel is given by $(R_p R_x \Delta t)(\exp - [t R_p])$, upon which will appear a peak of real coincidences centered about t_0 with integrated count rate $(R_p p)(\exp - [t_0 R_p])$. Here t_0 is the net delay of scattered-particle detection, including particle flight time, differential detector response times, and electronic delay, and Δt is the width of a single bin in the time spectrum. At high count rates, the dead time t_d of the time-to-amplitude converter (TAC) may become important, decreasing each of the above rates by the factor $\exp - [t_d R_x]$. Normally, $R_x \ll R_p$ and, since $t_0 < t_d$, one uses the x-ray detection to start the TAC. Further, unless R_p is very large, $\exp - [t_0 R_p]$ will be near unity and the randoms spectrum will be flat (see Fig. 5). The ratio of real coincidences N_t to randoms underlying the real peak N_r is given by $N_t/N_r = p/R_x T$, where T is the total width of the coincidence peak, the time resolution. In order to maintain a good peak to background, it is important to make p as large as possible, by maximizing the efficiency of and

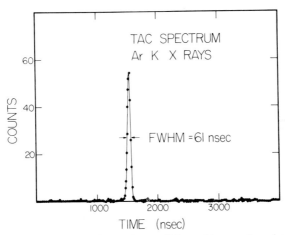

FIG. 5. Typical TAC spectrum taken under conditions of low randoms/reals (R. R. Randall, J. A. Bednar, B. Curnutte, and C. L. Cocke, *Phys. Rev. A* **13**, 204, 1976).

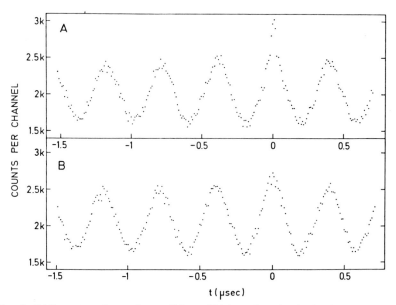

FIG. 6. TAC spectra taken under conditions of high randoms/reals showing structure due to time variation of the beam. The spectra are of (a) x-ray–particle time differences showing a coincidence peak of zero time delay, and (b) uncorrelated particle–particle time differences showing a small spurious peak at zero time delay due to high-frequency noise in the beam (J. U. Andersen, E. Laegsgaard, M. Lund, and C. D. Moak, *J. Phys.* **B9**, 3247, 1976).

solid angle subtended by the x-ray detector, and to keep R_x as low as practical, e.g., by minimizing the beam current consistent with an acceptable overall count rate. If the error bars in N_t and N_r are limited by counting statistics, one can show that, by increasing the beam current and thus R_x, one gains, in a time-limited measurement, enough statistical precision to more than compensate for a deteriorating N_t/N_r up to $N_t/N_r \sim 1$, thereafter approaching a limiting precision. We note that, by interchanging the roles of x-ray and particle detectors in the above discussion, one may alternatively obtain $N_t/N_r = p'/R_p T$. Here p' represents the probability that, given a detected x ray, the associated particle will be detected. Thus, one wishes to maximize the efficiency and solid angle of the particle detector as well, consistent with the necessary angular resolution. For large-angle scattering the small size of p' leads to deteriorating N_t/N_r and requires low values of R_p.

The assumption of steady beam current is not always justified, even for a DC machine such as a Van de Graaff accelerator. Andersen *et al.*[21]

[21] J. U. Andersen, E. Laegsgaard, M. Lund, C. D. Moak, and L. Kocbach, *J. Phys. B* **9**, 3247 (1976); and private communication.

7.2. INNER-SHELL VACANCY PRODUCTION

have found that strong systematic time structure in Van de Graaff beams is introduced by plasma oscillations in RF and possibly duoplasmatron ion sources. Under such conditions the randoms spectrum will develop structure characteristic of that of the instantaneous beam current. Figure 6 shows an example of such a time spectrum obtained under conditions of low N_t/N_r. It is clear that subtraction of random coincidences for such a case is at best treacherous, becoming increasingly difficult as the frequency of the beam oscillation becomes comparable to $1/T$. Such an effect is especially dangerous in low reals-to-randoms situations. For the case of protons on copper at 0.5 MeV, where low P_x required running at high count rates to obtain acceptable statistics, Andersen *et al.* report that an error of a factor of three can result from failure to deal adequately with structure in the randoms spectrum. Of particular concern is high-frequency "noise" in the beam, which may generate a false peak at zero time delay that is difficult to distinguish from a real coincidence peak.

7.2.3.2. Detectors. The value of T realized is generally dependent on the response time of the detector involved. If the experiment is not beam current limited, it is clearly advantageous to run R_x (and R_p) as high as possible, consistent with an acceptable R_xT product. The limiting factor may then become T, which should be minimized. Some general characteristics of commonly used detectors are summarized here.

7.2.3.2.1. X-RAY AND ELECTRON DETECTORS. The most commonly used x-ray detector is the Si(Li) detector. The time resolution that can be obtained from such a device, using leading-edge timing, is very dependent on the x-ray energy. A systematic study of this dependence has been made by Schmidt-Böcking *et al.*,[22] whose results are shown in Fig. 7. It is difficult to use the Si(Li) detector for timing below about 2 keV because the noise level in the fast-amplifier circuit is quite high. For a similar reason the slew-compensating electronics often used in γ-ray timing experiments are not practical for low-energy x rays. For high-energy x rays the Ge(Li) detector has been used by Burch *et al.*,[23] who achieved a time resolution of 7.3 nsec for lead K x rays (~74 keV).

For sufficiently high-energy x rays, a NaI detector may be used, coupled to a moderately fast photomultiplier tube. Such detectors may be made very large to obtain large $\Delta\Omega_x$. The timing is limited by the light output per unit time from the NaI crystal, which at low levels gives rise to a statistical distribution of time of emission of the photoelectrons. Andersen *et al.*[21] report a time resolution of ~60 nsec for the copper K x ray (~8 keV).

[22] H. Schmidt-Böcking, I. Tserruya, H. Zekland, and K. Bethge, *Nucl. Inst. Meth.* **120**, 329 (1974).
[23] D. Burch, W. B. Ingalls, H. Wieman, and R. Vandenbosch, *Phys. Rev. A* **10**, 1245 (1974).

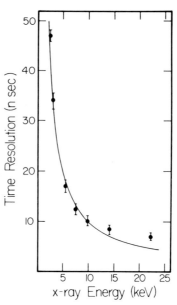

FIG. 7. Time resolution vs. x-ray energy for Si[Li] surface barrier coincidences obtained by H. Schmidt-Böcking, I. Tserruya, H. Zekl, and K. Bethge, *Nucl. Inst. Meth.* **120,** 329 (1974).

For lower-energy x rays a proportional counter may be used. Here the drift time between interaction of the primary x ray in the counter gas and the ultimate gas multiplication must be kept to a minimum.[24] Since drift velocities are typically of order 10^6 cm/sec, the primary interaction region should be kept as small as possible and at a uniform distance from the central wire, which is quite difficult to do for higher-energy x rays. A time resolution of 0.27 μsec (FWHM) was obtained by Stein *et al.*[7] for 4-keV x rays, and 30 nsec FWHM by Sackmann *et al.*[25] for ~1-keV x rays.

Better time resolution may be achieved if Auger electrons rather than x rays are detected. The most widely used electron detector is the channel electron multiplier, although fast timing is typical of most electron multiplier devices. Time resolutions at least as good as 8 nsec (FWHM) may straightforwardly be achieved.[26]

7.2.3.2.2. PARTICLE DETECTORS. The most commonly used particle detector is the surface barrier detector. Such a device is capable of sub-

[24] H. A. Staub, "Experimental Nuclear Physics" (E. Segré, ed.), Vol. 1, p. 30. Wiley, New York, 1953.
[25] S. Sackmann, private communication (1977); see also ref. 9.
[26] M. Rødbro, E. H. Pederson, C. L. Cocke, and J. R. Macdonald, *Phys. Rev. A* **19,** 1936 (1979).

7.2. INNER-SHELL VACANCY PRODUCTION

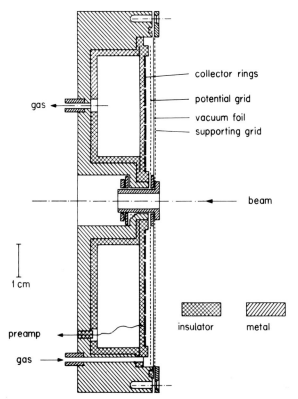

FIG. 8. Position-sensitive avalance detector used by G. Gaukler, H. Schmidt-Böcking, R. Schuch, R. Schulé, H. J. Specht, and I. Tserruya, *Nucl. Inst. Meth.* **141**, 115 (1977).

nanosecond timing and generates large enough instantaneous currents to be used with inductive pickoffs, which retain the total charge information. Since the x-ray detector is typically the slowest element in the system, it is seldom important to try to press the surface barrier detector to its ultimate timing capability. The major disadvantage of such detectors is that they deteriorate rapidly when exposed to the large-particle fluxes often encountered in x-ray–ion coincidence experiments. Damage by heavy ions is particularly severe; for example, a 25-MeV copper beam will seriously damage such a detector in a few minutes at 10^3 particles/mm²-sec.

Gaukler *et al.*[27] have described a position-sensitive avalanche detector capable of handling count rates up to 2×10^6 particles/sec with a time

[27] G. Gaukler, H. Schmidt-Böcking, R. Schuch, R. Schulé, H. J. Specht, and I. Tserruya, *Nucl. Inst. Meth.* **141**, 115 (1977).

resolution of 0.4 nsec (FWHM) (See Fig. 8). Such a detector does not suffer deterioration. It has poor energy resolution, but this does not generally present a problem in $P(b)$ experiments. By using simultaneous parallel detection on different annular sections of the collector plate, enormous gains may be made in the time needed to gather a sufficient number of coincidences and very small P_x experiments are possible.[28]

7.2.3.3. Target–Detector Arrangements. In order to optimize reals-to-randoms and to minimize data collection times, it is important to maximize the detection efficiencies of both x-ray and particle detector, i.e., p and p'. For the particle detector this means using an annular detector with as large a solid angle as is consistent with the desired precision in impact parameter. For the x-ray detector, one wishes to maximize the solid angle subtended. A typical scattering arrangement for a solid target is shown in Fig. 3. The beam is tightly collimated by a set of apertures following the last focusing element in the beam transport system. The impact parameter is changed by varying either the distance between the ion detector and the target or the radius of the annular aperture in front of the detector.

For higher-Z beams, P_x may depend sensitively on the electronic configuration of the projectile and thin gas targets become necessary to maintain controlled beam preparation. Because of the intrinsically low-count-rate nature of the coincidence experiment, it may be difficult to operate with thin enough targets to maintain true single-collision conditions. In studying inner-shell processes, however, it is generally the case that the configuration of the projectile's inner shells is critical. Targets thin enough to maintain core configuration purity may be considerably thicker than those needed for outer-shell charge state or excitation purity. Especially in the case of foil stripped beams, one must be careful that metastable systems with inner-shell vacancies such as the $(1s2s)2^3S_1$ state in heliumlike systems are not brought into the gas cell under the presumption that only ground-state ions are present.[29,30]

In Fig. 9 we show the gas target arrangement used first by Randall et al.[31] The target cell is surrounded by an intermediate vacuum pumped by a large diffusion pump. The flight path in the cell is minimized so that the

---

[28] R. Schuch, H. Schmidt-Böcking, R. Schulé, I. Tserruya, G. Nolte, and W. Lichtenberg, "Fith International. Conference Atomic Physics," (R. Marrus and M. H. Prior, eds.), p. 116. H. Shugart Berkeley, 1976.

[29] U. Schiebel, B. L. Doyle, J. R. Macdonald, and L. D. Ellsworth, *Phys. Rev. A* **16**, 1089 (1977).

[30] D. L. Matthews, R. J. Fortner, and G. Bissinger, *Phys. Rev. Lett.* **36**, 664 (1976).

[31] R. R. Randall, J. A. Bednar, B. Curnutte, and C. L. Cocke, *Phys. Rev. A* **13** 204 (1976); C. L. Cocke and R. R. Randall, *Phys. Rev. Lett.* **30**, 1016 (1973).

7.2. INNER-SHELL VACANCY PRODUCTION

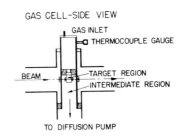

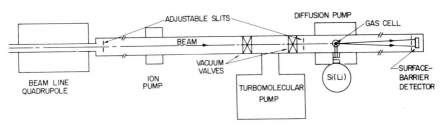

FIG. 9. Gas target used for $P(b)$ measurements by Randall *et al.*, *Phys. Rev. A* **13**, 204 (1976).

x-ray detector may view as much of the scattering region as possible. At pressures of 150 mTorr in the cell the pressure in the beam line rises only $\sim 2 \times 10^{-6}$ Torr. The exit apertures must be kept large enough to allow the scattering of interest to exit clearly, a requirement not difficult to meet for the very small angle scattering usually of interest ($\theta \lesssim 2°$). With the low density of a gas target, slit scattering may contribute significantly to R_p and must be minimized using beam scrapers or careful slit-edge choice.

Problems arise in evaluating the effective x-ray detection efficiency for a gas target. For the case of the target shown in Fig. 9, the average x-ray detection efficiency was evaluated by integrating $(4\pi/\Delta\Omega_x\epsilon_x)$ over the path through the gas cell.[19] Such a procedure is not completely justified because of pressure fringing effects, however. A more direct measurement of the efficiency using a known x-ray production cross section and Eq. (7.2.2) is desirable.[19,31] Alternatively, the $P_x(b)$ may be normalized to a known total cross section using Eq. (7.2.4).[9]

7.2.4. Experimental Results

The collisional production of inner-shell vacancies in target or projectile is most easily understood in two limiting cases. If Z_1/Z_2 is small compared to the scaled velocity v of the projectile [$v = E/1836\ U)^{1/2}$], direct ionization of the target electron by interaction with the time-varying field

TABLE I. Coincidence Measurements of $P_x(b)$ for Characteristic X-Ray and Auger Electron Production

Projectile	Target	E	Target type	Shell	Impact parameter range	Ref.
p	Ag, Se, Cu	1, 2 MeV	Solid	K	200–3000 fm	a
p	Al, Ca, Ni, Ag	0.3–3.0 MeV	Solid	K	Fixed $\theta = 5°$	b
p	Au, Se	2 MeV	Solid	K	30–450 fm	c
p, α, Li, O	Ni	1.25 and 2.19 MeV/amu	Solid	K	100–2200 fm	d
p	Ta, Au, Bi, U	4 MeV	Solid	K	40–700 fm	e
α	Pb	25, 40, 74 MeV	Solid	K	12–38 fm	f
p	Cu, Al, Ti	0.5–2 MeV	Solid	K	22–130 fm	g
α	Po	5.3 MeV	^{210}Po α-decay	K, L	0	h, i
p	Cu	0.5–2.5, 2 MeV/amu	Solid	K	100–3000 fm	j
p	Cu	0.5–2.0 MeV	Solid	K	$0° < \theta < 135°$	k
p	Ag	2 MeV	Solid	K	40–600 fm double differential	l
p	Pb	2, 3, 6 MeV	Solid	L	150–5000 fm	m
p	Au	1–2 MeV	Solid	L	200–2800 fm	n
O	Cu	25, 35, 43 MeV	Solid	K	400–3400 fm	o, p
F(5+, 9+); C(4+, 6+)	Ar	1.58 MeV/amu	Gas	K	800–11,000 fm	p
Cl	Pb	50–100 MeV	Solid	K	$\theta = 150°$–$174°$	q
					1–4 fm	
I	Te	18–46 MeV	Solid	L	500–11,000 fm	r
I	Au	60 MeV	Single crystal	M	Channeling expt. 0.1–1.1 Å	s
Ne$^+$	Ne	235, 363 keV	Gas	K	500–6000 fm	t
Na$^+$	Ne	420 keV	Gas	K	500–6500 fm	u
Cl	Al, Ti, Cu	21, 30 MeV	Solid	K	400–4400 fm	v
Cl(5+ to 11+)	Ar	15, 30 MeV	Gas	K	1000–7000 fm	w
O, Ne, S	Al	0.7–2.6 MeV	Solid	K	Fixed $\theta = 5.1°$	x
Ni	Mn, Sn, Pb	45, 94, 138 MeV	Solid	K, L	100–10,000 fm	y
Cu	Ni	50, 65.6 MeV	Solid	K	500–4000 fm	z
Cl	Cl, Ti, Ni	35 MeV	Solid	K	100–3200 fm	aa
p	Ag, Cu	1, 2 MeV	Solid	K	$110° < \theta < 15°$	bb
Ne$^+$	O	25, 385 keV	Gas	K*	500–8000 fm	cc

O⁺	Ne	308 keV	Gas	K*	500–7000 fm	cc
Mg⁺	Ne	380 keV	Gas	K*	1000–4000 fm	cc
O⁺	O	400 keV	Gas	K*	400–9000 fm	cc
N⁺	N	400 keV	Gas	K†	300–10,000 fm	cc
Ar⁺	Ar	4.5 MeV	Gas	K	100–1800 fm	dd
I	Ag	47, 100 MeV	Solid	K	100–2000 fm	ee
Ar	Al, Sn	5.2 MeV	Solid	K	100–400 fm	
α	Pb	5.3, 6.77, 8.78 MeV	Solid	K	^{210}Po decay (O); backscattering ($\theta = 90°, \sim 180°$)	ff
p	Cu, Ti, Sc, P, Si, Al, Mg, Cu, Al, Ti	1 MeV	Solid	K	Fixed $\theta = 8°$ (15° for Cu)	gg
		1–3.5 MeV				
Kr	Ge	118 MeV	Solid	K	25–800 fm	hh
p	Mo, Sn, Au	7 MeV	Solid	L	2–20 fm	ii

* Auger-electron and x-ray coincidence measurements.
† Auger-electron coincidence measurements only.

[a] E. Laegsgaard, J. U. Andersen, and L. C. Feldman, *Proc. Intl. Conf. Inner-Shell Ion Phenom.*, Atlanta (1972), 1019 (USAEC CONF—720404); and *Phys. Rev. Lett.* **29**, 1206 (1972).
[b] W. Brandt, K. W. Jones and H. W. Kraner, *Phys. Rev. Lett.* **30**, 351 (1973).
[c] M. Lund, Thesis, University of Aarhus (1974).
[d] R. Schulé, K. E. Stiebing, H. Zek, H. Schmidt-Böcking, I. Tserruya, and K. Bethge, Intl. *Conf. Phys. El. Atom. Coll. 9th*, 498, Seattle (1975); *Fifth Intl. Conf. Atom. Phys.*, 182, Berkeley (1976); and *J. Phys. B*, **10**, 2663 (1977).
[e] D. L. Clark, T. K. Li, J. M. Moss, G. W. Greenlees, M. E. Cage, and J. H. Broadhurst, *J. Phys. B* **8**, L378 (1975).
[f] R. J. Vader, A. van der Woude, R. J. de Meijer, J. M. Hansteen, and P. A. Amundsen, *Phys. Rev. A* **14**, 62 (1976).
[g] J. F. Chemin, J. Routurier, B. Saboya, Q. T. Thein, and J. P. Thibaud, *Phys. Rev. A* **11**, 549 (1975).
[h] J. P. Briand, P. Chevallier, A. Johnson, J. P. Roget, M. Tavernier, and A. Touati, *Phys. Rev. Lett.* **33**, 266 (1974).
[i] H. J. Fischbeck and M. S. Freedman, *Phys. Rev. Lett.* **34**, 173 (1975).
[j] J. U. Andersen, E. Laegsgaard, M. Lund, and C. D. Moak, *Nucl. Inst. Meth.* **132**, 507 (1976).
[k] J. U. Andersen, E. Laegsgaard, M. Lund, C. D. Moak, and L. Kocbach, *J. Phys. B* **9**, 3247 (1976).
[l] E. Laegsgaard, J. U. Andersen, and M. Lund, *Phys. Fenn.* **9**, Suppl. S1, 2 (1974).

(Continued)

TABLE I. (*Continued*)

[m] K. E. Stiebing, H. Schmidt-Böcking, R. Schulé, K. Bethge, and I. Tserruya, *Phys. Rev. A* **14**, 146 (1976).
[n] E. Laegsgaard, J. U. Andersen, and M. Lund, *Phys. Fenn.* **9**, Suppl. S1, 49 (1974).
[o] C. L. Cocke and R. R. Randall, *Phys. Rev. Lett.* **30**, 1016 (1973).
[p] R. Randall, J. A. Bednar, B. Curnutte, and C. L. Cocke, *Phys. Rev. A* **13**, 204 (1976).
[q] D. Burch, W. B. Ingalls, H. Wieman, and R. Vandenbosch, *Phys. Rev. A* **10**, 1245 (1974).
[r] H. J. Stein, H. O. Lutz, P. H. Mokler, and P. Armbruster, *Phys. Rev. A* **5**, 2126 (1972); H. J. Stein, H. O. Lutz, P. H. Mokler, K. Sistemich, and P. Armbruster, *Phys. Rev. Lett.* **24**, 701 (1970) and *Phys. Rev. A* **2**, 2575 (1970).
[s] R. Ambros, H. O. Lutz, and K. Reichelt, *Phys. Rev. Lett.* **32**, 811 (1974).
[t] S. Sackmann, H. O. Lutz, and J. Briggs, *Phys. Rev. Lett.* **32**, 805 (1974).
[u] N. Luz, S. Sackmann, and H. O. Lutz, *J. Phys. B* **9**, L15 (1976).
[v] C. L. Cocke, R. Randall, and B. Curnutte, *Int. Conf. Phys. Electr. Atom. Coll.*, 8th, 714, Belgrade (1974).
[w] C. L. Cocke, R. R. Randall, S. L. Varghese, and B. Curnutte, *Phys. Rev. A* **14**, 2026 (1976).
[x] K. W. Jones, H. W. Kramer, and W. Brandt, *Phys. Lett.* **57A**, 33 (1976).
[y] B. M. Johnson, K. W. Jones, and D. J. Pisano, *Phys. Lett.* **59A**, 21 (1976); B. M. Johnson, K. W. Jones, W. Brandt, F. C. Jundt, C. Guillaume, and T. H. Kruse, *Phys. Rev. A* **19**, 81 (1979).
[z] C. Annett, C. L. Cocke, and B. Curnutte, *Phys. Rev. A* **19**, 1038 (1979).
[aa] I. Tserruya, H. Schmidt-Böcking, and R. Schuch, *Phys. Rev. A* **18**, 2482 (1978).
[bb] J. F. Chemin, S. Adriamonje, S. Denagbe, J. Roturier, B. Saboya, and J. P. Thibaud, *Phys. Rev. A* **15**, 1851 (1977).
[cc] N. Luz, S. Sackmann, and H. O. Lutz, *J. Phys. B* **12**, 1973 (1979).
[dd] H. O. Lutz, W. M. McMurray, R. Pretorius, R. J. van Reenen, and I. J. van Heerden, *Phys. Rev. Lett.* **40**, 1133 (1978).
[ee] K. Jones, private communication (1977).
[ff] M. Lund, E. Laesgaard, J. U. Andersen, and L. Kocbach, *XICPEAC*, p. 40 (1977).
[gg] J. F. Chemin, private communication (1977).
[hh] D. Liesen, J. R. Macdonald, P. Mokler, and A. Warczak, *Phys. Rev. A* **17**, 897 (1978).
[ii] S. Röhl, S. Hoppenau, and M. Dost, *XICPEAC*, p. 76 (1977).

7.2. INNER-SHELL VACANCY PRODUCTION

of the projectile nucleus dominates. The process may be dealt with by taking initial and final electronic states centered on the target nucleus and treating the projectile nucleus as a perturbation. The plane-wave Born approximation (PWBA)[32] and semiclassical Coulomb approximation (SCA)[33-36] both use this approach. On the other hand, for very low v and nearly symmetric collisions the electrons try to adjust to the developing two-center potential by remaining in the same electronic orbitals of the transient molecule throughout the collision.[8] Transitions between orbitals result from their failure to complete the adjustment, occurring most readily when orbitals become degenerate or very nearly so. In this molecular-orbital regime the electronic structure of the projectile plays a central role in inner-shell processes.

Between these two limits lies a region of great theoretical difficulty. For Z_1/Z_2 not too large, small departures from the first Born theories may be dealt with in terms of increased binding energy and polarization effects.[37] For large Z_1/Z_2 ($\geq \frac{1}{3}$) electron capture may become important, especially if capture to an inner shell of the projectile is allowed. Perturbation theory treatments of this inner-shell vacancy transfer channel have been only partially successful.[38,39] It is perhaps in the study of electron transfer that the bridge between the Coulomb and molecular regimes may best be made.

We perform a division of the following section into discussions of Coulomb, molecular, and intermediate regimes recognizing that the dividing lines must be drawn somewhat arbitrarily. In Table I we summarize experimental $P(b)$ results from x-ray–ion coincidence experiments prior to approximately June, 1977. We include a partial summary of results presented through XICPEAC (Paris, July 1977), with a warning to the reader that a great deal of new material was presented there and appeared shortly thereafter in the literature.

7.2.4.1. Coulomb Regime: Point Projectiles. The SCA formulation of Bang and Hansteen[33] in which $P(b)$ is calculated directly, has served as

---

[32] E. Merzbacher and H. W. Lewis, in "Encyclopedia of Physics" (S. Flügge, ed.), Vol. XXXIV, p. 166. Springer-Verlag, Berlin and New York, 1958.

[33] J. Bang and J. M. Hansteen, *Mat. Fys. Medd. Dan. Vid. Selsk.* **31**, No. 13 (1959).

[34] J. M. Hansteen and O. P. Mosebekk, *Z. Phys.* **234**, 281 (1970); *Phys. Lett.* **29A**, 28 (1969).

[35] J. M. Hansteen and O. P. Mosebekk, *Nucl. Phys. A* **201**, 541 (1973).

[36] J. M. Hansteen, O. M. Johnsen, and L. Kocbach, *Atom. Nucl. Data Tables* **15**, 305 (1975).

[37] W. Brandt, R. Laubert, and I. Sellin, *Phys. Rev. Lett.* **21**, 518 (1966); G. Basbas, W. Brandt, and R. Laubert, *Phys. Rev. A* **7**, 983 (1973).

[38] A. M. Halpern and J. Law, *Phys. Rev. Lett.* **31**, 4 (1973).

[39] J. H. McGuire, *Phys. Rev. A* **8**, 2760 (1973).

a basis from which one may begin to test the perturbation theory predictions of differential cross sections with experiment. If the point projectile is assumed to move along a classical trajectory, an excellent approximation in most cases, the transition amplitude for ejection of an electron with energy E_f is given by

$$M_b(E_f) = \frac{1}{i\hbar} \int_{-\infty}^{\infty} e^{i\omega t} \langle f|V(t)|i\rangle \, dt, \qquad (7.2.5)$$

where $|i\rangle$ and $|f\rangle$ are initial and final electronic states, ω the transition frequency, and $V(t)$ the Coulomb potential between target electron and the projectile nucleus. If $\chi \gg 1$, the experimental $P(b)$ is to be compared to $\int_0^\infty dE_f |M_b(E_f)|^2$. It is pointed out by Hansteen[40] that the requirement that χ be large allows one to ignore diffraction effects in establishing the relationship between $M_b(E_f)$ and the differential cross section, and does not pertain directly to the validity of the total SCA cross section. It is well established[33] that the straight-line SCA and PWBA give the same total cross sections.

If χ is not large, a differential cross section must be obtained retaining the quantal characteristics of the scattering process. If only s-wave electronic states are involved and the collision is nearly elastic, the q-dependent transition amplitude $T_q(E_f)$ may be obtained from $M_b(E_f)$ by the Bessel transform[41,42]

$$T_q(E_f) = -iv \int_0^\infty b \, M_b(E_f) J_0(qb) \, db, \qquad (7.2.6)$$

where q is the transverse momentum transfer to the electron. The differential cross section is related to T_q by $d\sigma/d\Omega = (\mu^2/2\pi)|T_q|^2$, where v and μ are projectile velocity and the reduced mass. Physically, Eq. (7.2.6) allows amplitudes for collisions at different b to interfere with each other, the resulting differential cross section becoming the corresponding diffraction pattern. If the straight-line SCA amplitude for $M_b(E_f)$ is used, the resulting $T_q(E_f)$ is identical to that obtained directly in the PWBA approach.[42,43] Since the latter prescription neglects deflection of the projectile by the target nucleus, however, $|T_q(E_f)|^2$ will not give the correct angular distribution if q and θ are taken to be related by $q = 2\mu v \sin \theta/2$. This problem may be partially overcome by introducing into the integrand

[40] J. M. Hansteen, *Adv. Atom. Molec. Phys.* **11**, 299 (1975).
[41] R. McCarroll and A. Salin, *J. Phys. B* **1**, 163 (1968).
[42] H. Schiff, *Can. J. Phys.* **32**, 393 (1954).
[43] K. Taulbjerg, *J. Phys. B* **10**, L341 (1977).

7.2. INNER-SHELL VACANCY PRODUCTION

of Eq. (7.2.6) and appropriate phase factor accounting for distortion of the projectile wave by the Coulomb potential of the target.[44]

Approaching the problem from the other direction, $P(b)$ may be obtained by inverting the transform of Eq. (7.2.5) if T_q is known from, for example, a PWBA calculation. An approach of this type has been applied to K- and L-shell ionization by Kaminsky et al.[45] and by Beloshitsky and Nikolaev.[46]

The general formulation of the SCA problem was presented and calculations carried through for the K shell by Bang and Hansteen.[33] Calculations for L and M shells are discussed by Hansteen and Mosebekk.[34] The SCA was used to calculate several $P(b)$ curves by Hansteen and Mosebekk.[35] Tables of universal SCA probabilities and integrated cross sections are given in the straight-line approximation by Hansteen et al.[36] for K-, L- and M-shell ionization.

The binary encounter approximation (BEA) describes the ionization event as due to a classical encounter between the projectile and the free electron, whose initial velocity distribution is obtained from the target electronic wavefunction. The BEA total cross sections, which are similar to SCA and PWBA results, are easily expressed in a simple universal form[47] and have therefore been widely used. Developing the BEA in an impact-parameter-dependent form poses problems of principle, however, since it requires adopting a fundamentally incorrect position-dependent target velocity distribution. Unfortunately there does not seem to be an a priori favored way of choosing this distribution, to which $P(b)$ is quite sensitive. Prescriptions have been given by Hansen[48] and by McGuire[49] for calculating b-dependent BEA probabilities. McGuire gives extensive tables of $P(b)$ for several scaled velocities. At low velocities, his $P(b)$ curves tend to place more of the ionization at small b than does the SCA. Hansen incorporates some aspects of the uncertainity principle into his velocity distributions and obtains a $P(b)$ curve, for his "unconstrained" distribution, similar to the data for 2 MeV protons on selenium. It is probably not reasonable to expect detailed agreement between the classical theory and experiment for differential cross sections, nor is it particularly meaningful to ask whether the BEA gives better agreement with data than

---

[44] Dz. Belkic and A. Salin, *J. Phys. B* **9**, L397 (1976).

[45] A. K. Kaminsky, V. S. Nikolaev, and M. I. Popova, *Phys. Lett.* **53A**, 419 (1975); A. K. Kaminsky, and M. I. Popova, *Second Int. Conf. Inner-Shell Ioniz.*, p. 256, Frieburg, 1976.

[46] V. V. Beloshitsky and V. S. Nikolaev, *Phys. Lett. 51* **A**, 97 (1975).

[47] J. H. McGuire and P. Richard, *Phys. Rev. A* **3**, 1374 (1973).

[48] J. S. Hansen, *Phys. Rev. A* **8**, 822 (1973).

[49] J. H. McGuire, *Phys. Rev. A* **9**, 286 (1974).

does the SCA, since the former calculation rests on a somewhat arbitrarily chosen velocity distribution.

7.2.4.1.1. EXPERIMENTAL K-SHELL RESULTS. The earliest SCA-rated experimental results were those of Laegsgaard et al.[50] for 1–2 MeV protons on silver, selenium, and copper. The probabilities were so low in this experiment that it was necessary to use magnetic analysis of the scattered protons to isolate inelastic events and thus achieve a sufficiently low randoms rate. Part of these results are shown in Fig. 10, where good agreement, on an absolute scale, is seen with the SCA calculation. Maximum $bP(b)$ occurs, as predicted, near b equal to the so-called adiabatic radius $r_{ad} = (U + E_f)/\hbar v$ at which the maximum projectile angular velocity about the target nucleus matches the angular frequency associated with the ionization. Thus, as E decreases, the effective interaction distance scales roughly as r_{ad} and the ionization moves to smaller impact parameters. At higher energies, for which r_{ad} approaches r_k, the ionization comes primarily from $b \approx r_k$ and $P(b)$ does not continue to grow outward for higher bombarding energy.

Brandt et al.,[51] showed that the SCA gave a fair account of impact-parameter-dependent probabilities for low-energy proton K ionization of targets between silver and gold at a fixed proton scattering angle. They developed an analytic expression for the low-energy SCA $P(b)$, which departs from a full SCA result for b inside r_k, but is quite close to it for large impact parameters. They further suggested a procedure for introducing corrections to the impact-parameter-dependent theory to account for Coulomb deflection of the projectile and increased binding energy corrections, and found satisfactory fits to their data over a wide range of ionization probability.

For higher Z_2, knowledge of the necessary target fluorescence yield becomes better, but relativistic treatment of the target K shell becomes essential as well. Amundsen and Kocbach[52] performed relativistic SCA calculations (RSCA) for the K shell in the straight-line approximation for $l = 0$ final states and showed that for heavy targets the (RSCA) probabilities can be much higher than the (SCA) ones, although the $P(b)$ shape is not strongly affected. Physically, the RSCA probabilities are higher because the higher-momentum components in the relativistic wave functions facilitate the higher-momentum transfer to the electron needed for large energy transfers. Good agreement with the data of Lund[53] and

---

[50] E. Laegsgaard, J. U. Andersen, and L. C. Feldman, *Proc. Int. Conf. Inner-Shell Ion. Phenom. Atlanta,* 1972, 1019 (USAEC CONF-720404); and *Phys. Rev. Lett.* **29**, 1206 (1972).

[51] W. Brandt, K. W. Jones, and H. W. Kramer, *Phys. Rev. Lett.* **30** 351 (1973).

[52] P. A. Amundsen and L. Kocbach, *J. Phys. B* **8**, L122 (1975).

[53] M. Lund, Thesis, Univ. of Aarhus, 1974.

7.2. INNER-SHELL VACANCY PRODUCTION

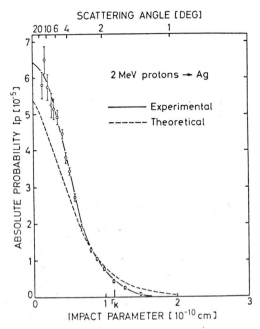

FIG. 10. Experimental values of $P(b)$ for K-shell ionization of silver by 2-MeV protons. The theoretical curve is an SCA calculation (E. Laegsgaard, J. U. Andersen, and L. C. Feldman, *Phys. Rev. Lett.* **29**, 1206, 1972).

Laegsgaard et al.[50] was found. Extension of the calculations to arbitrary final states and discussion of the general theory is given by Amundsen.[54] A prescription for correcting SCA cross sections and ionization probabilities to RSCA ones is given by Amundsen et al.[55] Clark et al.[56] have measured $P(b)$ for vacancy production in targets of bismuth for 4-MeV protons, and their results are shown in Fig. 11. The nonrelativistic SCA is seen to give approximately the correct shape for $P(b)$ but lies well below the data. The relativistic calculation (RSCA), is much closer to the experiment.

Experimental values for $P(b)$ for 25–74-MeV alpha particles on lead for near-zero impact parameters have been measured by Vader et al.[57] Their scattering angles were kept low enough to avoid large corrections due to

[54] P. A. Amundsen, *J. Phys. B* **9**, 971 (1976).
[55] P. A. Amundsen, L. Kocbach, and J. M. Hansteen, *J. Phys. B* **9**, L203.
[56] D. L. Clark, T. K. Li, J. M. Moss, G. W. Greenless, M. E. Cage, and J. H. Broadhurst, *J. Phys. B* **8**, L378 (1975).
[57] R. J. Vader, A. van der Woude, R. J. Meijer, J. M. Hansteen, and P. A. Amundsen, *Phys. Rev. A* **14**, 62 (1967).

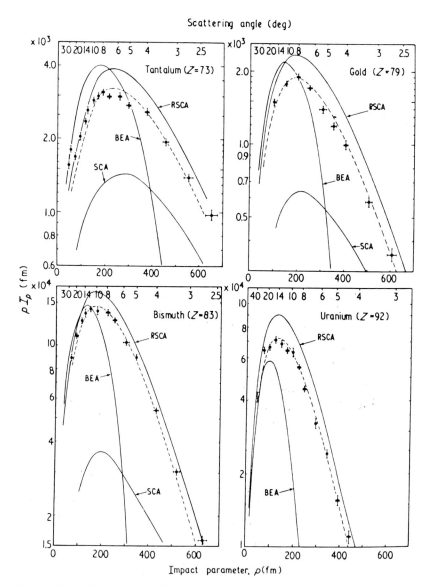

FIG. 11. Experimental values of $bP(b)$ for high-Z targets. The solid lines show absolute SCA, BEA, and RSCA calculations, while the dashed line shows the RSCA curve normalized to the experimental total cross section. (D. L. Clark, T. K. Li, J. M. Moss, G. W. Greenlees, M. E. Cage, and J. H. Broadhurst, *J. Phys. B* **8**, L378, 1975.)

7.2. INNER-SHELL VACANCY PRODUCTION

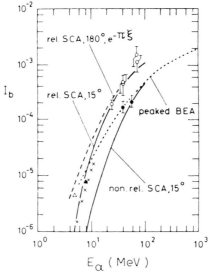

FIG. 12. Summary of data and calculations of $P(b)$ for the lead K shell at small b. Data are plotted vs. equivalent α-particle energy. Figure from R. J. Vader, A. van der Woude, R. J. Meijer, J. M. Hansteen and P. A. Amundsen, *Phys. Rev. A* **14**, 62, (1976). Data sources are ○, Vader *et al.*, α on lead; ●, G. Bissinger, T. H. Kruse, M. E. Williams, and W. Savin, *Int. Conf. Inner Shell Ion. Phen.*, 1036, AEC-Conf. 720404 (1972); △, H. J. Fischbeck and M. S. Freedman, *Phys. Rev. Lett.* **34**, 1973 (1975), α decay of polonium; ▲, M. Lund, Thesis, Aarhus, 1974, p on gold; X, D. Burch, W. B. Ingalls, H. Wieman, and R. Vandenbosch, *Phys. Rev.* A**10**, 1245 (1974), chlorine on lead. The data of Burch *et al.*, Bissinger *et al.*, and Lund are scaled according to J. H. McGuire, *Phys. Rev. A* **9**, 286, (1974). The solid SCA curves are calculated for a scattering angle of 15°, using the tangential approximation.

projectile deflections, while still allowing penetration to very small b. Their results, summarized in Fig. 12, again show the importance of treating the target K shell relativistically. The RSCA results are in good agreement with the data.

For low projectile velocities, the effects of Coulomb deflection of the projectile by the target nucleus become important. For the case of s to s transitions, Bang and Hansteen[33] showed that the major effect of deflection is to reduce the ionization probability by a factor of $e^{-\pi\xi}$ from the straight-line probability, where $\xi = \chi U/E$. Kocbach[58] has shown that this approximation gives results close to those found from requiring that the projectile move on a hyperbolic trajectory, so long as the correction is not too large. A somewhat more general correction may be made in the

---

[58] L. Kocbach, *Phys. Fenn.* **9**, Suppl. S1, 46 (1974).

tangential approximation, where b is replaced by the corresponding distance of closest approach and v by the true retarded velocity at that distance.[34,59] For scattering angles not too large, this approximation appears to be in reasonable agreement with the hyperbolic trajectory result for both dipole and monopole amplitudes.[60]

For very large scattering angles, effects of deflection may become quite large and the neglect of transitions to final electronic states with $l > 0$ is not justified.[60] Chemin et al.[61] reported a significant increase of $P(b)$, for 2-MeV protons on copper, as the scattering angle increased from 16 to 50°. These results were not confirmed by Andersen et al.,[21] who suggested that experimental problems due to time structure in the beam might be responsible for the earlier high probabilities. (More recent results of Chemin et al.,[62] are in agreement with those of the Aarhus group.) Andersen et al. did find a similar effect at lower bombarding energy (E_p = 0.5 MeV), however. Instead of remaining constant for b much less than r_{ad}, as the straight-line SCA calculation predicts, a rise in P for θ greater than 60° ($b \leq 100$ fm) was seen. This effect was explained as due to interference between dipole ionization amplitudes developed on incoming and outgoing parts of the projectile paths and a quantitative explanation of the data was obtained within the SCA framework. A similar effect was seen by Chemin et al.[62] in protons on silver, and discussed within a BEA framework with some success.

Such an interference phenomenon does not emerge naturally from a classical theory such as the BEA, however. Its importance in the interpretation of ionization probabilities deduced from measurements of ionization accompanying natural α decay was discussed by Ciocchetti and Molinari.[63] As a first approximation, one might consider such a process to be equivalent to one-half of the corresponding scattering event, but this is not necessarily correct. In Fig. 12, the values of $P(0)$ found by Fischbeck and Freedman,[64] who measured coincidences between α particles from a ^{210}Po source and the ensuing L and K x rays, as well as low-energy electrons, have been multiplied by 2 and included with the scattering data. As pointed out by Vader et al.,[57] interference effects may account for the departure of this point from the SCA curve, since terms in the transition amplitude that cancel for the full trajectory do not for the half-pass case.

[59] L. Kocbach, *Univ. Bergen, Sci. Tech. Rep.* **75** (1975).
[60] O. Aashamar and L. Kocbach, *Z. Phys. A* **279**, 237 (1976).
[61] J. F. Chemin, J. Roturier, B. Saboya, and Q. T. Dien, *Phys. Rev. A* **11**, 549 (1975).
[62] J. F. Chemin, S. Andriamonje, S. Denagbe, J. Roturier, B. Saboya, and J. P. Thibaud, *Phys. Rev. A* **15**, 1851 (1977).
[63] G. Ciocchetti and A. Molinari, *Nuovo Amento* **40B**, 69 (1965).
[64] H. J. Fischbeck and M. S. Freedman, *Phys. Rev. Lett.* **34**, 173 (1975).

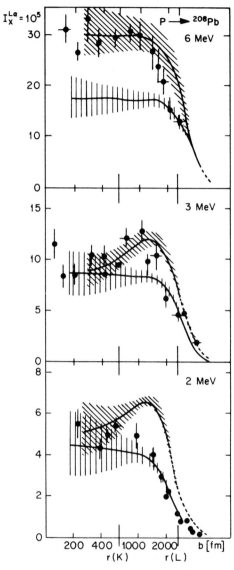

FIG. 13. Impact parameter dependences of lead L_α x-ray production by 2-, 3-, and 6-MeV protons. Theoretical curves are the SCA (dashed line) and BEA (solid line). Shaded areas indicate uncertainties in theoretical values due to uncertainties in the Coster–Kronig and fluorescence yields (K. Stiebing, H. Schmidt-Böcking, R. Schulé, K. Bethge, and I. Tserruya, *Phys. Rev. A* **14**, 146, 1976).

7.2.4.1.2. L-SHELL RESULTS. Whereas K-shell ionization probabilities tend to fall smoothly and monotonically to zero as b increases, L-shell ionization probabilities are predicted by both SCA and BEA theories to show b-dependent structure that reflects that of the 2s and 2p wavefunctions.[65,66] Only two $P(b)$ experiments on the L shell appear to have been reported. Laegsgaard et al.,[67] measured the total L-shell ionization probability of gold from proton bombardment between 1 and 2 MeV. While some of the structure predicted in $P(b)$ for individual subshells is thereby washed out, the overall shapes and sizes of the experimental and SCA curves were found to be in agreement. The SCA appears to place too much ionization at large b, and since distant encounters carry more weight in the total cross section, lead to too large total cross section. Very recently Amundsen[68] has extended the RSCA calculations to the L shell, and has found good agreement between theory and the 1-MeV data.

Stiebing et al.[69] resolved L_α and L_β transitions from 2–6-MeV protons on lead. The L_1 and L_2, L_3 subshells remain essentially unresolved, although some of the structure predicted by the SCA for the L_3 subshell does appear in the data (See Fig. 13). The SCA again predicts ionization probabilities that are too large at large b. The BEA probabilities, which do not predict the observed structure and which lie too low at small b, come close to the data at large b and thus give the better total cross sections. This situation is one in which, on the basis of total cross section data alone, one might be tempted erroneously to favor the BEA theory.

7.2.4.2. Molecular Regime. For $Z_1 \sim Z_2$ and projectile velocities small compared to those of the active electrons, the molecular orbital (MO) picture of Fano and Lichten[8] has been used with considerable success. In this model the electrons remain approximately in diabatic molecular orbitals during the collision, undergoing transitions between orbitals induced by the time-dependent radial and rotational motion of the vector internuclear distance.

The earliest applications of this model to impact parameter dependence data were to L-shell vacancy production. Referring to the schematic correlation diagram of Fig. 14, L-shell vacancies may be produced via promotion of the $4f\sigma$ orbital. As R, the internuclear distance, decreases, this orbital couples to a number of orbitals whose ultimate correlation is to higher-than-L shells. Thus, for collisions that penetrate inside some critical value of R, two L vacancies should be created almost every colli-

[65] J. M. Hansteen, O. M. Johnsen, and L. Kocbach, *J. Phys. B* **7**, L271 (1974).
[66] J. H. McGuire and K. Omidvar, *Phys. Rev. A* **10**, 182 (1974).
[67] E. Laegsgaard, J. U. Andersen, and M. Lund, *Phys. Fenn.* **9**, Suppl S1, 49 (1974).
[68] P. Amundsen, *J. Phys. B* **10**, 1097 (1977).
[69] K. E. Stiebing, H. Schmidt-Böcking, R. Schulé, K. Bethge, and I. Tserruya, *Phys. Rev. A* **14**, 146 (1976).

7.2. INNER-SHELL VACANCY PRODUCTION

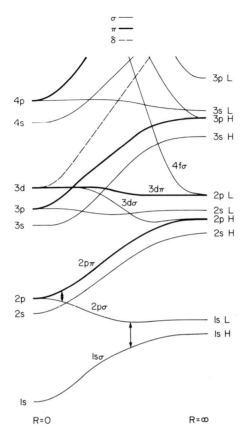

FIG. 14. Schematic correlation diagram for slightly asymmetric system. L and H refer to lighter and heavier collision partner, respectively.

sion, and $P(b)$ should take on a step function appearance. This indeed occurs experimentally,[1,2,70] until b, becomes sufficiently small that the 3d orbitals participate as well.

Until quite recently interpretations of experiments based on the MO model have been largely qualitative for L and higher shells. Apart from the results of Thomson et al.[5,70] (discussed in Chapter 7.3) and those of Stein et al.[7] most experimental information on these shells comes from IEL work. We thus confine this discussion to K-vacancy production, referring the reader to Part 4 and to several excellent review articles for discussion of higher shells.[4,71]

[70] G. M. Thomson, Phys. Rev. A **15**, 865 (1977).
[71] J. D. Garcia, R. J. Fortner, and T. M. Kavanagh, Rev. Mod. Phys. **45**, 111 (1973).

The best known K-vacancy production mechanism is the rotational transfer of vacancies from the $2p\pi$ orbital, which correlates to the 2p shell of the heavier atom, to the $2p\sigma$ orbital, which goes to the 1s of the lighter atom (see Fig. 14), provided $2p\pi$ vacancies are brought into the collision. A quantitative description of this process, based on scaling of d^+–d collisions, was presented by Briggs and Macek[72] for symmetric collisions. Taulbjerg, et al.[73,74] have extended this description to include heteronuclear cases and give universal $P(b)$ curves that may be scaled to arbitrary collision partners. The $P(b)$ curve is typically characterized by a broad adiabatic maximum with a value near 0.6 (per $2p\pi_x$ vacancy, where the collision is in the $x - z$ plane), which ultimately scales out in impact parameter roughly linearly with velocity, and a sharp kinematic peak going to unit probability for a C.M. scattering angle of 90°. The latter scattering transforms a $2p\pi_x$ orbital into a $2p\sigma$ one by simply rotating the internuclear axis within the inertia-bound electron wavefunction.

The total cross sections predicted by this model were found to be in good agreement with experiment for Ne–Ne collisions.[75,76] IEL work by Fastrup et al.[77] at Aarhus for first-row target–projectile combinations showed general agreement with the model and demonstrated that the K-vacancy production probabilities were greatly reduced when no $2p\pi$ vacancies were brought into the collision. The first full $P(b)$ curve displaying both adiabatic and the kinematic maxima was measured by Sackmann et al.[78] for Ne$^+$ on Ne, and is shown in Fig. 15. The evidence is thus overwhelming that the rotational coupling mechanism is the dominant one for the production of K vacancies at low v when that channel is open.

If there are no $2p\pi$ vacancies brought into the collision, this mechanism should be blocked. In such cases direct excitation of $1s\sigma$ and $2p\sigma$ orbitals to higher vacant orbitals and to the continuum might be important, as is discussed by Meyerhof et al.[79] However, it may also be the case that vacancies may be introduced into the $2p\pi_x$ orbital via a two-step process whereby $2p\pi_x$ vacancies are created by promotion processes at

[72] J. S. Briggs and J. Macek, *J. Phys. B* **5**, 579 (1972); *ibid.* **6**, 982 (1973).

[73] K. Taulbjerg and J. S. Briggs, *J. Phys. B* **8**, 1895 (1975); J. S. Briggs and K. Taulbjerg, *J. Phys. B* **8**, 1909 (1975).

[74] K. Taulbjerg, J. S. Briggs, and J. Vaaben, *J. Phys. B* **9**, 1351 (1976).

[75] R. K. Cacak, Q. C. Kessel, and M. E. Rudd, *Phys. Rev. A* **2**, 1327 (1970).

[76] N. Stolterfoht, D. Schnieder, D. Burch, B. Agaard, E. Boving, and B. Fastrup, *Phys. Rev. A* **12**, 1313 (1975).

[77] B. Fastrup, G. Herman, Q. Kessel, and A. Crone, *Phys. Rev. A* **9**, 2518 (1974); see also Ref. 74.

[78] S. Sackmann, H. O. Lutz, and J. Briggs, *Phys. Rev. Lett.* **32**, 805 (1974).

[79] W. E. Meyerhof, *Phys. Rev. A* **10**, 1005 (1974).

7.2. INNER-SHELL VACANCY PRODUCTION

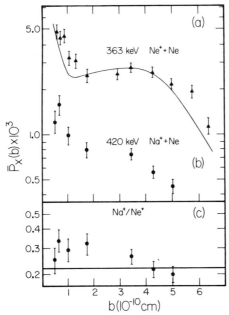

FIG. 15. Experimental K x-ray production probabilities for (a) Ne$^+$-Ne and (b) Na$^+$-Ne collisions. The solid curves are the rotational-coupling calculations. Both experiment and theory include contributions from recoil–x-ray coincidences. The lower curve (c) shows the ratio of the probabilities from (a) and (b) vs. impact parameter. The solid line in (c) shows the total cross-section ratio (S. Sackmann, H. O. Lutz, and J. S. Briggs, *Phys. Rev. Lett.* **32**, 805, 1974; N. Luz, S. Sackmann, and H. O. Lutz, *J. Phys.* B **9**, L15, 1976).

large R and carried into the inner region, there to be exchanged for $2p\sigma$ vacancies. Fastrup et al.[80] and Stolterfoht et al.[76] observed that the effective number of $2p\pi_x$ vacancies, N_π, as deduced from experiments on first-row collision partners, displayed a dynamic component. They found it possible to write $N_\pi = N_\pi^0 + N_\pi(v)$, and $\sigma = (N_\pi/6)\sigma_{\rm rot}$, where $\sigma_{\rm rot}$ is given by Taulbjerg et al.[74] Here the 6 assumes symmetric collisions, and N_π^0 is the $2p\pi_x$ occupation number brought into the collision. $N_\pi(v)$ represents vacancies produced in the first step. Since this step presumably takes place on an interaction radius much larger than r_K, it should be possible to write $P(b) = (N_\pi/6)P_{\rm rot}(b)$, where $P_{\rm rot}(b)$ is given by Tauljberg et al.,[74] and N_π is nearly b independent. Luz et al.[81] found this to be approximately the case, by measuring $P(b)$ for neon K x-ray production by Na$^+$ projectiles (Fig. 15). Here the rotational coupling mechanism

[80] B. Fastrup, E. Bøving, G. A. Larsen, and P. Dahl, *J. Phys.* B **7**, L206 (1974).
[81] N. Luz, S. Sackmann, and H. O. Lutz, *J. Phys.* B **9**, L15 (1976).

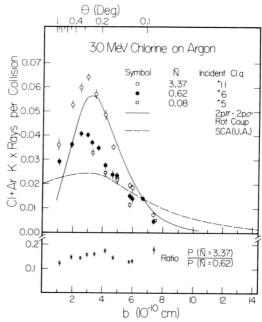

FIG. 16. The probability for production of chlorine or argon K x rays vs. b for 30-MeV chlorine on argon gas targets. The open and filled circles are for incident charge states +11 and +6. The parameter $\bar{N}$ represents approximately the average number of L vacancies borne by the projectile at the target center. The open squares are for +5 chlorine on a target of one-quarter normal pressure. The solid curve shows the rotational-coupling prediction, the dashed one the SCA prediction for the united-atom 2p ionization, each curve normalized to the experimental cross section for the higher incident charge. The lower figure shows the ratio between higher and lower incident charge data (C. L. Cocke, R. R. Randall, S. L. Varghese, and B. Curnutte, *Phys. Rev. A* **14**, 2026, 1976).

should be blocked, since Na$^+$ carries no 2pπ_x vacancies into the collision and the exit channel is closed. The similar shape of $P(b)$ for the Na$^+$ beam to that for Ne$^+$ projectiles, including evidence of a kinematic peak, indicates that the two-step process is partially opening the rotational coupling channel. This is an excellent example of the use of $P(b)$ data to identify a process whose importance could only be indirectly inferred from total cross section measurements.

The production of K vacancies in nearly symmetric collisions between higher-Z systems is not so clear-cut a situation. So long as the direct K-vacancy transfer channel is closed, proposed vacancy production mechanisms include the two-step process and direct 2pσ (and 1sσ) excitation. At higher nuclear charge, one would expect the probability of

7.2. INNER-SHELL VACANCY PRODUCTION

ejecting a 2p electron from the heavier system to decrease as its binding energy increases. The strong $2p\pi-2p\sigma$ coupling is so efficient, however, that even small values of $N_\pi(v)$ may cause the multistep process to dominate the K-vacancy production.

Measurements of $P(b)$ for chlorine K x-ray production in collisions of chlorine projectiles with aluminium, titanium, and copper[82] showed that, near matching Z, $P(b)$ tends to acquire a plateau at small b, qualitatively similar to the maximum in $P_{rot}(b)$. Similar results were seen in Ni + Mn collisions.[83] Both sets of experiments were done in solid targets, however, and the production of 2p vacancies in the projectile moving in the solid is expected.

In order to investigate the single-collision process one must go to either gas or *very* thin solid targets. Cocke *et al.*[84] found, in the bombardment of argon by chlorine projectiles in several charge states, the summed probability for chlorine and argon K-vacancy production followed nearly the shape expected for the rotational coupling mechanism (Fig. 16). This summed $P(b)$ presumably represents the $2p\sigma$ vacancy production probability, independent of the K-vacancy sharing ratio.[85] The measured $P(b)$ was found to be independent in shape of N_π^0, suggesting that outer-shell rearrangements so dominate the value of N_π brought to small R that, especially at higher bombarding energy, the exit channel effect is nearly destroyed.

It is possible that such two-step processes may dominate the K-vacancy production for much heavier systems. The $P(b)$ results of Annett *et al.*[86] for copper on nickel continue to show a shape qualitatively similar to the $P_{rot}(b)$ (see Fig. 17). A study of the target thickness dependence of $P(b)$ indicates that the same inner-shell process is operating under both single- and multiple-collision conditions. Since one expects the rotational-coupling process to dominate when the exit channel is open (latter condition), the two-step process for the former condition is strongly suggested. Quantitatively, the experimental curve is shifted to smaller b than is the theoretical one, $P_{rot}(b)$. It is possible that this may represent a breakdown of the scaling procedures given in Ref. 74. Departures from these scaling laws for very high Z, for which the spin–orbit Hamiltonian becomes important, are discussed by Anholt

---

[82] C. L. Cocke, B. Curnutte, and R. Randall, *Proc. Int. Conf. Phys. Elect. Atom. Coll., 8th (ICPEAC)* Belgrade, 1974, p. 714.
[83] B. M. Johnson, K. W. Jones, and D. J. Pisano, *Phys. Lett.* **59A**, 21 (1976).
[84] C. L. Cocke, R. R. Randall, S. L. Varghese, and B. Curnutte, *Phys. Rev. A* **14**, 2026 (1976).
[85] W. E. Meyerhof, *Phys. Rev. Lett.* **31**, 1341 (1973).
[86] C. Annett, C. L. Cocke, and B. Curnutte, *Phys. Rev. A* **19**, 1038 (1979).

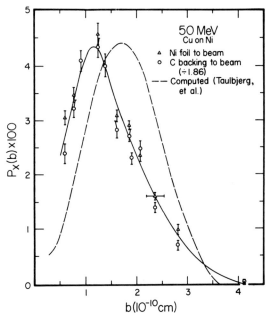

FIG. 17. $P_X(b)$ for nickel K_α x rays produced by 50-MeV Cu(8+) projectiles. The nickel foil thickness near 70 Å is roughly one-half that needed to achieve equilibrium L-vacancy population in the copper. The carbon-first data allowed the beam to pass the 4-μg/cm^2 carbon backing first. The solid curve is the rotational-coupling prediction, arbitrarily normalized. (Annett et al.[86])

et al.,[87] but will not be important for $Z \sim 29$. It is not impossible that it may represent the entrance of the direct processes into the picture. Unfortunately no reliable theoretical $P(b)$ curve for $2p\sigma$ direct ionization exists, although it seems unlikely that it should exhibit the strong decrease at small b seen experimentally.

As long as the two-step process is important, measurements of total K-vacancy production cross sections for nearly symmetric collisions measure only $N_\pi(v)$. That is, they shift the focus of the physics onto outer shells where explanations in terms of simple independent electron orbitals are likely to become more complicated if possible at all. It is at least clear that identification of the correct mechanism will not be possible with integral measurements alone, and that $P(b)$ measurements are essential.

Recent results of the Frankfurt–Heidelberg[88,89] group suggest that, for $Z_1 \sim Z_2 \sim 20$, solid target results may be in disagreement with the gas

[87] R. Anholt, W. E. Meyerhof, and A. Salin, *Phys. Rev. A* **16**, 951 (1877).
[88] H. Schmidt-Böcking, *Frankfurt Jahresbericht*, 70 (1975), unpublished.
[89] I. Tserruya, H. Schmidt-Böcking, and R. Schuch, *Phys. Rev. A* **18**, 2482 (1978).

7.2. INNER-SHELL VACANCY PRODUCTION

target Cl–Ar results. Their measurements of $P(b)$ for 35-MeV chlorine on solid targets of titanium, chlorine, and nickel are in good agreement with earlier results of Cocke et al.,[82] but do not show the characteristic rotational coupling shape at small b. Especially for the symmetric Cl–Cl case, the lack of any decrease in $P(b)$ for small b seems to be in clear disagreement with the gas target data, although the solid target data cover a range of b that barely overlaps the region of larger b studied in the Cl–Ar system. Tserruya et al.[89] suggest that ω_K may be b dependent for the gas target data. Such an effect has been seen previously in oxygen–neon collisions by Luz et al.,[90] who measured probabilities for both Auger electron and x-ray production. It would seem somewhat improbable that any b dependence of ω_K in the gas target experiments would be independent of projectile charge state. Further, the shape of $P(b)$ seen in the Cl–Ar system is qualitatively reproduced in the Cu–Ni system, for which the use of a solid target should remove any b dependence of ω_K. However, it is clear that the pioneering use by Lutz and collaborators of the Auger electron channel is important, and similar measurements for higher-Z symmetric systems would certainly help to clear up existing uncertainties there.

A rather different approach to K-shell ionization in collisions between complex systems has been taken by Brandt and Jones.[91] They treat the process as a statistical one in which K vacancies enact a random walk to the continuum on ladders of interacting level crossings. The collision is characterized by a diffusion constant D_K, and an effective interaction range R_0. By taking R_0 equal to the screening length, the single parameter D_K may be adjusted to fit the data. Reasonable fits to both total cross sections[92] and impact-parameter-dependent probabilities[89,92] may be obtained. Although it is probable that, for the systems studied thus far, specific mechanisms for K-vacancy production can be identified, for more complex situations the statistical approach may prove useful and necessary.

7.2.4.3. Intermediate Region. In the straight-line version of the SCA, it is possible to write $P(b)$ in terms of universal functions of a scaled velocity parameter ξ ($\equiv r_{ad}/r_n$) and a scaled impact parameter x ($\equiv b/r_{ad}$)[36,40,93].

$$P(b) = (Z_1^2/U)f_n(\xi, x). \tag{7.2.7}$$

---

[90] N. Luz, S. Sackmann, and H. O. Lutz, *Verhandl. DPG (VI)* **11**, 120 (1976); and *J. Phys. B* **12**, 1973 (1979).
[91] W. Brandt and K. W. Jones, *Phys. Lett.* **57A**, 35 (1976).
[92] K. W. Jones, H. W. Kramer, and W. Brandt, *Phys. Lett.* **57A**, 33 (1976).
[93] J. U. Andersen, E. Laegsgaard, M. Lund, and C. D. Moak, *Nucl. Inst. Mech.* **132**, 507 (1976).

The functions f_n depend on the shell (n) being ionized. As Z_1 approaches Z_2 the Coulomb field of the projectile becomes comparable to that of the target and the perturbation treatment and its corresponding scaling rules will break down. At low velocities, total cross sections for inner-shell vacancy production by heavy ions generally lie below those scaled from proton bombardment. This result was explained by Brandt et al.[37] as due to the increased binding energy of the target electron due to the presence of the projectile nuclear charge during the excitation process.

In order to follow the effect on $P(b)$ of raising the projectile charge, Cocke and Randall[31] measured the impact parameter dependence of the copper K-vacancy production for ^{16}O projectiles. The shapes of the $P(b)$ curves found were in good agreement with the SCA predictions and, by appropriate choice of scaling factor, are nearly identical[94] to the curves found by Laegsgaard et al.[50] for proton bombardment at the same velocity. The scaling needed is not $(Z_1)^2$, however, and the experimental cross sections fall increasingly below SCA for lower bombarding energy. A similar effect was found by Schmidt-Böcking et al.,[95] who measured $P(b)$ for nickel K-shell vacancy production by p, α, ^7Li, and ^{16}O ions at the same velocities. The light-ion data are in fair agreement with SCA, while the heavy-ion data fall progressively below SCA as Z_1 increases. By introducing a correction for increased binding into the SCA, they were able to obtain reasonably good agreement with the experiment.

This effect was studied in considerable detail by Andersen et al.,[93] who investigated the Z_1^2 scaling of $P(b)$ for hydrogen, beryllium, carbon, and oxygen ions on copper. Their approach was to start from the measured $P(b)$ curves for protons on the copper K shell, rather than the theoretical SCA $P(b)$ (which is similar, however), and to apply the SCA scaling laws to obtain $P(b)$ curves for higher Z_1. Such an approach allows a study of the scaling properties of $P(b)$ independent of the exact functional form of the probability. Their results again showed clearly that $Z_1^2 P(b)$ was not a universal function (see Fig. 18) but decreased strongly with increasing Z_1. This effect was quantitatively described as due to the increased binding effect and corrected scaled curves were found to be in good agreement with the data for small b. The effective binding energy used allowed relaxation of the target K shell during the collision. This relaxation, also considered by Schmidt-Böcking et al.,[95] is quite important to obtaining agreement with experiment.

[94] D. Burch, *Int. Conf. Phys. Elect. Atom. Coll. 8th (ICPEAC) Belgrade, 1974*, invited papers.
[95] H. Schmidt-Böcking, R. Schulé, K. E. Stiebing, K. Bethge, I. Tserruya, and H. Zekl, *J. Phys.* **B 10**, 2663 (1977).

7.2. INNER-SHELL VACANCY PRODUCTION

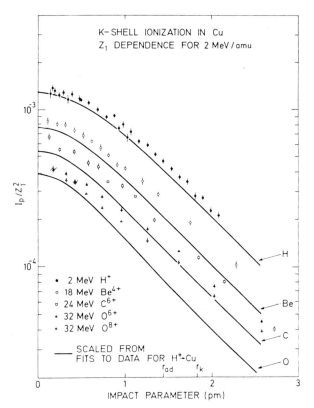

FIG. 18. Copper K-shell ionization probabilities, reduced by the factor $1/Z_1^2$, for p, Be^{4+}, C^{6+}, and O^{6+}, O^{8+} projectiles at 2 MeV/amu. The solid curves are scaled from protons on copper results, incorporating a "relaxed" binding energy correction (J. U. Andersen, E. Laegsgaard, M. Lund, and C. D. Moak, *Nucl. Inst. Meth.* **132**, 507, 1976).

Using a still higher Z_2 projectile, Burch *et al.*[96] measured cross sections and $P(b \simeq 0)$ for lead K-vacancy production by chlorine projectiles, at much lower scaled velocity. Although the total cross sections are almost exactly given by the simplest SCA or BEA calculations, it was pointed out that this is certainly due to a fortuitous cancellation of corrections due to relativistic effects and retardation–deflection corrections. The measured $P(0)$ (back-scattered chlorine particles were detected) lies about a factor of four below the SCA result. It is likely that a relativistic SCA calculation would be in better agreement. However, deflection correc-

[96] D. Burch, W. B. Ingalls, W. Wieman, and R. Vandenbosch, *Phys. Rev.* **A10**, 1245 (1974).

tions may be quite large for such large-angle scattering. Attempts to incorporate a binding correction into the perturbation theory results produced cross sections well below the experimental ones. We note that at chlorine energies as high as 100 MeV an appreciable fraction of the chlorine beam will bear K vacancies and account must be taken of the K-vacancy transfer channel discussed below.

Randall et al.[31] studied the impact parameter dependence of argon K-vacancy production by fluorine and carbon projectiles. By using gas targets they were able to study the dependence of $P(b)$ on the projectile charge state. With the K-vacancy transfer channel closed (C^{4+}, F^{5+} projectiles), vacancy production was seen to proceed with large probability well outside r_K, thus indicating immediately a breakdown of the SCA

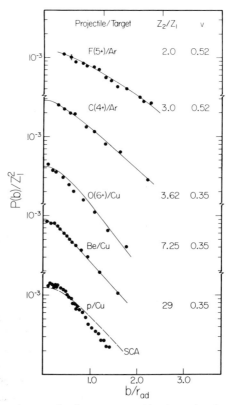

FIG. 19. $P(b)$ for various projectile–target combinations showing evolution out of the SCA region as Z_2/Z_1 is lowered. The scaled velocity is given as v. The SCA curve is shown for protons on copper, but is similar in shape for other systems. Data sources are: copper targets, Andersen et al., Nucl. Inst. Meth. **132**, 507, 1976; argon targets, Randall et al., Phys. Rev. A **13**, 204, (1976).

treatment. These data are shown in Fig. 19. We have included as well data from Andersen et al.,[93] in order to summarize the continuous evolution of $P(b)$ out of the perturbative region as Z_1/Z_2 is increased. All of these data represent cases for which direct 1s to 1s electron transfer is either blocked in the incoming channel or expected to be small. The data are plotted vs. x, since in the low-velocity regime the functional dependence of the SCA $P(b)$ on x is not velocity dependent.[33] This is still roughly the case, even after binding energy corrections are made, for scaled velocities in the vicinity of 0.4.[93] The SCA agrees well with the proton data, and $P(b)$ holds to the same shape up to $Z_1/Z_2 \sim 1/3$. Thereafter, $P(b)$ moves to larger impact parameters than would be expected from the perturbation treatment.[31] This may well be due to the opening, in a two-step process, of the K-vacancy transfer channel discussed below.

In a recent progress report, Laegsgaard et al.[97] survey the status of the SCA and its success in dealing with sufficiently asymmetric collision systems. Of particular interest is the tendency for binding and relativistic effects to cancel one another in many collision systems. For example, this occurs for the case of 2-MeV protons on silver, shown in Fig. 10, although neither effect is by itself small. This report may well describe the perturbation theory approach pressed as far in Z_1/Z_2 as will ultimately prove useful.

At high Z_1/Z_2, electron transfer from the target to vacant inner shells of the projectile may become important.[38,39] In the simplest perturbation treatment of electron capture, the Oppenheimer–Brinkman–Kramers approximation (OBK),[98] the probability for this process scales as $(Z_1/Z_2)^5$. Since the ionization probability for this process scales as $(Z_1/Z_2)^2$, the capture contribution, which is negligibly small for highly asymmetric collisions, becomes important and may dominate for $Z_1 \sim Z_2$ although the perturbation treatment itself ceases to be valid under such conditions. For example, the fraction of argon K-vacancy production by 3-MeV protons due to 1s–1s capture is only 0.5%.[99] The corresponding fraction for bare fluorine projectiles at the same velocity is near 70% (see Part 6).[100] For the case of K-vacancy production, the most important capture channel is usually that to the projectile K shell. Since the projec-

[97] E. Laegsgaard, J. U. Andersen, and M. Lund, *Int. Conf. Phys. Elect. Atom. Coll. 10th (ICPEAC), Paris, 1977,* invited paper.
[98] J. R. Oppenheimer, *Phys. Rev.* **31,** 349 (1928); H. C. Brinkman and H. A. Kramers, *Proc. Acad. Sci. Amsterdam* **33,** 973 (1930).
[99] J. R. Macdonald, C. L. Cocke, and W. W. Eidson, *Phys. Rev. Lett.* **32,** 648 (1974).
[100] J. R. Macdonald, L. M. Winters, M. D. Brown, T. Chiao, and L. D. Ellsworth, *Phys. Rev. Lett.* **29,** 1291 (1972); L. M. Winters, J. R. Macdonald, M. D. Brown, T. Chiao, L. D. Ellsworth, and E. Pettus, *Phys. Rev. A* **8,** 1835 (1973).

tile K shell must be open to enable the transfer, the K-vacancy production probability will become sensitive to the initial electronic configuration of the projectile.

In their study of binding energy corrections to the SCA, Andersen et al.[93] found a small dependence of $P(b)$ on the oxygen projectile charge state. A similar but much larger charge state dependence was first reported by Macdonald et al.[100] in the measurements with gas targets of total K-vacancy production cross sections in argon by carbon, oxygen, and fluorine projectiles. Projectiles that brought vacancies into the collision were found to be as much as three times as effective, per vacancy, in producing argon K vacancies as those which did not. (The validity of using a fluorescence yield slightly higher than the atomic one to extract vacancy production cross sections from x-ray production data in such experiments has been examined by Cocke et al.[101] Their measurements of K-Auger electron production cross sections in some of the same collision systems showed that the fluorescence yields used by Macdonald et al.[100] were essentially correct.) This K-vacancy transfer has now been observed by many experimenters (see Parts 5 and 6) and is well established.

The $P(b)$ data of Randall et al.[31] for bare fluorine and carbon projectiles on argon showed that K-vacancy transfer may take place with large probability at very large impact parameters (see Fig. 20). This behavior cannot be explained on the basis of an OBK charge exchange model, which predicts that $P(b)$ for electron capture is quite similar in width to the SCA curve for ionization. Although this calculation continues to be used to estimate total cross sections for inner-shell electron transfer between systems of high nuclear charge, the $P(b)$ data show clearly its inadequacy. This should in no way be surprising for such large Z_1/Z_2. These results show again the importance of differential measurements in assessing the validity of a theory. The K-vacancy transfer process is not yet fully understood theoretically, although better descriptions than the OBK one are available. Near Z symmetry, the process is the same as the $1s\sigma-2p\sigma$ vacancy sharing suggested by Meyerhof[85] to account for K-vacancy production in the heavier collision partner, except that both incoming and outgoing trajectories must be considered. The former process has received considerable theoretical study by Taulbjerg et al.[102] Randall et al.[31] attempted to treat the K-vacancy transfer between fluorine and argon in a molecular basis set of $1s\sigma$, $2p\sigma$, and $2p\pi$ orbitals,

---

[101] C. L. Cocke, Int. Conf. Phys. Elect. Atom. Coll., 9th (ICPEAC), Seattle, 1975; see also Ref. 31.

[102] K. Taulbjerg, J. Vaaben, and B. Fastrup, Phys. Rev. A **12**, 2325 (1975).

7.2. INNER-SHELL VACANCY PRODUCTION

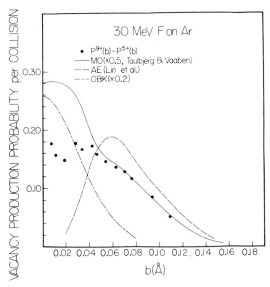

FIG. 20. The difference between argon K-vacancy production probabilities for F^{9+} and F^{5+} projectiles on argon, presumably representing roughly the probability for K-vacancy transfer as a function of b. Theoretical curves are OBK 1s–1s electron capture, reduced by a factor of 5; MO calculation of Taulbjerg and Vaaben, for a single initial $2p\pi_x$ vacancy; atomic expansion calculated by Lin et al. Data are from Randall et al., Phys. Rev. A **13**, 204, (1976).

using the coupling matrix elements calculated by Taulbjerg. A computational error led to incorrect and discouraging results in reference 31; however, a more recent calculation by Taulbjerg and Vaaben[103] shows that the electron capture probability calculated in such a model is extremely close in shape, although too high by a factor of two, to that of the experiment (see Fig. 20). Thus, it is likely that the K-vacancy sharing at large b seen in the experiment can be accounted for naturally within the molecular model.

More recently Lin and collaborators[104] have performed a coupled-channel calculation using only atomic hydrogenic 1s basis states and the formalism described by Bates.[105] At large impact parameters, their results are in good agreement with the data (see Fig. 20).

The asymmetry ratio of $\sim 1/3$, which might be considered from Fig. 19 roughly to define a boundary between atomic and molecular regimes, is

---

[103] K. Taulbjerg and J. Vaaben, private communication (1976).
[104] C. D. Lin, S. C. Soong, and L. N. Tunnell, private communication (1977).
[105] D. R. Bates, Proc. Roy. Soc. A **247**, 294 (1958).

also that at which the direct K-vacancy transfer becomes important. Continuing within the molecular basis discussed above, Vaaben and Taulbjerg[103] have made a preliminary investigation of the opening of this channel via $2p\pi-2p\sigma$ coupling on the incoming path. They find that if $2p\pi$ vacancies are present at small internuclear distances, this process may contribute importantly at large b to the ionization probability for the case of F^{5+} on argon given in Figs. 19 and 20. Thus, the two-step process discussed above for symmetric collisions may be competitive with direct ionization to the continuum even for such asymmetric collisions, and may be responsible for the shift of $P(b)$ to large b at large Z_1/Z_2.

7.2.4.4. Additional P(b) Results. Ambros et al.[106] have investigated the b dependence of iodine and gold M-shell x-ray production in an unusual coincidence experiment. Using an iodine beam on a single gold crystal, the transverse amplitude of oscillation of particles channeled in the (111) gold plane was identified from the iodine's energy loss. By detecting iodine and gold M x-rays in coincidence with the channeled ions, information on the x-ray production probability as a function of transverse oscillation amplitude was obtained. Larger amplitudes enable closer Au–I collisions. Although the actual structure of $P(b)$ cannot be easily unfolded from such data, approximate ranges for effective M x-ray production were deduced to be 0.12 and 0.25 Å for gold and iodine, respectively.

Schuch et al.[28] have measured the probability for production of two-electron–one-photon transitions in Ni–Ni collisions at very small b. Two K vacancies must be created in the collision in order that this radiation be generated. Although one might expect the probability for such a process to be proportional to the square of the probability for creating a single K vacancy, this was not observed to be the case. The $P(b)$ for the double-electron transition was found to continue to rise for impact parameters small enough that $P(b)$ for the characteristic x ray had reached a constant value. It is not clear to what this behavior is due.

A number of coincidence experiments on the b dependence of non-characteristic x radiation have been reported.[107–109] These and the double-jump experiment have in common the technical feature that measurements of very small P ($\sim 10^{-5}$) are necessary.

[106] R. Ambros, H. O. Lutz, and K. Reichelt, *Phys. Rev. Lett.* **32**, 811 (1974).

[107] I. Tserruya, H. Schmidt-Böcking, R. Schulé, K. Bethge, R. Schuch, and H. J. Specht, *Phys. Rev. Lett.* **36**, 1451 (1976); H. Schmidt-Böcking, I. Tserruya, R. Schulé, R. Schuch, and K. Bethge, *Nucl. Inst. Meth.* **132**, 489 (1976).

[108] F. Jundt, G. Guillaume, P. Fintz, and K. W. Jones, *Phys. Rev. A* **13**, 563 (1976).

[109] D. Burch, *Int. Conf. Phys. Elect. Atom. Coll., 9th (ICPEAC), Seattle, 1975*.

7.3. Coincidence Experiments Involving Charge-Analyzed Reaction Products

7.3.1. Related to L-Shell Vacancy Production

Much of the early coincidence work on L-shell vacancy production involved coincident detection of charge state analyzed projectile and targets. These experiments still represent much of our experimental information on differential cross sections for L-shell vacancy production in the MO regime. As most of this work is of the IEL type, we refer the reader to Part 4 and to Refs. 3 and 4 for a more comprehensive coverage. Here we restrict our discussion to experiments that grew out of the IEL ones involving the detection of Auger L electrons in coincidence with charge-analyzed scattered ions.

The IEL results for argon on argon had suggested that, for distance of closest approach (r_0) less than about 0.25 Å, two L vacancies were created per collision. If so, L-Auger electrons should be emitted immediately following the collision. Thomson et al.[5] were the first to detect such electrons in a coincidence experiment. Their apparatus is shown in Fig. 21. Argon L-Auger electrons, of energy near 200 eV, were detected in coincidence with charge-selected argon ions scattered at 21°. Coincident electron spectra are shown in the lower part of Fig. 21. The number of coincidences per collision was found to increase rapidly for $r_0 \lesssim 0.24$ Å, nicely confirming the IEL results. The charge state selection showed that the associated Auger electron energy was higher for higher parent ion charge states, a result expected from the energy level structure of the ions. Further, for low-charge states of $<2+$, coincidences per ion dropped off, as would be expected if the lower charge states are primarily generated by M-electron loss only, whereas the higher charge states result from inner-shell vacancy-producing collisions.

An extension of these results to higher bombarding energy and smaller r_0 has recently been reported by Thomson[110] who used essentially the same experimental approach. By making an absolute calibration of his Auger electron spectrometer he was able to confirm that the number of L vacancies per collision does saturate at two for $0.15 \leq r_0 \leq 0.24$ Å. This result is interpretable in terms of the Fano–Lichten model as due to saturation of the $4f\sigma$ promotion, which removes 2p electrons to higher levels with essentially 100% efficiency. In the saturation region one thus gets two L vacancies every collision, and the final charge state distribution for such collisions should be collision independent. This was observed as a

[110] G. M. Thomson, Phys. Rev. A **15**, 865 (1977).

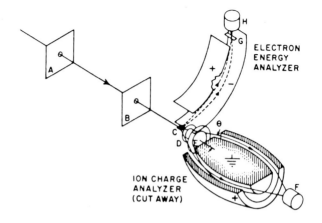

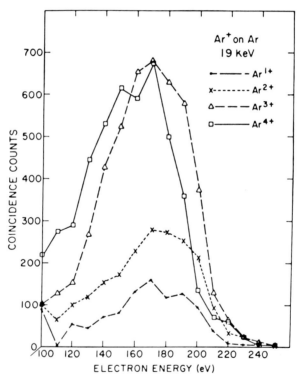

Fig. 21. Top: apparatus used for measuring electron–ion coincidences in Ar⁺–Ar collisions. Bottom: electron spectra measured in coincidence with charge-analyzed argon ions [G. M. Thomson et al., Phys. Rev. A **1**, 1439 (1970), G. M. Thomson, Phys. Rev. A **15**, 865, (1977)].

lack of correlation between coincident yield per collision (Y) and final ion charge state. For $r_0 \leq 0.15$ Å, Y was observed to rise sharply again, a result previously observed in IEL experiments.[3] This is attributed to further L-vacancy promotion via $3d\sigma-3d\pi-3d\delta$ rotational coupling producing up to ~6 L vacancies per collision. A correlation between outgoing argon charge state and Y was observed in this region, since the additional vacancy production does not saturate for $r_0 \geq 0.05$ Å. Collisions in which more L vacancies are created generally result in higher final argon charge states. Recent extension of this work to x-ray–ion coincidences[111] suggests a dramatic b dependence of ω_L in this region of impact parameters.

Schmid and Garcia[112] have carried out the first ab initio theoretical study of L-vacancy production within the context of the MO model for the Ar^+-Ar system. They calculate, within a comprehensive picture of the collision, Auger electron yields, inelastic energy losses, and charge state distributions. Their results confirm the earlier qualitative interpretation of the coincidence experiments and are even in good quantitative agreement with the data.

Studying the same collisional system, Thoe and Smith[113] detected argon L x rays in coincidence with charge-analyzed argon ions. Since the fluorescence yield of the argon L shell is quite sensitive to the M-shell occupation number, their results give information about not the vacancy production process but the fluorescence yields. A strong correlation between P_x and final charge state was found and reconciled with theoretical[114] fluorescence yields for the argon L shell. The analysis is complicated by the fact that ω_L is calculated for a given initial configuration, whereas the experiment detects the x ray in coincidence with the final charge state after Auger-dominated relaxation.

7.3.2. Coincident Charge State Analysis for Spectral Identification of X-Ray and Auger Electrons

The x-ray spectra and Auger spectra from systems whose inner shells have been collisionally vacated by heavy ions are generally rich with lines (see Part 8). Unless extremely high resolution spectrometers are used, it may be impossible to identify individual lines in the spectrum on the basis of energy resolution alone. Considerable simplification of the spectrum may be gained if the spectra are sorted according to the charge of the emitting ion, which may be achieved by detecting the emitted x radiation

---

[111] G. M. Thomson, private communication (1977).

[112] G. B. Schmid and J. D. Garcia, *Phys. Rev. A* **15**, 85 (1977).

[113] R. S. Thoe and W. W. Smith, *Phys. Rev. Lett.* **30**, 525 (1973).

[114] F. P. Larkins, *J. Phys. B.* **4**, L29 (1971); C. P. Bhalla and D. L. Walters, *Proc. Int. Conf. Inner-Shell Ion. Phenomena* (U.S.A.E.C. CONF 720404), p. 1572 (1972).

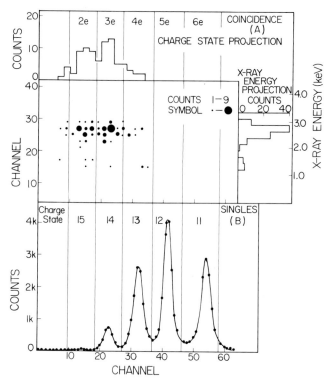

FIG. 22. (A) Two-dimensional spectra of downstream chlorine K x rays, following foil excitation, in coincidence with subsequently charge-analyzed chlorine ions. (b) Charge state spectrum taken without coincidence requirement. Metastable x-ray emitters are associated with charge states 14+ and 15+, although these are minor components of the total beam (C. L. Cocke, B. Curnutte, J. R. Macdonald, and R. Randall, *Phys. Rev. A* **9**, 57, 1974).

or Auger electrons in coicidence with the subsequent charge-analyzed parent ions.

This technique has been used to identify emitters produced by foil excitation. Cocke *et al.*[115] measured low-resolution (Si-[Li]) K x-ray spectra from foil-excited chlorine beams in coincidence with charge-state-analyzed chlorine ions. Here only the metastable emitters in the beam were studied, whose lifetimes were, however, too short to allow charge analysis before detection. The results, shown in Fig. 22, showed that the radiation came primarily from helium- and lithium-like emitters, which were identified as the $(1s2p)^3P_2$ and $(1s2s2p)^4P_{5/2}$ systems, each emitting M2 radiation to the ground state. Lifetime measurements made using the coincidence identification were confirmed by high-resolution measurements on the same beam.

[115] C. L. Cocke, B. Curnutte, J. R. Macdonald, and R. Randall, *Phys. Rev. A* **9**, 57 (1974).

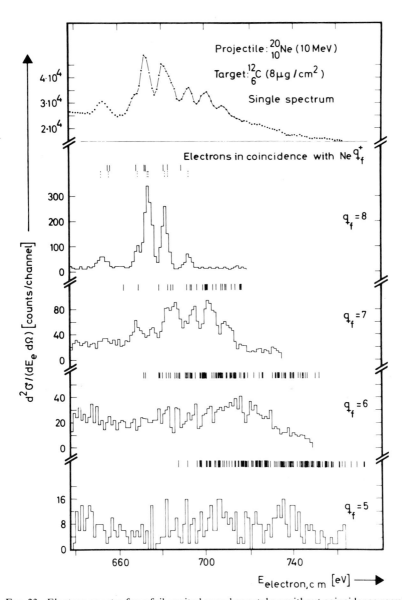

FIG. 23. Electron spectra from foil-excited neon beam taken without coincidence requirement (top) and in coincidence with subsequently charge analyzed Ne ions (bottom). Expected line locations are shown as short vertical bars (K. O. Groeneveld, G. Nolte, S. Schumann, and K. D. Sevier, *Phys. Lett.* **56A**, 29, 1976).

It is interesting to notice that such a technique does yield information on the probability that foil-excited systems emerge clothed with high-principal-quantum-number spectator electrons. The possibility that such systems might lead to erroneous beam foil lifetime measurements near the foil has often been discussed, most recently in connection with the observed nonexponential decay of the 2^3S_1 state in heliumlike systems.[116-118] Such spectator states are undetectable spectroscopically but should appear in the coincidence experiment by shifting a fraction of the primary coincident charge state spectrum to the next lower charge state, the fraction being the spectator electron formation probability. Figure 22 shows that such a fraction must be less than about 10% for few-electron systems in 48 MeV chlorine. Of course, if the principal quantum number is sufficiently high, the charge analysis may field-ionize the system and destroy the correlation.

Groeveld et al.[119] and Schumann et al.[120] have used coincident detection of Auger electrons from foil-excited neon beams and the subsequently charge-analyzed neon ions to simplify line identification in the Auger spectrum. They used a cylindrical-mirror electron analyzer but detected the electrons off the axis of the instrument, leaving clear the ion path through the system and into the following charge state spectrometer. Their spectra (Fig. 23) show that, even with charge state separation, the line density is too high to allow identification of individual transitions beyond lithium. The lithium spectrum is rather clear, however, and has been analyzed by Schumann et al.[120] to show that the relative intensities of the lines are consistent with a statistical population of the initial states, if the 2s and 2p excitation probabilities are equal. A clear analysis of the lithiumlike spectrum could never have been made from the singles spectrum alone.

7.3.3. Electron Capture from the K Shell of Heavy Targets

The realization that electron capture from inner shells of heavy targets may contribute importantly to inner-shell vacancy production has led to revived interest in such processes. Total electron capture cross sections

---

[116] J. A. Bednar, C. L. Cocke, B. Curnutte, and R. Randall, *Phys. Rev. A* **11**, 460 (1975).

[117] C. L. Cocke, *in* "Beam Foil Spectroscopy," (I. A. Sellin and D. J. Pegg, eds.), Vol. 1, p. 283. Plenum, New York, 1976.

[118] D. L. Lin and L. Armstron, *Phys. Rev.* **A16**, 791 (1977).

[119] K. O. Groeneveld, G. Nolte, S. Schumann, and K. D. Sevier, *Phys. Lett.* **56A**, 29 (1976).

[120] S. Schumann, K. O. Groeneveld, K. D. Sevier, and B. Fricke, *Phys. Lett.* **60A**, 289 (1977).

7.3. CHARGE-ANALYZED REACTION PRODUCTS

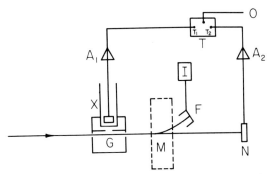

FIG. 24. Schematic arrangement for the observation of K x rays coincident with electron capture. G, gas cell; M, magnet; F, Faraday cup; I, integrator; N, surface barrier detector; X, Si[Li] detector; A, amplified; T, TAC. (J. R. Macdonald, C. L. Cocke, and W. W. Eidson, *Phys. Rev. Lett.* **32**, 648, 1974).

for targets much above helium tend to be dominated by capture from outer shells unless the projectile velocity is very high. However, it is possible to study capture from an isolated inner shell by detecting x radiation or Auger-electron emission from relaxation of the inner-shell vacancy in coincidence with the charge-changing event. Only the K shell has thus far been studied by this technique.

There is one problem of principle with such a technique. Although single-collision conditions may always be obtained by using a gas target of sufficiently low pressure, an ion capturing an electron from an inner shell may unfortunately find that even a single target atom represents a target of electron density sufficient to mask the results of the encounter with the inner shell. For highly charged incident ions, the probability P_i of L- (and, where appropriate, higher) shell capture may be sufficiently high, for impact parameters of the order of r_K, that the product $\sigma_K P_L$ is comparable to σ_{CK}. Here σ_K and σ_{CK} are cross sections for ionization of and capture from the K shell, respectively. Indeed, charge state distributions from single collisions have been reported to approach, for large scattering angles, those obtained in equilibrium from a thick target.[121,122] An event in which the K shell is ionized and L capture occurs is indistinguishable from direct K-shell capture. An estimate of $P_L \simeq \sigma_{CL}/\pi r_L^2$, where r_L is the radius of the L shell and σ_{CL} the cross section for L-shell capture, may be used to approximately correct the data for this process, provided the correction is not large.

[121] F. W. Martin and R. K. Cacak, *in* "Beam Foil Spectroscopy" (I. A. Sellin and D. J. Pegg, eds.), Vol. 2, p. 671. Plenum, New York, 1976.
[122] B. Rosner and D. Gur, *Phys. Rev. A* **15**, 70 (1977).

While the processes of most interest to study would be those in which a highly charged projectile captures K electrons from a target of similar nuclear charge, the above double-event process renders such experiments uninterpretable. Thus, experimental work to date has been confined to studying K-shell capture by protons on targets between $Z_2 = 7$ and 18. A typical experimental arrangement is that of Macdonald et al.[99] (Fig. 24). Here K x rays from an argon target gas were detected by a Si[Li] detector in coincidence with hydrogen atoms while the proton beam was deflected aside onto a Faraday cup. The neutralized particles were detected by a surface barrier detector whose surface area was sufficiently large to give a neutral detection efficiency of 100%. Under such conditions the ratio of coincidence events to singles x rays is directly the ratio of σ_{CK} to σ_{VK}, where σ_{VK} is the total K-vacancy production cross section. The x-ray detector efficiency and fluorescence yield do not enter. Thus, only a sepa-

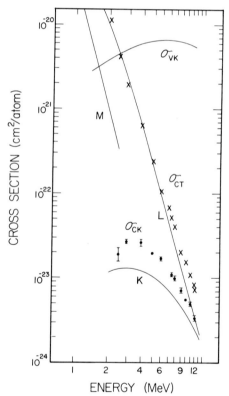

FIG. 25. Cross sections for total electron capture (σ_{CT}), K-vacancy production (σ_{VK}), and K-shell capture (σ_{CK}) for protons on argon. Solid lines for capture from individual shells are OBK calculations of Nikolaev (Macdonald et al., Phys. Rev. Lett. **32**, 648, 1974).

7.3. CHARGE-ANALYZED REACTION PRODUCTS

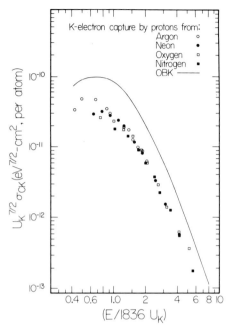

FIG. 26. A plot of K-shell capture cross sections for protons on various targets, multiplied by $(U_K)^{7/2}$, vs. the square of the scaled velocity. The OBK prediction for 1s to 1s capture is shown as the solid line (Cocke et al.[124]).

rate measurement of σ_{VK} is needed to complete the experiment. The results of Macdonald et al. are shown in Fig. 25, where it is seen that the contribution of K shell capture to K-vacancy production remains small for such an asymmetric system. The OBK calculations of Nikolaev[123] for capture from individual target shells, normalized to total capture cross sections at low energies, are in fairly good agreement with the data.

For lighter targets the fluorescence yields become so low as to make the detection of K x rays no longer feasible. Cocke et al.[124] have extended the measurements to targets of O_2, N_2, and Ne by replacing the Si[Li] detector used in the argon experiment by a cylindrical-mirror electron analyzer and detecting K-Auger electrons from target relaxation in coincidence with hydrogen atoms. Quite similar measurements have also been made by Rødbro et al.[26]

A summary of results for proton capture from K shells of higher-Z

---

[123] V. S. Nikolaev, Zh. Eksp. Teor. Fiz. **51**, 1263 (1966) [Sov. Phys.—JETP **24**, 847 (1967)].

[124] C. L. Cocke, T. R. Bratton, B. Curnutte, R. K. Gardner, and T. Saylor, Phys. Rev. **A16**, 2248 (1977).

targets is shown in Fig. 26, where $U_K^{7/2}\sigma_{CK}$ is plotted vs. the square of the scaled velocity. In the simplest OBK theory of charge exchange, data plotted in such a manner should fall on a universal curve. This is seen to be the case for scaled velocities greater than unity, although the absolute magnitude of the experimental curve is about a uniform factor of 3 below the OBK prediction for 1s–1s charge transfer. This serves to reinforce the recognized generalization that the OBK scaling of electron capture cross sections with projectile energy, from one nuclear charge to the next, or from one target shell to the next, seems to be useful while the absolute values are consistently too high. For proton bombardment, the contribution of capture to K-vacancy production is small, never exceeding 4% for the data of Fig. 26. The success of the OBK scaling rules illustrated by these data do tend to support conclusions drawn for higher Z_1, where the capture competes with ionization, based on the simple perturbation theory approach. More complete theoretical models[104] give good agreement with the data of Fig. 26, and certainly are to be preferred for the more nearly symmetric case over OBK.

The coincidence technique for detecting K-shell capture has been used to measure the differential cross section for electron capture by protons from the argon K shell as a function of the projectile scattering angle.[125] This experiment was partially motivated by the predictions in several first-Born calculations[126] of a node in the cross section at a scattering angle near the ratio of electron to projectile mass. In the theory, this zero result from a cancellation between OBK and nuclear amplitudes at small angles. The experimental result shows no sign of the predicted structure in $d\sigma/d\Omega$. Although the question arises whether it is fair to compare capture from the K shell of a neutral argon atom with a calculation of capture from an isolated K shell, a similar experimental result has recently been found for the case of protons on helium.[127] Both angular distributions may be accounted for[104,128] using an approach similar to that of Belkic and Salin.[44] It appears that the structure in the earlier theoretical angular distribution is an artifact of the formulation of the theory.

Acknowledgments

We thank C. D. Lin, J. H. McGuire, and K. Taulbjerg for communication of results prior to publication; L. Weaver for stimulating discussion; J. F. Chemin, G. W. Greenlees, K. O. Groeneveld, E. Laegsgaard, H. O. Lutz, I. Tserruya, H. Schmidt-Böcking, and G. Thomson for communication of current work.

[125] C. L. Cocke, J. R. Macdonald, B. Curnutte, S. L. Varghese, and R. Randall, *Phys. Rev. Lett.* **36**, 782 (1976).

[126] K. Omidvar, J. E. Golden, J. H. McGuire, O. L. Weaver, *Phys. Rev. A* **13**, 500 (1976).

[127] T. Bratton, C. L. Cocke, and J. R. Macdonald, *J. Phys.* **B10**, L517 (1977).

[128] S. R. Rogers and J. H. McGuire, *J. Phys.* **B10**, L497 (1977).

8. ION-INDUCED X-RAY SPECTROSCOPY*

By Forrest Hopkins

8.1. Introduction

The use of accelerators in atomic collision studies has facilitated the acquisition of a rapidly expanding base of information on ion–atom interactions. The range of ions and energies available has made accessible detailed investigations of inner-shell excitation and multiply excited states that are not possible with x-ray fluorescence or electron bombardment. The observation of the x rays emitted by excited ions has, along with Auger electron spectroscopy, served as a viable means of measuring various excitation probabilities as well as the energies and rates of transitions resulting from ion–atom collisions.

The breadth of energies of transitions discussed here varies from ~180 eV (68 Å), the boron K x-ray energy, up to several keV. K-shell excited states receive the majority of attention, simply because most available experimental and theoretical work concerns them. Several examples of L excitation are also pointed out. Incident energies of projectiles include ~1 keV/amu up to several MeV/amu. Collision partners as heavy as iodine are mentioned.

Experimental procedures are only briefly outlined, since to date they have been straightforward and can be found in the various experimental works cited. Similarly, collision mechanisms comprise a secondary source of interest here, although they are directly responsible for the production of the states involved. Total cross sections are touched upon only to the extent that they are related to specific spectroscopic information. Several comprehensive articles[1–3] cover in detail the modes of excitation occurring in these collisions. Further, numerous interesting

---

[1] Q. C. Kessel and B. Fastrup, in "Case Studies in Atomic Physics" (E. W. McDaniel and M. R. C. McDowell, eds.), Vol. III, p. 137. North-Holland, Amsterdam, 1973.
[2] J. D. Garcia, R. J. Fortner, and T. M. Kavanagh, *Rev. Mod. Phys.* **45**, 111 (1973).
[3] P. Richard, in "Atomic Inner Shell Processes," (B. Crasemann, ed.), Vol. I, p. 74. Academic Press, New York, 1975.

---

* See also Vol. 13A of this series, Chapter 3.2.

topics, e.g., radiative electron capture, metastable states, and molecular orbital x-ray emission, are omitted since they are covered in other parts of this volume.

The primary "experimental method" examined here is the classification and evaluation of an incredibly wide-ranging set of data made possible by the flexibility of modern accelerators. The accelerators themselves encompass a diverse group of machines, from high-voltage ion sources to Van de Graaff accelerators to cyclotrons. A basic understanding of the origin of x-ray emission[4] and the spectroscopic notation usually employed in delineating specific transitions[5] is assumed in the discussion to follow.

8.2. Detectors and Targets

8.2.1. Proportional Counters

Several types of gas-filled detectors are employed in the detection of x rays, including ionization chambers and Geiger–Müller counters. The third member of that group is the proportional counter, whose signal output is directly proportional to the energy of the photon. The latter is the only one discussed here, since it is the most widely used in ion-induced x-ray spectroscopy. It has served both as a poor resolution device for measuring total yields accurately and, even more importantly for the subsequent discussion here, as the photon-counting element in dispersive instruments such as crystal spectrometers.

The most common design[6,7,8] is a coaxial proportional counter with an ultrathin (~ 50 μm) center wire anode running along the axis of a cylindrical cathode of a few centemeters in diameter. The volume is filled with a chosen gas, which absorbs the photon, emitting a photoelectron. The photoelectron ionizes additional electrons, all of which are accelerated by a radial electric field established by applying a suitable voltage (1–2 keV) between anode and cathode. Under proper operating conditions of geometry, type and pressure of gas or gases, and voltage, an electron ava-

---

[4] J. H. Scofield, in "Atomic Inner Shell Processes" (B. Crasemann, ed.), Vol. I, p. 265. Academic Press, New York, 1975.

[5] A. E. Sandström, in "Handbuch der Physik " (S. Flügge, ed.), Vol. 30, p. 78. Springer-Verlag, Berlin and New York, 1957.

[6] S. C. Curran, in "Handbuch der Physik " (S. Flügge, ed.), Vol. 45, p. 174. Springer-Verlag, Berlin and New York, 1958.

[7] R. Jenkins and J. L. De Vries, "Practical X-Ray Spectrometry." Springer-Verlag, Eindhoven, The Netherlands, 1967.

[8] R. W. Fink, in "Atomic Inner Shell Processes " (B. Crasemann, ed.), Vol. II, p. 169. Academic Press, New York, 1975.

lanche is triggered in the high-field region adjacent to the anode wire. The voltage applied is sufficient to result in an amplification or gain from 1 to $\sim 10^6$, the amplitude of which depends upon the energy of the initial photoelectron and thereby of the photon. The signal resulting from the electron collection is processed through an RC circuit (preamp). The positive ions, which migrate to the cathode and emit photons upon recombination, leading to secondary electron emission and avalanche, are quenched either by the primary gas or a second gas added to the volume. Halogens and methane are frequently used as quench gases.

The resolution ΔE attainable for an x ray of energy E with a proportional counter is given by the expression[9]

$$[\overline{(\Delta E^2)}/E^2]^{1/2} = C/E^{1/2}, \qquad (8.2.1)$$

where $C = (\phi W)^{1/2}$, $\phi = F + f$, and W is the mean energy required to create an ion pair (20–30 eV). The quantity F is the Fano factor[10] and f is the relative variance of the avalanche size. Thus resolution, excluding practical contributions from nonuniform anode wire, end-field effects, and associated electronic conditioning of the signals, is limited by the statistics of the avalanche. Empirical resolution of $C = 0.146$ (keV)$^{1/2}$, nearly equivalent with the theoretical limit for an Ar–CH$_4$ mixture, has been reported.[11] Typical resolutions are between 15 and 30%, depending on, among other things, energy. Both the resolution and pulse amplitude are degraded by high count rates, but rates well above 10^4 counts/sec are feasible with some degradation.

The choice of gas is dictated by the photon energy and the experimental conditions. Propane, methane, and isobutane absorb soft x rays sufficiently, but photons with energy greater than ~ 2 keV require argon, krypton, or xenon. Published efficiency curves allow a wide range of choices.[7,12] The x rays enter the counter via a thin window of some material on the side. For sufficiently high-energy x rays, nonpermeable windows of several microns of beryllium or mica can be employed and the counter can be sealed with a fixed volume of gas. For softer x rays (<1.5 keV), very thin foils of usually carbonaceous material allow reasonable transmission.

The thickness (≤ 6 μm) of the latter requires that the detector be supplied with a continuous flow of gas in order to replace gas permeating the foil; hence the name flow proportional counter has been applied. Several

---

[9] G. D. Alkhazov, *Nucl. Instr. and Meth.* **89**, 155 (1970).
[10] U. Fano, *Phys. Rev.* **72**, 26 (1947).
[11] M. W. Charles and B. A. Cooke, *Nucl. Instr. and Meth.* **61**, 31 (1968).
[12] R. L. Kauffman and P. Richard, *in* "Methods of Experimental Physics: Spectroscopy" (D. Williams, ed.), Vol. 13, p. 148. Academic Press, New York, 1976.

358   8. ION-INDUCED X-RAY SPECTROSCOPY

TABLE I. Mass Absorption Coefficients μ of Several Carbonaceous Window Materials[a]

λ (Å)	Parylene N C_8H_8	Parylene C C_8H_7Cl	Polypropylene C_3H_6	Mylar $C_{10}H_8O_4$	Formvar $C_5H_8O_2$
10	0.117	0.157	0.088	0.223	0.190
20	0.784	0.975	0.593	1.357	1.156
40	4.457	5.220	3.374	3.997	3.350
60	0.500	2.611	0.379	0.999	0.851

[a] In units of $(\mu/\rho)\rho \times 10^{-4}$ (cm^{-1}) [taken from M. A. Spivak, *Rev. Sci. Instr.* **41**, 1614 (1970)].

such materials are listed in Table I. Stretched polypropylene and parylene have been successfully used down to thicknesses of 0.4 and 0.2 μm, respectively, where support grids are provided. The flow counter is normally operated at 1 atm, with a 90% Ar–10% CH$_4$ (P10) mixture commonly used. It should be noted that in principle most any thin solid foil can be used if durability and transmission are feasible and also that, in general, pressures in the counter can be much less than 1 atm (for both sealed and flow detectors). The plastic foils are frequently coated on the interior surface with several hundred angstroms of aluminum to prevent charge buildup. The detection of soft x rays ($\lambda > 10$ Å) especially requires thin-window techniques.[13]

8.2.2. Semiconductor Detectors

In recent years, solid-state detectors have been improved tremendously with regard to resolution and so have found increasing use in ion-induced x-ray spectroscopy. The two most successful versions are the lithium-drifted silicon and lithium-drifted germanium diodes. A newer member of the group is the intrinsic germanium detector. The semiconductor detector, a nondispersive device that registers all energies of photons simultaneously, allows a continual check on all regions of the spectrum of interest, in contrast to a dispersive counter, which selects out one energy or wavelength at a time.

The principle of operation is discussed in several sources.[14–18] Typi-

---

[13] B. L. Henke, in "Advances in X-Ray Analysis" (W. M. Mueller, G. Mallett, and M. Fay, eds.), Vol. 8, p. 269. Plenum Press, New York, 1964.

[14] G. Dearnaley and D. C. Northrop, "Semiconductor Counters for Nuclear Radiations." John Wiley Inc., New York, 1963.

[15] G. Bertolini and G. Restelli, in "Atomic Inner Shell Processes" (B. Crasemann, ed.), Vol. II, p. 124. Academic Press, New York, 1975.

[16] A. H. F. Muggleton, *Nucl. Instr. and Meth.* **101**, 113 (1972).

[17] D. A. Gedcke, *X-Ray Spectromet.* **1**, 129 (1972).

8.2. DETECTORS AND TARGETS

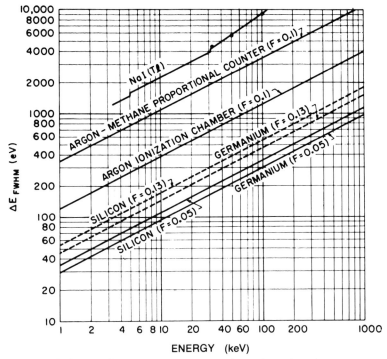

FIG. 1. Resolution as a function of photon energy for various dispersive detectors. For semiconductor detectors, the Fano factor F of 0.05 is the ideal case, while the curves with $F = 0.13$ represent resolution routinely attainable in practice [R. S. Frankel and D. W. Aitkens, *Appl. Spectrosc.* **24**, 557 (1970)].

cally, using silicon as an example, the diode is a cylindrical piece of p-type silicon crystal doped with lithium to mask impurities and to increase electrical resistivity. Diameters range from 4 to 16 mm. A layer of n-type silicon forms the boundary of the depletion (drifted) region, opposite the p-type. Two conducting electrodes, usually a few hundred angstroms of gold, complete the sandwich, one at either end of the cylinder. A reverse bias of 1–2 keV sweeps any free-charge carriers. Photons entering the front (p-type) end are absorbed in the depleted region, causing creation of a number of electron–hole pairs, given by (E/ϵ), where E is the photon energy and ϵ the energy required to create a pair. The latter quantity is ~3 eV for silicon and a little less for germanium. The charge is collected at the electrode and integrated in a charge-sensitive preamplifier. The output voltage pulse is proportional to the energy of the photon.

---

[18] R. S. Frankel and D. W. Aitken, *Appl. Spectrosc.* **24**, 557 (1970).

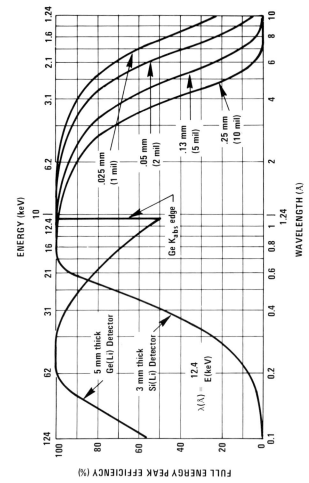

FIG. 2. Comparison of efficiencies as a function of energy for Ge(Li) and Si(Li) detectors [D. A. Gedcke, *X-Ray Spectromet.* **1**, 129 (1972)].

8.2. DETECTORS AND TARGETS

The ionization process is analogous to the creation of electron–ion pairs in a gas counter and the statistics of the mobile electrons determine the ultimate resolution in a similar way. The fact that a particular energy of photon yields about ten times more pairs in a semiconductor than in a gas (where pair production requires about 30 eV) means that resolution is vastly improved. Figure 1 presents the resolution of various detectors as a function of photon energy. The curves using a Fano factor of 0.13 for silicon and germanium represent current state of the art, while those with the value of 0.05 may be thought of as the theoretical limit, excluding other sources of line broadening (electronic, etc.).

The diodes are housed in a vacuum assembly with an entrance window typically made of beryllium of thicknesses from ~0.007 to 0.25 mm providing transmission of the x rays. A more recent design allows direct vacuum coupling of the crystal into the vacuum system containing the source, and avoids absorption in a window. The crystals must be cooled to liquid-nitrogen temperatures in order to prevent migration of the lithium and also to minimize noise due to thermal excitation. The cooling of them and the front end of the preamp is accomplished with various cryostat configurations coupled to nitrogen dewars. The crystals are from 3 to 5 mm thick.

Efficiency curves for Si(Li) and Ge(Li) detectors are shown in Fig. 2, for a variety of different entrance windows. Loss of efficiency at high energy is due to less than complete absorption; thus Ge(Li) is more efficient at higher energies. The loss at low energy is due primarily to attenuation in the entrance window. The crystals have a small "dead layer" of a fraction of a micron at the front surface, which may influence transmission and peak shape of the very low energy x rays ($<$ 1.5 keV). The detectors, with proper electronics, can sustain count rates of 10^4 without appreciable resolution degradation or gain shift. That stability is far superior to the performance of gas-filled proportional counters.

8.2.3. Single-Plane Crystal Spectrometers

Of various x-ray diffraction devices available, the most widely used in accelerator-based x-ray studies has been the Bragg crystal reflection spectrometer. The principles of operation have long been known and utilized in other fields, such as x-ray fluorescence and electron-induced excitation. Several excellent sources[7,19,20] furnish ample information of the

---

[19] J. L. Jones, K. W. Paschen, and J. B. Nicholson, *Appl. Optics* **2**, 955 (1963).

[20] V. Cauchois and C. Bonnelle, in "Atomic Inner Shell Processes " (B. Crasemann, ed.), Vol. II, p. 84. Academic Press, New York, 1975.

type discussed here concerning the application of the detector to high-resolution spectroscopy.

According to Laue interference conditions for diffraction from a three-dimensional point lattice, constructive interference takes place for angles satisfying the Bragg equation for diffraction order n,

$$n\lambda = 2d \sin \theta, \quad n = 1, 2, \ldots, \quad (8.2.2)$$

where λ is the wavelength of the photon, d the spacing of the atomic planes in the crystal, and θ the Bragg angle (complement of the angle of incidence). Although the equation holds for transmission as well as reflection, only situations of the latter type are discussed here. For very soft x rays, the effective value of d is dependent on n and is slightly different than the crystal spacing due to refraction upon entering and leaving the crystal. The detector is illustrated in Fig. 3, where O is the object (source) and F the focal point (detection), for a plane crystal. The de-

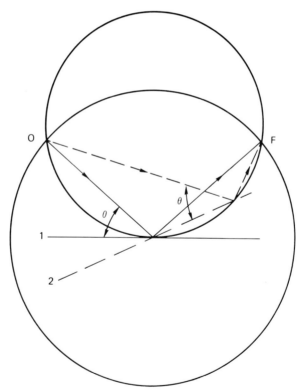

FIG. 3. Schematic of the Bragg plane crystal spectrometer (reflection geometry), with source at O and detector at F.

8.2. DETECTORS AND TARGETS

tecting element at F can be a photographic emulsion, a scintillator, solid state detector, channeltron, Geiger–Müller tube, or proportional counter. A vast majority of the ion-related data to date has been taken with the latter, which conveniently affords single-photon counting capability in conjunction with energy-dependent pulse amplitudes for line discrimination.

In practical application there is a fixed entrance slit that defines O or, alternatively, the source is limited to a reasonably small size. The crystal is rotated in the plane defined by the incident and reflected rays. For single-proton counting techniques, the detecting element (itself collimated to give best results) is rotated the proper angle to maintain a 2θ geometry for reflected rays. The angular dispersion is obtained by differentiating Eq. (8.2.2):

$$d\theta/d\lambda = n/(2d \cos \theta), \qquad (8.2.3)$$

which gives

$$dE/d\theta = (E \cos \theta)/n, \qquad (8.2.4)$$

where n as above is the order of diffraction and E the energy corresponding to wavelength λ. The linear dispersion $dl/d\lambda$ at the detector is given by

$$dl/d\lambda = (nL)/(2d \cos \theta), \qquad (8.2.5)$$

where L is the length from crystal to detector. From these equations, it can be seen that the resolution improves both for larger Bragg angles and for higher orders of diffraction.

Additional factors affect the ultimate empirical resolution $\Delta E/E$. The fact that x rays can penetrate a distance into the crystal before diffracting leads to a slightly different Bragg angle, although the transparency of the material usually limits the effect to shorter wavelengths. A second occurrence is vertical divergence (out of the plane in Fig. 3), which also leads to slightly different effective Bragg angles at a given setting of the crystal. Finite entrance and exit slits, where an extended source is involved, also contribute to the widths observed. Realizable resolution is a few parts in a thousand, with the proper choice of crystal.

In reality many crystals do not fulfill the requirements of a "perfect" crystal, that is they are composed of a group of small planes oriented in various directions giving rise to individual Bragg angles that exceed the range of reflection from a single plane. For random orientation, the net effect is that of a "mosaic" crystal and increases somewhat the angular range over which reflection can take place, a dependence known as the rocking curve.

TABLE II. Reflection Coefficients R_c for a Plane KAP Crystal[a]

Wavelength (Å)	R_c ($\times 10^{-5}$ rad)
8.34 (Al K_α)	8.4
18.3 (F K_α)	5.0
21.6 (Cr L_α)	4.3
23.6 (O K_α)	1.6

[a] From A. J. Burek, D. M. Barrus, and R. L. Blake, *Astrophys. J.* **191**, 533 (1974).

The reflecting power, the integral of the reflection coefficient over relevant values of θ, is of prime importance. The high resolution of crystal spectrometers is gained by the sacrifice of the high efficiency attainable with relatively poor resolution devices such as proportional counters and semiconductors. A series of measured reflection coefficients for a KAP crystal is contained in Table II. Total reflection efficiency of several percent is possible in first order, with further loss for higher orders. For accelerator experiments where time for data-taking is limited, the proper choice of crystal, based not only on resolution but also on reflective power, is essential.

As can be seen from Eq. (8.2.2), the region of wavelength accessible with a particular crystal is comparable to the $2d$ spacing. A list of several crystals with their associated $2d$ spacings is given in Table III. Energies (wavelengths) from about ~ 150 eV (80 Å) to $\sim$ keV (2 Å) can be covered

TABLE III. Crystal $2d$ Spacings[a]

Crystal	Plane	$2d$ Spacing (Å)	Useful wavelength region (Å)
LiF	(200)	4.027	0.351–3.84
NaCl	(200)	5.641	0.492–5.38
Si	(111)	6.271	0.547–5.98
Ge	(111)	6.532	0.570–6.23
SiO_2	(101)	6.687	0.583–6.38
PET	(002)	8.742	0.762–8.34
EDT	(020)	8.808	0.768–8.40
ADP	(101)	10.64	0.928–10.15
Mica	(002)	19.84	1.73–18.93
RAP	(100)	26.12	2.28–24.92
KAP	(100)	26.632	2.32–25.41
Lead stearate	—	100	8.72–95.4

[a] From "Handbook of Spectroscopy" (J. W. Robinson, ed.), Vol. 1, p. 239, CRC, Cleveland, 1974.

8.2. DETECTORS AND TARGETS

with substantial overlapping between different crystals. The term "crystal" is used to indicate any diffracting device, regardless of whether it is a true crystal or not. Limitations on $2d$ spacings of natural crystals have led to the use of pseudocrystals, where molecular layers with suitable spacing serve as the reflecting planes. A particularly successful method involves the accumulation of successive monolayers of metal stearates by means of the Langmuir–Blodgett dipping method, up to a total of about 100–200 layers. The lead stearate pseudocrystal in Table III has proved to be especially valuable in extending measurements of x rays as low in energy as the boron K line at 183 eV and even lower.

8.2.4. Single Cylindrical Curved-Crystal Spectrometers

The poor efficiency associated with plane crystals led to the development of various configurations that focus x rays. The two types to be discussed here are the Johann geometry[21] and the Johannson geometry,[22] both of which are pictured in Fig. 4. The former consists of a crystal bent to a radius R, which is twice the radius of the focusing or Rowland circle, leading to approximate focusing. The focusing defect indicated by the solid lines in Fig. 4a can be avoided by bending to radius R and then grinding to radius $R/2$, the radius of the focal circle. The latter arrangement, known as the Johannson condition and shown in Fig. 4b, leads to the effective use of the entire length of the crystal. In certain instances where grinding of the crystal is prohibited, as with the lead sterate pseudocrystal, the Johann geometry is the only choice.

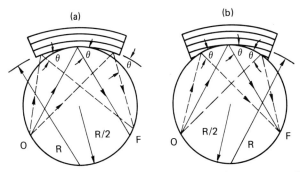

FIG. 4. (a) Johann curved-crystal spectrometer. (b) Johannson curved-crystal spectrometer.

[21] H. H. Johann, Z. Phys. **69**, 185 (1931).
[22] T. Johannson, Z. Phys. **82**, 507 (1933).

The angular dispersion for a curved crystal is the same as for a plane crystal in Eq. (8.2.2). The linear dispersion along the focal circle is given by

$$dl/d\lambda = (nR)/(2d \cos \theta), \qquad (8.2.6)$$

where R is the radius of curvature. This quantity and the resolution are again dependent upon the rocking angle, the angular range over which the mosaic structure allows a particular energy to be successfully reflected. A Gaussian will usually approximate the effective lineshape.

The many factors that make the reflective coefficient, and thereby the reflective power difficult to determine theoretically, include the actual diffraction pattern, and effects of geometry, imperfections in the crystal (mosaic). In addition, measures to increase resolution such as narrowing entrance and exit slits result in a concurrent reduction in usable intensity. Hence empirical checks on the intensity vs. resolution as a function of slit geometry and particular crystal used are an extremely valuable aid to the experimentalist. A sample of such measurements taken with crystals of 10 cm radius in the Johannson mode is presented in Figs. 5 and 6.[19] The

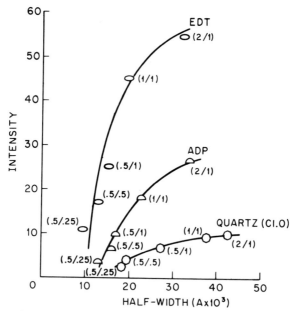

FIG. 5. Intensity–resolution function for Johannson-type crystals (radius of curvature 10 cm) at a wavelength of 8.339 Å (aluminum K_α). The figures in parentheses are the primary/secondary slit sizes in millimeters [J. L. Jones, K. W. Paschen, and J. B. Nicholson, Appl. Opt. **2**, 955 (1963)].

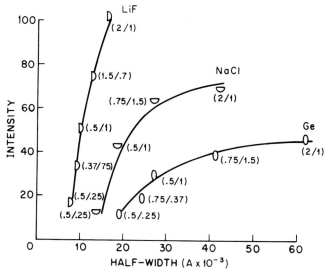

FIG. 6. Intensity–resolution function for Johannson-type crystals (radius of curvature 10 cm) at a wavelength of 3.360 Å (calcium K_α). The figures in parentheses are the primary/secondary slit sizes in millimeters [J. L. Jones, K. W. Paschen, and J. B. Nicholson, *Appl. Opt.* **2**, 955 (1963)].

half-width referred to is the Rayleigh criterion (width at half-maximum), which, although not indicative of the actual line profile, is a convenient signature of resolving power. The general trend is that, for a specific wavelength, the crystal that gives the best resolution for a particular choice of source (primary) and detector (secondary) collimation gives the best intensity also.

There are many background contributions that lead to extraneous yield in the region of a peak. The diffuse scattering of photons off the crystal surface into the detector is one. The fluorescence of the crystal by photons from the source can lead to characteristic lines of the crystal components entering the detectors. When the higher energies of projectiles from accelerators are used and nuclear reactions can occur with high probability, the flux of gamma rays produced may penetrate directly to the detector, requiring additional shielding. Compton scattering of the gamma ray in the counter gas can lead to pulses of the same amplitude as that produced by the x ray of interest. For small Bragg angles, ultraviolet radiation from the source can be reflected by the crystal into the detector. The latter problem can be reduced by placing appropriate filters somewhere in the photon path.

All of these problems make appealing the choice of a detector that out-

TABLE IV. Relative Discrimination Factors of Crystals for Orders Greater than $n = 1$[a]

Crystal	Plane	λ (Å)	Discrimination factor		
			$n = 2$	$n = 3$	$n = 4$
ADP	(101)	8.3 (Al K_α)	—	1.0	1.0
EDT	(020)		—	4.0	6.0
ADP	(101)	7.1 (Si K_α)	—	1.0	1.0
EDT	(020)		—	3.2	8.8
SiO_2	(01.0)		—	0.9	1.6
NaCl	(200)	4.7 (Cl K_α)	1.0	1.0	1.0
Ge	(111)		>10.3	0.12	0.3
SiO_2	(01.1)		0.4	6.2	>0.7
LiF	(200)	3.4 (Ca K_α)	1.0	1.0	1.0
NaCl	(200)		0.9	0.4	0.2
Ge	(111)		>60.0	1.1	0.1

[a] A high value indicates high rejection [J. L. Jones, K. W. Paschen, and J. B. Nicholson, *Appl. Opt.* **2**, 955 (1963)].

puts a variable-amplitude pulse (dependent on photon energy). For example, proportional-counter resolution is sufficient to allow the majority of these backgrounds to be rejected electronically by means of a single-channel-analyzer. Yet another superposition that can be avoided by such energy discrimination is the overlapping of higher-order reflections of shorter wavelengths with a line of interest. Although the Bragg condition does not select out two lines whose wavelengths happen to be integral multiples of each other, the choice of pulse height in the detector can. The relative strength of higher-order reflections is extremely dependent upon the type of crystal, as are the backgrounds listed above. Table IV demonstrates the great variety of that property as a function of crystal.

In addition to the plane and cylindrically curved crystal spectrometers discussed here, several other diffraction devices have been used to observe x rays produced in ion–atom collisions. These include the elastic continuously bendable crystals,[23] double-focusing spectrometer (concave spherical rather than cylindrical reflector),[24] and use of ruled gratings such as in the grazing incidence spectrometer.[25] The specialized functions of such devices will undoubtedly lead to greater use in the future, along with the transmission crystal geometry[26] optimal for shorter wavelengths. In

[23] R. L. Kauffman, L. C. Feldman, and P. J. Silverman, *Abstracts Int. Conf. Phys. X-Ray Spectra,* Gaithersburg, Md., 1976 (National Bureau of Standards).

[24] F. W. Martin and R. K. Cacak, *J. Phys. E* **9**, 662 (1976).

[25] D. J. Pegg, S. B. Elston, P. M. Griffin, H. C. Hayden, J. P. Forester, R. S. Thoe, R. S. Peterson, and I. A. Sellin, *Phys. Rev. A* **14**, 1036 (1976).

[26] Y. Cauchois, *J. Phys. Radium SVII* **4**, 61 (1933).

addition, various techniques facilitate the detection of long-wavelength (>10 Å) photons.[13]

8.2.5. Doppler-Tuned Spectrometer

A unique way of achieving high resolution has been devised by exploiting the Doppler shift of photons emitted by ions moving at high velocities. The shift is given by the expression

$$E = \gamma[1 - (v/c)\cos\theta]E_0, \qquad (8.2.7)$$

where E_0 is the intrinsic energy, θ the angle of detection with respect to the direction of the flight path of the ion, v the speed of the ion, c the speed of light, and $\gamma = [1 - (v/c)^2]^{-1/2}$.

The instrument is pictured in Fig. 7 and operates on the following principle.[27,28] An absorber is placed between the emitter and the detector, which has an absorption edge just below the energy of the photon. This is often and most conveniently simply the edge of the appropriate electronic shell in the neutral element itself. The angle θ is varied through the region where the Doppler-shifted energy matches and then surpasses the threshold necessary for absorption. For proper choice of thickness, the absorber is essentially transparent (or nearly so) at angles below a critical angle and opaque (within a few percent) above that angle. Thus the intensity of the line of interest will be drastically curtailed when its energy matches or exceeds the absorption edge energy. The total yield in the detector, such as a proportional counter, will reflect that loss.

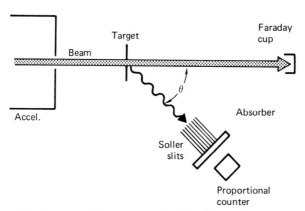

FIG. 7. Schematic of the geometry for a Doppler-tuned spectrometer.

[27] R. W. Schmeider and R. Marrus, *Nucl. Instr. and Meth.* **110**, 459 (1973).
[28] C. L. Cocke, B. Curnutte, and R. Randall, *Phys. Rev. A* **9**, 1823 (1974).

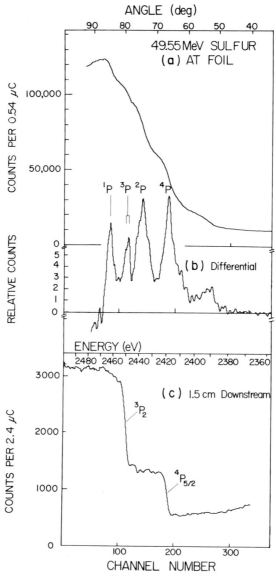

FIG. 8. Spectra of sulfur K x rays taken with the Doppler-tuned spectrometer. (a) Detector viewing a thin foil directly, (b) a differentiation of the spectrum in (a), and (c) viewing the beam 1.5 cm beyond the foil. The angle θ is the angle expressed in Eq. (8.2.7) with corresponding energy scale indicated [C. L. Cocke, B. Curnutte, and R. Randall, *Phys. Rev. A* **9**, 1823 (1974)].

8.2. DETECTORS AND TARGETS

A spectrum taken[28] with a Doppler-tuned spectrometer (DTS) is shown in Fig. 8. The absorber used was a polystyrene–sulfur material of thickness 0.025 mm, providing the sulfur K absorption edge. Sulfur K x rays emitted by ~50-MeV sulfur ions passing through a thin carbon foil constitute the total yield in the top part of the figure, recorded as a function of detector angle. The dips correspond to the loss of specific lines. A differential of the total yield curve with respect to θ produces the more familiar representation of the x rays as peaks. The bottom part of the figures presents the decay of metastable states where the ions were viewed downstream from the exciter foil. The geometrical resolution of the DTS is given by differentiation of Eq. (8.2.7):

$$\Delta E/E_0 = \gamma(v/c) \sin \theta \; \Delta\theta. \tag{8.2.8}$$

Thus it is limited by x-ray angular divergence, which is to say, the degree of collimation of the emitted photons (assuming that beam divergence is negligible). The collimation can be accomplished either by two thin collimators some distance apart or, more efficiently, with Soller slits, a series of thin parallel metal planes with spacing chosen to limit emission to the desired angle of acceptance. Resolution on the order of 0.1% can be reached without restricting counting rates severely.

In fact, one of the great advantages of the DTS is that it retains for the most part the efficiency characteristic of a nondispersive detector, which is far superior to that of a crystal spectrometer. Disadvantages include the fact that only a narrow-energy region can be scanned with a given absorber and that the number of possible absorption edges is limited. An obvious restriction is that only emission from fast-moving ions (projectiles) can be viewed. Additional problems include the nonuniformity of the thickness of absorbers and the fact that absorption edges are in reality not step functions but have threshold shapes. However, it is a very useful approach made possible by the energies attainable with modern accelerators.

8.2.6. Targets

The various considerations involved in the use of solid targets in conjunction with heavy projectiles are similar for both dispersive and nondispersive detectors. The three main effects that have to be accounted for are beam charge state equilibrium, beam energy degradation, and self-absorption in the target of any photons emitted by either target or projectile ions.

The latter two processes can be minimized by the use of thin solid foils rather than infinitely thick solids (slabs) in which the beam slows down and eventually stops. The optimum thickness depends upon the appropriate absorption coefficient and the stopping power of the ion in the particular solid medium. Ideally both the absorption[29] and energy loss are kept to a sufficiently small amount as to be negligible, a situation that still allows reasonable counting rates even for low-efficiency, nondispersive detectors. Of course, the angle of a foil or a slab with respect to the beam and the target–detector axis determines the effective absorption and thereby the visible (to the detector) part of the flight path over which energy loss is important. Where use of a slab is unavoidable, thick-target corrections[30] can be made to obtain approximate cross sections for the production of the x rays from either partner involved in the collision.

An unavoidable uncertainty associated with the penetration of a heavy ion through a solid is the ambiguity introduced by the characteristic charge state distribution of the ion.[31] Typically for a given situation involving heavy ions ($Z > 3$), the beam consists of significant fractions of several different charge states, the equilibrium distribution of which is reached in many cases within a few hundred angstroms of solid material. Even very thin solid foils (≤ 5 μm/cm^2) of most materials may be thick enough for a beam to reach or at least move in the direction of the equilibrium distribution. As discussed subsequently, inner-shell vacancy cross sections can depend very strongly on the charge state and specific configuration of the projectile. The yield of a target x ray must be treated in general as a composite yield from collisions involving the different charge states of the projectile. It represents an average cross section for excitation of a state or group of states.

The geometry of an experimental configuration is not overly critical where nondispersive detectors are utilized, aside from the obvious need to subtend solid angles sufficient to provide usable counting rates. Likewise the need to vacuum-couple the detector into the target system or to filter the photon emission with various absorbers is usually apparent, depending upon the specific experiment. However, the use of crystal spectrometers further entails choosing a particular orientation of the crystal and a specific way of defining the source size, i.e., alignment is much more critical than for a Si(Li) detector. The active area of the target should be placed at the correct source location of a crystal in order to ensure maximum reflected intensity and resolution. Source definition can be attained either by a single entrance slit or by a Soller entrance slit on

---

[29] R. D. Dewey, R. S. Mapes, and T. W. Reynolds, in "Progress in Nuclear Energy, Series IX-Analytical Chemistry " (H. A. Elion and D. C. Stewart, eds.), Vol. 9, p. 307. Pergamon Press, New York, 1969.

[30] E. Merzbacher and H. W. Lewis, in "Handbuch der Physik " (S. Flügge, ed.), Vol. 34, p. 166. Springer-Verlag, Berlin, 1958.

[31] A. B. Wittkower and H. D. Betz, Atom. Data 5, 113 (1973).

8.2. DETECTORS AND TARGETS

the spectrometer. Yet another alternative is to collimate the beam to an appropriate source size before it encounters the target, precluding the need for entrance slits on the detector.

The specific orientation of the crystal with respect to the target becomes most important when the x rays of interest are polarized. The reflection efficiency of the crystal depends upon the direction of plane polarization, being greater for those vectors parallel to the axis of rotation than for those perpendicular to the axis of rotation.[32] Where polarization is negligible, the crystal can be placed with axis rotation parallel or perpendicular to the beam axis according to individual requirements of operation. Resolution for an extended source such as a beam spot on a solid is similar for the two orientations, provided appropriate slit sizes are used.

The angle between beam axis and target–detector axis is also a matter of choice, but certain consequences should be considered. Particularly for x-ray emission from fast-moving projectiles, the Doppler shift described by Eq. (8.2.7) can be substantial. The effect is minimized when the detector is at 90° to the direction of flight of the emitting ion. Thus 90° is often a desirable angle between beam axis and target–detector axis. For example, the shift at 90° for an ion with 1 MeV/amu energy is only ~0.1%. Of course, where energy shifts facilitate the separation of overlapping lines, such as the case where the same line is observed in a symmetric collision from projectile and target, the angle of detection may be varied to advantage. If x rays are polarized, a nonisotropic angular distribution of emission results and corrections for both dispersive and nondispersive detectors are necessary to relate observed yields to excitation probabilities.

The use of thin gas targets circumvents all three of the above problems. The primary limitation on usable gas pressure (density) is usually the maintaining of charge state purity of the beam. Beam ions undergo electron capture and loss in the active region of the target as well as any pressure gradient region prior to the active region. It is imperative that the beam pass through no windows (thin solid foils), if collisions involving individual charge states are to be studied. Accordingly, the establishment of a pressure gradient must be accomplished by differential pumping. One method is to define the central target region with apertures of a few millimeters diameter, which are the only connection between the vacuum of the target and the rest of the system. In addition, a second set of collimators may be placed at each end of the primary gas cell, defining an intermediate pumping region between beamline and gas cell. The latter configuration is referred to as doubly differentially

---

[32] B. E. Warren, "X-Ray Diffraction," p. 334. Addison-Wesley, Massachusetts, 1969.

pumped. Details of geometry and ensuring single-collision conditions require careful measurements of charge change of the beam[33] in a given situation.

Distances between apertures must be chosen so that cumulative effects of beam charge change for practical gas pressures are minimal, a restriction that depends on the magnitude of beam collision cross sections and thus on the particular projectile and target gas being used. Target lengths on the order of 1 cm are feasible. Primary target densities up to ~ 300 μm can be attained, provided sufficient pumping of the external region(s) is provided.

One consequence of the use of such a gas cell with crystal spectrometers is the extension of the source size. Very dilute gases usually yield much lower counting rates than solid targets and hence density is increased by increasing path length of the beam in the gas. The orientation of the crystal with axis of rotation parallel to the beam axis conveniently utilizes the full width of the crystal for Bragg reflection.

For those situations where beam purity is not a requirement, an enclosed gas cell provides a means of attaining much higher gas pressures and hence counting rates. Thin metal foils or plastic windows of a few hundred μg/cm^2 thickness allow penetration of heavy beams without prohibitive energy loss and pressures up to an atmosphere. They can be conveniently placed over the apertures of a system such as the differentially pumped gas cell. The viewing port on the cell for a Si(Li) or a crystal spectrometer can also be capped with a thin plastic window, the thickness of which is chosen to allow the maximum possible transmission of the x ray of interest.

8.3. Multiple Inner-Shell Vacancy Production

8.3.1. Multiple Vacancies and the K_α Satellite Structure

Early experiments[34,35,36] performed with proportional counters and lithium-drifted silicon solid-state detectors revealed shifts in the centroid energies and increases in the widths of the characteristic x-ray peaks induced in solid targets by bombardment with various heavy ions. Specht[34]

---

[33] L. M. Winters, J. R. Macdonald, M. D. Brown, T. Chiao, L. D. Ellsworth, and E. W. Pettus, *Phys. Rev. A* **4**, 1835 (1973).

[34] H. J. Specht, *Z. Phys.* **185**, 301 (1965).

[35] P. Richard, I. L. Morgan, T. Furuta, and D. Burch, *Phys. Rev. Lett.* **23**, 1009 (1969); D. Burch and P. Richard, *Phys. Rev. Lett.* **25**, 983 (1970).

[36] M. J. Saltmarsh, A. van der Woude, and C. A. Ludemann, *Phys. Rev. Lett.* **29**, 329 (1972).

8.3. MULTIPLE INNER-SHELL VACANCY PRODUCTION

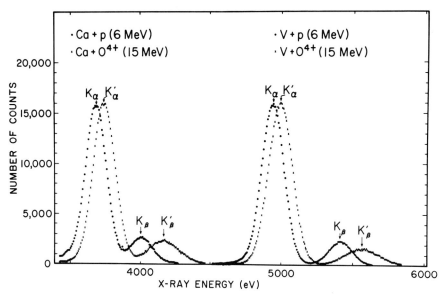

FIG. 9. Shifted K x rays of calcium and vanadium produced by heavy-ion bombardment as recorded with a Si(Li) detector. The unprimed symbols K_α and K_β refer to the proton-induced spectra, while the primed symbols designate peak centroids for 15-MeV oxygen-induced spectra [D. Burch and P. Richard, *Phys. Rev. Lett.* **25**, 983 (1970)].

first reported shifts in the energies of radiation induced by collisions with fission fragments as compared to photon-induced x rays. A comparison between spectral features in the calcium and vanadium K_α and K_β emission resulting from collisions with protons and 15-MeV oxygen is shown in Fig. 9. Shifts of ~50 and ~160 eV for the K_α and K_β lines, respectively, for the heavier projectile suggested multiple L vacancies created within the same collision producing the K hole. The magnitude of the shifts and the increased widths of both peaks were consistent with Hartree–Fock–Slater calculations for multiply ionized ions, for which a multitude of states within each peak would be unresolved by the ~280 eV resolution of the Si(Li) detector.

Similar effects were observed in the L x ray spectra of even heavier collision partners, as for example, I + Mo and I + Kr at ≲ 0.4 MeV/amu.[37] Energy shifts beyond the single-vacancy diagram line energies for both target and projectiles were interpreted again as being due to multiple vacancies, in these cases configurations involving one L and multiple M vacancies. The use of low-resolution detectors similarly precluded more specific information about the states of ionization.

[37] P. H. Mokler, *Phys. Rev. Lett.* **26**, 811 (1971).

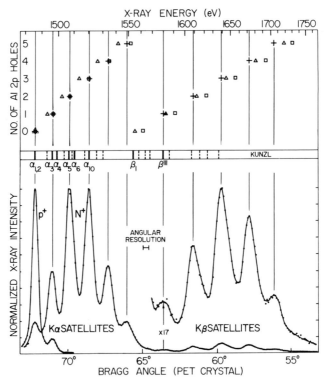

FIG. 10. Proton- and nitrogen-induced aluminum K_α x-ray spectra. At the top are calculated x-ray energies obtained with a Herman–Skillman program (squares) and a screening-constant program (triangles). Spectra taken with a PET crystal [A. R. Knudson, D. J. Nagel, P. G. Burkhalter, and K. L. Dunning, *Phys. Rev. Lett.* **26**, 1149 (1971)].

The use of crystal spectrometers with vastly improved resolution precipitated a new and much more precise series of experiments dealing with line energies and intensities. Concurrent with the low-resolution work mentioned above, the utility of the dispersive devices was demonstrated in a paper[38] that discussed energy shifts of argon L x rays produced in $Ar^+ + Ar$ collisions at incident energies from 50 to 330 keV. The observed effects of multiple outer-shell excitation were accessible only through the superior resolution provided by Bragg reflection. Most subsequent work has been done on K_α and K_β emission from various systems, although additional information on the L-shell x-ray spectra has been recorded and is mentioned briefly in this chapter. The complexity of the spectral features of the latter and a corresponding lack of theoretical calculations of energies and rates are contrasted with the situation for

[38] M. E. Cunningham, R. C. Der, R. J. Fortner, T. M. Kavanagh, J. M. Khan, C. B. Layne, E. J. Zaharis, and J. D. Garcia, *Phys. Rev. Lett.* **24**, 931 (1970).

8.3. MULTIPLE INNER-SHELL VACANCY PRODUCTION

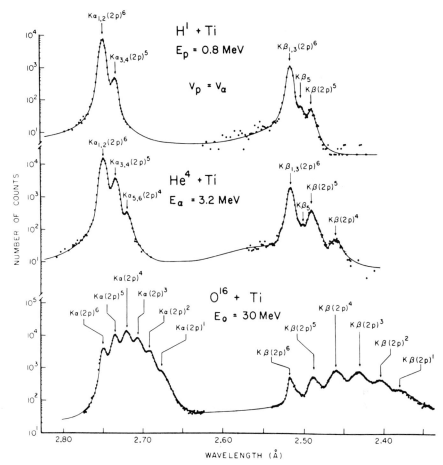

FIG. 11. Thick target titanium K x rays produced by proton, alpha, and oxygen bombardment and taken with a LiF crystal [C. F. Moore, M. Senglaub, B. Johnson, and P. Richard, *Phys. Rev. Lett.* **40A**, 107 (1972).

a wide-range K-shell excited states, for which the foundation is more substantial and which furnishes the focus of the discussion here.

Resolved K_α and K_β x-ray structures produced in heavy-ion collisions were first recorded with crystal spectrometers for the elements iron[39] and aluminum.[40] The aluminum spectra for proton and nitrogen bombardment are presented in Fig. 10. Each of the peaks in the K_α group induced by nitrogen was attributed to a set of transitions involving initial states belonging to configurations with one K vacancy in addition to a certain

[39] D. Burch, P. Richard, and R. L. Blake, *Phys. Rev. Lett.* **26**, 1355 (1971).

[40] A. R. Knudson, D. J. Nagel, P. G. Burkhalter, and K. L. Dunning, *Phys. Rev. Lett.* **26**, 1149 (1971).

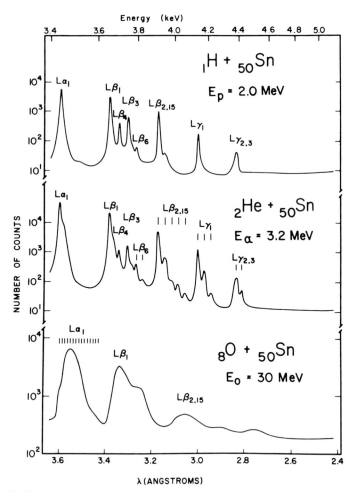

FIG. 12. The L x-ray emission induced by protons, alphas, and oxygen ions. Vertical lines indicate Hartree–Fock–Slater energies for multiple M-shell ionization. Crystal used was a LiF [D. K. Olsen, C. F. Moore, and P. Richard, *Phys. Rev. A* **7**, 1244 (1973).

number of L (2p) vacancies, ranging from 0 to 5, as indicated by theoretical calculations. The K_β structure reflects a similar spread in energies due to increasing degrees of ionization. The theoretical energies are average values for the states arising from a given configuration, an appropriate quantity for comparison since the resolution of the detector as indicated by the diagram line in the hydrogen spectrum is insufficient to resolve individual transitions.

Subsequent experiments revealed similar structures for magnesium[41]

[41] D. G. McCrary, M. Senglaub, and P. Richard, *Phys. Rev. A* **6**, 263 (1972).

and silicon.[42] The dramatic dependence of the degree of target multiple ionization upon the atomic number (Z value) of the projectile involved in the collision is demonstrated in Fig. 11, taken from work[43] on the titanium K_α and K_β spectra resulting from bombardment by hydrogen, helium, and oxygen. An extremely comprehensive body of such high resolution information for a variety of collision partners at a wide range of relative velocities has been accumulated, for both projectiles and target ions in gases and solids, and constitutes the bulk of the data discussed here.

The phenomenon of extensive multiple ionization has also been observed in several instances in L x-ray spectra.[44-46] One example of that is depicted in Fig. 12 for the L shell of antimony and projectiles hydrogen, helium, and oxygen. It is evident that discrete structural information for multivacancy configurations produced in the heavier-ion collision is difficult to extract, presumably due to the superposition of a prohibitively large number of states. Thus, although total L x-ray yields can in certain cases be related to specific configurations, in general the complexity of the situation limits accessibility to the type of information available in the K x-ray spectra.

8.3.2. Multiple-Ionization Formulation

A theoretical description of the process of producing L vacancies within the same collision leading to a K vacancy has been advanced and has proved to be a useful basis for comparing various collision systems. A major assumption in the theory is that the ionization of a particular electron is not correlated with the events involving other electrons, that is, the process is independent of other excitations within the same atom. A major stimulus for that approach was the observation[47] that the distribution of strengths in the various satellite peaks followed closely that predicted by a simple binomial distribution. The development of the following treatment is attributed to several different sources[48-52] although most

---

[42] D. G. McCrary and P. Richard, *Phys. Rev. A* **5**, 1249 (1972).
[43] C. F. Moore, M. Senglaub, B. Johnson, and P. Richard, *Phys. Lett.* **40A**, 107 (1972).
[44] R. C. Der, R. J. Fortner, T. M. Kavanagh, J. M. Khan, and J. D. Garcia, *Phys. Lett.* **36A**, 239 (1971).
[45] R. J. Fortner, *Phys. Rev. A* **10**, 2218 (1974).
[46] D. K. Olsen, C. F. Moore, and P. Richard, *Phys. Rev. A* **7**, 1244 (1973).
[47] D. Burch, *Proc. Int. Conf. Inner Shell Ionization Phenom. Future Appl.*, p. 1464. Atlanta, Georgia, 1972, (Natl. Tech. Information Service, U.S. Dept. of Commerce, Springfield, Va., 1972).
[48] J. M. Hansteen and O. P. Mosebekk, *Phys. Rev. Lett.* **29**, 1361 (1972).
[49] R. L. Kauffman, J. H. McGuire, P. Richard, and C. F. Moore, *Phys. Rev. A* **8**, 1233 (1973).
[50] J. H. McGuire and P. Richard, *Phys. Rev. A* **8**, 1374 (1973).
[51] J. S. Hansen, *Phys. Rev. A* **8**, 822 (1973).

of the notation employed below is taken directly from one of them.[52] The important experimental quantities are the x-ray yields of the individual satellite peaks. In most cases, accurate values of the yields can be obtained by least-squares fitting the data with a series of gaussians superimposed on a linear or quadratic background, as demonstrated in Fig. 13.

Assuming the electrons are not correlated, the cross section for producing one K vacancy and n L vacancies simultaneously in an atom with initially N_K K electrons can be written as

$$\sigma^I_{1K,nL} = N_K \int_0^\infty 2\pi b P_K(b) P_{nL}(b) \, db, \qquad (8.3.1)$$

where $P_K(b)$ and $P_{nL}(b)$ are the respective probabilities per unit area for creating a single K vacancy and n L vacancies as functions of the impact parameter b. A further distinction can be made for the L shell, e.g., i vacancies in the 2s subshell and j vacancies in the 2p subshell ($i + j = n$),

$$\sigma^I_{1K,ij} = N_K \int_0^\infty 2\pi b P_K(b) P_{ij}(b) \, db. \qquad (8.3.2)$$

The quantity P_{ij} can be expressed as

$$P_{ij} = C \binom{2}{i} [P_{2s}(b)]^i [1 - P_{2s}(b)]^{2-i}$$
$$\times C \binom{6}{j} [P_{2p}(b)]^j [1 - P_{2p}(b)]^{6-j}, \qquad (8.3.3)$$

where $P_{2s}(b)$ and $P_{2p}(b)$ are the individual probabilities for removing 2s and 2p electrons, respectively. The binomial coefficients $C\binom{m}{n}$ reflect a statistical distribution. It should be noted that states with the same value of n but different populations of the 2s and 2p subshells cannot usually be revolved.

The relationship between the vacancy production cross section and the observed x-ray spectrum is given by

$$\sigma^X_{K,nL} = \sum_{ij} \sigma^I_{K_{ij}} \omega_{ij}, \qquad i + j = n, \qquad (8.3.4)$$

where ω_{ij} is the average fluorescence yield for the configruation (Kij). The fluorescence yield is given by

$$\omega_{ij} = \Gamma_X/(\Gamma_X + \Gamma_A), \qquad (8.3.5)$$

where Γ_X and Γ_A are the radiative and Auger rates, respectively.

A simplification results from a consideration of the range of impact

---

[52] F. Hopkins, D. O. Elliott, C. P. Bhalla, and P. Richard, *Phys. Rev. A* **8**, 2952 (1973).

8.3. MULTIPLE INNER-SHELL VACANCY PRODUCTION

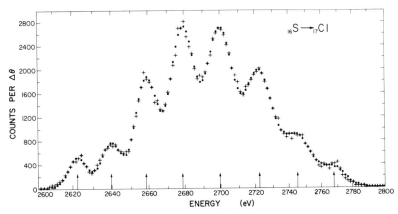

FIG. 13. Least-squares fit to a high-resolution chlorine K_α spectrum (NaCl crystal). Crosses are experimental points; dots represent the fit. Arrows indicate peak centroids [R. L. Watson, F. E. Jensen, and T. Chiao, *Phys. Rev. A* **10**, 1230 (1974)].

parameters that is significant. The dominant contribution is from impact parameters small compared to the L-shell radii, that is, P_K approaches zero for b much greater than the K-shell radius. As can be seen in Fig. 14, calculations[49] indicate that $P_{2s}(b)$ and $P_{2p}(b)$ are nearly constant over that range of b. Hence $P_{2s}(0)$ and $P_{2p}(0)$ can be taken as constant "average" probabilities for the region where the integrand of Eq. (8.3.2) is nonzero and hence factored out of the integral, leaving the quantity

$$\sigma^I_{1K,ij} = N_K P_{ij}(0) \int_0^\infty 2\pi b P_K(b) \, db = N_K P_{ij}(0) \sigma^I_K, \quad (8.3.6)$$

where σ^I_K is the total K-shell vacancy cross section.

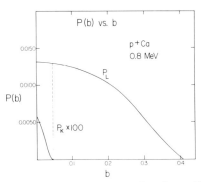

FIG. 14. BEA calculations for $P_K(b)$ and $P_L(b)$ as functions of impact parameters b for case of ionization of calcium by 0.8 MeV protons [R. L. Kauffman, J. H. McGuire, P. Richard, and C. F. Moore, *Phys. Rev. A* **8**, 1233 (1973)].

The ratio of $\sigma^x_{1K,nL}$ to the total K_α x-ray yield is given by

$$R^x_n = \sigma^x_{1K,nL}/\sigma^x_K = \sum_{ij} P_{ij}\, \omega_{ij} \bigg/ \sum_{n=0}^{7} \sum_{ij} P_{ij}\, \omega_{ij}. \quad (8.3.7)$$

Thus the available experimental values of R^x_n can be fit simultaneously by varying the two parameters $P_{2s}(0)$ and $P_{2p}(0)$. Values for ω_{ij} can be taken from theoretical calculations or estimated from sets of values for ions neighboring in Z value. In fact, qualitative trends can be noticed by simply taking ω_{ij} to be constant and equal for all configurations and thereby reducing the ratio to one involving just probabilities. Further, in practice, the probabilities extracted are typically not very sensitive to differences between P_{2s} and P_{2p} and thus it is often sufficient to allow for a single average L-shell ionization probability $P_L(0)$,

$$R^x_n \approx P_n = \binom{8}{n} [P_L(0)]^n [1 - P_L(0)]^{8-n}. \quad (8.3.8)$$

The neglecting of fluorescence yield differences is appropriate so long as the degree of ionization is not high, as seen in following sections.

The probabilities resulting from analyses of the above type can be related[50] to L-shell ionization cross sections through

$$P_{2p}(0) = \frac{\sigma_{2p}/N_{2p}}{2\pi\langle r^2_{2p}\rangle}, \quad (8.3.9)$$

where σ_{2p} is the ionization cross section, N_{2p} the number of 2p electrons, and $\langle r^2_{2p} \rangle$ the square of the rms radius of the subshell, usually as provided by theoretical calculations. Comparisons between experimental and theoretical cross sections are presented in Section 8.5.1.

A consideration that has been neglected in most multiple-ionization analyses to date but that must clearly be taken into account for a complete picture is the possibility of alteration of L-vacancy distributions by L-shell cascading prior to the K x-ray transitions. Recent developments in this area are covered in Chapter 8.6.

8.3.3. Collision Mechanisms

Although the primary purpose of this part is to consider experimental methods and results for spectroscopic investigations of various excited ionic states, the collisional processes leading to those states should be considered since they account for the strengths observed. Only a brief summary is given here, as several informative review articles[1,2] as well as additional parts in this volume dwell at length on this topic. The importance of the specific collision mechanisms is pointed out in subsequent discussions of empirical results.

8.3. MULTIPLE INNER-SHELL VACANCY PRODUCTION

The first process to be considered is ionization, whereby an electron from one of the colliding partners is elevated to a continuum state, creating an inner-shell vacancy. One theoretical approach is to treat the interaction as a "binary" event between the electron excited and the projectile's nuclear charge (or effective nuclear charge if screening effects are included). The problem can then be solved classically (binary encounter approximation),[53,54] semiclassically,[55] or quantum mechanically (plane wave Born approximation).[56] The three calculations give qualitatively similar results for ionization of inner shells by light ions. The behavior of a particular vacancy production cross section as expressed in the BEA is representative and is given by the expression

$$\sigma_n^I = \frac{\sigma_0 N_n Z_1^2}{U_n^2} G\left(\frac{V_1}{V_2}\right), \quad (8.3.10)$$

where Z_1 is the atomic number of the projectile, N_n the number of electrons in the shell or subshell, U_n the binding energy of the shell of subshell, $\sigma_0 = 6.56 \times 10^{-14}$ cm^2 eV^2, V_1 the velocity of the projectile, and V_2 the velocity of the bound electron. The incident energy dependence is given by the function $G(V_1/V_2)$, which depends solely on the ratio of the velocities. The function peaks around velocity matching, i.e., a ratio of 1.0, and falls off slowly above that value and rather rapidly below it. In simplest form, these theories all constitute a "sudden approximation" in the sense that the bound electron does not respond to the charge of the projectile but is taken to be frozen in its initial orbit until it is ionized. Note also that the cross section for all three theories is proportional to the square of Z_1.

Within the context of the PWBA, modifications[57] have been made to account for the change of orbit of the electron in response to the presence of the charged projectile. For deep penetration of the electronic shell, i.e., small impact parameters, the binding energy of the electron increases and the cross section for ionization decreases, as can be seen from the U_n^{-2} dependence in Eq. (8.3.10). For distant collisions, those for which the impact parameter is large compared to the shell radius, the shell is polarized in the direction of the projectile and, due to the decreased interaction distance, the ionization cross section increases. The net effect on the total cross section due to these two compensating phenomena de-

---

[53] M. Gryzinski, *Phys. Rev.* **138**, A336 (1965).
[54] J. D. Garcia, *Phys. Rev. A* **4**, 955 (1971).
[55] J. M. Hansteen and O. P. Mosebekk, *Nucl. Phys. A* **201**, 541 (1973).
[56] G. S. Khandelwal, B. H. Choi, and E. Merzbacher, *Atom. Data* **1**, 103 (1969).
[57] G. Basbas, W. Brandt, R. Laubert, A. Ratkowski, and A. Schwarzschild, *Phys. Rev. Lett.* **27**, 171 (1971); G. Basbas, W. Brandt, and R. Laubert, *Phys. Rev. A* **7**, 983 (1973).

pends strongly on the value of the ratio of the velocities. Ionization cross sections can deviate substantially from the Z_1^2 dependence, particularly for heavy ions, due to such effects.

Attempts have also been made[58] to allow for a screening of the projectile's nuclear charge by its own attached electrons. A reduction in the effective charge seen by a target atom depresses the ionization cross section. For highly asymmetric collisions ($Z_1 \ll Z_2$), inner-shell ionization should not be very sensitive to screening but outer-shell ionization (of loosely bound electrons) can depend in a significant way on the degree of screening.

Two other modes of excitation are available within the Coulomb or "single-particle" excitation scheme. An electron can be elevated to a higher unoccupied shell within the same atom or it can be transferred to the projectile. The former process, Coulomb excitation, to bound states, has been studied extensively[59] for light ions such as hydrogen and helium but has only recently received attention[60] for heavier ions, i.e., more tightly bound inner shells. The latter process, charge exchange or electron capture, again is a long standing subject of investigation insofar as light ions and the outer shells of heavier ions are concerned. However, data[60,61] on the dependence of target vacancy production on the electronic structure of highly stripped high-Z ($Z_1 \geq 6$), made available through the use of Van de Graaff accelerators, have stimulated application of Coulomb charge exchange theories[62,63] to inner-shell pickup.

All of the channels of excitation discussed above can be described with an alternative mechanism, the molecular orbital (MO) or "quasi-molecular" treatment.[2] In contrast to the sudden approximations above, the MO scheme[64] allows for the orbitals of the colliding partners to adjust to the time-varying two-center field. As the internuclear distances decrease, the various shells of the combined system (quasi-molecule) are occupied according to the Pauli exclusion principles and the initial binding

---

[58] G. Basbas, *Proc. Ninth Int. Conf. Physics Electron. Atom. Coll.*, p. 502. University of Washington Press, Seattle, 1975.

[59] F. J. de Heer, "Atomic and Molecular Processes," p. 327. Academic Press, New York, 1966.

[60] F. Hopkins, R. Brenn, A. Whittemore, N. Cue, V. Dutkiewicz, and R. P. Chaturved *Phys. Rev. A* **13**, 74 (1976); F. Hopkins, A. Little, and N. Cue, *Phys. Rev. A* **14**, 1634 (1976).

[61] M. D. Brown, L. D. Ellsworth, J. A. Guffey, T. Chiao, E. W. Pettus, L. M. Winters, and J. R. Macdonald, *Phys. Rev. A* **10**, 1255 (1974).

[62] J. H. McGuire, *Phys. Rev. A* **8**, 2760 (1973); A. M. Halpern and J. Law, *Phys. Rev. Lett.* **31**, 4 (1973).

[63] G. Lapicki and W. Losonsky, *Phys. Rev. A* **15**, 896 (1977); A. L. Ford, E. Fitchard, and J. F. Reading, *Phys. Rev. A* **16**, 133 (1977).

[64] U. Fano and W. Lichten, *Phys. Rev. Lett.* **14**, 627 (1965).

8.3. MULTIPLE INNER-SHELL VACANCY PRODUCTION

energies of the isolated ion and atom. For example, the K shell of the combined system is occupied by the two K electrons of the heavier partner, which track the 1sσ molecular orbital.

Several consequences arise from the modulation of the binding energies. Levels corresponding to different initial subshells can cross, with electrons (or vacancies) switching from one to the other. Hence the outgoing channel will be described as excitation to bound states if both subshells are in the same partner, or as charge exchange if they reside in the different partners. Theoretical developments for level crossings have been made for several collision systems and energies and have been applied to inner- as well as outer-shell excitation for symmetric ($Z_1 = Z_2$) and asymmetric collisions. A classic example is the successful description[65] of the neon K-vacancy production cross sections for Ne$^+$ + Ne at energies from ~50 to 200 keV. The nuclear rotational coupling of the 2pπ to the 2pσ molecular states leads to a transfer of a 2p (2pπ) vacancy to the 1s shell (2pσ) of the less tightly bound neutral neon atom (with the possibility of subsequent vacancy sharing as discussed below). Calculations of those probabilities for a variety of systems are feasible.[66]

In addition to level crossings, which occur when certain critical internuclear distances are reached, electrons (vacancies) can be shared between levels even when they do not reach the crossing points. Such a vacancy-sharing mechanism was proposed[67] to account for the behavior of K-shell vacancy cross sections. A simple expression for K-vacancy sharing based on a charge exchange model was given as the ratio

$$S = \exp(-\alpha/v), \qquad (8.3.11)$$

where $\alpha = \pi[(2I_H)^{1/2} - (2I_L)^{1/2}]$, with I_H and I_L the K-shell binding energies of the heavier and lighter partners, respectively. Thus at intermediate internuclear distances for both the incident and exit channels, the correlation between molecular levels can lead to vacancy creation, in this case by sharing a vacancy produced early in the same collision or in a previous collision and carried into a second collision. Similar sharing can occur between higher shells.

It should be noted that the molecular picture should be particularly valid in the adiabatic regime, that is, where orbital velocities of the bound electrons are much larger than the relative velocity between nuclei. The Coulomb excitation theories discussed above are, in their simplest forms, appropriate in the "high-velocity" limit where nuclear velocities exceed

---

[65] J. S. Briggs and J. H. Macek, *J. Phys. B* **5**, 579 (1972); *ibid.* 982 (1972).
[66] K. Taulbjerg, J. S. Briggs, and J. Vaaben, *J. Phys. B* **9**, 1351 (1976).
[67] W. E. Meyerhof, *Phys. Rev. Lett.* **31**, 1341 (1973).

orbital electron velocities. The intermediate region, for which values of the velocity ratio range from ~0.2 to 1.0, constitutes a transition stage that demands a unified theory incorporating both the direct Coulomb excitation and the molecular electron promotion processes.

8.4. Spectroscopy of Individual States

8.4.1. Line Energies

Perhaps the most direct information furnished by high-resolution spectroscopy is the transition energies of the x rays. The resolution of crystal spectrometers of a few parts in a thousand allows separation of transitions according to configurations and, in the case of few-electron ions, ac-

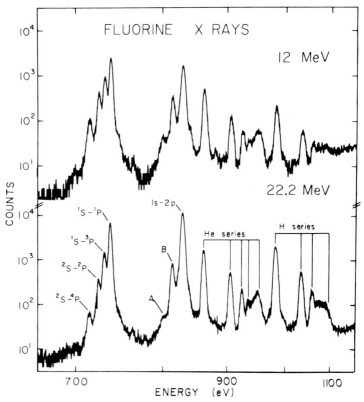

FIG. 15. K x rays emitted by beam of fluorine ions passing through a thin carbon foil at 12 and 22 MeV energy. RAP crystal used [R. L. Kauffman, C. W. Woods, F. F. Hopkins, D. O. Elliott, K. A. Jamison, and P. Richard, *J. Phys. B* **6**, 2197 (1973)].

8.4. SPECTROSCOPY OF INDIVIDUAL STATES

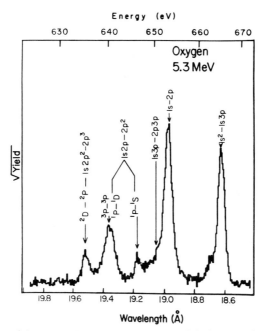

FIG. 16. Region of the oxygen K x-ray spectrum around the hydrogenic Lyman α (1s–2p) transition, which involves a number of multiply excited states. Data taken with an RAP crystal [D. L. Matthews, W. J. Braithwaite, H. H. Wolter, and C. F. Moore, *Phys. Rev. A* **8**, 1397 (1973)].

cording to individual states. Although collisions using gas targets have provided some of the available data, the most prolific source of high charge states and thereby simplified decay schemes is the beam–foil interaction. Stripping at tandem Van de Graaff energies ($\geq$ MeV/amu) in thin solid foils makes one-, two-, and three-electron states available even in ions as heavy as chlorine, while larger accelerators accommodate still higher Z. In addition to profuse production of optical and ultraviolet radiation, x rays and in particular K x rays can be obtained in sufficient amount to be easily accessible to low-efficiency detectors such as plane- and curved-crystal spectrometers.

Spectra of fluorine[68] and oxygen[69] beam K x rays are shown in Figs. 15 and 16, respectively. The former includes all of the K x rays seen, ranging from three-electron (lithiumlike) to one-electron (hydrogenlike)

---

[68] R. L. Kauffman, C. W. Woods, F. F. Hopkins, D. O. Elliott, K. A. Jamison, and P. Richard, *J. Phys. B* **6**, 2197 (1973).

[69] D. L. Matthews, W. J. Braithwaite, H. H. Wolter, and C. F. Moore, *Phys. Rev. A* **8**, 1397 (1973).

TABLE V. Three-, Two-, and One-Electron K-Transition Energies (eV)

		Fluorine (Ref. 68)		Sulfur (Ref. 28)		Chlorine (Ref. 72)	
Initial state	Final state	Expt.	Theor.	Expt.	Theor.	Expt.	Theor.
$1s2p^2(^4P)$	$1s^22p(^2P)$	—	—	2416.5	2419.6	—	—
$1s2p^2(^2P)$	$1s^22p(^2P)$	—	—	2435.3	2438.2	—	—
$1s2s2p(^4P)$	$1s^22s(^2S)$	715.1	713.1	2426	2416.8	2741	2732
$1s2s2p(^2P)$	$1s^22s(^2S)$	724.8	726.0	—	—	2762	2751
$1s2p(^3P)$	$1s^2(^1S)$	731.1	731.4	2448	2447.7	2775	2765
$1s2p(^1P)$	$1s^2(^1S)$	737.1	737.7	2461.8	2461.1	2789	2779
$1s3p(^1P)$	$1s^2(^1S)$	857.5	857.5	—	—	3271	3260
$1s4p(^1P)$	$1s^2(^1S)$	899.7	899.7	—	—	3442	3429
2p	1s	827.4	827.4	—	—	2961[a]	2961
3p	1s	981.0	980.6	—	—	3511	3508
4p	1s	1034.9	1034.2	—	—	3708	3700

[a] Empirical value normalized to theoretical value.

transitions. The heliumlike and hydrogenlike structures are very similar, with the primary 1s–2p line followed by the 1s–3p, 1s–4p, . . . , 1s–np series. The hydrogen series is conveniently made more prominent by simply raising the beam energy. A magnified portion of the corresponding oxygen spectrum (Fig. 16) reveals lines involving multiply excited states and also demonstrates that overlapping occurs even for highly stripped situations. In addition, the fine structure, i.e., the various possible J values indicated in the $^{(2S+1)}L_J$ spectroscopic notation, where S, L, and J are spin, orbital, and total angular momenta of the system, cannot be separated in this region with splittings expected to be far less than the experimental resolution of ~3 eV.

Similar measurements have been performed for aluminum,[70] silicon,[71] sulfur,[28,72] and chlorine.[72,73] The latter two cases were investigated with both crystal spectrometers and Doppler-tuned spectrometers. A sampling of various line energies is tabulated in Table V, along with the relevant theoretical numbers, which come from a variety of sources as noted in the individual references. In general, agreement is very good and thus provides confidence in delineating classification schemes for neighboring ions using theoretical values as a guide.

[70] J. R. Mowat, K. W. Jones, and B. M. Johnson, *Phys. Rev. A* **14**, 1109 (1976).
[71] S. L. Varghese, C. L. Cocke, B. Curnutte, and G. Seaman, *J. Phys. B* **9**, L387 (1976).
[72] P. Richard, C. F. Moore, D. L. Matthews, and F. Hopkins, *Bull. Am. Phys. Soc.* **19**, 570 (1974).
[73] C. L. Cocke, B. Curnutte, J. R. Macdonald, and R. Randall, *Phys. Rev. A* **9**, 57 (1974).

8.4. SPECTROSCOPY OF INDIVIDUAL STATES

TABLE VI. Theoretical Energies and Fluorescence Yields for Neon K X-Ray Transitions[a]

KL^n				Final state $1s^22s^m2p^n$					
n	m	n	$^{(2S+1)}L$	m	n	L	E_X (eV)	100ω	$100\langle\omega\rangle$
5	2	1	^1P	2	0	^1S	895.6	9.57	2.39
			^3P				—	0.0	
	1	2	^1D	1	1	^1P	888.2	1.60	
			^3P			^3P	892.3	35.5	
			^3D			^3P	893.3	2.88	
			^1S			^1P	895.1	2.32	
			^1P			^1P	895.2	20.8	14.8
			^3S			^3P	900.1	6.07	
			$^3P_+$			^3P	904.2	0.48	
			^5P			^3P	878.4	28.8	
	0	3	^1P	0	2	^1S	889.6	3.74	
			^3D			^3P	889.6	2.19	
			^1D			^1D	891.4	6.35	
			^3S			^3P	891.8	100.0	10.8
			^3P			^3P	894.2	3.02	
			^1P			^1D	895.9	4.78	
			^5S			^3P	879.8	2.84	
6	1	1	$^2P_-$	1	0	^2S	906.6	46.8	
			$^2P_+$			^2S	914.0	1.22	23.8
			^4P			^2S	894.9	23.6	
	0	2	^2D	0	1	^2P	904.0	3.15	
			^2P			^2P	905.7	100.0	
			^2S			^2P	911.2	7.65	24.4
			^4S			^2P	895.9	7.13	

[a] Taken from Ref. 74.

It is apparent that spectroscopy of K_α and K_β structures of lower charge states, e.g., beryllium-like (equivalent to KL^5 for an ion whose neutral state includes a filled L shell), becomes increasingly more difficult. An illustration of the degree of overlapping, based on theoretical calculations,[74] is presented in Table VI for the KL^5 and KL^6 configurations of neon. The region from ~895 to ~905 eV includes transitions arising not only from different multiplets within a given subshell configuration, e.g., 1s 2s 2p, but also from different subshell and total electron configurations (KL^5 vs. KL^6). In addition, the fluorescence yields, the relative probability that a given state x-ray decays rather than Auger decays, fluctuates tremendously. Obviously some assumption must be made about initial

[74] C. P. Bhalla, *Phys. Rev. A* **12**, 122 (1975); C. P. Bhalla, *J. Phys. B* **8**, 1200 (1975).

population probabilities, which in principle may depend on the mode of excitation such as beam–foil or gas collisions, in order to relate fitted line energies from spectra such as in Fig. 10 to calculated values. A common procedure is to assume statistical populations of multiplets $[=(2S + 1)(2L + 1)]$ within a given subshell configuration. Those quantities are folded in with explicit calculations of the type in Table V to yield average energies for all transitions belonging to a given 1s $2s^m$ $2p^n$ configuration. Further averaging over subshell population is accomplished by using binomial probabilities for the distribution of n vacancies between the 2s and 2p subshells. It should be kept in mind that the latter procedure is at best an approximation and may not be strictly valid for a given situation.

A compilation[12] of available data and theoretical values for K_α x-ray transition energies as a function of Z of the emitting ion is given in Fig. 17. Hartree–Fock and Hartree–Fock–Slater calculations give very similar results and exhibit reasonably good agreement, with only a slight underes-

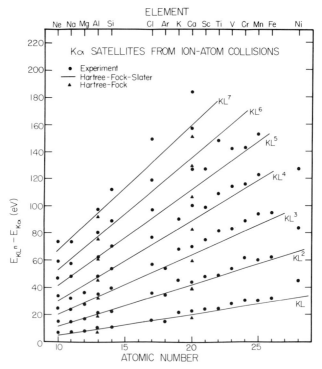

FIG. 17. A comparison of measured K_α satellite x-ray energies with theoretical values, expressed as increase above the K_α diagram energy [R. L. Kauffman and P. Richard, in "Methods of Experimental Physics" (D. Williams, ed.), Vol. 13, p. 148. Academic Press, New York, 1976].

8.4. SPECTROSCOPY OF INDIVIDUAL STATES

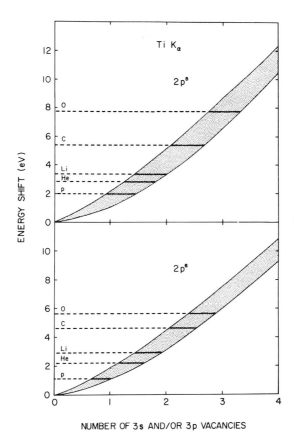

FIG. 18. Shifts in titanium K_α x-ray energies with increasing numbers of 3s or 3p vacancies. The bounded gray area represents Hartree–Fock calculations, while horizontal lines represent empirical shifts for the projectiles indicated [K. W. Hill, B. L. Doyle, S. M. Shafroth, D. H. Madison, and R. D. Deslattes, *Phys. Rev. A* **13**, 1334 (1976)].

timation for the higher satellites. In fact, part of that discrepancy may be due to varying M-shell populations, which were ignored in the calculations, where full M shells are assumed. The effect of allowing M vacancies is demonstrated in Fig. 18 for titanium,[75] with and without an L vacancy. The shifts in energy of several electron volts observed in the spectra as a function of projectile indicate significant probabilities for an incomplete M shell at the time of the transitions. The magnitude of the shifts has been found to be almost negligible for silicon.[76]

[75] K. W. Hill, B. L. Doyle, S. M. Shafroth, D. H. Madison, and R. D. Deslattes, *Phys. Rev. A* **13**, 1334 (1976).

[76] R. L. Kauffman, K. A. Jamison, T. J. Gray, and P. Richard, *Phys. Rev. Lett.* **36**, 1074 (1976).

8. ION-INDUCED X-RAY SPECTROSCOPY

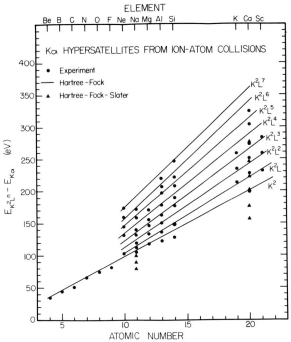

FIG. 19. A comparison of measured (dots) K_α hypersatellite x-ray energies with Hartree–Fock–Slater predictions (triangles) and Hartree–Fock calculations (solid lines) [R. L. Kauffman and P. Richard, in "Methods of Experimental Physics" (D. Williams, ed.) Vol. 13, p. 148. Academic Press, New York, 1976].

The less complete results[12] for K_α hypersatellites, involving initial states with two K-shell vacancies, are presented in Fig. 19. The Hartree–Fock calculation gives considerably better agreement with experimental values, all of which represent target x rays produced in various collisions. Note that the K^2L^7 line for $Z = 10$ is equivalent to the 1s–2p hydrogen-like transition produced in oxygen and fluorine by beam–foil excitation discussed previously.

8.4.2. Fluorescence Yields

The interpretation of x-ray spectra must include an assessment of the appropriate fluorescence yields for the various states observed. The fluorescence yield is the relative probability of x-ray decay $[\Gamma_x/(\Gamma_x + \Gamma_A)]$ where Γ_x and Γ_A are the radiative and Auger widths (rates), respectively, for a particular transition of a particular state. Those rates are extremely sensitive to the specific configuration and spin. The great variance for neon K_α transitions can be seen in Table VI. Regarding K

8.4. SPECTROSCOPY OF INDIVIDUAL STATES

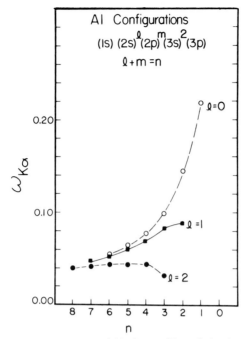

FIG. 20. Theoretical K_α fluorescence yields for satellites of aluminum, where l indicates the number of 2s electrons, m the number of 2p electrons, and $n = l + m$ [C. P. Bhalla and P. Richard, *Phys. Lett.* **45A**, 53 (1973)].

and L transitions, the only values available for most elements are the theoretical or empirical single-vacancy fluorescence yields.[77] Early attempts to account for multiply ionized situations consisted of a statistical scaling of the rates.[78] Subsequently, the need for a more exact procedure led to extensive calculations for a few selected cases.[52, 74, 79, 80, 81] An example[82] of the critical dependence of ω_n (the satellite fluorescence yield) on the presence of 2s vacancies is shown in Fig. 20. In addition, total K x-ray and Auger electron yield measurements have been made for several ions highly stripped by collisions, including neon[83] and argon,[84] providing some test cases for theory.

[77] W. Bambynek, B. Crasemann, R. W. Fink, H. U. Freund, H. Mark, C. D. Swift, R. E. Prince, and P. V. Rao, *Rev. Mod. Phys.* **44**, 716 (1972).
[78] F. P. Larkins, *J. Phys. B* **4**, L29 (1971).
[79] M. H. Chen and B. Crasemann, *Phys. Rev. A* **12**, 959 (1975).
[80] C. P. Bhalla, *J. Phys. B* **8**, 2792 (1975).
[81] C. P. Bhalla, *Phys. Lett.* **45A**, 19 (1973).
[82] C. P. Bhalla and P. Richard, *Phys. Lett.* **45A**, 53 (1975).
[83] D. Burch, N. Stolterfoht, D. Schneider, H. Wieman, and J. S. Risley, *Phys. Rev. Lett.* **32**, 1151 (1974).

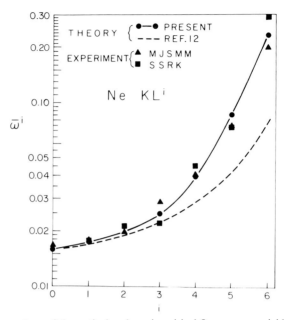

FIG. 21. Comparison of theoretical and semiempirical fluorescence yields for neon KL^n configurations. Empirical values are from D. L. Matthews, B. M. Johnson, L. E. Smith, J. J. Mackey, and C. F. Moore, *Phys. Lett.* **48**, 93 (1974) and N. Stolterfoht, D. Schneider, P. Richard, and R. L. Kauffman, *Phys. Rev. Lett.* **33**, 1418 (1974). The dashed line is due to C. P. Bhalla, N. O. Folland, and M. A. Hein, *Phys. Rev. A* **8**, 649 (1973) [M. H. Chen, B. Crasemann, and D. L. Matthews, *Phys. Rev. Lett.* **34**, 1309 (1975)].

As with a comparison of x-ray energies, the assessment of the relative intensities of the satellites in a spectrum in order to obtain a total average fluorescence yield requires some assumptions about initial populations. Assuming a statistical $(2S + 1)(2L + 1)$ weighting of multiplets leads to the average quantities $\langle \omega \rangle$ contained in the last column of Table VI. Again taking 2s and 2p subshell populations to be equally likely, the values ω_n for each satellite configuration can be derived. Given that theoretical basis, the average *total* fluorescence yield can be expressed in semiempirical form as[85]

$$\bar{\omega} = 1 \bigg/ \sum_n (R_n/\omega_n), \qquad (8.4.1)$$

where R_n is the observed intensity of the KL^n x-ray satellite. Con-

---

[84] R. R. Randall, J. A. Bednar, B. Curnutte, and C. L. Cocke, *Phys. Rev. A* **13**, 204 (1976).
[85] R. L. Kauffman, F. Hopkins, C. W. Woods, and P. Richard, *Phys. Rev. Lett.* **31**, 621 (1973).

8.4. SPECTROSCOPY OF INDIVIDUAL STATES

versely, providing that the total fluorescence yield has been measured along with high-resolution x-ray and Auger electron data, the experimental values for ω_n can be determined,[86,87]

$$\omega_n = \bar{\omega} R_n / Q_n, \qquad (8.4.2)$$

where Q_n are the observed Auger electron satellite intensities corresponding to the KL^n configuration. A comparison of experimental quantities ω_n obtained in this fashion for neon with theoretical values[88] is presented in Fig. 21. The various spectra utilized were produced by using oxygen and chlorine as projectiles, resulting in highly stripped neon. Considerable care must be exercised in taking the overlapping of states into account, as discussed in the previous section. It should be noted that certain states for which x-ray or Auger channels are closed in the LS coupling scheme can be represented in the intermediate coupling scheme, which allows for mixing of states (see, e.g., the ^5P–^3P x-ray transition at 878.4 eV in Table VI).

In certain situations where the measurement of both the x rays and Auger electrons needed to define $\bar{\omega}$ is impractical, e.g., with solid targets where emitted electrons are absorbed, the use of Eq. (8.4.1) can give valuable estimates on total fluorescence yields. A high-resolution x-ray spectrum thus furnishes a basis for comparing, for example, similar collisions in gaseous and solid environments and deducing the relevant inner-shell vacancy cross sections.[76]

A vast majority of the experimental and theoretical work to date has been concerned with K transitions. However, the rates and fluorescence yields for certain multiply ionized L-shell excited configurations have been treated,[89,90] although in a more limited manner.

8.4.3. Multielectron Transitions

In the filling of inner-shell vacancies, the radiative transitions that predominate and that constitute the essential subject of this work are due to single-electron rearrangement. Fully allowed transitions proceed ac-

---

[86] D. L. Matthews, B. M. Johnson, L. E. Smith, J. J. Mackey, and C. F. Moore, *Phys. Lett.* **48A**, 93 (1974).

[87] N. Stolterfoht, D. Schneider, P. Richard, and R. L. Kauffman, *Phys. Rev. Lett.* **33**, 1418 (1974).

[88] M. H. Chen, B. Crasemann, and D. L. Matthews, *Phys. Rev. Lett.* **34**, 1309 (1975).

[89] C. P. Bhalla, *Proc. Int. Conf. Inner Shell Ionization Phenom. Future Appl.*, p. 1572. Atlanta, Georgia, 1972, (Natl. Tech. Information Service, U.S. Department of Commerce, Springfield, Va., 1972).

[90] M. H. Chen and B. Crasemann, *Phys. Rev. A* **10**, 2232 (1974).

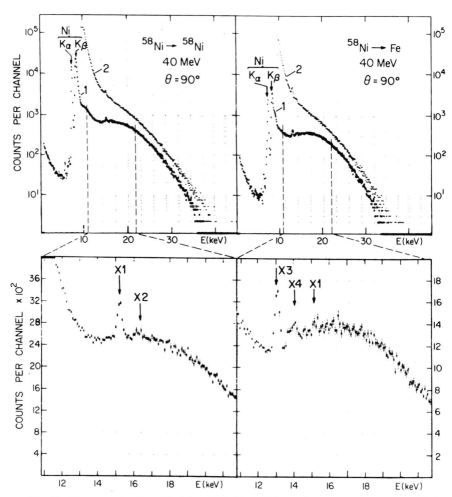

FIG. 22. X-ray spectra from Ni–Ni and Ni–Fe collisions. Lower spectra are magnified portions of the corresponding upper ones. The peaks marked X1 and X2 represent two-electron one-photon K_α and K_β transitions, respectively, in nickel. The peaks marked X3 and X4 are the same transitions in ion. [W. Wölfli, Ch. Stoller, G. Bonani, M. Suter, and M. Stöckli, *Phys. Rev. Lett.* **35**, 656 (1975)].

cording to electric dipole (E1) selection rules,[91] while various metastable states can decay by other multipolarities (e.g., M1, E2). However, the emission of a photon accompanied by multielectron rearrangement is also permissible, governed by those same selection rules. There is substantial

[91] B. W. Shore and D. H. Menzel, "Principles of Atomic Spectra," John Wiley and Sons, Inc., New York, 1968.

8.4. SPECTROSCOPY OF INDIVIDUAL STATES

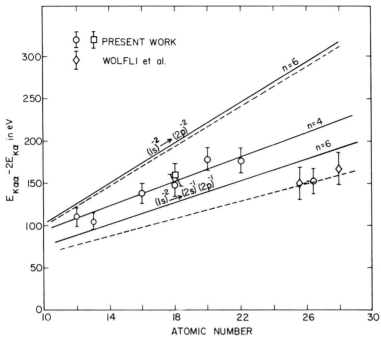

FIG. 23. Theoretical and measured $K_{\alpha\alpha}$ energy shifts. Solid and dashed lines represent Hartree–Fock calculations without and with multiplet splitting. The top pair are for an $E2$ $K_{\alpha\alpha}$ transition, the lower lines for $E1$ transitions with n L vacancies at the time of deexcitation. The diamonds are data points from W. Wölfli, Ch. Stoller, G. Bonani, M. Suter, and M. Stöckli, Phys. Rev. Lett. **35**, 656 (1975) [A. R. Knudson, K. W. Hill, P. G. Burkhalter, and D. J. Nagel, Phys. Rev. Lett. **37**, 679 (1976)].

evidence from several recent experiments for that mode of correlated deexcitation of highly excited states produced in ion–atom collisions.

One such example is demonstrated in Fig. 22, spectra taken with a Si(Li) detector of the x-ray emission produced in Ni → Ni and Ni → Fe collisions at 40 MeV incident energy.[92] The peaks marked X1 and X2 at roughly double the energy of the characteristic nickel K x rays correspond to transitions where a fully vacant K shell (i.e., hypersatellite) is filled simultaneously with two outer electrons and only a single photon is emitted. Calculations[93] of energies indicate agreement with an $E1$ transition, $(1s)^{-2} \to (2s)^{-1}(2p)^{-1}$ where the L shell may contain additional spectator vacancies, for the X1 peak. The X2 line corresponds to a tran-

---

[92] W. Wölfli, Ch. Stoller, G. Bonani, M. Suter, and M. Stöckli, Phys. Rev. Lett. **35**, 656 (1975).

[93] J. P. Briand, Phys. Rev. Lett. **37**, 59 (1976); W. Wölfli and H. D. Betz, ibid., p. 61; T. Åberg, K. A. Jamison, and P. Richard, ibid., p. 63.

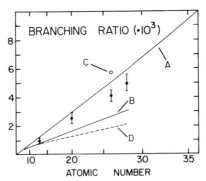

FIG. 24. Comparison of measured and calculated branching ratios for $K_{\alpha\alpha}$ and K_α transitions. A. J. P. Vinti, *Phys. Rev.* **42**, 632 (1932); B. T. Åberg, K. A. Jamison, and P. Richard *Phys. Rev. Lett.* **37**, 63 (1976); C, H. P. Kelley, *Phys. Rev. Lett.* **37**, 386 (1976); D. M. Gavrila and J. E. Hansen, *Phys. Lett.* **58A**, 158 (1976); are theoretical calculations [Ch. Stoller, W. Wölfli, G. Bonani, M. Stöckli, and M. Suter, *Phys. Rev. A* **15**, 990 (1977)].

sition where an L and M electrons fill the K vacancies. The respective energies are given roughly by the expressions

$$E_{K\alpha\alpha} = 2E_{K\alpha} + \Delta E_s, \quad (8.4.3)$$

$$E_{K\alpha\beta} = E_{K\alpha} + E_{K\beta} + \Delta E_s, \quad (8.4.4)$$

where $E_{K\alpha}$ and $E_{K\beta}$ are the ordinary single-vacancy K_α and K_β x-ray energies, respectively, and ΔE_s is an adjustment for the reduction in screening due to the vacant K shell.

A summary[94] of data on $E_{K\alpha\alpha}$ energies for these and several other ions along with Hartree–Fock calculations is presented in Fig. 23. The solid lines for the $(1s)^{-1} \rightarrow (2s)^{-1}(2p)^{-1}$ transitions allow for either an initially filled L shell ($n = 6$) or an initially doubly vacant L shell ($n = 4$). Notice that the $(1s)^{-2}(2p)^{-2}$ E2 transition lies too high in energy to account for the lines. All of these data were obtained from either projectiles or targets involved in heavy ion–atom collisions.

Equally interesting are the transition rates for the above two-electron decays as compared to the conventional one-electron decays. The branching ratio $N(K_\alpha^h)/N(K_{\alpha\alpha})$, where $N(K_\alpha^h)$ is the relative probability for deexcitation of a given two-K-vacancy state by two sequential individual one-electron transitions (the first a K_α hypersatellite, the second a K_α satellite), is shown in Fig. 24 as a function of Z of the ion.[95] The various theories appear to give an adequate description of both the magnitude and the Z dependence of the ratio.

[94] A. R. Knudson, K. W. Hill, P. G. Burkhalter, and D. J. Nagel, *Phys. Rev. Lett.* **11**, 679 (1976).
[95] Ch. Stoller, W. Wölfli, G. Bonani, M. Stöckli, and M. Suter, *Phys. Rev. A* **15**, 990 (1977).

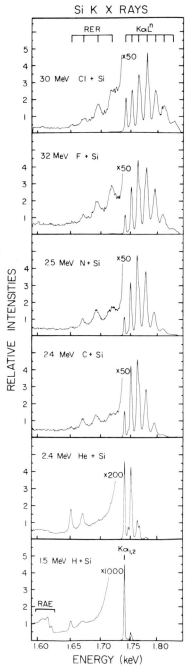

FIG. 25. Silicon K x-ray spectra produced by chlorine, fluorine, nitrogen, carbon, helium, and hydrogen bombardment [K. A. Jamison, J. M. Hall, J. Oltjen, C. W. Woods, R. L. Kauffman, T. J. Gray, and P. Richard, *Phys. Rev. A* **14**, 937 (1976)].

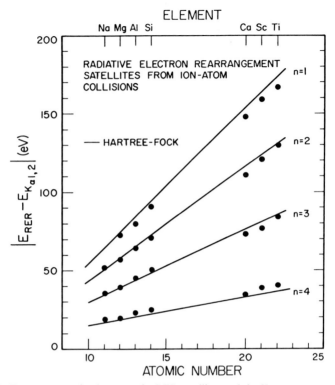

FIG. 26. Energy separation between the RER satellites and the $K\alpha_{1,2}$ energy as a function of atomic number. Data represented by points and Hartree–Fock calculations by solid lines [K. A. Jamison, J. M. Hall, and P. Richard, *J. Phys. B* **8**, L458 (1975)].

Yet another type of two-electron one-photon transition has been proposed[96,97] to account for the group of low-energy x-ray lines observed in high-resolution spectra[98,99] below the $K_{\alpha_{1,2}}$ diagram line of various elements. An example of the spectral features resulting from bombardment of silicon by different ions is depicted in Fig. 25. The peak at ~1.65 keV is the first of a series, the higher members of which become more prominent when the degree of ionization increases (reflected in the production of the higher satellites when heavier ions are used). It has been shown that the energy separation between the low-energy peaks and the $K_{\alpha_{1,2}}$ is

[96] K. A. Jamison, J. M. Hall, and P. Richard, *J. Phys. B* **8**, L458 (1975).
[97] K. A. Jamison, J. M. Hall, T. Oltjen, C. W. Woods, R. L. Kauffman, T. J. Gray, and P. Richard, *Phys. Rev. A* **14**, 937 (1976).
[98] P. Richard, C. F. Moore, and D. K. Olsen, *Phys. Lett.* **43A**, 519 (1973).
[99] J. McWherter, J. E. Bolger, H. H. Wolter, D. K. Olsen, and C. F. Moore, *Phys. Lett.* **45A**, 57 (1973).

consistent with the Hartree–Fock calculations for the sequence $(1s)^{-1}(2p)^{-n}$ to $(2s)^{-2}(2p)^{-n+1}$ for $n = 1-4$. Thus for a configuration with one K and one 2p vacancy, the transition involves the filling of the K vacancy by a 2s electron simultaneously with the "shakeup" of the other 2s electron to the initial 2p vacancy. The energy of the photon emitted is the usual energy minus the 2s–2p binding energy difference. The agreement with theory extends over a wide range of ions, as depicted in Fig. 26. The other peaks in the series correspond to the same process for states with increasing numbers of L vacancies. Hence each peak is related to a particular K_α satellite, e.g., its strength relative to the satellite is independent of the mode of excitation.

The origin of the proposed radiative electron rearrangement (RER) is due to final state mixing. Considering the $KL^1 (1s)^{-1}(2p)^{-1}$ initial configuration, decay to the $(2p)^{-2}$ configuration with multiplets ^1S, ^1D, and ^1P is possible. The final configuration $(2s)^{-2}$ ^1S will be mixed by the electron–electron interaction with the $(2p)^{-2}$ ^1S state. The mixing coefficient, along with the inclusion of multiplet fluorescence yields, determines the branching ratio. The latter agrees well with theory, on the order of 0.2% for the $(KL^1RER)-KL^1$ branch in silicon.

Yet another feature of cooperative electron rearrangement is visible at the start of hydrogen-induced spectrum in Fig. 25. The structure denoted RAE (radiative Auger effect) consists of transitions whereby an electron is ejected in conjunction with photon emission.[100] In this particular case, the high-energy endpoint is similar to that of the usual KLL Auger electron energy, indicating both the ejected electron and the one filling the K vacancy belonged to the L shell. Transitions form a band below that energy. These structures have been seen for a variety of targets and projectiles.[101]

8.5. Single-Collision Phenomena

8.5.1. Multiple L-Shell Ionization Probabilities

The majority of multiple ionization data has been taken using x-ray emission from target atoms in solids. Although in certain low-energy, high-projectile-Z, low-target-Z collisions there is a possibility of interaction with other atoms in the solid prior to deexcitation, most cases discussed here involve minimal contact with neighbors due to the low recoil

---

[100] T. Åberg, *Phys. Rev. A* **4**, 1735 (1971).
[101] P. Richard, J. Oltjen, K. A. Jamison, R. L. Kauffman, C. W. Woods, and J. M. Hall, *Phys. Lett.* **54A**, 169 (1975); G. Presser, *Phys. Lett.* **56A**, 273 (1976).

velocities produced. The role of the valence shell interaction between adjacent atoms, to be discussed at length in Chapter 8.6, is neglected for the moment. Consequently the vacancy-producing event can be treated as a single collision and appropriate comparisons with theory can be directly made.

The simplest case to consider is one in which the ionization occurring in addition to the creation of the K vacancy is limited for the most part to one or at most two L vacancies. This is the situation encountered when light ions, such as protons and alpha particles, are used as projectiles, as

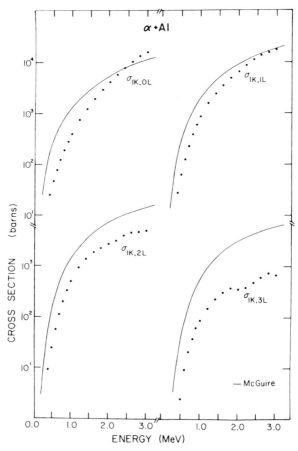

FIG. 27. Multiple-ionization cross sections for aluminum K_α configurations with one K vacancy and 0,1,2,3, L vacancies, produced by alpha bombardment. BEA results are shown as solid curves [P. Richard, R. L. Kauffman, J. H. McGuire, C. F. Moore, and D. K. Olsen, *Phys. Rev. A* **8**, 1369 (1973)].

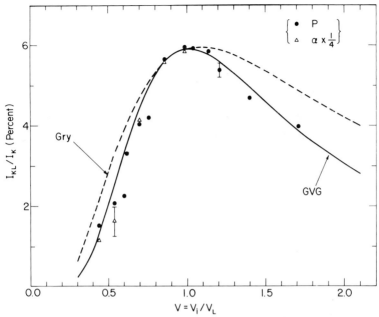

FIG. 28. Ratios of the double-ionization cross section σ_{KL} to single-ionization cross section σ_K, for ionization of vanadium by protons (p) and alphas (α), as a function of the velocity ratio (projectile to bound electron). The alpha data have been divided by a factor of 4 ($= Z_\alpha^2/Z_p^2$) [N. Cue, V. Dutkiewicz, P. Sen, and H. Bakhru, Phys. Lett. **46A**, 151 (1973)].

can be seen in Fig. 11. The dependence[102] of the individual aluminum K_α satellite cross section measurement is shown in Fig. 27. The BEA approximation is seen to give trends correctly, although discrepancies do exist in the various magnitudes. The ratio σ_{KL}/σ_K is a convenient quantity with which to make comparisons, since systematic errors in both experiment and theory are eliminated. The double- to single-ionization ratios[103] for protons and alphas incident on vanadium in Fig. 28 are well reproduced by the classical theories. Both the peaking around velocity match and the quadratic scaling in projectile Z predicted in Eq. (8.3.10) are evident in the experimental ratio.

As indicated previously, the degree of multiple ionization increases substantially for a wide range of targets when heavier ions are used as projectiles. In Fig. 29, the K_α x-ray satellite intensities[49] from elements calcium through manganese induced by 30-MeV oxygen ions are still

---

[102] P. Richard, R. L. Kauffman, J. H. McGuire, C. F. Moore, and D. K. Olsen, Phys. Rev. A **8**, 1369 (1973).
[103] N. Cue, V. Dutkiewicz, P. Sen, and H. Bakhru, Phys. Lett. **46A**, 151 (1973).

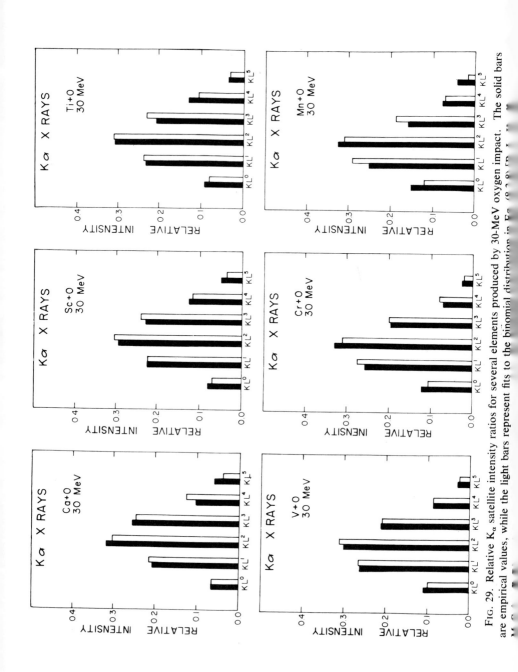

FIG. 29. Relative K_α satellite intensity ratios for several elements produced by 30-MeV oxygen impact. The solid bars are empirical values, while the light bars represent fits to the binomial distribution.

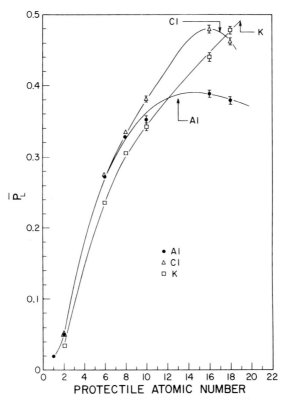

FIG. 30. Best-fit single K-, multiple L-shell ionization probabilities P_L for a variety of 1.7-MeV/amu projectiles incident on aluminum, chlorine, and potassium. Arrows indicate symmetric collisions [R. L. Watson, F. E. Jensen, and T. Chiao, *Phys. Rev. A* **10**, 1230 (1974)].

remarkably well fit by the binomial distribution from Eq. (8.3.8). The incident energy dependence of the quantity $P_L(0)$ extracted from data for oxygen bombardment of aluminum has been observed from below to well above velocity match with the Al L electrons.[104] Agreement with theory is not as good as with light ions; a somewhat arbitrary increase in the effective L-shell binding energy yields some improvement.

The dependence of the multiple ionization[105] upon projectile Z is shown in Fig. 30. The expectation of a higher degree of multiple L-shell ionization for the lighter target aluminum, based on the Coulomb excitation theories, is not supported by the data. A significant role by molecular or-

[104] D. K. Olsen, C. F. Moore, and R. L. Kauffman, *Phys. Lett.* **44A**, 109 (1973).
[105] R. L. Watson, F. E. Jensen, and T. Chiao, *Phys. Rev. A* **10**, 1230 (1974).

TABLE VII. Multiple L-Shell Probabilities

Target	Projectile	Energy (MeV)	$P_L^{expt.}$	$P_L^{theor.}$ (BEA)	Ref.
^{13}Al	^{1}H	1.7	0.019	0.142^a	105
	^{2}He	1.7	0.050	0.423^a	105
	^{8}O	1.7	0.328	0.967^a	105
	^{8}O	1.9	0.328	1.125^b	52
	^{17}Cl	1.1	0.414	8.474^b	52
	^{18}Ar	1.7	0.379	1.00^a	105
^{19}K	^{2}He	1.7	0.034	0.127^a	105
	^{8}O	1.7	0.305	0.841^a	105
	^{18}Ar	1.7	0.468	0.996^a	105
^{20}Ca	^{1}H	0.8	0.0115	0.0134^b	49
	^{2}He	0.8	0.044	0.0537^b	49
	^{8}O	1.9	0.289	0.918^b	49
^{55}Mn	^{1}H	0.8	0.0028	0.0057^b	49
	^{2}He	0.8	0.0101	0.0228^b	49
	^{8}O	1.9	0.233	0.493^b	49

^{a} Computed using Eq. (8.3.8).
^{b} Computed using Eq. (8.3.7).

bital processes, i.e., level crossing and vacancy sharing, in the L vacancy creation is indicated. It should also be kept in mind that the data for the high values of P_L appropriate for a high degree of stripping are somewhat suspect in light of the possibility of L-hole filling prior to K-hole filling. However, trends observed with the solid targets should still be of value in attempts to describe the configurations created initially in the collisions.

A sampling of P_L values for various systems, both experimental and theoretical, is summarized in Table VII. One problem arises in the simplified approach presented in Section 8.3.2 for heavier ions. The values of P_L, which is a probability per unit area, exceed 1.0, the maximum physically real number. Apparently the assumptions regarding the limitation of impact parameters to the size of the K-shell radius are in error. Indeed, an explicit integration of Eq. (8.3.1) using computed values for $P_K(b)$ and $P_{nL}(b)$ exhibits the expected ceiling of 1.0 but still substantially overestimates the experimental numbers.[105]

All of the above analyses were made with the assumption of statistical population of magnetic substates and consequently unpolarized x-ray emission. Recent evidence[106] for polarization of KL^1 transitions produced in aluminum by protons and alphas calls for a fresh examination of the subject. Specifically the $K_{\alpha'}(^1P_1 \rightarrow {}^1S_0)$, $K_{\alpha 3}(^3P_{2,1,0} \rightarrow {}^3P_{2,1,0})$, and

[106] K. A. Jamison and P. Richard, *Phys. Rev. Lett.* **38**, 484 (1977).

$K_{\alpha_4}(^1P_1 \to {}^1D_2)$ transitions are seen in Fig. 31 to vary in intensity depending on the orientation of the crystal, in this case an ADP. The intensities $I_\perp$ (Rowland circle coplanar with the beam axis) and $I_\parallel$ (Rowland circle normal to the beam axis) can be used to obtain the polarization fraction P,

$$P = \frac{I_\parallel - I_\perp}{I_\parallel + I_\perp} \frac{1 + \cos^2(2\theta)}{1 - \cos^2(2\theta)}, \qquad (8.5.1)$$

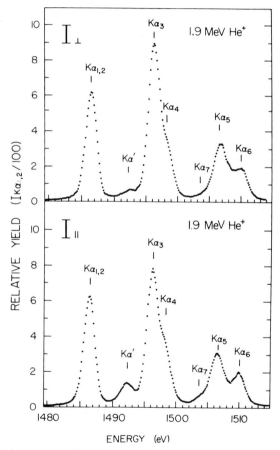

FIG. 31. Aluminum K_α satellite spectra induced by 1.9-MeV alphas. Data taken with an ADP crystal. Upper spectrum taken with axis of crystal rotation perpendicular to the beam axis (Rowland circle coplanar with beam axis) and lower spectrum taken with rotation axis parallel to beam axis (Rowland circle perpendicular to beam axis) [K. A. Jamison and P. Richard, *Phys. Rev. Lett.* **38**, 484 (1977)].

where the second term is a correction at Bragg angle θ due to not being at 45°. Polarizations as large as 52% ($K_{\alpha'}$ using 1.9-MeV helium) were observed, indicating a preferential population of the $m_l = 0$ substate over the $m_l = \pm 1$ substates in the 2p shell. The importance of such effects should decrease for the more highly ionized configurations produced by heavier ions, since the large number of initial and final states tends to decrease the magnitude of polarization fractions.

8.5.2. Hypersatellite Production

Many of the features inherent in the K_α satellite spectra and total x-ray yields can be studied in a second set of satellites also collisionally induced, the hypersatellites.[107,108] The latter, as stated above, consist of groups of transitions for which the initial states include two K vacancies

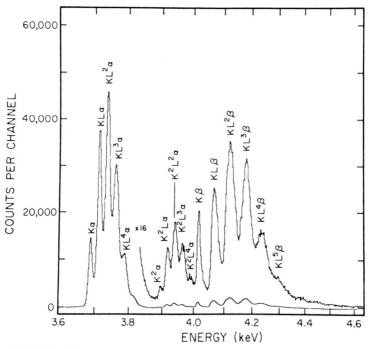

FIG. 32. Calcium K x rays produced by 48-MeV oxygen bombardment, including the K^2L^n hypersatellite transitions. Spectrum taken with a LiF crystal [D. K. Olsen and C. F. Moore, *Phys. Rev. Lett.* **33**, 194 (1974)].

[107] P. Richard, W. Hodge, and C. F. Moore, *Phys. Rev. Lett.* **29**, 393 (1972).
[108] D. K. Olsen and C. F. Moore, *Phys. Rev. Lett.* **33**, 194 (1974).

8.5. SINGLE-COLLISION PHENOMENA

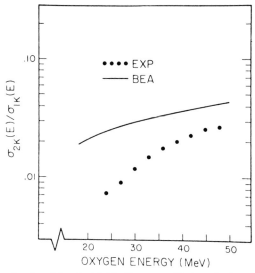

FIG. 33. Measured ratio of the double K- (hypersatellite) to single K-vacancy cross section resulting from oxygen bombardment, compared to a BEA prediction [D. K. Olsen and C. F. Moore, *Phys. Rev. Lett.* **33**, 194 (1974)].

in addition to some number n of L vacancies. Fluorescence yields for these configurations are, with a few exceptions,[109] not available and hence some uncertainty exists in obtaining vacancy cross sections from the x-ray yields. Nonetheless, relative yields of that group made available by the high resolution of crystal spectrometers provide valuable information on the double K-hole creation process.

One such piece of information is the ratio of the double to single K-hole production for calcium bombarded by oxygen ions. A spectrum is shown in Fig. 32. The energy dependence of that ratio, as seen in Fig. 33, is qualitatively described by a BEA calculation of the form of Eq. (8.3.1), where a double-hole probability $P_K^2(b)$ is utilized. In addition, fits to the L-vacancy distributions of both satellites and hypersatellites produced similar values of $P_L(0)$, implying that the assumption of no correlations for events involving K and L electrons is a valid one for this system.

Yet another interesting dependence of the double K- to single K-hole cross section ratio is upon the atomic number or Z value of the target atom.[110] The behavior displayed in Fig. 34 for 30-MeV chlorine is indicative of level crossing, in this case a transfer of the chlorine beam L va-

---

[109] C. P. Bhalla, *J. Phys. B* **8**, 2787 (1975).

[110] C. W. Woods, F. Hopkins, R. L. Kauffman, D. O. Elliott, K. A. Jamison, and P. Richard, *Phys. Rev. Lett.* **31**, 1 (1973).

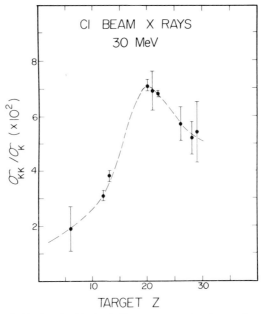

FIG. 34. Ratio of chlorine double K-shell vacancy production to single K-shell vacancy production for 30 MeV ^{35}Cl incident on a variety of targets [C. W. Woods, F. Hopkins, R. L. Kauffman, D. O. Elliott, K. A. Jamison, and P. Richard, *Phys. Rev. Lett.* **31**, 1 (1973)].

cancies expected due to the equilibrium charge state distribution to the chlorine K shell via the $2p\pi$–$2p\sigma$ rotational coupling. Above $Z \sim 20$, the filled-target L shell is more tightly bound than the L shell of chlorine stripped to charge 10+ or 11+ and hence occupies the $2p\pi$ level.

8.5.3. Neon as a Case Study

Many of the uncertainties and ambiguities inherent in the measurement of x rays emitted from solid targets can be avoided by the use of gas targets. Self-absorption in the target can be eliminated. Especially important for heavier ions, the charge state purity of the beam can be maintained, a critical control since inner-shell vacancy cross sections are in many cases sensitive to the electronic structure of the projectile. The fact that ions moving in solids are characterized by an equilibrium charge distribution consisting of a variety of charge states dictates that x-ray yields from solids can only be treated as average quantities, i.e., converted into mean cross sections. Although incident energy degradation effects can be minimized by using thin solid foils, the thin gas targets conveniently avoid that problem also.

8.5. SINGLE-COLLISION PHENOMENA

A veritable wealth of information has been accumulated for multiply ionized neon. Neon provides additional advantages beyond those listed above. Its being a noble gas precludes various possible molecular effects that can enter into both the collisional excitation and the subsequent deexcitation. The absence of M-shell electrons, a condition met unless electrons are excited to that shell in the collision, simplifies the relaxation

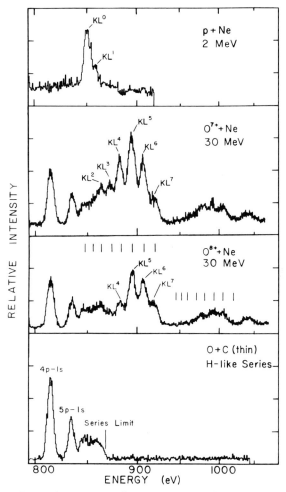

FIG. 35. Dependence of neon K_α satellite structure upon the charge state of 30-MeV incident oxygen ions. Top portion is a proton-induced spectrum. RAP crystal used. Bottom portion exhibits hydrogenic oxygen lines produced by excitation in a thin carbon foil. Two groups of vertical lines indicate calculated K_α satellite and hypersatellite energies [R. L. Kauffman, F. Hopkins, C. W. Woods, and P. Richard, *Phys. Rev. Lett.* **31**, 621 (1973)].

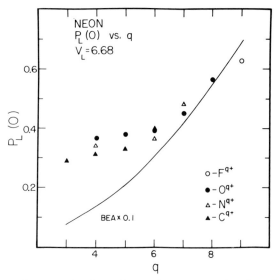

FIG. 36. A plot of measured values of $P_L(0)$ for neon as a function of charge state of various ions with 1.5-MeV/amu incident energy. The solid curve is the Z^2-dependent BEA prediction [R. L. Kauffman, C. W. Woods, K. A. Jamison, and P. Richard, Phys. Rev. **11**, 872 (1975)].

scheme. Coster–Kronig transitions cannot occur (2s vacancies are frozen) and also the L-S coupling possibilities are more limited. Theoretical treatments of x-ray energies and rates, as discussed in Chapter 8.4, are made more feasible by all these reasons. Likewise the collisional aspects of the excitation become manageable. Accordingly, the many high-resolution x-ray and Auger electron studies performed to date constitute a classic case study in ion-induced spectroscopy.

The dramatic dependence of the relative intensities of the neon K_α x-ray statellites on the charge state of a heavy projectile is well established.[85,111] As can be seen in Fig. 35, the degree of stripping increases noticeably when the charge of a 30-MeV oxygen ion is increased from 5+ to 8+ (fully stripped). The substantial amount of L-shell ionization is in contrast to the predominant KL^0 strength in the proton- and electron-induced spectra.[112] The latter are nearly equivalent at $\geqslant 2.0$ MeV/amu incident energy, approaching a limiting value for the KL^1 strength commensurate with electron shakeoff (rather than collisional ionization of the L shell).

[111] D. L. Matthews, B. M. Johnson, and C. F. Moore, Phys. Rev. A **10**, 451 (1974).
[112] D. L. Matthews, C. F. Moore, and D. Schneider, Phys. Lett. **48A**, 27 (1974).

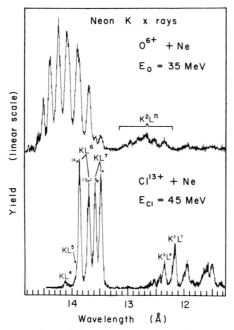

FIG. 37. K x-ray spectra of neon induced by oxygen and chlorine ions. Mica crystal used [D. L. Matthews, B. M. Johnson, G. W. Hoffman, and C. F. Moore, *Phys. Lett*, **49A**, 195 (1974)].

A survey[113] of $P_L(0)$ values for a variety of projectiles and charge states is contained in Fig. 36. The similarity of values for ions of different nuclear charge but the same ionic charge implies nearly full screening due to attached projectile K electrons, insofar as the vacancy creation in the L shell is concerned. However, the attached L electrons are ineffective and affect the multiple ionization only minimally. The incident energy dependence for fixed charge state was found[113] to closely approximate the ln E/E high-energy dependence for L-shell ionization predicted by PWBA, rather than the more rapid $1/E$ falloff characteristic of the BEA theory. This regime, i.e., projectile velocities being much higher than orbital velocities of the neon L electrons, is one in which direct Coulomb excitation should be dominant over molecular processes.

Stripping of neon down to primarily two- and three-electron states, where coupling schemes are simplified and the spectroscopy of individual

[113] R. L. Kauffman, C. W. Woods, K. A. Jamison, and P. Richard, *Phys. Rev. A* **11**, 872 (1975).

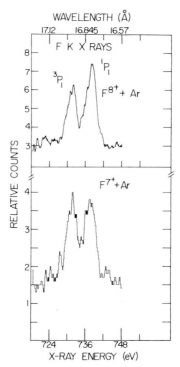

FIG. 38. Fluorine K x rays in the region 712–750 eV produced by F^{7+} and F^{8+} at 35 MeV energy in collisions with argon. Taken with an RAP crystal [J. R. Macdonald, P. Richard, C. L. Cocke, M. Brown, and I. A. Sellin, *Phys. Rev. Lett.* **31**, 684 (1973)].

states is made more feasible, has been accomplished[114,115] with highly charged chlorine and argon ions at ~1–2 MeV/amu energies. The spectrum in Fig. 37 demonstrates this accessibility, with the two-electron $1s2p(^1P)-1s^2(^1S)$ and metastable $1s2p(^3P)-1s^2(^1S)$ decays clearly visible. The use of a mica crystal afforded the increased resolution necessary to resolve the transitions, which were only partially resolved in the previous studies utilizing an RAP crystal. Further, the overlapping of multiplets from different configurations is evident in the decreased linewidth of the peak attributed to three-electron (KL^6) ^4P states in the lower of the two spectra as compared to the upper spectrum where four-electron (KL^5) strength leads to a single broadened peak. An important experimental finding is that the ^4P peak intensity relative to the remainder of the KL^6

---

[114] D. L. Matthews, B. M. Johnson, G. W. Hoffman, and C. F. Moore, *Phys. Lett.* **49A**, 195 (1974).

[115] J. R. Mowat, R. Laubert, I. A. Sellin, R. L. Kauffman, M. D. Brown, J. R. Macdonald, and P. Richard, *Phys. Rev. A* **10**, 1446 (1974).

8.5. SINGLE-COLLISION PHENOMENA

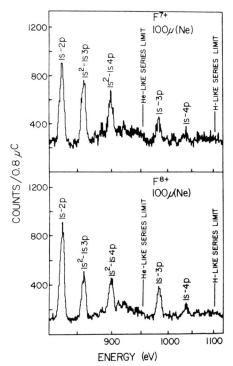

FIG. 39. High-energy portion of fluorine K x-ray spectrum (RAP crystal) produced by F^{7+} and F^{8+} bombardment of neon at 36 MeV [F. Hopkins, R. L. Kauffman, C. W. Woods, and P. Richard, *Phys. Rev. A* **9**, 2413 (1974)].

group is indicative of a preferential population of the 1s2s2p configuration over the 1s2p² configuration.[116]

Although neon and other gases have been utilized primarily as the neutral partners to be excited in various heavy-ion collisons, they also serve in effect as projectiles to excite the incident ions under single-collision conditions. The same x-ray spectroscopy can be applied to transitions taking place in beam ions. In fact, for a great many of the few-electron states formed, which are solely photon emitting due to the absence of an Auger channel, it is the only means available to study the collision. The transitions arising from the $^3P_1$ and $^1P_1$ states visible in Fig. 38 are the same decays referred to in Fig. 36 except they were observed[117] in a two-electron fluorine ion passing through an argon gas target. In the case

---

[116] D. L. Matthews, R. J. Fortner, D. Schneider, and C. F. Moore, *Phys. Rev. A* **14**, 1561 (1976).

[117] J. R. Macdonald, P. Richard, C. L. Cocke, M. D. Brown, and I. A. Sellin, *Phys. Rev. Lett.* **31**, 684 (1973).

of the incident F^{8+}, they represent x rays emitted following electron capture to the 2p subshell (or a higher n state that led through cascading to the 2p). For F^{7+}, they correspond to some form of excitation of the ion, such as direct Coulomb excitation of a $1s^2(^1S_0)$ ground state to the $1s2p(^1P_1)$ state or, as will be discussed in Chapter 8.7, perhaps elevating an initial $1s2s(^3S_1)$ metastable component of the beam into the $1s2p(^3P_1)$ state. The complexity of such a collision becomes apparent when the various possible hydrogenic transitions arising from F^{7+} or F^{8+} in neon[118] are considered (see Fig. 39). Direct one-step excitation would explain the one-electron transitions for F^{8+}. However, starting initially with a ground-state F^{7+} would require ionization of one K electron plus excitation of the other to an np level or loss of both K electrons plus electron capture from the neon. Projectile x-ray spectroscopy is obviously a useful means of investigating multiple events (other than just multiple ionization) occurring within a single collision.

8.6. Chemical Effects

8.6.1. Effects of Outer-Shell Relaxation on X-Ray Spectra

In general the relative intensities of x rays in a high-resolution spectrum do not directly reflect the primary vacancy production. A correction for the fluorescence yield must be made and, as has been discussed previously, that quantity can vary tremendously depending on the degree of ionization of the state as well as the particular configuration and multiplet being considered. A second effect is the possibility of outer vacancy filling prior to the transition filling the inner vacancy, leading to an alteration of the vacancy distribution created initially in the collision. Regarding K_α and related satellite structure, both the intensities of the satellites and the mean energies can be influenced[39] by such processes involving the L- and M-shell vacancies, respectively. Early experiments[119,120] attempted to characterize peak centroid energies in terms of the valence electron density available in specific compounds. In general, the cases considered in this section involve x-ray emission from target atoms that travel only short distances (< 1 Å) due to recoil energy prior to deexcitation.

---

[118] F. Hopkins, R. L. Kauffman, C. W. Woods, and P. Richard, *Phys. Rev. A* **9**, 2413 (1974).

[119] P. G. Burkhalter, A. R. Knudson, D. J. Nagel, and K. L. Dunning, *Phys. Rev. A* **6**, 2093 (1972).

[120] J. McWherter, D. K. Olsen, H. H. Wolter, and C. F. Moore, *Phys. Rev. A* **10**, 200 (1974).

8.6. CHEMICAL EFFECTS

The consequences upon the L-vacancy distribution observed at the time of K x-ray emission due to the competition between K- and L-vacancy filling rates can be described schematically as[121]

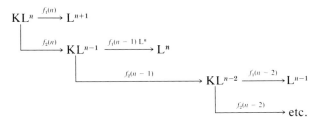

As represented above, KL^n is the initial configuration (one K and n L vacancies), $f_1(n)$ the fraction decaying directly by K_α x-ray emission, and $f_2(n)$ the fraction transformed to configuration KL^{n-1} due to L-hole filling. Allowing for varying fluorescence yields ω_n, the fraction $f_1(n)$ is given by

$$f_1(n) = [\Gamma_K^n/(\Gamma_K^n + \Gamma_L^n)]\omega_n, \qquad (8.6.1)$$

where Γ_K^n and Γ_L^n are the total K and L widths, respectively, for the nth configuration. The value ω_n is here normalized so that the total yield of K_α x rays is given by 1.0 and satellite intensities are represented by fractions. The fraction $f_2(n)$ is simply the ratio $\Gamma_L^n/(\Gamma_K^n + \Gamma_L^n)$. The relationship between the observed satellite intensity N_0^n and the configurations N_c^n created in the collision is given by

$$N_0^n = f_1(n)[N_0^n + f_2(n+1)N_c^{n+1} + f_2(n+1)f_2(n+2)N_c^{n+2} + \cdots]. \qquad (8.6.2)$$

Thus a given satellite reflects not only its own creation probability but also contributions from higher-order configurations that cascade into it via L-shell transitions.

The various widths of multiply ionized states incorporated into f_1 and f_2 are, with a few exceptions, unmeasured and also unknown theoretically. However, a plot of the quantity $(\Gamma_K + \Gamma_L)/\Gamma_K$ using the available single-vacancy widths is informative and also appropriate for situations where limited multiple ionization takes place. Figure 40 demonstrates those values as a function of Z of the ion, based on both theoretical and empirical L_3 rates. It is apparent that even doubly ionized (KL^1) states should exhibit significant rearrangement effects. It is to be expected that as L electrons are depleted, the L rate will compete more favorably with the K rate (the majority of which is due to L–K transitions). A major uncertainty in specifying the extent of the rearrangement is the determina-

[121] T. K. Li, R. L. Watson, and J. S. Hansen, *Phys. Rev. A* **8**, 1258 (1973).

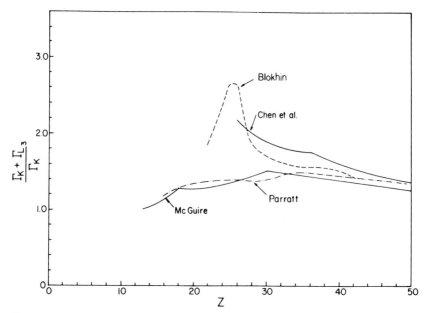

FIG. 40. The rearrangement factor $(\Gamma_K + \Gamma_{L3})/\Gamma_K$ obtained using single-vacancy theoretical widths from E. J. McGuire, *Phys. Rev. A* **3**, 587 (1971) and M. H. Chen, B. Crasemann, and V. O. Kostroun, *Phys. Rev. A* **4**, 1 (1971) and experimental widths from M. A. Blokhin, "The Physics of X-Rays," 2nd ed., Nauka, Moscow, 1957, Atomic Energy Tech. Rep. No. 4502, and L. G. Parratt, *Rev. Mod. Phys.* **31**, 616 (1959). Taken from T. K. Li, R. L. Watson, and J. S. Hansen, *Phys. Rev. A* **8**, 1258 (1973).

tion of M-shell populations, which can at best be inferred from energy shifts. Those populations directly affect the various rates, such as the L cascading discussed here.

One study[122] of the proton-induced KL^1 strength for a variety of elements in solids, including a few compounds, attributed the discrepancy between observed relative intensities and predicted intensities as being due to L-hole filling. Values for $(\Gamma_K + \Gamma_{L3})/\Gamma_K$ were deduced in this way and found to lie reasonably close to the solid (McGuire) curve in Fig. 40.

8.6.2. Dependence Of Solid Target Spectra on Environment

The highly stripped target ions produced in collisions involving heavy projectiles were first used to determine chemical shifts (of centroid energies) of the KL^n peaks in metals as compared to nonmetals. Including

[122] V. Dutkiewicz, H. Bakhru, and N. Cue, *Phys. Rev. A* **13**, 306 (1976).

8.6. CHEMICAL EFFECTS 419

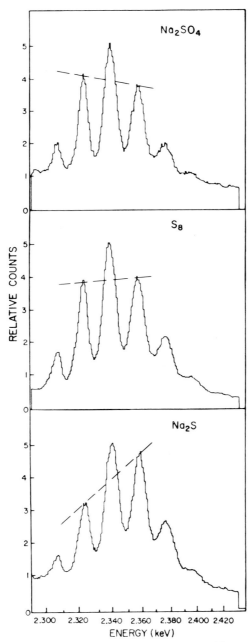

FIG. 41. Sulfur K_α x-ray spectra (NaCl crystal) from several different solid sulfur targets, produced by 32 MeV oxygen [R. L. Watson, T. Chiao, and F. E. Jensen, *Phys. Rev. Lett.* **35**, 254 (1975)].

information on the more pronounced nature of K_β satellite shifts, data for aluminum x rays from Al, Al_2O_3, and AlN[119] and silicon x rays from Si and SiO_2[120] indicated primarily that the degree of ionization is important in determining the magnitude of the shifts. Higher-order K_β configurations (e.g., KL^4) exhibited smaller shifts than the KL^2 and KL^3 groups, all in the range 5–10 eV. The attraction of local neighboring valence electrons was a possible explanation for the neutralization of the positive shifts, themselves indicative of the absence of M electrons.

A noticeable dependence of the intensities of K_α satellites on the nature of the compound containing the element has been observed for a series of solid components from aluminum to chlorine.[123,124] As shown for sulfur in Fig. 41, the spectra show a shifting of strength from compound to compound. The trends are conveniently gauged by extracting P_L values of the type discussed at length above. A summary of those values derived from 32.4-MeV oxygen-induced spectra is presented in Fig. 42, as a function of the total valence electron density. The latter is taken as consisting of all electrons in the outermost, unfilled shell of each bonding partner, with the assumption that they are delocalized within the region defining the molecular volume. In a fairly consistent fashion, the higher P_L values correspond to lower densities. In addition most M electrons associated with the ion initially are expected to be ionized in such a violent collision. Significant contributions to L-hole filling from valence electrons of neighboring atoms are indicated, i.e., interatomic transitions must be considered. The specific form of the process is undetermined, although Auger cascades (as opposed to radiative) should predominate in any event. It may consist of a transition of a neighboring valence electron directly into the L vacancy with an additional (Auger) electron ejected from the same shell or perhaps the ion's own remaining valence shell.[125] Alternatively, the neighboring valence electrons may be attracted into the valence (M, in these cases) shell of the ion and then undergo ordinary intraatomic transitions.

All of the above analysis is based on the assumption that the ionization probabilities are independent of compound and the initial configurations are on the average the same. However, differences in cross sections due to the changes in L-shell binding energies from material to material may account in part for the variation in P_L.[23] The striking anomaly in all of the data is the lack of an effect for aluminum, which has been verified by

---

[123] R. L. Watson, T. Chiao, and F. E. Jensen, *Phys. Rev. Lett.* **35**, 254 (1975).

[124] R. L. Watson, A. K. Leeper, B. I. Sonobe, T. Chiao, and F. E. Jensen, *Phys. Rev. A* **15**, 914 (1977).

[125] P. H. Citrin, *J. Electron. Spectrosc. Relat. Phenom.* **5**, 273 (1974).

8.6. CHEMICAL EFFECTS

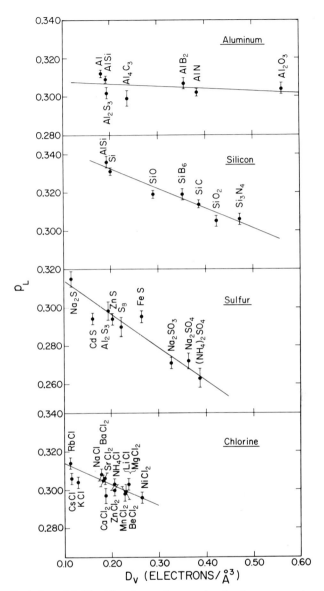

FIG. 42. Variation of $P_L(0)$ with valence electron density for compounds of several different elements, all resulting from 32-MeV oxygen bombardment [R. L. Watson, A. K. Leeper, B. I. Sonobe, T. Chiao, and F. E. Jensen, *Phys. Rev. A* **15**, 914 (1977)].

additional work.[126] Both valence electron and binding energy arguments would predict a change commensurate with that in the silicon series.

The extent to which rearrangement occurs, an unknown in the analyses above, can be judged by comparing spectra from gas and solid targets. In Fig. 43, the centroid energies of the silicon K_α satellite peaks induced by chlorine ions are shifted upward to a degree in the SiH_4 spectra consistent with a fully ionized M shell.[76] The measured (calculated, assuming no M electrons) values for the KL^2, KL^3, KL^4, KL^5, and KL^6 centroid energies were 5(6.3), 7(8.0), 7(9.7), 11(11.4), and 11(13) eV, respectively. The experimental shifts are taken relative to the solid silicon energies, which match up well with the calculated Hartree–Fock values for a silicon ion with a filled M shell. Since the absence of M electrons is indicated for the gas, that spectrum should be free of L-shell relaxation effects and hence, with proper corrections for fluorescence yields, be representative of the L vacancy distribution created in the collisions. It is apparent that the L cascading in the solid alters the distribution by ~1 to 2 vacancies prior to K_α emission. A similar effect has been observed[127] for neon ions in gaseous form vs. being implanted in solid hosts. An important consequence of these findings is the knowledge that the average fluorescence yields discussed in Section 8.4.2 can be noticeably larger in gases than solids, and thus care must be exercised when poor-resolution total yields from gases and solids are compared.

The configurations of the valence shells are uncertain above and beyond the crude indication given by screening effects on the L to K transitions. The rate at which vacancies in the valence shell itself are filled determines the effective L rates to be used in Eq. (8.6.1). A more direct investigation of the rearrangement in the outer shells of ions has been made[128] by observing the K x rays from very low Z ions, for which the L shell is the valence shell. Rather than the centroid energies as for the higher-Z ions, the relative intensities of the K_α reflect valence shell population. For detection of target boron and carbon K x rays produced by projectiles as heavy as oxygen at ~1 MeV/amu energy, the Bragg crystal spectrometer used consisted of a double-focusing (spherical lens) lead stearate pseudocrystal with a channeltron as the photon-counting element. For thin boron and carbon solid foils, the major finding was the lack of higher satellites, e.g., KL^2 and KL^3 (in the case of carbon), expected for such violent collisions. One possible explanation[128] is L-hole filling within the K-hole lifetimes ($\geqslant 2 \times 10^{-14}$ sec for boron and

---

[126] F. Hopkins, J. Sokolov, and A. Little, *Phys. Rev. A* **14**, 1907 (1976).
[127] R. J. Fortner and D. L. Matthews, *Phys. Rev. A* **14**, 2357 (1976).
[128] F. W. Martin, W. Freeman, and J. Sauls, to be publ. in *J. Phys. B*.

8.6. CHEMICAL EFFECTS

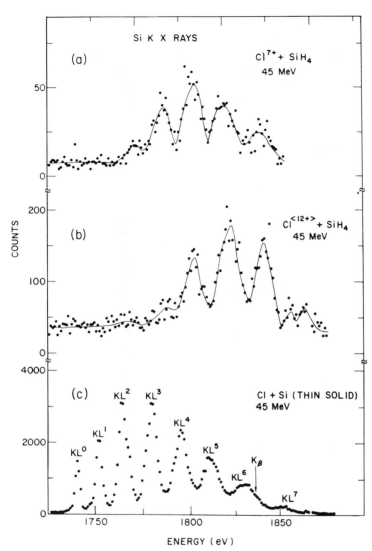

FIG. 43. Silicon K_α x-ray spectra (EDT crystal) from (a) SiH_4 gas induced by 45-MeV Cl^{7+}, (b) SiH_4 gas induced by 45-MeV Cl^{12+} prepared by inserting a prefoil before the gas cell, and (c) thin solid silicon foil induced by 45 MeV chlorine [R. L. Kauffman, K. A. Jamison, T. J. Gray, and P. Richard, *Phys. Rev. Lett.* **36**, 1074 (1976)].

$\geq 10^{-14}$ sec for carbon). Such quenching could occur within the plasma response times of $\sim 1.4 \times 10^{-15}$ sec for the valence electrons of boron and carbon. Thus it would appear that the valence shell itself in solids may experience fast rearrangement. This is supported by the observation that for the higher-Z projectiles, the silicon satellite energies observed from the solid are predicted by Hartree–Fock calculations assuming a full M shell.

8.6.3. Outer-Shell Rearrangement in Gaseous Molecules

Quenching effects similar to those for solids have also been observed in gases, where the rearrangement is restricted to a single molecule.[129] The results are summarized in Fig. 44 for silicon and sulfur K_α x-ray emission induced by 53-MeV chlorine ions. The "nearly atomic" targets of SiH_4 and H_2S yield satellite strength centered substantially above that for the corresponding spectra from both "heavy" molecules and solids (see Fig. 44b, e). The energy shifts relative to the solids indicate that most M electrons are ionized from the SiH_4 and H_2S, while population in both the multivalence electron SiF_4 and SF_6 is greater, i.e., is more similar to that in the solids.

An analysis of the type indicated by Eq. (8.6.1) produced the curves in Fig. 44c, f. The vacancy distribution was obtained by correcting the x-ray distribution for the fluorescence yields ω_n scaled from available theoretical numbers

$$R^n = \frac{R_x^n}{\omega_n} \bigg/ \sum_{n=0}^{7} \frac{R_x^n}{\omega_{n'}} \tag{8.6.3}$$

where values for R_x^n from $SiH_4(H_2S)$ were used as indicative of those resulting initially from the collision. Then allowing for L-hole quenching in the heavier molecules, which contain a much larger number of valence electrons before and presumably after the collision, "relaxed" distributions were calculated. The L widths were approximated by assuming a statistical scaling of the rates, i.e., $\Gamma_L^n = n\Gamma_L^1$, where Γ_L^1 is the L width for the configuration KL^1. The K rates were scaled from calculations for neighboring ions. In order to reproduce the data, values for the quantity Γ_L^1 of twice the single L vacancy rate were used. The solid curve in Fig. 44c is based on no depletion of M electrons (M vacancies are filled immediately as might be the case with the solid), while the dashed curve allows for depletion and the corresponding reduction of the L rates. Those curves plus similar ones for sulfur (Fig. 44f) coincide reasonably well with

---

[129] F. Hopkins, A. Little, N. Cue, and V. Dutkiewicz, *Phys. Rev. Lett.* **37**, 1100 (1976).

8.6. CHEMICAL EFFECTS

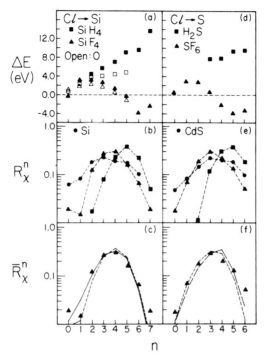

FIG. 44. The centroid energy shifts and relative intensities of the K_α satellites from several gases compared to same for solids. Shifts in (a) and (d) are differences between gases and solids. Intensities in (b) and (e) are in fractional form. Theoretical curves in (c) and (f) are described in the text [F. Hopkins, A. Little, N. Cue, and V. Dutkiewicz, *Phys. Rev. Lett.* **37**, 1100 (1976)].

the distributions for the heavy molecules. It is evident that the bonding partners are supplying electrons for the quenching of the L vacancies, probably directly through the valence shells (a reasonable hypothesis where covalent bonds are involved).

Further work[130] suggests that the valence electrons readjust to the stripped ions valence shell and then undergo ordinary L transitions. Comparisons of chlorine K_α spectra from HCl, CH_3Cl, and CCl_4 produced by 50-MeV silicon ions demonstrate a lower-energy structure for CCL_4 than for HCl and CH_3Cl, which are similar. Since the central carbon atom is present in both the heavier molecules, the difference can be attributed to participation by electrons from the other peripherial chlorine bonding partners with the carbon valence shell serving as an intermediate location.

[130] N. Cue, W. A. Little, and F. Hopkins, *Bull. Am. Phys. Soc.* **22**, 655 (1977).

8.7. Multiple-Collision Phenomena

8.7.1. Collisional Quenching

The spectroscopic methods outlined above have been applied for the most part to x-ray emission from "target" ions, that is, ions with low velocities. In solid environments, the lifetimes of the excited states created are usually shorter than interaction times with nearest-neighboring nuclei, although effects due to local electron densities can be manifested as outer-shell rearrangement. As a result, the relative intensities of the various lines represent to a large degree the vacancy distributions and states produced in single collisions. The opposite of this situation is true for inner-shell vacancies in ions moving at rapid velocities in dense media, particularly solids. The lifetimes due to radiative and Auger decay can be much longer than typical collision times, so that in multiple collisions given states are altered or quenched in some way.[131] The cross sections are expected to be largest for outer-shell rearrangement via capture, loss, and excitation to higher bound states. Thus, following creation of a state with an inner-shell vacancy, the projectile may undergo several such "soft" collisions before the inner vacancy is filled, possibly enough to reach an equilibrium population of states. In such a multiple-collision regime, the spectrum is representative to some extent and perhaps for the most part of that equilibrium population characteristic of a specific ion moving at a certain velocity in a solid rather than the distribution of states resulting from the "hard" collision producing the inner hole.[132]

The quenching of states by collisional processes rather than by the ordinary relaxation experienced by isolated excited ions can be significant whenever the normal (static) lifetime exceeds collisional times. An extreme example of quenching in very dilute gases has been reported[133] for the ^{3}P metastable state referred to in Fig. 38, for fluorine in various gases. The ratio of ^{3}P yield to ^{1}P yield is shown in Fig. 45a at 20 MeV incident energy as a function of gas pressure. The nonlinearity is attributed[133] to collisional quenching of the long-lived ($\tau \approx 0.5$ nsec) $^3P_1$ state following formation in the exciter gas. The $^1P_1$ state with a lifetime of $\sim 2 \times 10^{-13}$ sec decays quickly enough so that its yield is directly proportional to its creation probability and thus is essentially independent of target pressure. In its simplest form, the quenching cross section σ_Q can be specified as

$$\sigma_Q = (nv\tau)^{-1}, \tag{8.7.1}$$

---

[131] H. D. Betz, *Rev. Mod. Phys.* **44**, 465 (1972).
[132] H. O. Lutz, H. J. Stein, S. Datz, and C. D. Moak, *Phys. Rev. Lett.* **28**, 2 (1972).
[133] D. L Matthews, R. J. Fortner, and G. Bissinger, *Phys. Rev. Lett.* **36**, 664 (1976).

8.7. MULTIPLE-COLLISION PHENOMENA

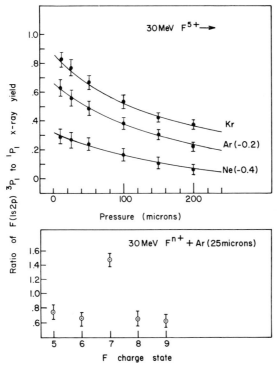

FIG. 45. Upper part depicts the ratio of the fluorine $^3P_1$ yield to $^1P_1$ yield as a function of target gas pressure for 30 MeV F^{5+} incident on krypton, argon, and neon. Argon and neon offset 0.2 and 0.4 units for display purposes. Bottom part presents the dependence of the ratio upon initial charge state [D. L. Matthews, R. J. Fortner, and G. Bissinger, *Phys. Rev. Lett.* **36**, 664 (1976)].

where n is the particle density of the medium, v the velocity of the ion, and τ the average interaction or collision time for alteration of the state. Values of this quantity were extracted from the data in Fig. 45 based on a set of rate equations and found to be roughly equal to the geometric size of the target gas atoms. A recent calculation[134] of the soft collision probability has confirmed the agreement.

A second interesting point is evident in the lower part of Fig. 45. The excitation cross section is anomalously high for the two-electron F^{7+} ion. One explanation is the possible presence of a $1s2s(^3S_1)$ component in the beam that is excited to the $1s2p(^3P_1)$ state, which subsequently decays. The cross section for that process could be much larger than for ground-state excitation. The $^3S_1$ state is metastable, having a lifetime in

[134] J. I. Gersten, *Phys. Rev. A* **15**, 940 (1977).

fluorine much longer than the flight time from stripper foil on the accelerator to target cell, for the indicated experiment. A consideration of such metastable components of accelerator-produced ion beams is an important aspect of ion-induced collision studies, since the results can be affected so much by even small quantities ($\sim 1\%$) of the metastables if the relevant interaction is highly probable.

The various production and destruction (quenching) probabilities for an ion traversing a solid yield an equilibrium distribution of states that emit x rays during penetration of and upon emergence from solids. Spectra of the type in Fig. 15 have been utilized to determine excitation probabilities for ions that emit after they have emerged, i.e., undergo postfoil deexcitation. Those population probabilities can be correlated with empirical information on emerging charge state distributions to give estimates of the degree of excitation within specific charge states.[135,136]

8.7.2. Ions Moving in Solids

The identification of x-ray lines emitted by ions traversing solid media is based on the same notation describing emission from stationary ions. Figure 46 contains the relative K_α x-ray satellite strengths for a 3-MeV aluminum beam as a function of the Z of the host solid. The fluctuation in relative intensities is attributed to a varying probability of L-shell population established after the collision producing the K vacancy.[137] A distribution is characteristic of the equilibrium L population resulting from a balance between electron capture and loss processes, the cross sections for which depend significantly upon the specific environment. Studies of this type[138-140] have been made for a variety of ions and energies.

In addition to the usual relative measurements, the resolution furnished by a crystal also makes possible total yield measurements of groups of lines lying close in energy. Those yields can be related to probabilities of finding vacancies in a particular shell of a projectile. The separation of argon L structure from carbon K emission for argon ions moving in graphite at relatively low energies (≤ 200 keV) allowed the determination of the

[135] F. Bell, H. D. Betz, H. Panke, W. Stehling and E. Spindler, *J. Phys. B* **9**, 3017 (1976).
[136] F. Hopkins, in "Fourth Conference on Scientific and Industrial Applications of Small Accelerators," Denton, Texas, 1976, (J. L. Duggan and I. L. Morgan, eds.), p. 37. IEEE Publ. No. 76 CH 1175-9NPS (1976).
[137] A. R. Knudson, P. G. Burkhalter, and D. J. Nagel, in "Atomic Collisions in Solids," (S. Datz, B. R. Appleton, and C. D. Moak, eds.), Vol. 1, p. 421. Plenum, New York, 1975.
[138] H. Panke and F. Bell, *Phys. Lett.* **43 A**, 351 (1973).
[139] A. R. Knudson, P. G. Burkhalter, and D. J. Nagel, *Phys. Rev. A* **10**, 2118 (1974).
[140] R. J. Fortner, D. L. Matthews, and J. D. Garcia, *Phys. Lett.* **53A**, 464 (1975).

8.7. MULTIPLE-COLLISION PHENOMENA

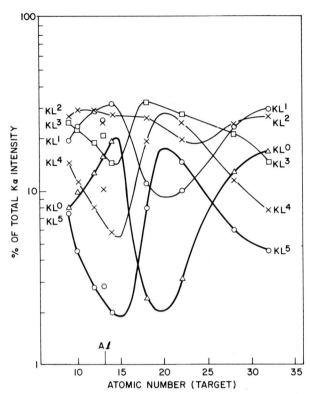

FIG. 46. Fractional intensity of KL^n components in aluminum K_α spectra from 3.0-MeV aluminum projectiles as a function of target Z [A. R. Knudson, P. G. Burkhalter, and D. J. Nagel, in "Atomic Collisions in Solids," (S. Datz, B. R. Appleton, and C. D. Moak, eds.), Vol. 1, p. 421. Plenum Press, New York, 1975.

probabilities for argon L vacancies as a function of penetration depth.[141] As seen in Fig. 47, the fraction of ions with an L vacancy reaches a maximum value of about 200 Å and then steadily decreases as the ion loses energy and slows down, with the magnitude dependent on initial incident energy. Further, the centroid of the argon L x-ray spectrum served as an indication of the M-shell population at the time of emission, which should reflect M equilibrium reached well before deexcitation. The measurement of specific quenching cross sections of the type discussed above is very difficult to do accurately, even provided high-resolution spectra. The overlapping of lines along with uncertainties about multiplet

---

[141] R. J. Fortner and J. D. Garcia, *Phys. Rev. A* **12**, 856 (1975).

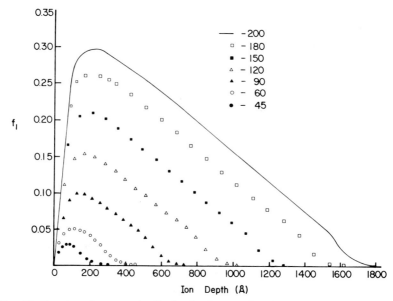

FIG. 47. Fraction of argon projectiles having L vacancies plotted as a function of depth of penetration into graphite for several different initial energies (keV) [R. J. Fortner and J. D. Garcia, *Phys. Rev. A* **12**, 856 (1975)].

strengths and fluorescence yields makes the analysis a macroscopic one at best, with information on specific individual states simply unattainable.

One very interesting feature of x-ray emission from projectiles in solids is that in certain cases the line profile is noticeably broadened. An example is given in Fig. 48 for the case[142] of 90-keV neon. The electron excited and Ne$^+$ + Ne gas spectra show the discrete satellite peaks characteristic of K$_\alpha$ emission. However, the structure from neon ions moving in copper is smooth, with no indication of even partially resolved satellites. Doppler effects due to multiply scattered beam, with the concomitant spread in emission angles, have been estimated to be negligible for this and the following cases. Similar effects have been observed for neon at 4.0 MeV penetrating a variety of solids.[136] The effective widths of lines needed to produce such effects is substantial, on the order of 10–20 eV. A more recent investigation[143] involved a specific line, the $1s^2(^1S_0)-1s2p(^1P_1)$ transition in fluorine, at higher incident energies, ≳1 MeV/amu. The observed broadening depends critically upon both

---

[142] R. J. Fortner, D. L. Matthews, J. D. Garcia, and H. Oona, *Phys. Rev. A* **14**, 1020 (1976).

[143] F. Hopkins and D. L. Matthews, *Bull. Am. Phys. Soc.* **22**, 550 (1977).

8.7. MULTIPLE-COLLISION PHENOMENA

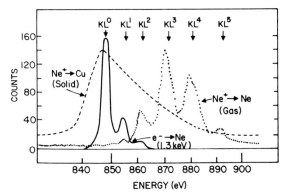

FIG. 48. Comparison of neon K_α x-ray spectra (mica crystal) produced by electron bombardment of neon gas (solid line), 90-keV neon ions on neon gas targets (dotted curve), and 90-keV neon ions on copper (dashed curve) [R. J. Fortner, D. L. Matthews, J. D. Garcia, and H. Oona, *Phys. Rev. A* **14**, 1020 (1976)].

atomic number and the structural nature of the solid, ranging from less than 1 eV for graphite to more than 10 eV for diamond, zinc, and germanium. The explanations for all of these broadened features are currently not clear. A possibility is Heisenberg broadening of the intrinsic linewidths due to very short collision times. Another is binding energy modulations due to the fact that the ions' flight is through a sea of electrons punctuated periodically by encounters with nuclear centers. Effects upon transition energies due to screening by valence electrons[144] have been indicated by measurements of energy shifts of few-electron sulfur transitions from within solids.[145] Yet another phenomenon being considered is photon wave train perturbation due to elastic scattering of target valence electrons by the moving ion.[91]

It is evident that the use of x-ray spectroscopy to determine states of ions passing through solid media is one of the few direct methods for doing so. Undoubtedly a large part of future research at accelerator facilities will be involved in this area. The capabilities of accelerators in providing different types and energies of ions make them natural tools in ion–solid interaction studies.

Acknowledgments

The author is indebted to Robert L. Kauffman and Dennis L. Matthews for helpful comments concerning the manuscript.

---

[144] W. Brandt, *in* "Atomic Collisions in Solids," (S. Datz, B. R. Appleton, and C. D. Moak, eds.), Vol. 1, p. 261. Plenum, New York, 1975.
[145] F. Bell, H. D. Betz, H. Panke, and W. Stehling, *J. Phys. B* **9**, L443 (1976).

9. ION-INDUCED AUGER ELECTRON SPECTROSCOPY*

by Dennis L. Matthews

9.1. Introduction

In ion-induced Auger electron spectroscopy (IAES) we study the electronic structure and formation dynamics of multiply ionized atoms. For example, Fig. 1 depicts the principle of IAES. A beam of ions having $Z \geq 1$ (not necessarily fully ionized) collisionally produces an inner shell vacancy in either or both the target and beam. These vacancies can be filled by a process known as the Auger effect,[1,2] first observed by Pierre Auger in 1923. The Auger effect is simply described as atomic relaxation via the emission of an electron having a discrete energy. We measure the energy, linewidths, intensities, and direction of emission of the Auger electrons. From these measurements we ascertain the probability of formation, the degree of multiple ionization, binding energies, quantum numbers, and lifetimes for highly excited states. From these results responsible ionization mechanisms can be studied.

Applications of IAES are numerous. For instance, determination of fluorescence yields and lifetimes for excited states of highly ionized atoms provides information pertinent to solving containment problems for Tokamak-type thermonuclear reactors. The determination of the magnitude of radiative cooling of the plasma core due to contaminant ions having high fluorescence yields and long lifetimes would not have been possible without both Auger and x-ray measurements on these highly ionized atoms. Furthermore, Auger yield measurements can accurately determine inner-shell vacancy fractions in ions traversing foils or dense gases. Such studies scrutinize the possibility of producing x-ray wavelength lasing from the stimulated decay of ion beams that have population

---

[1] P. Auger, *Compt. Rend.* **177**, 169 (1923).
[2] E. H. S. Burhop, "The Auger Effect and Other Radiationless Transitions." Cambridge University Press, 1952.

---

* Work performed under the auspices of the U.S. Energy Research and Development Administration under Contract No. W-7405-Eng-48.

9. ION-INDUCED AUGER ELECTRON SPECTROSCOPY

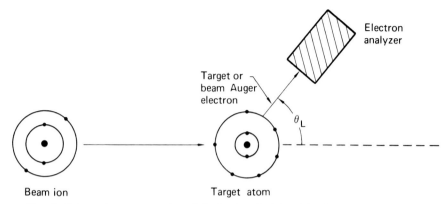

FIG. 1. Schematic representation of the basic principle of ion-induced Auger electron spectroscopy (IAES).

inversions between excited- and ground-state configurations. Spectroscopy information such as linewidths, energies, and direction of emission is also relevant to the fields of astrophysics and conventional plasma physics.

The energy range of electrons studied in this work spans ~50 to 3000 eV. Practical high-dispersion measurements in this range require electrostatic analyzers and a host of other special experimental techniques. Reviews of the Auger effect produced by both e⁻ and ion bombardment as well as general techniques for electron spectrometry have been given by Burhop,[2] Burhop and Asaad,[3] Parillis,[4] Carlson,[5] Krause,[6] Rudd and Macek,[7] Sevier,[8] Garcia et al.,[9] Richard,[10] Burch,[11] and Stolterfoht.[12,13] In this work we restrict ourselves to Auger electrons produced

---

[3] E. H. S. Burhop and W. N. Asaad, in "Advances in Atomic and Molecular Physics " (D. R. Bates and I. Esterman, eds.), Vol. 8, p. 164. Academic Press, New York, 1972.

[4] E. C. Parillis, "The Auger Effect," Acad. Sci. Uzbek SSR, Taschkent, USSR.

[5] T. A. Carlson, "Photo-Electron and Auger Spectroscopy." Plenum Publishing Corporation, New York, 1975.

[6] M. O. Krause in "Atomic Inner Shell Processes " (B. Crasemann, ed.), Vol. II, p. 33. Academic Press, New York, 1975.

[7] M. E. Rudd and J. H. Macek, in "Case Studies in Atomic Physics " (E. W. McDaniels and M. C. McDowell, eds.), Vol. 3, p. 48. North-Holland, Amsterdam, 1972.

[8] K. D. Sevier, "Low Energy Electron Spectrometry." Wiley-Interscience, New York, 1972.

[9] J. D. Garcia, R. J. Fortner, and T. M. Kavanagh, Rev. Mod. Phys. **45**, 111 (1973).

[10] P. Richard in "Atomic Inner-Shell Processes " (B. Crasemann, ed.), Vol. I, p. 73. Academic Press, New York, 1975.

[11] D. Burch, Proc. VIIIth Int. Conf. Electron. and Atom. Coll., p. 97. Invited Lectures and Progress Reports, Institute of Physics, Beograd, 1973.

[12] N. Stolterfoht, in Reference 11, page 117.

9.1. INTRODUCTION

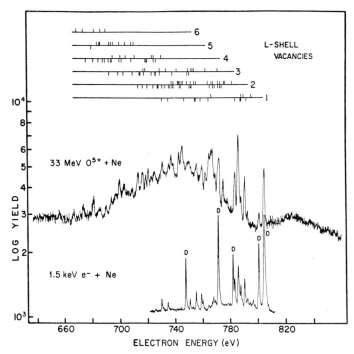

FIG. 2. Comparison of Auger electron spectrum produced by ion and electron bombardment (from D. L. Matthews, B. M. Johnson, J. J. Mackey, L. E. Smith, W. Hodge, and C. F. Moore, *Phys. Rev. A* **10** 1177, 1974; Reproduction courtesy of The Physical Review).

following ionization primarily by heavy-ion bombardment ($Z > 1$). Work in this area is relatively new and has greatly expanded our understanding of both Auger decay and heavy-ion collision mechanisms. Figure 2 best illustrates the differences in Auger spectra produced by conventional techniques (i.e., e^- bombardment) and by violent heavy ion–atom collisions. As opposed to e^- impact, the heavy ion inflicts substantial outer-shell ionization in addition to the inner-shell vacancy while also imparting significant recoil energy to the target atom. In a heavy-ion-produced spectrum, these phenomena are characterized by increased line density, decrease in line energies, and substantial line broadening (owing to the target recoil). We concentrate on relating these and other observations to the study of the theoretical descriptions of the ionization phenomena.

This part is separated into two main chapters. In Chapter 9.2 the tech-

---

[13] N. Stolterfoht, *in* "Topics in Current Physics" (I. A. Sellin, ed.), Vol. 5, Chap. 5. Springer-Verlag, Berlin, 1978.

niques behind the production and measurement of Auger electrons are discussed. For instance, a highly detailed description is given for two types of electrostatic analyzers used personally by the author. Other analyzers are only briefly treated since they have been adequately discussed elsewhere. Some discussion is also given to describing the eccentricities of ion accelerators with regard to production of ion beams suitable for our measurements. The general intent of the experimental section is to familiarize the reader with the "black magic" of simultaneous heavy-ion production and low-energy electron measurement. We hope that unnecessary guesswork can be avoided in future endeavors as a result of some of the detail given in this section. Chapter 9.3 is more conventional in that it reviews Auger emission theory and then describes a panacea of experiments, some of which are directly and others "abstractly" related to theoretical studies of ionization schemes.

9.2. Techniques for Production and Detection of Auger Electrons

9.2.1. Chambers and Detectors

To perform IAES measurements, a multitude of equipment is necessary. We simply cannot discuss each in detail, and so summarization will be the rule. In the sections that follow we will peruse the various experimenters' schematics of equipment and paraphernalia necessary for a given type of measurement.

Before discussing the key element in IAES measurements, namely, the electron analyzer, we should briefly address the problem of e^- detection and scattering-chamber design. Considering electron detection first, the most common method of counting single electrons in the 0.05–3.0 keV energy range is the continuous-channel electron multiplier. Conveniently sized and ruggedly constructed units are commercially available (Galileo Electrooptics, EMI, and Johnson Laboratories). Toburen[14] reports measured efficiencies for these windowless detectors of nearly 100% for low-energy electrons. Other investigators have also considered the efficiencies of these devices over various energy ranges. Figure 3 summarizes their results. Be sure to note that these efficiencies are much higher than those reported by Frank[15] in Sevier's[8] book on electron spectrometry. Other operational characteristics of these devices worth noting include very high gain (10^6–10^8 electrons out per incident electron)

[14] L. H. Toburen, e.g., *Phys. Rev. A* **3**, 216 (1971).
[15] L. A. Frank, *Rev. Sci. Inst.* **40**, 685 (1969).

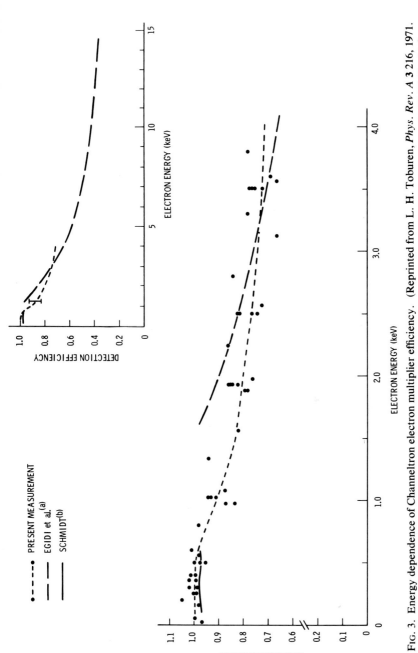

FIG. 3. Energy dependence of Channeltron electron multiplier efficiency. (Reprinted from L. H. Toburen, *Phys. Rev. A* **3** 216, 1971. Courtesy of The Physical Review.) (a) From A. Egidi, R. Marconero, G. Pizzella, and F. Sperli, *Rev. Sci. Instr.* **40** 88 (1969); (b) from K. C. Schmidt, Bendix Electro-Optics Division of Technical Applications Note 9803, 1969 (unpublished).

and very low dark-current count rate (usually less than 1 counts/sec). However, electron multipliers cannot be operated for extended periods in poor vacuum or at high counting rates without serious loss of gain, see Sevier[8] (and Part 2).

For most measurements the scattering chamber design is relatively simple. However, when using high-energy ion accelerators, beam time is usually expensive (timewise and costwise). Therefore, it becomes practical to have a multifunctional chamber. In addition, because of the small rigidity of low-energy electrons, drastic magnetic shielding of the electrostatic analyzer must be incorporated into the chamber design. It is

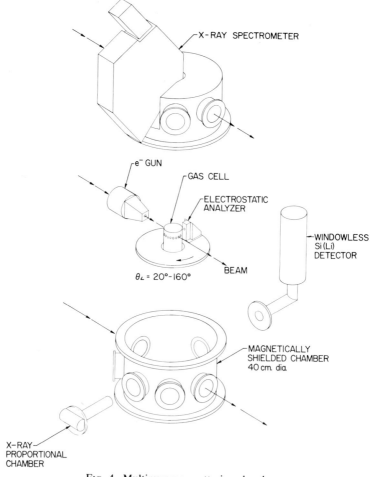

FIG. 4. Multipurpose scattering chamber.

9.2. PRODUCTION AND DETECTION OF AUGER ELECTRONS

often convenient to perform the magnetic shielding with commerically available, machinable, paramagnetic material such as permalloy, or netic–konetic (Perfection Mica Company). A versatile scattering chamber for performing simultaneous x-ray and Auger measurements using ion bombardment is shown in Fig. 4.

9.2.2. Description of Electron Spectrometers*

It is most important to describe the basic component of any electron intensity and energy measurement, namely, the electrostatic analyzer. We restrict ourselves to the basic types of electrostatic deflection analyzers. Apart from their simplicity and unlike more exotic transmission types (see Sevier[8]), deflection analyzers have the unique advantage of optically masking the e^- detector from the e^- source, thus permitting only negatively charged electrons to be analyzed. This characteristic greatly enhances the signal to background capability since ion–atom collisions copiously produce photons and high-energy charged particles, which can inundate an in-line detector.

There are four basic configurations of electrostatic deflection analyzers: the parallel-plate, cylindrical mirror, spherical, and cylindrical analyzer. Their basic design and characteristics have been extensively reviewed by Sevier,[8] Krause,[6] Eland,[16] Wannberg et al.,[17] and Burhop and Asaad.[3] We only briefly review their operating characteristics here. However, modifications that make some of them particularly appropriate for IAES are rigorously discussed.

9.2.2.1. Parallel-Plate Analyzer with Retardation Lens.
The basic parallel-plate or plane-mirror analyzer (PMA) was first described by Yarnold and Bolton[18] and later more rigorously by Harrower.[19] In its basic configuration (see Fig. 5 and ignore lenses) the PMA consists of two parallel plates held at different potentials such that electrons of the proper energy will be electrostatically reflected before reaching the back plate. Electrons of the proper energy enter the entrance slit at 45° to the plane of the plates and then follow a parabolic trajectory, returning again to the front plate and then exiting through another slit. However, electrons of improper energy follow the wrong trajectory to make it out the exit slit, and so energy selection is performed. The equation for distance between entrance and exit slits for proper analysis of electrons of energy E en-

---

[16] J. D. Eland, "Photoelectron Spectroscopy." John Wiley and Sons, New York, 1974.
[17] B. Wannberg, U. Gelius, and K. Siegbahn, *J. Phys. E* **7**, 149 (1974).
[18] G. D. Yarnold and H. C. Bolton, *J. Sci. Instr.* **26**, 38 (1949).
[19] G. A. Harrower, *Rev. Sci. Instr.* **26**, 850 (1955).

---

* See also Vol. 4A, Section 1.1.8.2; as well as Vol. 7B, Section 9.2.2.1.2.

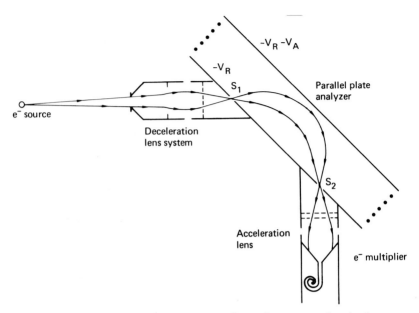

FIG. 5. Parallel-plate analyzer with energy-retarding and energy-accelerating lens setup.

tering at 45° and with angle β relative to plane of paper is

$$l = \{[2\, dE \sin(2 \cdot 45°)]/eV_p\} [1 - (\sin^2 \beta)/(\sin^2 45°)]^{1/2}, \quad (9.2.1)$$

where d is the slit separation and V_p the potential on back plate. An electron having perfect energy and trajectory (i.e., $\beta = 0°$) then travels the characteristic length $l_0 = 2\, dE/eV_p$. The analyzer constant relating potentials on analyzer plates to analyzed electron energy is given by $C = 2d/l_0$. The analyzer resolution function is given by

$$\Delta E/E = [(S_1 + S_2)/d] + A\alpha^2, \quad (9.2.2)$$

where S_1, S_2 are the entrance and exit slit widths, α the deviation of incident electron trajectory from 45°, and A a constant. Since there are no terms linear in α in Eq. (9.2.2) the PMA is said to exhibit only weak first-order focusing.

The PMA has been extensively used to measure electron production following ion–atom collisions.[13,20–23] Its principal advantages include the

[20] D. J. Volz and M. E. Rudd, *Phys. Rev. A* **2**, 1395 (1970).
[21] R. McKnight, *Rev. Sci. Instr.* **46**, 98 (1975); R. McKnight and R. G. Rains, *Phys. Rev. A* **14**, 1389 (1976).
[22] A. K. Edwards, e.g., *Phys. Rev. A* **12**, 1830 (1975).
[23] N. Lee and A. K. Edwards, *Phys. Rev. A* **11**, 1768 (1975).

9.2. PRODUCTION AND DETECTION OF AUGER ELECTRONS

ultimate simplicity of construction, compact size, and simple transmission determination. Its correctable disadvantage is that good resolution requires small acceptance angles ($\pm 3°$), which when combined with one-dimensional focusing implies very poor efficiency. Proper deceleration and focusing of incident electrons, however, will result in high efficiency and resolution. Volz and Rudd[20] first introduced a simple retarding lens system for the PMA that resulted in high-resolution measurements of argon LMM Auger electrons following H$^+$ bombardment. N. Stolterfoht and students at the Hahn-Meitner Institut in Berlin have since developed a high-efficiency electrostatic retarding lens capable of reducing incident electron energies to as low as 3 eV while actually increasing solid angle. This author has also developed a similar lens by combining the technique of computer simulation and experimental trial and error. In the following, a complete PMA system utilizing retarding lenses is described. It represents a useful and practical instrument, which has already been used by the Berlin group to perform numerous important measurements. The lengthy compilation of lens characteristics is intended for those wishing either to copy this system or to learn enough to modify their own system.

The main purpose for deceleration prior to electrostatic analysis is to take advantage of the PMA linear resolution function, i.e., $\Delta E = RE$, to gain resolving power. More specifically, since the analyzer resolution R is a constant, reducing the incident electron energy E correspondingly reduces the instrumental contribution to linewidth ΔE.

One is correct in assuming that retarding e$^-$ energies certainly does not require anything so elaborate as an electrostatic lens. In fact, a negatively charged transparent (80–90% transmission) gridded mesh placed perpendicular to the trajectory of a beam of electrons will reduce their energy by an amount equal to the potential in the mesh. However, if the electron trajectory is not perpendicular to the mesh, as will be the case for a pencil beam diverging from a source, then an enhancement in the divergence of the beam will result. Figure 6 demonstrates this phenomenon. For parallel grids separated by distance R the change in velocity due to the applied retarding force $\mathbf{F} = -e\,\mathbf{E}$ is given by

$$\mathbf{V}_f = \left(V_x^2 - \int_0^R \frac{2eE_x}{m_e}\,dx \right)^{1/2} \hat{x} + V_y\,\hat{y}. \tag{9.2.3}$$

Therefore only the velocity component antiparallel to the applied electric field is reduced. As the retardation increases toward the critical value, $V_x^2 = \int_0^R (2eE_x/m_e)\,dx$, the electron deflects more and more toward 90°. At the critical value the electron shoots off at $\mathbf{V}_f = V_y\,\hat{y}$. The obvious solution to this dilemma is to establish two-dimensional retardation, thus

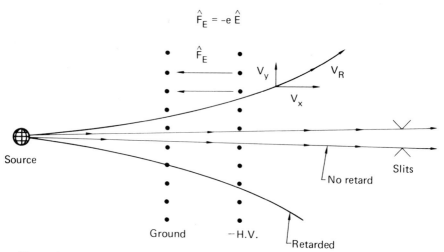

FIG. 6. Illustration of divergence of a pencil beam of electrons in a uniform retarding electric field.

creating a deceleration and focusing electrostatic lens. In fact, the development of electrostatic focusing lenses is well described by Spangenberg.[24] In addition, there are numerous accounts of combining rather elaborate electrostatic lens setups with deflection analyzers.[25,26]

For our purposes, however, a simple two-element cylindrical lens will suffice. It is also necessary to have an acceleration lens for focusing electrons exiting the analyzer into the e⁻ multiplier. Since electron multipliers can be obtained with large acceptance cones, then the acceleration focusing lens characteristics are much less crucial. As a consequence, the acceleration lens parameters are only listed so as to allot space to discuss the design of the deceleration lens system.

Referring again to Fig. 5, this time concentrate on the lens setup instead of the analyzer. The principal operating characteristics of the system are as follows. Electrons emanating from a point source are retarded in energy and focused by the deceleration lens (held at V_R) onto analyzer slits S_1. After energy analysis the electrons are then accelerated from the exit slit onto the entrance cone of the e⁻ multiplier. When a spectrum is taken, the analyzer voltage V_A is held fixed while the retardation voltage V_R is scanned. This method of voltage scanning is unique in that all de-

[24] K. R. Spangenberg, "Vacuum Tubes," McGraw-Hill, New York, 1948.
[25] N. Oda, S. Tahira, F. Nishimura, and F. Koike, *Phys. Rev. A* **15**, 574 (1977).
[26] G. E. Chamberlain, S. R. Mielczarek, and C. E. Kuyatt, *Phys. Rev. A* **2**, 1905 (1970).

9.2. PRODUCTION AND DETECTION OF AUGER ELECTRONS

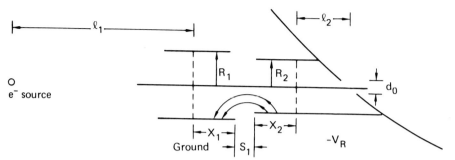

FIG. 7. Definition of relevant lens parameters: analyzer slit to e⁻ source distance remains fixed; d_0 is the slit width, l_1 the source to first grid distance, l_2 the second grid to slit distance (remains fixed), X_1 the active length of grounded element of retarding lens, X_2 the active length of high-voltage element, R_1 the radius of grounded element, R_2 the radius of high-voltage element, and S_1 is the lens element spacing (~1 mm).

flection analysis takes place at a fixed energy; hence the spectrometer contribution to linewidth is constant although energy is scanned by retardation. Note that the use of gridded lenses is very important (as will be seen shortly) for they precisely shape and clamp electrostatic field lines so as to prevent stray fields in the analyzer and other sensitive areas.

A schematic of the deceleration lens system is shown in Fig. 7. We demonstrate with computer simulation that a large enhancement in the normal solid angle intercepted by the analyzer slits can be obtained while also retarding the incident electron energy. For our setup the distance $l_1 = 83.9$ mm, $l_2 = 8.4$ mm, and $S_1 = 1.0$ mm are fixed mainly for convenience (S_1 should remain small[24]). An understanding of the operation of this lens can be qualitatively described by following the electrostatic forces exerted on an electron moving off axis. In Fig. 7 the curved electric field force lines act in a direction labeled by the arrows. As is obvious from the figure, there are regions in the lens where the electron encounters divergent and convergent lens action (i.e., electrical forces, respectively, away from and toward the axis of symmetry. We control the extent of divergence to convergence by changing lens parameters, such as spacing between grids or element radii. The proper ratio of divergence to convergence will then result in imaging a large solid angle from the source onto the analyzer entrance slit.

Figure 8 represents a computer simulation[27] of electron trajectories when retarding from 200 to 48 eV in a lens having parameters (in milli-

---

[27] This computer code, which solves for potentials and plots electron trajectories, is available from the author upon request.

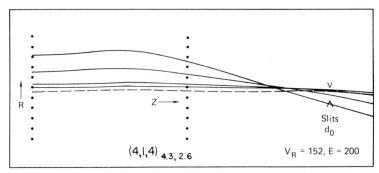

FIG. 8. Example of the focusing action of a lens having elements of different radii. The lens parameters are $(X_1, S_1, X_2)_{R_1,R_2}$ (mm) with incident electrons of energy 200 eV retarded by 152 V to 48 eV. For this and other lens trajectory plots we only show the region from first grid to analyzer slits. The trajectories shown are for divergence from a point source at angles of 1×, 3×, 5×, 11×, and 15× the normal acceptance half-angle of analyzer (0.09°).

meters) $(X_1, S_1, X_2)_{R_1,R_2} = (4,1,4)_{4.3,2.6}$. These trajectories use (r, θ, z) as coordinates for five independent initial trajectories. The trajectories are 1×, 3×, 5×, 11×, and 15×, vertical acceptance half-angle (0.09° when using 0.25-mm entrance slit), and are spaced in that order radially outward on the figure). To reduce the computer memory necessary for the calculation the trajectory plots are started at the first grid (in the direction from source to slit) in the lens and continued to the entrance slits. Actually it is unnecessary to calculate the trajectory except between the two grids since the e^- is otherwise in a field-free region and should follow a straight line trajectory. Notice that all but the 15× trajectory make it into the analyzer entrance slit. Also note that the lens action is slightly too convergent since the trajectories cross the $r = 0$ symmetry axis slightly before the entrance slit. The lens parameters chosen for this simulation are identical to those used in an actual lens. In experiments with the actual lens, ~200 eV electrons from the argon LMM–Auger transition (L_3–$M_{2,3}$, $^1S_0$ at 201.1 eV; see Werme et al.[28]) were detected with and without 152 V of retardation. Thus, electrostatic analysis was made at 48 eV when retarding as in the computer simulation. We made the observation that the counting rate with retardation exceeded that without by a factor of ~2. If we take account of the E^{-1} transmission factor for this analyzer[8] this implies that $(200/48) \times 2 = 8.4$ times the normal (as retarding) solid angle is being focused into the analyzer entrance slits when retarding. The calculated trajectories in Fig. 8 agree reasonably well with

[28] L. O. Werme, T. Bergmark, and K. Siegbahn, *Phys. Sci.* **8**, 149 (1973).

experiment, since the 11× trajectory is the last to make it into the analyzer slit. Actually the agreement may have been better since an aperture present between the analyzer and target region collimated out some of the larger acceptance angles. Of course, the simulated trajectories are only estimates since uncertainties in the field calculation as well as our inherent inability to truly simulate the electron environment (e.g., no stray magnetic or electric fields are included in the calculation) always remains a problem.

Making the assumption that the previous agreement between experiment and simulation is nonfortuitous, we used the computer simulation to design a set of optimum lenses for a variety of experimental requirements. For example, Fig. 9 shows the trajectories for a lens designed to retard electrons to 5 eV from an initial energy of 200 eV. Note that trajectories up to 11× are focused into the entrance slit. The lens parameters are $(4,1,3.5)_{4.1,4.1}$ where both lens elements have the same radii. The reason for this is discussed later. This lens configuration has also been checked experimentally. Ten percent of the no retardation counting rate was observed at $V_R = 196$ V, again in reasonable agreement with the solid angles predicted by the computer simulation.

The lens shown in Fig. 9 is very sensitive to the second lens element dimensions. In addition, it cannot be used with equal success for retarding electrons to 5 eV that have energies much different than 200 eV (± 50 eV before poor focusing reduces counting rate). A dramatic demonstration of the effects of different second grid to lens gap spacing, X_2, for this lens is shown in Fig. 10. In summary for a given retarding voltage the effect of *decreasing* X_2 is to *decrease* the focusing power of the lens (i.e., increase focal length). In addition, the effect of having dissimilar lens element

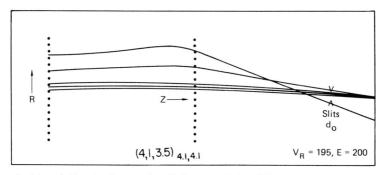

FIG. 9. A special lens having equal radii elements designed for retardation to very low energies with maximum efficiency. Note that all but the 15× normal trajectory make it into analyzer slit.

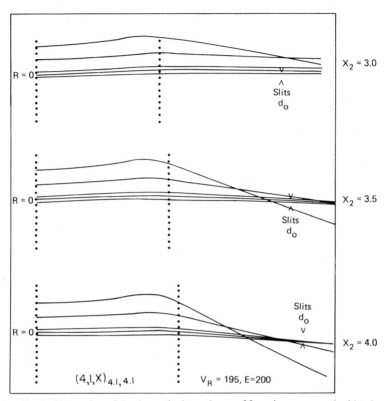

FIG. 10. An illustration of the dramatic dependence of focusing on second grid to lens gap spacing X_2. In general, a smaller X_2 results in a weaker lens.

radii is shown in Fig. 11. For a given retarding voltage $R_2 \leq R_1$ for convergence (in our experimental setup!). In general, the smaller R_2 to R_1 ratio the more convergent is the lens action.

Except for the lens described in Fig. 8 we have only discussed a lens capable of reducing electron energies to very small values. Moreover, the higher the initial unretarded energy the more efficiency we lose with retardation since the solid angle gained cannot compensate for the change in analyzer transmission function (remember that transmission is proportional to analyzed energy). Fortunately, most ion–atom collisions result in kinematical broadening of Auger lines (see Dahl,[29] Matthews et al.,[30]

[29] P. Dahl, M. Rødbro, B. Fastrup, and M. E. Rudd, *J. Phys. B* **9**, 1567 (1976).
[30] D. L. Matthews, B. M. Johnson, J. J. Mackey, L. E. Smith, W. Hodge, and C. F. Moore, *Phys. Rev. A* **10**, 1177 (1974).

9.2. PRODUCTION AND DETECTION OF AUGER ELECTRONS

Garcia *et al.*[9]). It therefore becomes practical to design a moderate resolution analyzer system with very high efficiency, a needed characteristic for studying some ion–atom collisions where beam currents are small (such as with a tandem Van de Graaff). A lens system that retards to ~20 eV seems practical since spectrometer contribution to linewidth will then be (0.8%) (20 eV) = 160 meV, which is reasonable and certainly much less than the kinematically broadened linewidths (~1 eV) for the collisions of interest. In addition, we can achieve very high efficiency for retarding electrons of incident energies from 100 eV to 1 keV.

Figure 12 shows a compilation of computer-designed high-efficiency lenses for reducing 100–800 eV electrons to 20 eV. At the lowest incident energies, considerably more focusing power is needed because of the

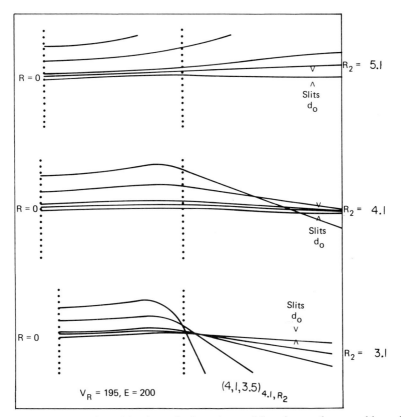

FIG. 11. An illustration of the dramatic dependence of focusing on the second lens element radius. In general, the smaller the R_2/R_1 ratio, the more convergent is the lens. These three trajectory plots are not plotted on the same horizontal scale.

small field gradients. Therefore, the second lens radius was reduced to 3.0 mm for the 100 eV incident energy. At the highest energies just the opposite problem exists, so that X_2 was decreased to curtail the focusing power of the lens. The results of this design give a set of lenses that have phenomenal focusing ability over the energy range of interest ($15\times$ normal in $V_R = 80$–180 V range and $\sim 12\times$ normal at the $V_R = 480$–780 V range). Unfortunately, to obtain maximum efficiency these lenses do require in one case (at $V = 80$) the changing of the second element radius and the distances X_1, X_2 at some other retarding voltages. However, the lens elements can be constructed such that this change re-

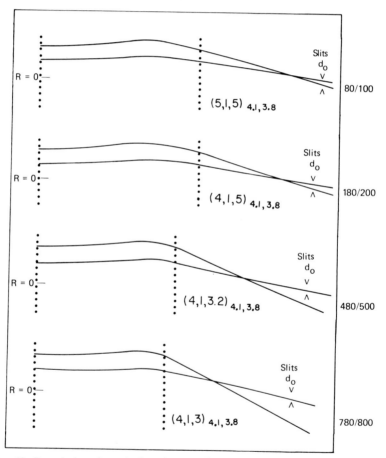

FIG. 12. Best designs for reducing electrons from 100, 200, 500, and 800 eV to 20 eV. Only the $11\times$ and $15\times$ trajectories are plotted.

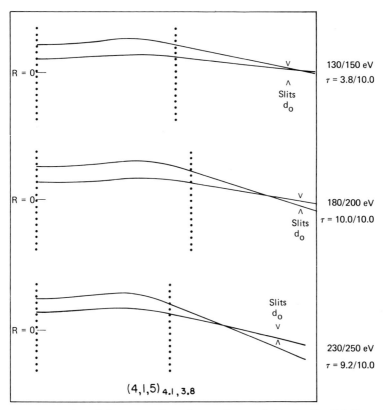

FIG. 13. Illustration of the energy dependence of transmission for the $(4,1,5)_{4.1,3.8}$ lens when used to retard electrons from 150, 200, and 250 eV to 20 eV. Only the 11× and 15× normal acceptance trajectories are shown. The transmission τ is tabulated. Note the sharp decrease for lower energies than optimum as compared to higher energies.

quires a minimum of effort. In addition, for most high-resolution Auger spectra the energy spanned in a spectrum is never more than 100 eV. Figure 13 shows the $(4,1,5)_{4.1,3.8}$, which is optimized for 200 eV operated at $V_R = 150$ and 250 V so as to bracket the optimum by 100 eV. For this case the efficiency only varies by $\sim 60\%$ over the entire energy range. In fact, the biggest variation occurs at the lowest energies retarded. Surprisingly, for the lens optimized for 800 eV there is little change in transmission between 500 and 1000 eV. For these higher energies this relatively flat transmission has also been confirmed experimentally[31] when reducing neon K-Auger electrons from 800 to 20 eV.

[31] N. Stolterfoht, H. Gabler, and U. Leithauser, *Phys. Lett.* **45A**, 351 (1973).

As promised earlier, we briefly discuss the acceleration lens system. First, however, it might also be useful to just quote the computer-designed lens parameters necessary for high efficiency when analyzer slit size is increased, e.g., from 0.23 to 1.0 mm. This change to larger analyzer slits may be desirable for obtaining more total solid angle, i.e., efficiency. For retarding 200 eV electrons to 20 eV a lens with parameters $(5,1,5)_{8.1,6.5}$ focuses well up to 11×; however, 15× is grossly overfocused. Note the main changes in the lens are the radii of both elements of the grid. The lens for trajectories less than 15× is fairly weak, and so it will possibly work better for higher retarding voltages or for reducing the energy of electrons having higher initial (unretarded) energies. Little experimental development of a lens of this size has taken place at out laboratory, and so one may have to juggle the parameters somewhat to optimize its performance.

The postanalysis acceleration lens for our analyzer (using 0.25-mm slits) is shown in Fig. 14. Its focusing characteristics are similar to the deceleration lens, except that the second half of the lens is actually the entrance cone to the e^- multiplier. Fortunately, the field lens necessary to focus electrons into the cone is attractive and therefore there is little problem directing electrons onto the 4.5-mm radius of the multiplier collector cone. We have also added a "clean-up grid" that is biased slightly negative (2–3 V) relative to the analyzer slits. This prevents low-energy secondary electrons produced at the exit slits (and elsewhere) from reaching the electron multiplier.

Before demonstrating the performance of the plane-mirror analyzer it is convenient to show a schematic of the electronics necessary for its opera-

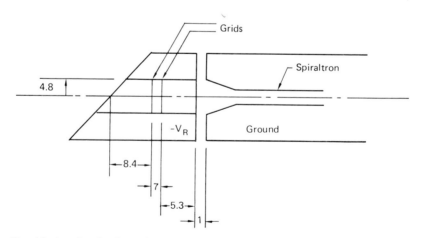

FIG. 14. Acceleration lens with relevant dimensions (in mm) and characteristics.

9.2. PRODUCTION AND DETECTION OF AUGER ELECTRONS

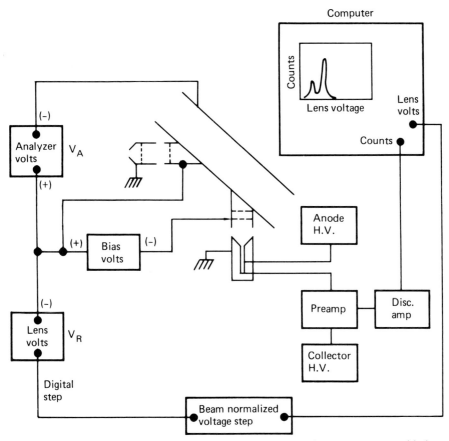

FIG. 15. Schematic illustration of electronics necessary to make a measurement with the PMA.

tion. Figure 15 is such a representation. The process of measuring an electron energy spectrum is straightforward. First, the fixed analysis voltage V_A is set to pass some electron energy E_A (eV) $\sim C V_A$ (V), where $C = 1.6$ for our analyzer. If there were no retardation, then electrons of energy E_0 would be incident upon the analyzer entrance slit. However, prior to arriving at the entrance slits the desired energy electrons are retarded from E_0 to a constant energy E_{0R} by a retarding voltage V_R. Thus E_A is set such that $E_A = E_{0R}$. The maintaining of a constant E_A can only be achieved by scanning V_R. In practice, V_R is scanned by digital to analog power supply. The resultant measured electron counting rate per increment ΔV_R is then stored in the memory of a computer. Display of the data reveals a histogram of total measured electron counts per increment

452   9. ION-INDUCED AUGER ELECTRON SPECTROSCOPY

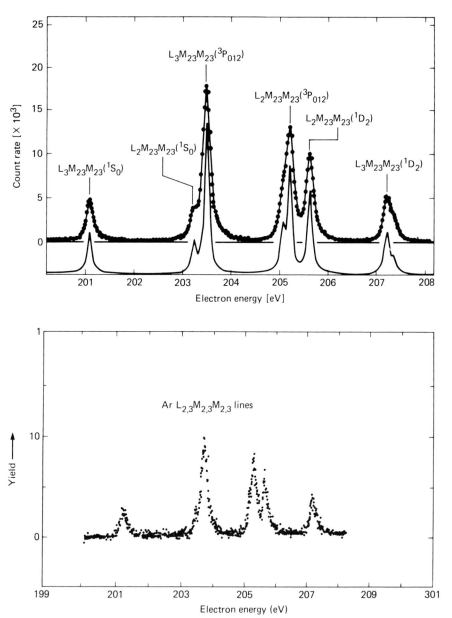

Fig. 16. Comparison of the argon $L_{2,3}M_{2,3}M_{2,3}$ lines measured with the Hahn–Meitner group's PMA analyzer (top) and the one developed by this author.

9.2. PRODUCTION AND DETECTION OF AUGER ELECTRONS   453

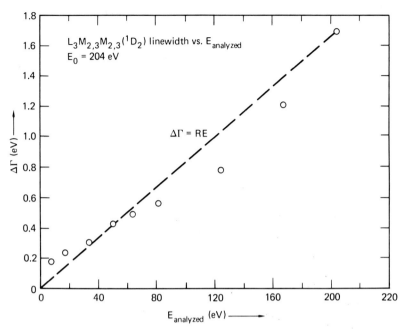

FIG. 17. Measured variation in instrumental linewidth (open circles) with retarding voltage as compared to theoretical prediction $\Delta\Gamma = RE$.

of kinetic energy. The amount of time spent at each energy interval is normalized to the number of beam particles that have passed through the collision region. The yield of electrons at a given energy is then a measure of the probability of producing them per beam ion.

Now we can consider the performance of the parallel-plate analyzer system. Using a $(4,1,3.5)_{4.1,4.1}$ lens and 0.25 mm slits the linewidths of the argon $L_2-M_{2,3}^2$ and $L_3-M_{2,3}^2$ normal Auger lines were measured following 1.5-keV e^- bombardment. The electrons were retarded to 5 eV before analysis. Figure 16 compares our spectrum with one taken by Ridder and co-workers[32] using the Hahn–Meitner retarding lens PMA system (at 3-eV analysis energy for their spectrum). The instrumental contribution to linewidth is small ($\sim 50-70$ MeV) in both cases; therefore, the spectral linewidths are mostly due to the natural linewidth. This spectrum serves to demonstrate the ultimate resolving power capabilities for the system.

In Figure 17 linewidths are plotted for the argon $L_3-M_{2,3}^2$ ($^1D_2$) Auger line at 203.47 eV (from Ref. 28) as a function of analyzed energy. The deviation from the constant resolution line $\Delta E_s = RE$ is troublesome. The deviation at the lowest analyzing energies (highest retarding voltages) is

[32] D. Ridder, J. Dieringer, and N. Stolterfoht, *J. Phys. B* **9**, L1 (1976).

ignored because the natural linewidth is dominating. However, above ~60 eV analyzed energy the measured linewidth is at all times smaller than predicted by the known resolution function. This effect probably occurs because focused trajectories do not resemble the trajectories integrated by the slits in the unretarded case. Evidently the smaller spread in incident angles that accompanies the well-formed e⁻ trajectories intercepted in narrower than predicted lineshapes (see Rudd in Sevier[8] for a discussion of normal lineshapes). By the same process a grossly overfocused set of e⁻ trajectories can produce spectral lineshapes larger than predicted. Hence due caution should be exercised in extrapolating analyzer linewidths outside regions where good focusing is known to exist.

High-resolution measurements of neon K-Auger electrons have also been performed with the $(4,1,4)_{4,1,4,1}$ lens. The linewidths associated with ~800-eV Auger lines are even less than those reported by Matthews and co-workers[30] using an analyzer having an intrinsic resolution of 0.02% or $\Delta E_s = 160$ meV at these energies. The efficiency is very poor at 800 eV; however, as can be expected from the earlier computer simulation of the $(4,1,4)_{4,1,4,1}$ lens. Future measurements will be performed with one of the higher-efficiency lenses presented earlier. In fact, the Hahn–Meitner group has already reported (see Ref. 31) higher-efficiency measurements when retarding 800 eV neon K-Auger electrons to 20 eV.

9.2.2.2. Cylindrical-Mirror Analyzer. The cylindrical mirror analyzer (CMA) shown schematically in Fig. 18 has been used in many measurements[33-37] of Auger electrons produced in ion–atom collisions. This analyzer's popularity stems from its high luminosity (or efficiency) since almost the entire azimuthal angle 2π is accepted. For measurements where extremely high transmission and luminosity are required at the expense of resolution, the CMA is the ideal choice of analyzer.

The CMA was first introduced by Blauth[38] and Mehlhorn.[39] Its focusing properties have been rigorously studied by Zashkvara et al.,[40] Sar-el[41] and Aksela.[42] Referring to Fig. 18, in operation a negative potential ΔV_0 is applied between cylinder r_2 and r_1 so as to reflect and image

---

[33] C. W. Woods, R. L. Kauffman, K. A. Jamison, N. Stolterfoht, and P. Richard in *Phys. Rev. A* **13**, 1358 (1976).

[34] C. W. Woods, R. L. Kauffman, K. A. Jamison, N. Stolterfoht, and P. Richard, *Phys. Rev. A* **12**, 1393 (1975).

[35] K. O. Groeneveld, G. Nolte, S. Schumann, and K. D. Sevier, *Phys. Lett.* **56A**, 29 (1976).

[36] I. A. Sellin, *in* "Topics in Current Physics" (I. A. Sellin, ed.), Springer, Berlin, 1978.

[37] R. Bruch, Doctoral Dissertation, University of Berlin, July 1976.

[38] E. Blauth, *Z. Phys.* **147**, 228 (1957).

[39] W. Mehlhorn, e.g., *Z. Phys.* **160**, 247 (1960); *Z. Phys.* **208**, 1 (1966).

[40] V. V. Zashkvara, M. I. Korsunskii, and O. S. Kosmachev, *Zh. Tekhn. Fiz.* **36**, 132 (1966). *Sov. Phys.-Tech. Phys.* **11**, 96 (1966).

9.2. PRODUCTION AND DETECTION OF AUGER ELECTRONS

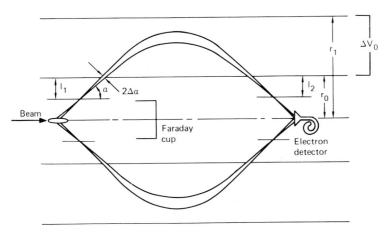

FIG. 18. Cylindrical mirror analyzer schematic.

electrons defined by slits S_1 onto the detector slit S_2. Electrons of energy E_0 that leave the slit S_1 at angle α reach S_2 after a distance L given by[42]

$$L/r_0 = [(l_1 + l_2)/r_0] \cot \alpha + 2(K\pi)^{1/2} \cos \alpha \exp(\sin^2 \alpha) \exp(K^{1/2} \sin \alpha), \quad (9.2.4)$$

where $K = (E_0/\Delta V_0) \ln(r_1/r_2)$. This device has second-order focusing when $\partial^2 L/\partial \alpha^2 = 0$, which occurs for $l_1 + l_2 = 2r_0$, $K = 1.3098$, and $\alpha = 42.3°$. The dispersion function D, which is related to energy resolution, is given by $D = E_0(\partial L/\partial E)$. The base energy resolution $R_b = \Delta E_b/E$, which is the maximum energy spread at a given E accepted by the analyzer, was determined by Risley[43] to be

$$R_b = [(S_1 + S_2)/D] + [\Delta L(\pm \Delta \alpha)/D] + (r_c/r_i)^2, \quad (9.2.5)$$

where r_c is the average radius of the source. The three terms in R_b are, respectively, the instrumental broadening due to finite source and detector length, finite angular acceptance, and off-axis electrons having nonzero angular momenta. The analyzer transmission function is given by $T = \sin \alpha \, \Delta \alpha$.

Unfortunately, the full-sized CMA has several disadvantages that restrict its user to Auger cross section determination or total yield measurements. As usual, high resolution and high efficiency are not compatible. A resolution of only about 1.4% seems to be all that is practical for ion–atom collision experiments. As with the parallel-plate analyzer, preretardation of incident electrons can be applied[44] to improve ef-

---

[41] H. Sar-el, *Rev. Sci. Instr.* **38**, 1210 (1967); *Rev. Sci. Instr.* **39**, 533 (1968).
[42] S. Aksela, *Rev. Sci. Instr.* **42**, 810 (1971).
[43] J. S. Risley, *Rev. Sci. Instr.* **43**, 95 (1972).

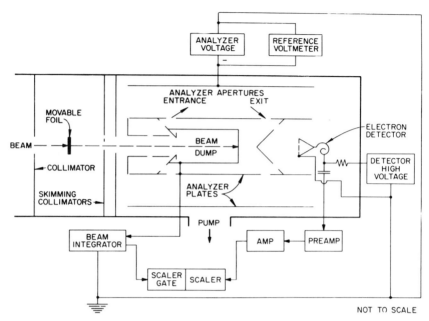

FIG. 19. Cylindrical mirror analyzer experimental setup from I. Sellin, *Nucl. Instr. Meth.* **110**, 477 (1973) (courtesy of North-Holland Publ. Co.).

fective resolution. However, the transmission is reduced since the special geometry of the analyzer makes it awkward to design a retardation lens to match the emittance from a small source to the enormous acceptance angle of the spectrometer. Another disadvantage is that the target is submerged within the analyzer and the takeoff angle fixed at 42.3°. This means that angular distribution measurements cannot be performed. The measurement of Auger emission at the forward scattering angle of 42.3° is also jeopardized by the large secondary electron background present in the forward hemisphere. This is especially true for heavy-ion bombardment, where electrons "knocked off"[45] the projectile further contribute to the secondary electron background using a special geometry. Woods and co-workers[33] have used a CMA to detect at back angles but the take-off angle is not single-valued since it spans from 90 to 174°. Therefore, angular distribution measurements are still not possible. One further disadvantage is encountered when ion beams are stopped inside the analyzer. This practice can create nuclear or bremsstrahlung radiation, which subsequently produces electrons, thus creating a significant background in the electron detector. However, Schumann and Groene-

[44] K. Maeda, "Electron Spectroscopy," p. 177. North-Holland, Amsterdam, 1972.
[45] D. Burch, H. Wieman, and W. B. Ingalls, *Phys. Rev. Lett.* **30**, 823 (1973).

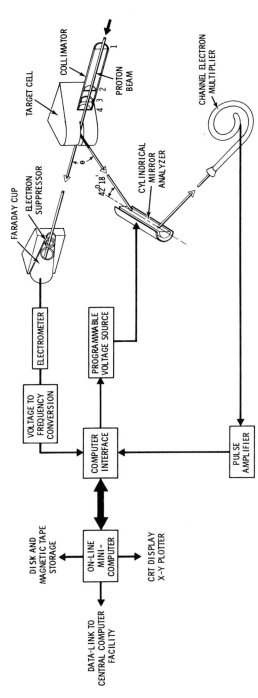

FIG. 20. Half-cylindrical mirror analyzer setup for performing angular distribution measurements from L. H. Toburen, *Phys. Rev. A* **3**, 216 (1971) (courtesy of The Physical Review).

veld[46] at Frankfurt have circumvented this problem by making an annular ring of e⁻ multipliers, thus allowing the beam to pass through the analyzer.

Beam–foil Auger measurements have particularly benefitted from the standard CMA. Figure 19 is a schematic of the beam–foil Auger apparatus used by Sellin and co-workers[47] at Oak Ridge National Laboratory. Similar systems have also been developed by Bruch[37] and Groeneveld.[46] The on-axis geometry of the CMA particularly suits the beam–foil experimentalist's need for a movable exciter foil geometry as well as detection of electrons emitted in the forward hemisphere (at 42.3°). The forward angle position of measurement is often necessary when viewing the in-flight Auger decay of an ion beam. If the Auger electron velocity is less than the velocity of the ion then no electrons will be observed for angles larger than a derivable angle (see Sellin[47] for details).

It must be conceded that Auger electron angular distribution measurements can be performed with the CMA if only half the cylindrial volume is used. Spectrometers composed of only half-cylindrical electrodes have been successfully used by Risley,[43] Burch et al.,[48] and Toburen.[14] Figure 20 shows a system used by Toburen to measure electrons ejected at angles from 20 to 130°. The input acceptance angle is also reduced to 5° so that much of the original efficiency is not utilized.

9.2.2.3. The Spherical-Sector Analyzer. The spherical analyzer was originally developed by Purcell.[49] It is essentially a half-spherical capacitor that provides second-order focusing when the electrons transverse 180°. The resolution function is given by

$$\Delta E/E = [(S_1 + S_2)/2\bar{r}] + \alpha^2. \qquad (9.2.6)$$

Thus, at a given slitwidth, the resolution is proportional to $1/\bar{r}$, where $\bar{r}$ is the average radius of curvature for the electron path through the analyzer. For S_1 and S_2 of unreasonable size the resolution can be made as good as desired simply by increasing the electrode sizes, hence increasing $\bar{r}$. Striving for high resolution has typically led to production of a gigantic analyzer that is clumsy to use in a crowded laboratory. Angular distribution measurements are also seriously impaired by the physical dimension and geometry of the analyzer.

Although extensively used in other types of research, the spherical analyzer has only recently been used in ion–atom collision experiments.[50]

---

[46] S. Schumann and K. Groeneveld, Frankfurt University, private communication.
[47] I. Sellin, *Nucl. Instr. Meth.* **110**, 477 (1973).
[48] D. Burch, W. B. Ingalls, J. S. Risley, and R. Heffner, *Phys. Rev. Lett.* **29**, 1719 (1972).
[49] E. M. Purcell, *Phys. Rev.* **54**, 818 (1938).
[50] M. M. Duncan and M. G. Menendez, e.g., *Phys. Rev. A* **13**, 566 (1976); M. C. Menendez and M. M. Duncan, *Phys. Lett.* **54A**, 409 (1975).

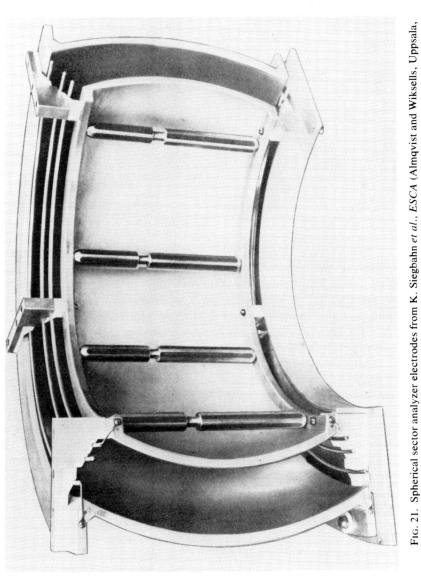

Fig. 21. Spherical sector analyzer electrodes from K. Siegbahn et al., ESCA (Almqvist and Wiksells, Uppsala, 1967) (courtesy of Almqvist and Wiksells Publ. Co.).

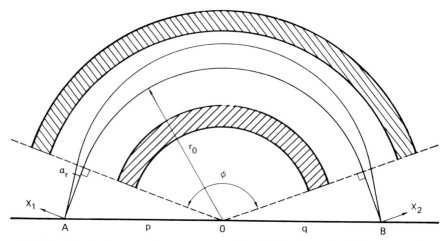

FIG. 22. Spherical sector analyzer parameters from B. Wannberg, U. Gelius, and K. Siegbahn, *J. Phys. E* **7**, 149 (1974) (with permission, The Institute of Physics, Techno House, Bristol).

Interestingly, a simple variation on the spherical capacitor type known as the spherical sector analyzer has been extensively used for Auger electron measurements[30,51,52] following ion–atom collisions. From the work of Siegbahn et al.,[53] and Wannberg et al.,[17] respectively, Fig. 21 shows a picture of the spherical sector electrodes, while Fig. 22 portrays a schematic view of the relevant geometry and parameters.

The spherical sector analyzer differs from the normal spherical analyzer only in the ϕ angle circumscribed by the electrodes. It was discovered that the use of smaller electrodes did not seriously affect the resolution available with the full spherical analyzer. Siegbahn et al.[53] have popularized this instrument by demonstrating its outstanding general performance. Typical orbit radii $\bar{r}$ in units operating have been 36 cm with angle ϕ optimized at 157.5°. A commercial unit available from the McPherson Inst. Co. having $\Delta E/E = 0.0002$ has been used by Matthews et al.,[30] Mackey et al.,[51] Johnson,[52] and Schneider et al.,[54] who all performed measurements at the University of Texas. The Texas groups performed their measurements with the vertical acceptance slit perpendicular to the ion beam axis, thus only viewing a small segment of the beam–target interaction path. Even so, healthy counting rates

---

[51] J. J. Mackey, L. E. Smith, B. M. Johnson, C. F. Moore, and D. L. Matthews, *J. Phys. B* **7**, L447 (1974).
[52] B. M. Johnson, Doctoral Dissertation, University of Texas at Austin, 1975.
[53] K. Siegbahn et al., ESCA, Almquist and Wiksells, Uppsala, 1967.
[54] E.g., D. Schneider, C. F. Moore, and B. M. Johnson, *J. Phys. B* **9**, L153 (1976).

9.2. PRODUCTION AND DETECTION OF AUGER ELECTRONS

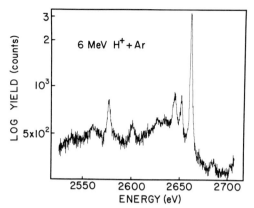

FIG. 23. Argon KLL–Auger spectrum from the work of J. J. Mackey, L. E. Smith, B. M. Johnson, C. F. Moore, and D. L. Matthews, *J. Phys. B* **7**, L447 (1974) (with permission, The Institute of Physics, Techno House, Bristol).

were observed with the small-particle currents (~10 nA) available on tandem Van de Graaffs. Figure 23 shows a dramatic test of the efficiency and resolution of this analyzer. This K-Auger spectrum[50] was taken with only a few micro-Coulombs of proton beam ionizing a dilute gas target. It demonstrates that even at high electron energies this analyzer is capable of good resolution and efficiency. Another advantage of

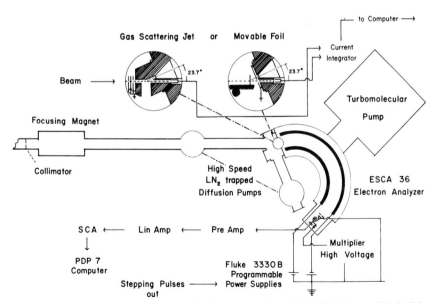

FIG. 24. Experimental setup for spherical sector analyzer from B. Johnson, Ph.D. Dissertation, Univ. Texas, Austin 1975.

this type of analyzer is realized when performing beam Auger measurements. With slit operated perpendicular to beam direction the spherical sector analyzer has a very small horizontal acceptance angle—thus limiting kinematical broadening. The principal disadvantages of the analyzer stem not only from its monstrous size and geometry (eliminating angular distribution measurements) but also the inability to turn off high resolution so a quick scan can be made of spectral features such as total intensity. Its physical drawbacks at angles other than $\theta_L = 90°$ can be overcome with a clever experimental setup. Figure 24 shows such a system by Johnson[51] for performing beam–gas or beam–foil Auger measurements at $\theta_L = 23.7°$.

Generally speaking, the spherical sector analyzer is the best possible choice for purely spectroscopic measurements at one observation angle (θ_L usually 90° by design) owing to its inherent high resolution and efficiency. Thus there is no need of preretardation lenses or other complicated electron optics.

9.2.2.4. Cylindrical Analyzer. Probably the oldest form of electrostatic analyzer still in use is the cylindrical analyzer first developed by Hughes and Rojansky[55] in 1929. It is also called the 127° or $\pi/\sqrt{2}$ ana-

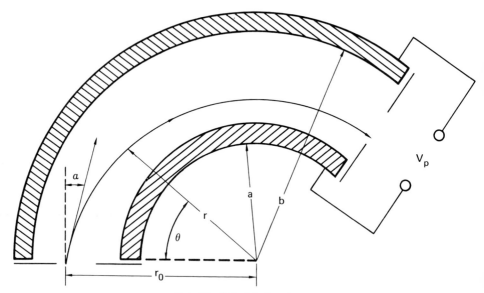

FIG. 25. Cylindrical analyzer.

[55] A. L. Hughes and V. Rojansky, *Phys. Rev.* **34**, 284 (1929).

lyzer since its developers took advantage of the fact that first-order focusing of electrons takes place every 127°17' in a radial inverse first-power electric field. A schematic of this analyzer is shown in Fig. 25. The analyzer resolution function[8] is given by

$$\Delta E/E = [(S_1 + S_2)/\bar{r}] + \tfrac{4}{3}\alpha^2 + \beta^2, \qquad (9.2.7)$$

where β is the angular divergence perpendicular to analysis plane. Since the resolution depends on $1/\bar{r}$ instead of $1/2\bar{r}$ as in the case of the spherical analyzer, then it requires twice as large an analyzer to get the same brute-force resolution. This necessitates development of retarding lenses to achieve what was done for the PMA. The principal advantages of this type of spectrometer are its simplicity and compact size ($\bar{r}$ typically 5–10 cm). Its lack of double focusing (resulting in poor efficiency) is its only shortcoming. In fact, Rudd and co-workers (see review in Ref. 7) initiated the field of IAES by using this type of analyzer to study proton-induced Auger spectra.

9.2.2.5. *Efficiency Function for Deflection Analyzers.* One common characteristic of all electrostatic deflection analyzers is their efficiency variation with analyzer energy. This subject was discussed by Rudd (in Sevier[8]). However, it is important to again state that any electrostatic deflection analyzer (having *linear* dispersion) will have a transmission function that varies in proportion to the analyzer energy. This phenomenon stems from the fact that all these analyzers have constant resolution with energy $\Delta E/E$ = const. More explicitly, to determine electron production cross section σ_A at a given lab angle, the expression for electron yield per unit E is

$$Y(E) = \frac{d\sigma_A}{dE}\left[\frac{\Delta\Omega}{\Omega}\,TlPl\right]\Delta E, \qquad (9.2.8)$$

where T is analyzer transmission, $\Delta\Omega/\Omega$ the solid angle intercepted by the analyzer, for target length l at pressure P with incident beam current I (particle/sec), and ΔE the energy window (transmission function) of the analyzer at a given E. Since $\Delta E = RE$ when R is the resolution constant and letting $U = [TlPl\,\Delta\Omega/\Omega]$, then $Y(E) = U\,(d\sigma/dE)RE$, which implies that the measured count rate depends on E. A cross section σ_A is then determined from the expression

$$\sigma_A = \frac{1}{UR}\int^E \frac{Y(E)\,dE}{E} \cong \frac{\Delta E_i}{UR}\sum_i \frac{Y(E_i)}{E_i}, \qquad (9.2.9)$$

provided dY/dE is a slowly varying function compared to the energy interval ΔE between successive analysis energies E_i in a spectrum.

9.2.3. Fast-Ion Accelerator Peculiarities

It is instructive to review some of the characteristics of the common types of electrostatic ion accelerators. The discussion is confined to the three most commonly used ion accelerators: (1) high-voltage ion sources, (2) Van de Graaff accelerator, and (3) tandem Van de Graaff accelerator. A thorough review of some of these accelerators has been given in Vol. 5, Part 1, of this treatise, as well as a recent review article,[56] and so we merely address those characteristics that influence IAES.

A schematic of an IAES experiment is shown in Fig. 26. For our measurements we need a versatile ion beam source capable of producing a geometrically small beam of ions of atomic number Z, having charge Q, energy E, and intensity I for injection into the scattering chamber. The choice of type of accelerator depends on the parameters Z, Q, E, and I necessary for a given measurement. Shortly the optimum accelerator for various regimes of those parameters is discussed. First, however, it is important to discuss the vacuum and magnet requirements for transporting and bending heavy ion beams. Particularly for all ions heavier than protons, it is important to maintain very high vacuum in beam transport tubes to avoid changing the desired charge Q of the projectile due to capture and loss collisions. Charge changing before the point of magnetic momentum analysis (see Fig. 26) can also reduce the beam current available for a given charge Q. This occurs because analysis depends on $m\mathbf{v}/QB$, where $m\mathbf{v}$ is particle momentum and B is applied (transverse to $m\mathbf{v}$) magnetic field. The percentage change in beam current associated with charge Q goes as $I/I_0 \simeq n\sigma x$, where n is ambient molecular density in vacuum, σ the charge-changing cross section, and X the path length the ion traverses in vacuum. As an example, for a typical good vacuum pressure of 1×10^{-6} Torr and $\sigma \simeq 10^{-16}$ cm^2 (geometric for heavy ions) 0.3% of the beam is lost per meter of beam path. This is significant since most beam transport lengths are much larger than 1 m (10–100 m) and pressures are sometimes more like 1×10^{-5} Torr in the actual beam line.

In addition to good vacuum, every ion-scattering experiment needs a good beam transport system. For heavy ions of small Q special problems are presented when it comes to magnetic momentum analysis and magnetic focusing via quadrupole doublet lenses.[57] Relative to that used for protons, very strong quadrupole doublet lenses will be needed to focus these ions since their magnetic rigidity mv/Q is large compared to an equal-velocity proton. For the same reason a huge deflection magnet will

---

[56] For an excellent review of Van de Graaff accelerator technology see *Nucl. Instr. Meth.* **30**, 621 (1973).

[57] H. A. Enge, *Rev. Sci. Instr.* **30**, 248 (1959).

9.2. PRODUCTION AND DETECTION OF AUGER ELECTRONS

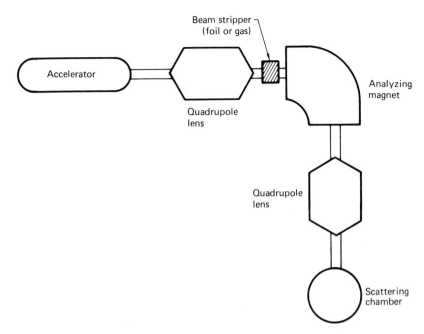

FIG. 26. Experimental setup for ion-induced Auger spectroscopy laboratory.

be needed if momentum analysis through a full 90° is necessary. As an example, for momentum analysis of a 3-MeV Kr⁺ beam, a magnetic field is required that is 80 times stronger than that needed to bend an equal velocity proton. Generally, the bending magnet problem is avoided by accepting smaller dispersion by deflecting through angles less than 90°.

Some experiments call for varying the incident charge state at a given beam energy. Within the confines of the charge state distribution available at a given E for ions "stripped" by foil or gas targets, a preanalyzing magnet beam stripper can be used (Fig. 26). The analyzing magnet can then be used to select both the energy (momentum) and charge state of the beam. This idea has been commonly used with tandem Van de Graaff experiments.[58,59]

Table I summarizes some of the basic parameters achievable from the various types of accelerators discussed here. The tandem Van de Graaff deserves some special discussion since it is particularly difficult to maintain a stable beam intensity during the course of acquiring an Auger elec-

[58] R. L. Kauffman, F. F. Hopkins, C. W. Woods, and P. Richard, *Phys. Rev. Lett.* **31**, 621 (1973).

[59] J. R. Mowat, I. A. Sellin, P. M. Griffin, D. J. Pegg, and R. S. Peterson, *Phys. Rev. A* **9**, 644 (1974).

TABLE I. Electrostatic Accelerator Characteristics

Model	Typical energy range E (MeV)/Q	Maximum current I (mA)	Atomic number of ions Z (amu)	Charge states Q
High-voltage power supplies	~0.01–1.5 with air insulation 3.0 for Dynamatron	10–50 (H$^+$)	Anything producable by ion source Maximum current usually H$^+$	Singly ionized species dominate intensity
Normal Van de Graaff	0.2–5.0	~0.5	Usually has an RF source that effectively produces only rare gas ions and H$^+$	Singly ionized dominates
Tandem Van de Graaff	$(1.0Q' + 1.0)$ MeV to $(20.0Q' + 20.0)$ MeV Q' is ion charge after terminal stripping	0.02 (H$^+$) All other ions typically much less because production of negative ions very inefficient	$Z = 1$–36 for fully stripped ions	Upper and lower limits depend sensitively on terminal voltage

9.2. PRODUCTION AND DETECTION OF AUGER ELECTRONS

tron spectrum. The tandem accelerator is charged positive at the center relative to its entrance and exit ports. Negative ions are injected into the tandem and where they are electrostatically attracted to positive terminal. When they reach the terminal the surplus electrons are "stripped" away in collisions either with a thin foil or gaseous target, thus leaving a positive ion that is repelled from the terminal and further accelerated to the exit port. Foil strippers yield the highest charge status for a given energy and are necessary to achieve complete stripping of the ion. Unfortunately the foils tend to break rather quickly under irradiation by heavy ions. In fact, Dobberstein and Henke[60] have shown that the dose for foil breakage follows the simple law

$$D = C/S_v, \qquad (9.2.10)$$

where S_v is the stopping power, and $C = NAn$, where N is foil density, A the foil area irradiated, and n the average energy transferred per atom to our foil. Obviously one should use a gas stripper where possible to get the Z, Q, E, I desired. (See Y. L. Intema in Ref. 56 for more detail on foil problems.)

9.2.4. Optimum Experimental Setup for a Given Type of Auger Electron Measurement

The optimum experimental setup for a given type of measurement depends sensitively on the accelerator necessary for producing the Z, Q, I, E parameters of choice. For moderate to high current accelerators (first and second type discussed in Table I) the best all-around setup involves the use of the PMA with retarding lens described in Section 9.2.2.1. Such a setup as used by the Hahn–Meitner group in Berlin (see Stolterfoht[61]) is shown in Fig. 27. This system allows one to measure angular distributions from 20 to 160°, perform low- or high-resolution spectral measurements, and to measure ionization cross sections using gas, vapor, or foil targets.

For measurements using tandem Van de Graaffs or other low-particle-current accelerators, the PMA becomes less practical because of its inherent low efficiency (particularly when not retarding). In this case three different types of analyzers would be necessary for optimum results. A CMA for cross-section measurements, a spherical sector for high-resolution spectroscopy, and a half-CMA for angular distribution measurements. Obviously a well-rounded experimental setup for low-particle-current machines has not been developed. Future analyzer design should be guided in this direction.

[60] P. Dobberstein and L. Henke, *Nucl. Instr. Meth.* **119**, 611 (1974).
[61] N. Stolterfoht, *Z. Physik* **248**, 81 (1971); D. Schneider, Ph.D. Dissertation, Free University of Berlin (1975).

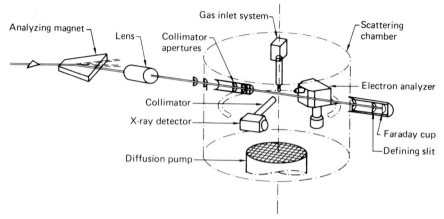

FIG. 27. Plane-mirror analyzer system from the Hahn–Meitner Institut in Berlin, courtesy of Dieter Schneider.

9.3. Studies of Ion–Atom Collision Phenomena Using Auger Spectroscopy

9.3.1. Theory of Auger Electron Emission

Simply stated, Auger electron emission is the deexcitation of an ionized (or excited) atom via electron emission. Studies of the physical mechanism for producing such emission along with descriptions of the characteristics of observed spectra have prompted much theoretical work (see Burhop,[2] Burhop and Asaad,[3] Walters and Bhalla,[62] McGuire,[63] Kelly,[64] Chen and Crasemann[65]). Theoretical studies of multiply ionized atoms (i.e., multiple outer-shell vacancies present in addition to inner-shell) have occurred only recently.[65,66] Such descriptions are still limited mainly to describing the K-Auger transitions from neon. In this section we emphasize the phenomena associated with the Auger decay of multiply ionized atoms.

9.3.1.1. *Basic Description of Auger Lines.* The basic description of Auger decay can best be obtained by studying Fig. 28. Taking for an example the K-Auger process (although a vacancy in any shell can be filled

---

[62] D. L. Walters and C. P. Bhalla, *Phys. Rev. A* **3**, 1919 (1971).

[63] E. J. McGuire *in* "Atomic Inner-Shell Processes" (B. Crasemann, ed.), Vol. I, p. 293. Academic Press, New York, 1975.

[64] H. P. Kelly in Reference 62, page 331.

[65] E.g., M. H. Chen and B. Crasemann, *Phys. Rev. A* **12**, 959 (1975).

[66] E.g., C. P. Bhalla, *Phys. Rev. A* **12**, 122 (1975).

9.3. STUDIES OF ION–ATOM COLLISION PHENOMENA

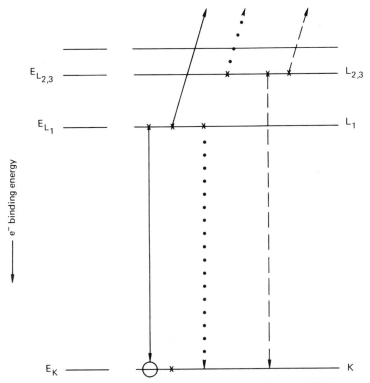

FIG. 28. Schematic of K-Auger electron process. Solid lines are KL_1L_1 transitions, dotted lines are $KL_1L_{2,3}$, and dashed lines are $KL_{2,3}L_{2,3}$.

by an Auger process) we note the occurrence of three possible types of lines. The types of lines depend upon the origination of the electrons involved in the transition. For instance, the KL_1L_1 (also labeled 1s2s2s) line involves the movement of a 2s electron into the 1s vacancy while another 2s electron is ejected. Incidentally, the electron movements may or may not be simultaneous (see D. Shirley[67] for more detail). Similarly, the $KL_1L_{2,3}$ (1s2s2p) involves vacancy filling with the 2s while the 2p is ejected, and the $KL_{2,3}L_{2,3}$ (1s2p2p) involves movement of the two outermost electrons. It can be guessed that these different types of transitions will have different energies and intensities since the different orbitals have different binding energies, quantum numbers, and electron availability.

9.3.1.2. Line Energies for Characteristic and Satellite Auger Decay.

Characteristic Auger lines are those formed by electron or photon bombardment, where the initial ionization results mainly in inner-shell va-

[67] D. A. Shirley, *Phys. Rev. A* **7**, 1520 (1973).

cancy production. Thus, the lines observed in the spectrum can be attributed to initial states having only a inner-shell vacancy and no simultaneous outer-shell vacancies. Satellite Auger lines are those lines identifiable as coming from states having some degree of outer shell excitation or ionization also present at the time of K-vacancy filling. Calculations of the characteristic KLL-Auger transition energies have been catalogued for a wide range of atoms by Asaad and Burhop[3] and Sevier.[8] The mechanism for characteristic Auger decay is evident from the expressions for the energies of various types of transitions. Using some examples from Ref. 3 assuming pure L,S coupling, we obtain

$$E_A(KL_1L_1, {}^2S \longrightarrow {}^1S) = |E_B(K) - 2E_B(L_1) - F^0(2s,2s)|,$$
$$E_A(KL_1L_{2,3}, {}^2S \longrightarrow {}^1P) = |E_B(K) - E_B(L_1) - E_B(L_{2,3})$$
$$- F^0(2s,2p) - \tfrac{1}{3} G^1(2s2p)|, \quad (9.3.1)$$
$$E_A(KL_1L_{2,3}, {}^2S \longrightarrow {}^3P) = |E_B(K) - E_B(L_1) - E_B(L_{2,3})$$
$$- F^0(2s,2p) + \tfrac{1}{3} G^1(2s,2p)|,$$

where E_B is the binding energy for the various shells. The terms $F^\nu(nl, nl')$ and $G^\nu(nl, nl')$ are the radial matrix elements expressing the electrostatic interaction between electrons in the (nl) and (nl') orbitals. These elements are also called the Slater F and G integrals.[68] The energy obtained by the Auger electron emitted in the transition is whatever remains after subtracting from the K-shell binding energy (B.E.) the B.E. of the two L shells and their electrostatic interaction energy. Figure 29 is the KLL-Auger spectrum of neon following 1.5-keV electron bombardment. In accordance with the relevant binding energies, notice that the $KL_{2,3}L_{2,3}$ transitions are much higher in energy than the KL_1L_1. The observed satellite energies and the spectral intensities and linewidths are discussed shortly. Definite selection rules also exist for the Auger transition process, which is discussed in the next section.

Auger transition energies for satellite lines arising from initial states having an *arbitrary* number of vacancies have been calculated by Larkins,[69] Bhalla *et al.*,[70] Kauffman,[71] Schumann[72] and Matthews *et al.*[73] Working with binding energies is cumbersome for multiply ionized atoms since there are only a few experimental measurements for them. Instead,

[68] J. C. Slater, "Quantum Theory of Atomic Structure," Vol. I. McGraw-Hill, New York, 1960.

[69] F. P. Larkins in Reference 62, page 377.

[70] C. P. Bhalla, N. O. Folland, and M. A. Hein, *Phys. Rev. A* **8**, 649 (1973).

[71] R. L. Kauffman, C. W. Woods, K. A. Jamison, and P. Richard, *Phys. Rev. A* **11**, 873 (1975).

[72] S. Schumann, private communication.

[73] D. L. Matthews, B. M. Johnson, and C. F. Moore, *At. Data and Nucl. Data Tables* **15**, 41 (1975).

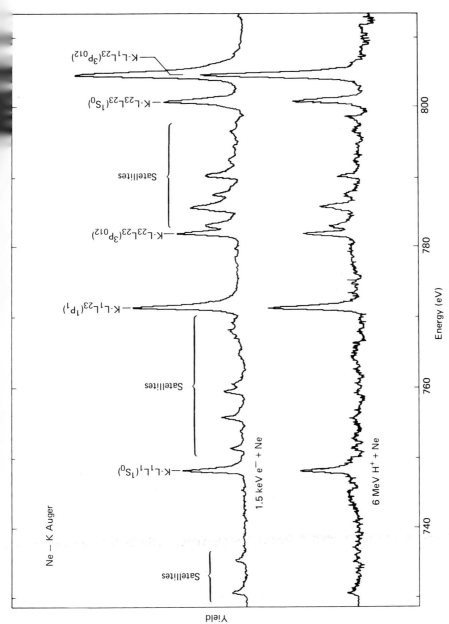

FIG. 29. Comparison of K-Auger electron spectra produced by electrons and fast protons.

computer calculations using the Hartree–Fock method result in a computation of total energies of initial and final states. General expressions for KLL-Auger transitions of neon having initial configuration $1s^l 2s^m 2p^n$ are presented by Matthews and co-workers[73]:

$$\begin{aligned}
KL_1^m L_1^n &= E_A(lmnss, {}^{2s+1}L_j - {}^{2s'+1}L'_{j'}) \\
&= E_T(1s^l 2s^m 2p^n, {}^{2s+1}L_J) - E_T(1s^{l+1} 2s^{m-2} 2p^n, {}^{2s'+1}L'_{J'}), \\
KL_1^m L_{2,3}^n &= E_A(lmnsp, {}^{2s+1}L_J - {}^{2s'+1}L'_{J'}) \\
&= E_T(1s^l 2s^m 2p^n, {}^{2s+1}L_J) \\
&\quad - E_T(1s^{l+1} 2s^{m-1} 2p^{n-1}, {}^{2s'+1}L'_{J'}), \\
KL_{2,3}^m L_{2,3}^n &= E_A(lmnpp, {}^{2s+1}L_J - {}^{2s'+1}L'_{J'}) \\
&= E_T(1s^l 2s^m 2p^n, {}^{2s+1}L_J) - E_T(1s^{l+1} 2s^m 2p^{n-2}, {}^{2s'+1}L'_{J'}),
\end{aligned}$$ (9.3.2)

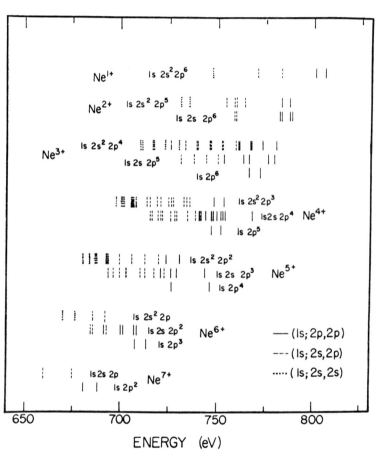

FIG. 30. Calculated neon KLL-Auger transition energies. (From R. L. Kauffman, Ph.D. Dissertation, Kansas State Univ., Manhattan 1975.)

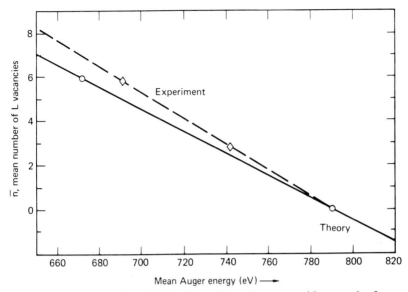

FIG. 31. Comparison of theoretical and experimental mean transition energies for neon K-Auger decay. Experiment values from N. Stolterfoht, D. Schneider, D. Burch, B. Asgaard, E. Bøving, and B. Fastrup, *Phys. Rev. A* **12**, 1313 (1975) (courtesy of The Physical Review).

where E_T is the total energy of a particular term ($^{2s+1}L_J$) of a configuration and is determined from the corresponding Hartree–Fock wavefunction.

Following Slater,[68] E_T is defined as

$$E_T(^{2s+1}L, nl, n'l') + \overline{E} + \sum_k a_{nl,n'l',k} F^k(nl, n'l')$$

$$+ \sum_k b_{nl,n'l',k} G^k(nl, n'l'), \qquad (9.3.3)$$

where, in this case, n is the principal quantum number and l the orbital quantum number, $\overline{E}$ is the total energy for the average configuration, $a_{nl,n'l',k}$ and $b_{nl,n'l',k}$ are coefficients depending on the particular term being calculated in a given configuration (see Slater[68] for tabulations of a, b). F^k and G^k are electrostatic electron–electron interaction integrals.

From the work of Kauffman[71] Fig. 30 represents the results of such calculations for neon. Evidently a large number of transitions are obtained, presenting a complicated spectrum like that observed in Fig. 2. The general features of Auger satellite lines are their decrease in transition energy with increasing number of initial outer-shell vacancies. This phenomenon occurs because the binding energies of the final state outer-shell electrons increase faster with increasing vacancy number than does the B.E.

of the initial-state K shell. Thus from expression (9.3.1) the transition energy can only decrease. Even though Auger spectra from multiply ionized atoms are complicated, empirical techniques can be used to predict number of outer-shell electrons missing. Stolterfoht[74] has developed a curve for neon relating average Auger transition energy (experimentally determined) to number of L shell vacancies present at time of Auger emission. Figure 31 presents his work along with a comparison to a theoretical determination.[75] A very interesting empirical result follows from the graph, namely, there is a linear relationship between mean outer-shell vacancy number and average Auger electron energy. This is a very valuable result since the average Auger transition energy is easily measured but the mean L vacancy is often impossible to determine from very complicated spectra. The differences between the experimental and theoretical curve for high charge states are not presently known.

9.3.1.3. Transition Rates, Relative Line Strengths, and Linewidths. The expression for the Auger transition rate was originally derived by Wentzel[76] using first-order perturbation theory. The expression for the Auger transition rate is given by Bhalla[66]

$$T_A(i \longrightarrow f) = 2\pi\rho_F \bar{\Sigma} |\langle \psi_f(LSJM) | \sum_{i,j} \left(\frac{e^2}{r_{ij}}\right) | \psi_i(L'S'J'M')\rangle|^2, \quad (9.3.4)$$

where ρ_F is the density of final states, r_{ij} the distance between i and j electrons involved in transition, and ψ_i, ψ_f are the initial and final state wavefunction in the L,S,J,M quantum number representation. It is obvious that since the electrostatic operator e^2/r_{ij} is a scalar operator, then $L = L'$, $S = S'$, $J = J'$, $M = M'$, or else expression (9.3.4) vanishes. These conditions invoke the selection rules for the Auger effect, namely, $\pi_i = \pi_f$, $\Delta L = \Delta S = \Delta J = 0$ between initial and final states. Calculations of Auger transition probabilities within various coupling schemes have been reviewed by Asaad and Burhop.[3] It turns out that these rates are constant with Z and are $\sim 10^{15}$ sec^{-1}. This constant rate probably results from the fact that the outer-shell electrons involved in the trnasition are completely screened from the nuclear field (Z dependent). This lack of Z dependence is an important result when it comes to understanding fluorescence yields as discussed in Section 9.3.1.4.

Even though the total Auger transition rate (sum of KL_1L_1, $KL_1L_{2,3}$, $KL_{2,3}L_{2,3}$ rates) is calculated to be constant with Z, the individual rates for the different types of transitions are by no means equal (nor constant with

---

[74] N. Stolterfoht, *Proc. Fourth Conf. Sci. Industr. Appl. Small Accelerators, IEEE Cat.* #76CH1175-9 NPS, page 311 (1976).
[75] D. L. Matthews, unpublished.
[76] G. Wentzel, *Z. Phys.* **43**, 524 (1927).

9.3. STUDIES OF ION-ATOM COLLISION PHENOMENA

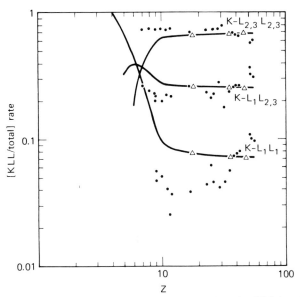

FIG. 32. Comparison of theoretical and experimental values for KLL/total Auger decay rate as a function of Z; from D. L. Walters and C. P. Bhalla, *Phys. Rev. A* **3**, 1919 (1973) (courtesy of The Physical Review).

Z as measured from experiment). Figure 32 from the work of Walters and Bhalla[62] shows a comparison of theoretical and experimental relative transition rates as a function of Z. For $20 \geq Z \geq 10$ (calculations break down for $Z < 10$) we see that the intensities $KL_{2,3}L_{2,3}$, $KL_1L_{2,3}$, KL_1L_1 are in proportion 15:4.5:1. This tells us a basic fact about Auger decay: it favors ejection of weakly bound and high l-value electrons. Another point that is obvious from Fig. 32 is that below some Z value the theoretical values diverge sharply from those of experiment. As discussed by Asaad and Burhop,[3] this problem could not be resolved by including the effects of configuration interaction. Kelly,[64] however, using many-body perturbation theory has had good luck in calculating relative Auger rates for the neon atom.

Auger transition rates for multiply ionized atoms have mainly been performed by Chen and Crasemann,[65] Bhalla,[66] and McGuire.[63] These calculations have to be very accurate since the Auger rates may change drastically for initial states whose multiplet terms only differ by their spin (even by their J value[77]). Table II demonstrates this effect in the calculated Auger rates[78] for the first satellite lines of argon LMM-Auger decay

---

[77] M. Chen and B. Crasemann, to be published.
[78] M. Chen and B. Crasemann, *Phys. Rev. A* **10**, 2232 (1974).

(single L, single M vacancies). Auger rates always depend on the number of outer-shell electrons present at the time of Auger emission. Figure 33 shows this variation with L-vacancy number in the K-Auger rate for neon having one K vacancy and M L vacancies. We have used the calculated rates of reference 64, but have statistically averaged them over the multiplet states [(i.e., weighted according to $(2L + 1)(2S + 1)$] before plotting. It is obvious that the Auger rate monotonically decreases with increasing charge state. A simple-minded argument for this effect can be presented. Expression (9.3.4) is really just a determination of $\langle 1/r_{ij}\rangle^2$, i.e., the reciprocal square of the average electron interaction length. This average length must increase with decreasing L-shell electron population, and thus the rate decreases.

The fact that Auger transition linewidths are proportional to the total decay rate for the inner-shell vacancy provides a means for checking absolute transition rates with experiment. The natural intensity distribution of the Auger line (with energy E_0) for the transition i → f is given by the Lorentzian distribution

$$I(E) \sim \Gamma_{i,f}/[(E - E_0)^2 + \tfrac{1}{4}\Gamma_{i,f}^2], \qquad (9.3.5)$$

where $\Gamma_{i,f} = \Gamma_i + \Gamma_f$ is the total decay rate of initial and final states. For most cases of interest the rate Γ_f is zero since the ground state is infinitely long-lived. Thus the Auger linewidth gives a measure of the total initial vacancy decay rate. This rate is related to the total Auger rate through the expression $\Gamma_{\text{Auger}} = \Gamma_i(1 - \omega)$, where ω is the fluorescence yield (probability of x-ray decay) of the initial vacancy state. For characteristic Auger lines in low-Z atoms ($Z \leq 20$), $\omega \ll 1$ so that the total transition rate essentially equals the total Auger rate. In this case, a more severe test of theoretical rates can be made than with relative rates since

TABLE II. Calculated Argon Auger Rates (10^{-3} a.u.) for Decay of $1s^22s^22p^53s^23p^5$ Configuration[a]

Initial state	Final state		
	$3p^{-3}$	$3s^{-1}3p^{-2}$	$3s^{-2}3p^{-1}$
^1S	6.4	2.7	0.07
^3S	0	0.07	0.07
^1P	0.2	0.2	0.07
^3P	8.0	2.9	0.07
^1D	7.4	2.7	0.07
^3D	0.2	0.07	0.07

[a] From M. H. Chen and B. Crasemann, Ref. 78.

9.3. STUDIES OF ION–ATOM COLLISION PHENOMENA

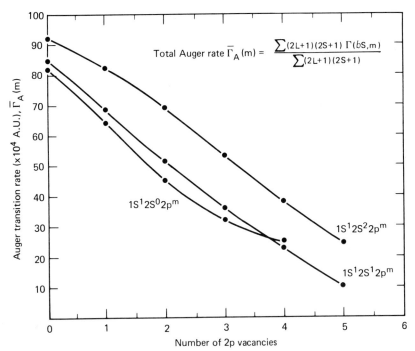

FIG. 33. Variation in total Auger rate with number of 2p vacancies for neon as adapted from M. H. Chen and B. Crasemann, *Phys. Rev. A* **12**, 959 (1975) (courtesy of The Physical Review).

the ratios of two numbers tend to cancel errors resident in the absolute values of the two. A comparison of theoretical and experimental absolute rates for the neon K shell is given in the work of Kelly.[79] Ironically, the discrepancy between theory and experiment proved to be the fault of experiment (see later measurements by Gelius *et al.*[80] and Ridder *et al.*,[32] which agree with theory).

9.3.1.4. Fluorescence Yields for Singly and Multiply Ionized Atoms.

As briefly mentioned in Section 9.3.1.3, Auger decay is not the only means for deexciting a state having an inner-shell vacancy. In fact, some states have an Auger transition rate of zero but still have relatively fast decay rates since they are able to go by photon emission (x-ray decay). It is outside the scope of this part to discuss x-ray decay, but the concept of fluorescence yield is important to understanding the relative intensities of Auger lines and relating Auger yields to vacancy production. In addition,

---

[79] H. P. Kelly, *Phys. Rev. A* **11**, 556 (1975).
[80] N. Gelius, *et al.*, *Chem. Phys. Lett.* **28**, 1 (1974).

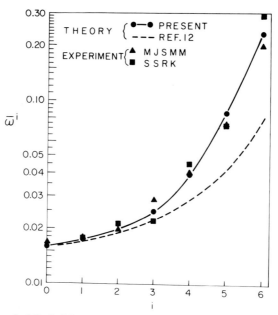

FIG. 34. Theoretical K-shell fluorescence yields for neon KL^i-charge states from M. Chen and B. Crasemann, *Phys. Rev. A* **12**, 959 (1975) (courtesy of The Physical Review). Reference 12 refers to work by C. P. Bhalla, N. O. Folland, and M. A. Hein, *Phys. Rev. A* **8**, 649 (1973). MJSMM refers to Ref. 114, while SSRK refers to Ref. 113.

the determination of fluorescence yields require Auger electron measurements.

The fluorescence yield or probability of radiative decay of a state having quantum numbers L,S is given by

$$\omega(LS) = \Gamma_X(LS)/[\Gamma_X(LS) + \Gamma_A(LS)], \quad (9.3.6)$$

where Γ_X is the radiative rate and Γ_A is the Auger rate. Comprehensive reviews of fluorescence yield work are available from Bambynek et al.,[81] Burhop and Asaad,[3] and Sevier.[8] As an example of how the K-shell fluorescence yield (single K vacancies, no multiple ionization) scales with Z, consider that the radiative rate $\Gamma_{KX} \sim Z^4$ while recalling that $\Gamma_{K,A} \approx$ const. Thus, $\omega_K \sim (1 + b_K Z^{-4})^{-1}$, where b_K is a constant ($\sim 7.5 \times 10^5$). Walters and Bhalla[62] have calculated ω_K vs. Z and shown that it varies from $\omega_K \sim 0.02$ at $Z = 10$ to $\omega_K \simeq 0.9$ at $Z = 55$. As already mentioned for $Z \leq 20$, then $\omega_K \ll 1$. This is an important result since it allows the derivation of vacancy production cross section σ_T from measurements of

---

[81] W. Bambynek, B. Crasemann, R. W. Fink, A. Freund, H. Mark, C. D. Swift, R. E. Price, and P. V. Rao, *Rev. Mod. Phys.* **44**, 716 (1972).

Auger production cross section (see Glupe and Mehlhorn[82]) σ_A, since $\sigma_T = \sigma_A/1 - \omega \simeq \sigma_A$ for $\omega \ll 1$.

Fluorescence yields for multiply ionized atoms may be very large even for low-Z atoms (see Table II). These large fluorescence yields directly influence the relative intensities of Auger line observed in measured spectra since $I(LS) \sim [1 - \omega(LS)]I_0$ (see Ref. 83) where $I(LS)$ is the intensity of an Auger transition labeled by quantum number L,S. For the same reason, the Auger production cross section $\sigma_A \neq \sigma_T$ since ω is not small compared to 1. Bhalla[66,84] and Chen and Crasemann[65] are mostly responsible for calculations of K- and L-shell fluorescence yields for multiply ionized atoms such as nitrogen[84] and neon[65,66] and Argon,[78] sulfur, and chlorine.[77]

Figure 34 from Chen and Crasemann[65] demonstrates the variation of $\bar{\omega}$ (statistically averaged over L,S and configuration) with number of L-shell vacancies for the neon K shell. Conceptually speaking, the fluorescence yield increases with increasing vacancy number because the radiative transition rate decreases more slowly than the Auger rate. The variation with M vacancy number in the case of argon L,ω values is even more dramatic (see Chen and Crasemann[78]).

9.3.1.5. Angular Distribution of Auger Electron Emission.

One final bit of Auger emission theory concerns the emission pattern or angular distribution to be expected for Auger decay. For characteristic lines Cleff and Mehlhorn[85] are responsible for developing the theory and making the first measurements of the small anisotropies produced in electron impact ionization of the argon L shell. As will be seen in Section 9.3.3.3, large anisotropies exist when ionization takes place using ion bombardment. It is therefore worth reviewing the basics of Auger emission patterns.

Following Cleff and Mehlhorn[85] we note that Auger electrons can be emitted anisotropically provided that the initial state has quantum number $j > 1/2$ and its magnetic sublevels M_J are populated nonstatistically. Within the intermediate coupling scheme (atomic states labeled by $LSJM$) this nonisotropic emission of electrons relative to a quantization axis (beam direction) amounts to a nonequal population $P(LSJM)$ of the magnetic sublevels during inner-shell ionization. If we assume that the target atoms are randomly oriented, then the nonequal population of the magnetic sublevels is entirely due to the ionization process. In fact, due to the unpolarized nature of the incident beam, the cross section for removal

---

[82] G. Glupe and W. Mehlhorn, *Phys. Lett.* **25A**, 274 (1967).
[83] D. L. Matthews, R. J. Fortner, D. Schneider, and C. F. Moore, *Phys. Rev. A* **14**, 1561 (1976).
[84] C. P. Bhalla, *J. Phys. B* **8**, 2792 (1975).
[85] B. Cleff and W. Mehlhorn, *J. Phys. B* **7**, 593 (1974).

of an inner-shell electron depends only on the value $|m_j|$, or considering the atom as a whole, then the population $P(JM)$ of substate JM depends only on $|m|$. Under these conditions the angular distribution of Auger electrons is given by

$$\omega(\theta)\, d\Omega = (2J + 1)^{-1} \sum_M \sum_{M'',m_s} [P(LSJM)\, P(LSJM \longrightarrow L''S''J''M'',E,m_s,\theta)]\, d\Omega, \quad (9.3.7)$$

where we have summed over final states M'' and m_s while averaging over initial states M and multiplied the population probability $P(LSJM)$ by the probability of emission $P(LSJM \to L''S''J''M'',E,m_s,\theta)$ of an electron of energy E, spin projection m_s (relative to beam), into a solid angle $d\Omega$ with direction θ to incident beam. Equation (9.3.7) breaks the theory of anisotropic Auger emission into the theory of (1) population and (2) emission.

The theory of nonequal population of substates is identical to that treated in connection with the observation of polarized X-ray emission (McFarlane[86], Jamison and Richard[87]). As mentioned earlier, the cross sections for population of different (M) states are not necessarily equal. From McFarlane[86] the population probabilities are related to the ionization cross section through the expression

$$P(SLJM) \equiv \sigma(sljm): \sum_{m_s ml} [j] \begin{pmatrix} s & l & j \\ m_s & ml & -m \end{pmatrix}^2 \delta(slj_sm_l), \quad (9.3.8)$$

assuming an unpolarized beam. Calculation of these partial cross sections has been performed by McFarlane,[86] Jamison and Richard,[87] Fano and Macek,[88] and Percival and Seaton.[89] As an example, assume we are looking at the M_L dependence on cross section for an $l = 1$ initial state. The possible M_L values are $M_L = 0, \pm 1$ or $|M_L| = 0, |M_L| = 1$. For nonequal populations to occur, then $\sigma(|M_L| = 0) \neq \sigma(|M_L| = 1)$. Normally these partial ionization cross sections are equal except for special collisions. Three types of collisions do exist where they are different. When using electrons for ionization Cleff and Mehlhorn[85] found that when the impact energy was at threshold then by definition $\sigma(|M_L| = 1) = 0$ since the ionized electron could carry away no orbital angular momentum. Furthermore, in the high-energy impact limit, $E_p > 100 E_B$, where E_B is the binding energy of the target electron; then the Bethe approximation is valid. Under these circumstances the transition matrix element for ioni-

---

[86] S. C. McFarlane, *J. Phys. B, Atom. and Molec. Phys.* **5**, 1906 (1972).
[87] K. A. Jamison and P. Richard, *Phys. Rev. Lett.* **38**, 484 (1977).
[88] U. Fano and J. H. Macek, *Rev. Mod. Phys.* **45**, 553 (1973).
[89] I. C. Percival and M. J. Seaton, *Philos. Trans. Roy. Soc. London, Ser. A* **251**, 113 (1958).

9.3. STUDIES OF ION-ATOM COLLISION PHENOMENA 481

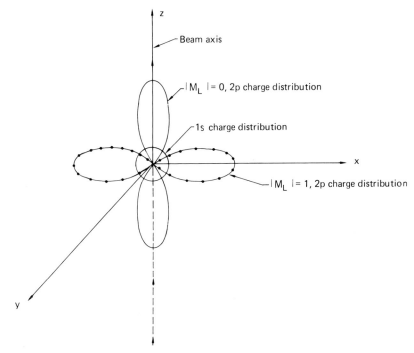

FIG. 35. Alignment mechanism for Coulomb ionization. Radial charge distributions for $|M_L|$ state of a 1s2p electronic configuration. Note that an ion beam traveling on axis Z has a greater probability of interacting with both 1s and 2p orbitals when $(M_L)_{2p} = 0$.

zation is essentially the dipole matrix element; thus $\Delta M_l = \pm 1$ is the selection rule. This would also be the case for photoionization. The third example of $\sigma(0) \neq \sigma(1)$ deals with multiple-ionization phenomena and thus is of great interest to the majority of the measurements discussed here. Jamison and Richard[87] observed that $\sigma(0) \neq \sigma(1)$ because of the polarization they observed for initial single K-, single L-vacancy states formed in aluminum by H^+ or He^+ bombardment. They used the Hansteen and Mosebekk[90] formulation of multiple-vacancy production via simultaneous K and L shell vacancy formation. In this formulation, the ionizing particle is required to cross both the K and L shell; thus the range of available impact parameters is restricted. In essence, because of the M_L dependence of the geometry of overlapping spatial charge distributions for the 1s and 2p shells, the probability for simultaneous K,L ejection for $M_L = 0$ is greater than for $|M_L| = 1$. Figure 35 schematically portrays this effect since for $|M_L| = 1$ projections, the probability of projectile in-

[90] J. M. Hansteen and O. P. Mosebekk, *Phys. Rev. Lett.* **29**, 1361 (1972).

teraction with the 1s and 2p electrons is much less than for the $|M_L| = 0$ projection.

As explained later, electron capture and electron promotion also significantly contribute to the ionization and excitation of inner-shell electrons. At present, there are no theoretical descriptions of the $|M|$ dependences of these mechanisms, although Auger experiments indicate anisotropies when these are the responsible ionization mechanisms.

The probability of Auger emission into a given angle depends not only on the population but also the transition amplitude and phase of the different partial waves for the emitted Auger electron. In general, the relative angular intensity $\hat{\omega}(\theta) = 4\pi\omega(\theta)/\omega_{\text{tot}}$ is given by

$$\hat{\omega}(\theta) = 1 + \sum_{n=1}^{\infty} A_{2n} P_{2n}(\cos \theta), \qquad (9.3.9)$$

where P_{2n} are the even-order Legendre polynomials and A_{2n} are the general anisotropy coefficients, which are functions of the population probability transition amplitude and phase. For the particular case where the final state $J'' = 0$, then A_{2n} are only functions of the population probability. In general cases, however, the A_{2n} are probably small because of depolarization effects.[87] Future theoretical endeavors should attempt to match ionization mechanisms to experiments determining anisotropies.

9.3.2. Ion–Atom Collision Mechanisms*

Auger electron production can be used to study ion–atom reaction mechanisms primarily by measurements of total vacancy production. However, the extent of formation of multiply ionized states, the production of high spin states, and the formation of states that decay by anisotropic Auger emission are also signatures of certain types of reaction mechanisms. In this section, we briefly discuss three main reaction or collision mechanisms and some illustrative Auger measurements that have been used to study them.

Three categories of collision mechanisms are (1) electron promotion, which is described via the Fano–Lichten[91] or molecular orbital (MO) model; (2) direct Coulomb excitation (DCE) or ionization described usually with the plane-wave Born approximation (PWBA)[92] to Coulomb scattering; (3) e⁻ capture or charge exchange, again explained within PWBA although allowing electron pickup by the projectile. Numerous

[91] U. Fano and W. Lichten, *Phys. Rev. Lett.* **14**, 627 (1965).
[92] E. Merzbacher and H. W. Lewis *in* "Handbook der Physik," (S. Flügge, ed.), Vol. 34, p. 166. Springer-Verlag, Berlin, 1958.

* See also Vol. 4A, Part 4.

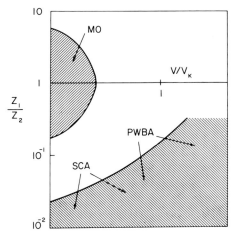

FIG. 36. Regions of validity for various approximation schemes as a function of Z_1/Z_2 and collision velocity V (in units of the K-shell electron velocity V_K). Shaded regions are shown for plane-wave Born approximation (PWBA), semiclassical approximation (SCA), and the quasi-molecular electron promotion model (MO). From D. Madison and E. Merzbacher *in* "Atomic Inner Shell Processes" (B. Crasemann, ed.), Vol. 1. Academic Press, New York, 1975.

reviews and journal articles consider these mechanisms and their theoretical description in detail (see Garcia *et al.*,[9] Richard,[10] Madison and Merzbacher,[93] Stolterfoht,[13] Garcia *et al.*,[94] McGuire,[95] Halpern and Law[96]). Ignoring charge exchange for the moment, the classification of ranges of importance of mechanisms (1) and (2) depends on the projectile nuclear charge Z_1, the target nuclear charge Z_2, the velocity of the projectile V_1, and the orbital velocity of the inner-shell electron V_2, to be ionized. We must also state that conceptually (at least for most collisions) there is no difference between removing target or projectile electrons via these reaction mechanisms. This is an important result since often it is necessary to describe experiments that study postcollision states of both projectile and target.

In terms of the parameters Z_1, Z_2, V_1, and V_2 (K-shell orbital velocity in this case) Fig. 36 from Ref. 93 best summarizes the regions of validity for theoretical descriptions of K-shell ionization via the MO model and the PWBA. It is unfortunate to relate that many measurements are done in the unshaded regions where combinations of mechanisms can be

---

[93] D. H. Madison and E. Merzbacher in Reference 62, page 2.
[94] J. D. Garcia, E. Gerjuoy, and J. Welker, *Phys. Rev.* **165**, 66 (1968).
[95] J. H. McGuire, *Phys. Rev. A* **8**, 2760 (1973).
[96] A. M. Halpern and J. Law, *Phys. Rev. Lett.* **31**, 4 (1973).

responsible for the ionization and most theoretical descriptions are invalid. Obviously, much theoretical work remains in order to fill in the shaded regions.

9.3.2.1. Electron Promotion. The basic characteristics of the e^- promotion class of ion–atom collisions is that they are usually slow where $V_1 \ll V_2$ and involve collision partners whose bound states reasonably overlap in energy. The usual explanation invoked for the phenomenon is that during the slow collision an electron in an inner shell is promoted to a higher level by a dynamic interaction between transient molecular orbitals (see Lichten,[97] Briggs and Macek,[98] Eichler[99]). Figure 37 provides a skeletal display of the movement along transient molecular orbitals of a neon 1s electron during the course of a Ne–Ne collision. Considered in a stepwise manner the transition occurs (following the path of the e^- vacancy) as $[2p \rightarrow 2p\pi] \rightarrow [2p\sigma \rightarrow 1s]$. The brackets indicate that this transition occurs in two stages: (1) $2p \rightarrow 2p\pi$ vacancy transfer, which occurs in proportion to the vacancy occupation number $N_\pi(i)$, where i denotes the number of initial 2p vacancies in the projectile; and (2) rotational coupling of $2p\pi$ to $2p\sigma$ (then transfer to 1s upon separation), thus transferring vacancy to 2p with cross section σ_{ROT}. The resulting total K-vacancy production for incident ions having i vacancies then goes as $\sigma_T{}^i = \sigma_{ROT} N_\pi(i)$. Briggs and Macek[98] showed theoretically that the cross sections scaled as $\sigma_T(E_A) = \frac{1}{6} N Z_s^{-2} \sigma(E_D)$ for symmetric collision systems involving first-row atoms, where $N/6$ is the probability that the $2p\pi$ MO has a vacancy, Z_s is the separated atom Z, $\sigma(E_D)$ is the cross section for $D^+ \rightarrow D$ 1s vacancy formation at bombarding energy E_D, and $E_A = M_A Z_s^2 E_D$, where M_A is the reduced mass number of atom A. The velocity dependence of electron promotion exhibits a well-defined threshold $[E_T \approx 2Z_s^3 \text{(a.u.)}]$ for ionization followed by a sharp rise to near maximum cross section—the cross section then very slowly increases with V to the maximum. As discussed later, experiment has shown that N_π is velocity dependent, and so σ_T from experiment increases faster with energy than predicted by theory.

Multiple ionization can be produced in MO collisions. For this to happen the electron promotion process governs inner-shell vacancy formation and ionization simultaneously produces outer-shell vacancies. However, no theoretical descriptions of multiple ionization exist at this time.

The colliding systems need not have states of identical binding energy

[97] W. Lichten, *Phys. Rev.* **164**, 131 (1967).
[98] J. S. Briggs and J. Macek, *J. Phys. B* **5**, 579 (1972); *J. Phys. B* **6**, 982 (1973). Also J. S. Briggs, *Proc. IXth Int. Conf. Phys. Electron. Atom. Coll.*, p. 384, Univ. of Washington Press, Seattle, 1975.
[99] J. Eichler, U. Wille, B. Fastrup, and K. Taulbjerg, *Phys. Rev. A* **14**, 707 (1976).

9.3. STUDIES OF ION–ATOM COLLISION PHENOMENA   485

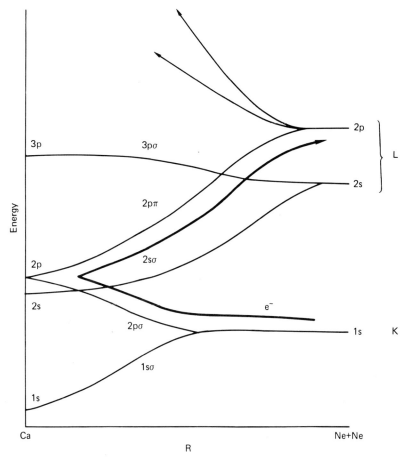

FIG. 37. Correlation diagram for molecular orbital energy levels produced in slow Ne → Ne collisions. Promotion path of 1s electron is plotted by an arrow.

for vacancy formation to occur. In fact, in an asymmetric collision system ($Z_1 \neq Z_2$) the inner-shell vacancy is normally produced in the lighter (smaller 1s binding energy) collision partner. However, there exists a finite probability for vacancy formation in the heavier partner. The ratio of heavy to light partner vacancy formation is known as the vacancy-sharing ratio (Meyerhof,[100] Stolterfoht[13]) and has been experimentally studied. Meyerhof described this phenomenon on the basis of Dempkov "coupling"[101] or charge transfer. The sharing ratio is simply

[100] W. E. Meyerhof, Phys. Rev. Lett. 31, 1341 (1973).
[101] Y. N. Dempkov, Zh. Eksp. Teor. Fiz. 45, 195 (1963) transl. Sov. Phys.-JETP 18, 138 (1963).

$S = \exp(-\alpha/v)$, where v is ion velocity and $\alpha = \pi[(2E_H)^{1/2} - (2E_L)^{1/2}]$, where E_H and E_L are the binding energy of 1s electrons in heavy and light particles, respectively. This model works remarkably well except for certain deviations for low-Z collisions, which can be explained by adjustment of the relevant binding energies because of two-step transitions plus relaxation (see Stolterfoht and Leithäuser[102]).

9.3.2.2. Direct Coulomb Ionization or Excitation. Direct excitation to bound or continuum states of inner-shell electrons proceeds via an electrostatic interaction with a charged incident particle. Madison and Merzbacher[93] have reviewed the most common theoretical approximations used to describe Coulomb scattering. Coulomb ionization is usually described within the theoretical framework of the (1) plane-wave Born approximation (PWBA),[93,103] (2) semiclassical approximation (SCA),[104] or (3) binary encounter approximation (BEA).[105-108] The regions of validity for approximations (1) and (2) were shown in Fig. 36. McGuire and Richard[109] noted that some of the main features of Coulomb ionization can be pointed out by the BEA formulation for ionization cross section

$$\sigma(E_i) = [NZ_1^2\sigma_0/U^2]G(V), \qquad (9.3.10)$$

where E_i is the incident ion energy, Z the projectile nuclear change, N the number of possible electrons that can be ionized in a given shell ($N = 2$ for K shell), $\sigma_0 = \pi e^4$ (cm^2 eV^2), U the binding energy of electron to be ionized, and $G(V)$ a function of v_1/v_2, i.e., a function of projectile to orbital electron velocity, which has been tabulated by numerous investigators.[106-110] Often we see universal cross section curves where $G(V) = (\sigma/\sigma_0)U^2/NZ_1^2$ is plotted vs. $V = v_1/v_2$ or $E/\lambda u$, where $\lambda = m_i/m_e$ (i.e., v^2). The salient features of Coulomb scattering are described by expression (9.3.10), namely, the cross section $\sigma(E_i)$ scales as Z_1^2, U^{-2}, and depends on v_1/v_2 such that a maximum is obtained for $v_1/v_2 \simeq 1$. The same scaling features are predicted by the SCA and PWBA models (see Ref. 9). For $v_1/v_2 \ll 1$, Basbas et al.[110] have shown that the

---

[102] N. Stolterfoht and U. Leithäuser, Phys. Rev. Lett. **36**, 186 (1976).

[103] W. Henneberg, Z. Phys. **86**, 592 (1933).

[104] J. Bang and J. M. Hansteen, Kg. Dan. Vidensk. Selsk. Mat.-Fys. Medd. **31**, No. 13 (1959).

[105] M. Gryzinski, Phys. Rev. **138**, A305, A322, and A336 (1965).

[106] E. Gerjuoy, Phys. Rev. **148**, 54 (1966).

[107] J. D. Garcia, Phys. Rev. A **1**, 280 and 1402 (1970) also Phys. Rev. A **4**, 955 (1971).

[108] L. Vriens, Proc. Roy. Soc. London, **90**, 935 (1966); Case Stud. Atom. Phys. **1**, 6 (1971).

[109] J. H. McGuire and P. Richard, Phys. Rev. A **3**, 1374 (1973).

[110] G. Basbas, W. Brandt, and R. Laubert, Phys. Rev. A **7**, 983 (1973); G. Basbas, W. Brandt, and R. H. Ritchie, Phys. Rev. A **7**, 1971 (1973).

PWBA must be corrected for "Coulomb deflection" and "increased binding energy" effects. The effect of Coulomb deflection is to steer the normal trajectory (distance of closest approach) of the incident particle so as to obstruct deep penetration of the inner shell. This reduces the probability of ionization. The amount of deflection is proportional to particle momentum, and so for the same velocities Coulomb deflection is not significant for heavy projectiles. In general, Coulomb deflection is usually unimportant for $v_1/v_2 > 0.5$ or for ionizing particles whose M_1 is much greater than protons or alpha particles. Increased binding energy effects are produced by the fact that ionizing particles traveling at slow velocities can penetrate an inner shell and reside within it for a time long compared to the characteristic period of the orbiting inner-shell electron. The electronic state from which ionization occurs can then be characterized by a new binding energy $U' = U_0[(Z_1 + Z_2)/Z_2]^2$, where Z_2 is the target nuclear charge. The vacancy production cross section is then reduced since $\sigma \sim (U')^{-2}$. The correction approaches unity for $v_1 > v_2$.

The Coulomb ionization mechanism can produce multiply ionized atoms since the projectile can produce ionization in every shell it traverses. Generally, the multiple-ionization probability is much higher with heavy-ion bombardment owing to the greater strength of the Coulomb interaction since $U(r) \sim Z_1 Z_2 e^2/r$.

Attempts to theoretically predict multiple ionization usually resort to the impact parameter formulation.[90,109,111] The fundamental assumption in descriptions of multiple ionization is that it can be expressed as a binomial distribution of the single ionization probability. From Richard[10] we see that the multiple-ionization cross section can be expressed as

$$\sigma_{1K,nL} = \sum_{i,j} \int_0^\infty 2\pi b P_K(b) P_{ij}(b) \, db, \quad (9.3.11)$$

where $\sigma_{1K,nL}$ is the cross section for simultaneous one K- and n L-vacancy production, b the impact parameter, $P_K(b)$ the probability of ionizing a K-shell electron at b, and $P_{ij}(b)$ the probability of putting i vacancies in the 2s orbital and j vacancies in the 2p orbital, where $i + j =$ total number of L-shell vacancies n. If we assume the probability of producing 2s and 2p vacancies is independent, then

$P_{ij}(b)$

$$= \binom{2}{i} [P_{2s}(b)]^i [1 - P_{2s}(b)]^{2-i} \binom{6}{i} [P_{2p}(b)]^j [1 - P_{2p}(b)]^{6-j}, \quad (9.3.12)$$

where P_{2s} and P_{2p} are the probabilities of producing a single 2s or 2p va-

[111] J. S. Hansen, *Phys. Rev. A* **8**, 822 (1973).

cancy. Expression (9.3.11) can be greatly simplified since $P_{ij}(b)$ is fairly constant over the small range of b to which $P_K(b)$ contributes (see Kauffman et al.[112]). Thus

$$\frac{\sigma_{1K,nL}}{\sigma^{TOT}} = \frac{\Sigma_{ij} P_{ij}}{\Sigma_{n=0}^{\infty} \Sigma_{ij} P_{ij}} \qquad (9.3.13)$$

expresses the relative multiple-ionization cross section. This expression has been successfully used to predict x-ray statellite spectra[10] provided one takes account of the fluorescence yield for each configuration. This same expression has also been successfully used to fit Auger satellite spectra.[113,114]

9.3.2.3. Electron Capture or Charge Exchange. The final ionization mechanism considered here is that of electron capture of target inner-shell

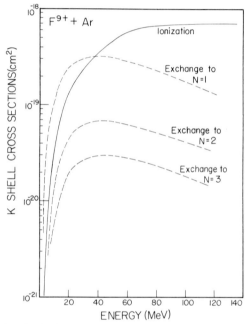

FIG. 38. Theoretical ionization and charge-exchange cross sections for $F^{9+} \rightarrow Ar$. At lower energies the cross section may be too large. From J. H. McGuire, *Phys. Rev. A* **8**, 2760 (1973) (courtesy of The Physical Review).

[112] R. L. Kauffman, J. H. McGuire, P. Richard, and C. F. Moore, *Phys. Rev. A* **8**, 1233 (1973).

[113] N. Stolterfoht, D. Schneider, P. Richard, and R. L. Kauffman, *Phys. Rev. Lett.* **33**, 1420 (1974).

[114] D. L. Matthews, B. M. Johnson, L. E. Smith, J. J. Mackey, and C. F. Moore, *Phys. Lett.* **48A**, 93 (1974).

9.3. STUDIES OF ION–ATOM COLLISION PHENOMENA

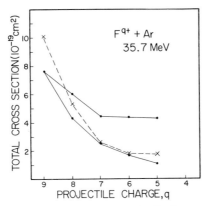

FIG. 39. Dependence of total cross section on projectile charge state. Upper solid curve represents calculations using nuclear charge of fluorine (9+) whereas lower solid curve represents calculations using the atomic charge $(q+)$ of fluorine. The dashed curve represents data of McDonald et al., Phys. Rev. Lett. **29** 1291 (1972). Figure from J. H. McGuire, Phys. Rev. A **8**, 2760 (1973) (courtesy of The Physical Review).

electrons into empty orbitals of the projectile (see Halpern and Law,[96] Nikolaev,[115] Garcia et al.,[94] and McGuire.[95]). A commonly used approximation to the capture process is called the Brinkman–Kramers (BK) approximation (Halpern and Law,[96] Nikolaev[115]). It is simply the PWBA for a point charge ion (bare nucleus) including the normally omitted function for electron pickup by the projectile. The BK approximation is only valid for $v_1 \gg v_2$, although it appears to work well as low as $v_1 \sim v_2$. In general, the electron capture probability is at a maximum when target and projectile have some matched energy level and the velocity of the projectile matches the velocity of the bound electron to be captured. Above its maximum value, the capture cross section drops fairly fast (E^{-6} for $v_1 \gg v_2$) with bombarding energy (see Ref. 113). McGuire[95] has shown for small Z_1, where $U_1 < U_{\text{target}}$, that the charge exchange from target K shell to bare nucleus projectile scales as Z_1^4 and N^{-2}, where N is the principal quantum number of the capturing level in the projectile. As an example for $F^{9+} \to Ar$ collisions, Fig. 38 from McGuire shows the bombarding energy and N dependence for charge exchange in comparison to that for ionization. Of course, for nonbare nucleus bombardment the charge exchange from target to ion K shell depends upon whether or not a vacancy exists in the projectile K shell. Figure 39 demonstrates projectile charge dependence on charge exchange. Much theoretical work remains to be done to characterize charge exchange by non-point charge ions, particularly at low energies and for $Z_1 \sim Z_2$.

[115] V. S. Nikolaev, Sov. Phys.-JETP **24**, 847 (1967).

9.3.2.4. Some Experimental Studies of Basic Mechanisms.

Numerous Auger electron measurements have greatly contributed to the understanding of the previously discussed collision mechanisms. We restrict ourselves to discussing a few illustrative experiments for each collision mechanism.

Studies of Auger electron production from collisions described by the electron promotion model have been restricted primarily to cross section or yield measurements because of the severe kinematical broadening.[116] For comparison with theoretical descriptions, Auger spectroscopy techniques have been used to study the Ne → Ne collision system by McCaughey et al.,[117] Fastrup et al.,[118] Cacak and co-workers,[119] and Stolterfoht et al.[116] As an example of what IAES can provide, we review some of the results of Stolterfoht and co-workers.[116] They measured (1) Auger electron production cross sections for $Ne^{0,1,2+}$ → Ne collisions in the energy range from 45 keV to 2.2 MeV, (2) angular and relative intensity of target (T) and projectile (P) Auger decay, and (3) average charge state of P and T after collision via Auger average energy (see Section 9.3.1.2). Figure 40 shows their results for total K-vacancy production cross section in Ne^+ → Ne collisions. The theoretical curve of Briggs and Macek[98] apparently agrees with experiment below 200-keV bombarding energy but significantly underestimates the vacancy production probability at higher energies. This disagreement results from a breakdown in the conventional two-state approximation for producing K vacancies (see Section 9.3.2.1). For collision velocities corresponding to ions having energy above 200-keV, N_i, the vacancy occupation number becomes a function of velocity.[118] This phenomenon results from the fact that at higher energies there is an increased probability for outer-shell vacancy production (hence rising N_i) prior to reaching the K-shell radius, where 1s electron promotion takes place. The incident ion charge state dependence of the cross section vs. energy measurements is also a good test of $N_i(v)$. According to theory[91,98] the vacancy production cross section should scale as $0:1:2$ where $i = 0, 1, 2$ and $N_i = 0, \frac{1}{6}, \frac{1}{3}$. Figure 41 shows the experimental data for this regime. For all the charge states the breakdown is again attributed to the velocity dependence of N_i. Stolterfoht and coworkers empirically determined that N_i varies as $e^{-1/v}$, in agreement with a model proposed by Barat and Lichten[120] and Meyerhof.[100] Stolter-

---

[116] N. Stolterfoht, D. Schneider, D. Burch, B. Asgaard, E. Bøving, and B. Fastrup, *Phys. Rev. A* **12**, 1313 (1975).

[117] M. P. McCaughey, E. J. Knystantas, H. C. Hayden, and E. Everhart, *Phys. Rev. Lett.* **21**, 65 (1968).

[118] B. Fastrup, E. Bøving, G. A. Larsen, and P. Dahl, *J. Phys. B* **7**, L206 (1974).

[119] P. K. Cacak, Q. C. Kessel, and M. E. Rudd, *Phys. Rev. A* **2**, 1327 (1970).

[120] M. Barat and W. Lichten, *Phys. Rev. A* **6**, 211 (1972).

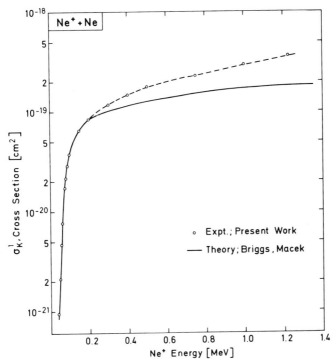

FIG. 40. Total cross section σ_K^1 for single-K vacancy production in Ne$^+$ → Ne collisions as a function of incident-ion energy from work of N. Stolterfoht, D. Schneider, D. Burch, B. Asgaard, E. Bøving, and E. Fastrup, *Phys. Rev. A* **12**, 1313 (1975) (courtesy of The Physical Review).

foht et al.[116] also looked at the vacancy production for the projectile and found that target and projectile yields were identical and isotropic. In addition, the mean charge states of the target and projectile were found to be equal. This indicates that the incident ion reaches an equilibrium outer-shell charge state distribution in the course of interacting with one target atom!

Measurements by Stolterfoht[13] and Woods and co-workers[33] have provided K-vacancy sharing ratios over a wide range of collision partners and interaction velocities. Figure 42 summarizes this work. The only deviations from the theoretical ratio (see Section 9.3.2.1) are for light projectile collision systems such as B$^+$ → CH$_4$ and Li$^+$ → He. These discrepancies were explained by *two* electron transitions from the 2pσ to 2pπ MO. Bøving[121] has also recently considered these deviations for B + C and O + Ne collisions.

[121] E. Bøving, *J. Phys. B* **10**, L63 (1977).

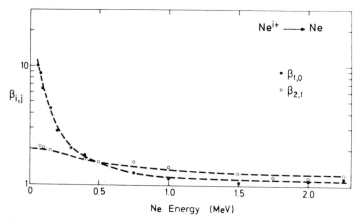

FIG. 41. Cross-section ratio $\beta_{1,0} = \sigma_A^{i=1}|_{\sigma_A^{i=0}}$ and $\beta_{2,1} = \sigma_A^{i=2}|_{\sigma_A^{i=1}}$ as a function of incident-particle energy. σ_A^i denotes the Auger-electron production cross section for the Ne^{i+} + Ne collision with $i = 0, 1,$ and 2. From N. Stolterfoht, D. Schneider, D. Burch, B. Asgaard, E. Bøving, and B. Fastrup, *Phys. Rev. A* **12**, 1313 (1975) (courtesy of The Physical Review).

Numerous reviews have been given for the study of Coulomb ionization by Auger electron measurements (see Stolterfoht,[13] Rudd and Macek,[7] Richard,[10] Garcia et al.,[9] and Ogurtsov[122]). In discussing some of the Coulomb ionization measurements it is convenient to separate yield or cross section determinations from the more spectroscopic measurements such as determination of satellite structure and linewidths. Auger yield measurements have been used to determine vacancy production cross sections (singly and multiply ionized) as a function of Z_1, v_1/v_2, and atomic charge Q of incident particle. High-resolution Auger measurements have been used to determine single to double K-vacancy production, the average degree of multiple outer-shell ionization, linewidths, line energies, line intensities, and fluorescence yields of multiply ionized atoms.

Considering vacancy production measurements with low-Z projectiles, we know that $\omega_K \ll 1$ for these collisions; hence the Auger production cross section essentially equals that for ionization (i.e., $\sigma_T \sim \sigma_A$). As an example, Stolterfoht and Schneider[123] have measured K-shell ionization cross sections for proton bombardment of gaseous targets such as CH$_4$, N$_2$, and Ne. These measurements, which are shown in Fig. 43, provide a good test of the BEA formalism discussed earlier (Section 9.2.2.2). Simi-

---

[122] G. N. Ogurtsov, *Rev. Mod. Phys.* **44**, 1 (1972).
[123] N. Stolterfoht and D. Schneider, *Phys. Rev. A* **11**, 721 (1975).

9.3. STUDIES OF ION–ATOM COLLISION PHENOMENA

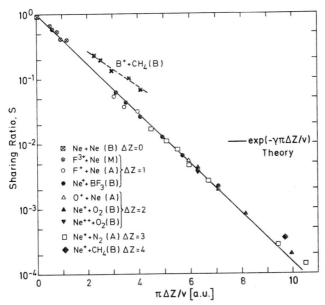

FIG. 42. Ratio S for K-vacancy sharing as a function of $\pi \Delta Z/v$, where v is the projectile velocity and ΔZ the difference in the atomic numbers of the collision partners. Labels A, B, and M refer to measurements at Aarhus by Fastrup et al. (Ref. 131, p. 1058), at Berlin by Stolterfoht et al. (Ref. 131, p. 1060), and at Manhattan by Woods et al. (Ref. 34). The theoretical results with $\gamma = 0.87$ a.u. are obtained using the formalism by Meyerhof (Ref. 100) (courtesy of N. Stolterfoht).

lar measurements have also been performed by Toburen[14] and Watson and Toburen.[124] The discrepancy between theory and experiment is with the uncertainty of measurement except at low-collision velocities, where the inclusion of increased binding energy and Coulomb deflection should be added. Argon L-vacancy production as a function of proton and He^+ bombarding energy has also been tested experimentally (see Volz and Rudd,[20] Crooks and Rudd,[125] Rudd,[126] Watson and Toburen[124] and Stolterfoht[127]). The Z_1^2 of the ionization cross section as a function of v_1/v_2 has also been tested by Stolterfoht and Schneider[113] for H^+, He^{2+} bombardment of CH_4, N_2, Ne targets and by Watson and Toburen for D^+, He^{2+} bombardment of C_2H_4. In both sets of measurements the proper ratio $\sigma(He^{2+}) = (Z_1 = 2)^2/\sigma(H^+)$ is only obtained for $v_1/v_2 \sim 1$. Again, in-

---

[124] R. L. Watson and L. H. Toburen, Phys. Rev. A 7, 1853 (1973).
[125] G. B. Crooks and M. E. Rudd, Phys. Rev. Lett. 25, 1599 (1970).
[126] M. E. Rudd, Phys. Rev. A 10, 518 (1974).
[127] N. Stolterfoht, D. Schneider, and P. Ziem, Phys. Rev. A 10, 81 (1974).

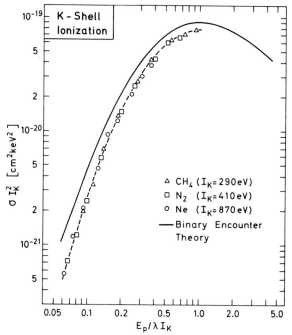

FIG. 43. Scaled cross sections for K-shell ionization of carbon, nitrogen, and neon. Scaling parameters used are the K binding energy I_K and mass ratio λ for protons and electrons. From N. Stolterfoht and D. Schneider, *Phys. Rev. A* **2**, 721 (1975). Theory from J. D. Garcia, *Phys. Rev. A* **1**, 1402 (1970), and *ibid.* **4**, 455 (1971) (courtesy of The Physical Review).

creased binding and Coulomb deflection prevents the proper scaling for lower velocity ratios.

For heavy-ion bombardment the PWBA or BEA are not expected to hold since the requirement of $Z_1 \ll Z_2$ (i.e., a point charge) is not met. Many measurements of x-ray production cross sections have been done in this region. To the contrary, Auger measurements are few in number, and primarily restricted to the Cl → Ne K-Auger experiments of Burch *et al.*,[128] the F → Ne work of Woods *et al.*,[34] and the F → Ar L-Auger measurements of Matthews and Fortner.[129] Unfortunately, these Auger measurements are performed on collisions where the Coulomb ionization process is not the exclusive excitation mechanism (see Woods[34]). In addition, owing to the high degree of multiple ionization present, states with

[128] D. Burch, N. Stolterfoht, D. Schneider, H. Wieman, and J. S. Risley, *Phys. Rev. Lett.* **32**, 1151 (1974).
[129] D. L. Matthews and R. J. Fortner, unpublished.

9.3. STUDIES OF ION-ATOM COLLISION PHENOMENA

TABLE III. Neon K-Auger Electron Production Cross Sections[a,b]

Projectile	$V = v_t/v_e$	Energy (MeV)	Charge state, $\sigma_A \times 10^{19}$ (cm^2)						
			1	4	5	6	7	8	9
H	0.97	1.5	1.11	—	—	—	—	—	—
H	1.12	2.0	1.18[c]	—	—	—	—	—	—
H	1.37	3.0	1.11	—	—	—	—	—	—
H	1.58	4.0	0.94	—	—	—	—	—	—
H	1.77	5.0	0.84[d]	—	—	—	—	—	—
O	0.97	24.0	—	32	32	45	80	113	—
O	1.09	30.0	—	—	41	56	83	138	—
O	1.17	35.0	—	—	—	52	74	121	—
F	0.97	28.5	—	—	37	34	60	91	135

[a] From Ref. 130.
[b] Relative errors are estimated to be ±15%.
[c] Cross sections are normalized to 2.0-MeV H$^+$ as given by Schneider et al. (1973) (Ref. 132).
[d] Burch et al. (Ref. 128) obtain a cross section of 0.86 × 10^{-19} cm^2, in good agreement with the present value.

large fluorescence yields are formed; therefore, the approximation $\sigma^T \sim \sigma_A$ is no longer valid. Finally, the atomic charge state of the ion can significantly influence the resultant ionization cross section so that comparison with theory is problematic. Table III from the work of Woods et al.[130] demonstrates that Z_1^2 scaling is not observed for H,O,F → Ne collisions. The Auger production cross sections for neon appear to be ~50% low for the small projectile charge states. This same trend has been noticed by Matthews and Fortner for argon L-vacancy production by 1–2 MeV/amu fluorine ions. Figure 44 (Ref. 130) shows the dependence of neon K-vacancy production on bombarding oxygen charge state. The previously mentioned failure of $Z_1^2 \sigma_p(H)$ scaling is shown by the dashed line, where σ_p is the prediction of H$^+$ → Ne K-vacancy production. The full line is $q^2 \sigma_p$, which supposedly implies maximum possible screening of the projectile nucleus by its bound electrons. According to calculations by Basbas,[131] this large degree of screening is somewhat unphysical. Basbas calculates that screening effects are small for projectile velocities less than that of the target K-shell electron except for screening by the projectile 1s electrons. Outer-shell screening effects by the 2s and 2p electrons become more appropriate for ion velocities like ~20 MeV/amu

[130] C. W. Woods, R. L. Kauffman, K. A. Jamison, C. L. Cocke, and P. Richard, *J. Phys. B* **7**, L474 (1974).
[131] G. Basbas, *Proc. IXth Int. Conf. Physics Electron Atom. Coll.*, p. 502. University of Washington Press, Seattle, 1975.

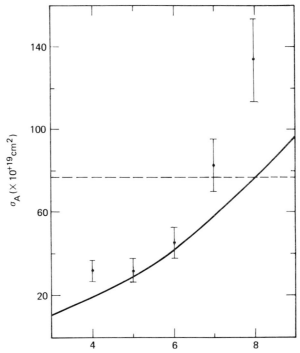

FIG. 44. Charge state dependence of neon K-shell Auger electron production by 24-MeV oxygen. The broken curve is $Z^2\sigma_p$, where σ_p is the Coulomb ionization prediction for 1.5-MeV $H^+ \to Ne$. The full curve is $q^2\sigma_p$. From C. W. Woods, R. L. Kauffman, K. A. Jamison, C. L. Cocke, and P. Richard, *J. Phys. B* **7**, L474 (1974) (with permission, The Institute of Physics, Techno House, Bristol, England).

in the calculated cases of $O^{5+} \to Al$ K-shell ionization. Again, referring to Fig. 44, the dramatic increase in vacancy production for oxygen projectile 1s electrons missing is probably due to electron capture of neon 1s electrons into empty projectile 1s orbitals. This effect is discussed later. In another projectile charge state, dependence of neon K-vacancy production, namely $Cl^{n+} \to Ne$ collisions, Burch and co-workers[128] attributed variations in total vacancy production to screening effects. Here again they assume 0.75 charge units of screening for each projectile electron—a somewhat unphysical value. More likely is electron capture of target K-shell electrons by the highly ionized chlorine L-shell.

Now we discuss some Auger spectroscopy measurements that relate to Coulomb ionization studies. Figure 45 from a review paper by Stolterfoht[74] best summarizes the neon K-Auger satellite production dependence on the projectile Z. The extent of multiple ionization drastically in-

9.3. STUDIES OF ION–ATOM COLLISION PHENOMENA

creases with increasing projectile Z. This phenomenon is demonstrated by a general decrease in average Auger line energy as discussed in Section 9.3.1.2. For 4.2-MeV $H^+ \to$ Ne bombardment the neon K-Auger spectrum is relatively simple. However for O^{5+} and Cl^{12+} (see Matthews et al.[30,83]) bombardment, a complicated Auger spectrum emerges due to overlaps from many Auger lines of different charge states. Fortunately, the spectrum returns to simplicity for the violent $Xe^{31+} \to$ Ne collisions, where only the Auger decay of lithium- and helium-like (doubly excited, i.e., double K vacancies) configurations are observed. These few-electron neon Auger lines provide us with the only opportunity to study individual multiplet states produced in ion–atom collisions. In a later section we return to a detailed study of the lithiumlike neon studies formed by Cl^{12+} and Xe^{31+} bombardment.

Stolterfoht and co-workers[113] using the binomial distribution recipe for

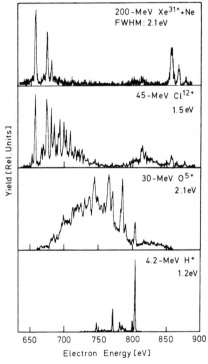

FIG. 45. Neon K-Auger spectra excited by various projectiles. Measured linewidths shown for each spectrum. From N. Stolterfoht, *Proc. Fourth Conf. Sci. Indust. Appl. Small Accel.* (J. Duggan and I. L. Morgan, eds.), p. 311. *IEEE Catalog #76CH 1175-9 NPS* (1976). Reprinted with permission of the IEEE, New York.

TABLE IV. Peak Widths vs. Proton Bombardment in
H$^+$ + Ne K-Auger Measurements[a]

Proton energy (MeV)	ΔE FWHM KL$_{2,3}$L$_{2,3}$(^{1}D$_2$) line (eV)
0.15	1.06
0.20	0.81
0.40	0.73
0.50	0.67
0.60	0.64
6.0	0.51

[a] From Ref. 30.

multiple ionization (Section 9.3.2.2) have determined the charge state distribution and hence individual cross sections for O^{5+} → Ne collisions. A description of this work is given in Section 9.3.3.

Figure 45 also demonstrates that the measured linewidths (including spectrometer transmission function) vary, depending on Z_1. The widths of these lines are determined by the recoil velocity distribution of the target ion. This means that the widths should be proportional to projectile momentum or energy. Matthews and co-workers[30] as shown in Table IV have demonstrated this effect for p → Ne K-Auger linewidth measurements. The linewidths obtained with e$^-$ bombardment and 6.0-MeV H$^+$ projectiles are essentially the same, indicating as expected that momentum transfer at high impact velocities is negligible.

Even with proton bombardment, the probability of multiple ionization can be appreciable (see Hansteen and Mosebekk[90]). Figure 46 (from Ref.

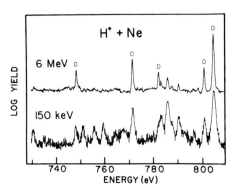

FIG. 46. Comparison of neon KLL-Auger spectrum produced by 6-MeV and 150-keV proton bombardment. Letter D labels diagram lines. All other lines of appreciable strength are attributed to initial single K-, single L-vacancy states. From D. L. Matthews, B. M. Johnson, J. J. Mackey, L. E. Smith, W. Hodge, and C. F. Moore, *Phys. Rev. A* **10**, 1177 (1974) (courtesy of The Physical Review).

9.3. STUDIES OF ION–ATOM COLLISION PHENOMENA

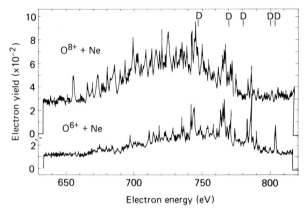

FIG. 47. Neon K-Auger spectrum produced by 35-MeV O^{6+} and O^{8+} bombardment. Letter D marks position of lines from single K-vacancy to L-vacancy states. The strong low-energy lines (~655–670 eV) seen with O^{8+} bombardment stem from formation of lithiumlike neon vacancies. From C. F. Moore, J. Mackey, L. Smith, J. Bolger, B. Johnson, and D. Matthews, *J. Phys. B* **7**, L302 (1974) (with permission, The Institute of Physics).

20) shows the neon K-Auger spectrum from 6-MeV H^+ and 0.15-MeV H^+ bombardment. Only the lines marked D come from singly ionized atoms—all others come from atoms having one and two L-shell vacancies in addition to the K vacancy. The phenomenon of enhanced satellite production by low-energy protons has also been studied by Schneider *et al.*[132] Their measurements of the energy dependence of the effect also indicates its disappearance with high bombarding energies. The enhancement at low collision velocities can be attributed to a variation in the ratio of L- to K-vacancy production probability over the available range of impact parameters which are bombarding-energy dependent.

Hansteen and Mosebekk[90] predicted a Z_1^2 scaling in the first satellite (one K, one L vacancy) to normal line production for velocity- matched low-Z ions. Matthews *et al.*[133] observed this effect for 1.0-MeV He^+ and 0.25-MeV H^+ bombardment of neon. The q^2 (atomic charge) dependence of Auger satellite production was first tested by Moore *et al.*[134] in the $O^{6+}, O^{8+} \rightarrow$ Ne K-Auger measurements. Figure 47 demonstrates their results. Note the decreasing centroid energy with increased projectile

---

[132] D. Schneider, D. Burch and N. Stolterfoht, *Proc. 8th Int. Conf. Physics Electron Atom. Coll.*, p. 729. Institute of Physics, Beograd, 1973.

[133] D. L. Matthews, B. M. Johnson, J. J. Mackey, and C. F. Moore, *Phys. Lett.* **45A**, 447 (1973).

[134] C. F. Moore, J. J. Mackey, L. E. Smith, J. Bolger, B. M. Johnson, and D. L. Matthews, *J. Phys. B* **7**, L302 (1974).

TABLE V. Experimental Energies and Relative Intensities for Initial ^{3}P and ^{1}P States with Configuration $1s^12s^22p^5$ of Neon[a]

Energy (eV)	Final state configuration	Relative intensity				
		Ref. (14)	Electron	Proton	Alpha	Oxygen
Triplet						
730.4	$2s^02p^5(^2P)$	0.12	0.16	0.17	0.17	0.29
751.0	$2s^12p^4(^2P)$	0.11	0.14	0.16	0.16	0.14
753.6	$2s^12p^4(^2S)$	0.02	—	0.02	0.02	—
759.2	$2s^12p^4(^2D)$	0.17	0.20	0.26	0.26	0.19
767.9	$2s^12p^4(^4P)$	0.03	—	0.12	0.12	0.16
783.1	$2s^22p^3(^2P)$	0.31	0.42	0.44	0.45	0.55
785.7[b]	$2s^22p^3(^2D)$	0.63	0.70	1.28	1.29	1.00
		Sum 1.62	2.45	2.47	2.33	
Singlet						
734.9	$2s^02p^5(^2P)$	0.09	0.10	0.11	0.11	0.20
755.6	$2s^12p^4(^2P)$	0.22	0.21	0.24	0.23	0.21
787.6	$2s^22p^3(^2P)$	0.23	0.23	0.20	0.20	0.21
790.2	$2s^22p^3(^2D)$	0.46	0.46	0.45	0.46	0.38
		Sum = 1.0				

[a] From Ref. 135.
[b] This line is probably unresolved.

charge states, signifying greater multiple ionization. The first appearance of the lithiumlike neon Auger lines observed clearly with Cl^{12+} and Xe^{31+} was first noticed in the O^{8+} spectrum. As expected, the greater multiple ionization is produced by ions with an open shell into which electron capture can take place.

A special test of the statistical nature of these heavy-ion–atom collisions is presented later in this part. For now, however, there exists a measurement that should be mentioned in connection with production of statistical population of multiplet states. Bhalla et al.[135] examined the production of $1s^12s^22p^5$ (^{1,3}P)-Auger emitting states for oxygen, helium, and hydrogen bombardment of neon. Their results are summarized in Table V. It is interesting to note that the proper statistical population ratio $^3P/^1P = 3$ is more nearly approached with ion bombardment than with e^- impact. Curiously, high-resolution x-ray measurements have shown statistical populations for helium-like states in neon. This certainly indicates the need for further study of the $1s^12s^22p^5$ neon KLL-Auger line strengths.

Auger experiments demonstrating the charge exchange mechanism are

[135] C. P. Bhalla, D. L. Matthews, and C. F. Moore, *Phys. Lett.* **46A**, 336 (1974).

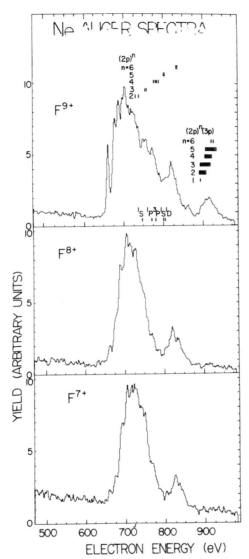

Fig. 48. Projectile charge state dependence of neon Auger electrons by fluorine at 25 MeV. The hypersatellite peak at 915 eV is present only in the F^{9+} case. The calculated positions for $2p^n 3p$ initial configurations are shown. The spectrum filling near 750 eV is attributed to hypersatellite Auger electrons from the initial configurations $(2p)^n$, as shown. The intensity of the lithiumlike configurations is also dramatically increased for the F^{9+} case. From C. Woods, R. L. Kauffman, K. Jamison, N. Stolterfoht, and P. Richard, *Phys. Rev. A* **12**, 1393 (1975) (courtesy of The Physical Review).

somewhat limited. Figure 44, which shows the q dependence of neon K vacancy production by oxygen ions,[35] is at present the only Auger yield measurement with some reasonable evidence for K-shell to K-shell charge exchange. Recall that the cross section was greatly enhanced when the oxygen K shell was open (i.e., charge states $q = 7, 8$). This behavior agrees with the prediction of McGuire[95] for charge exchange in $F^{8,9+}$ collisions with argon (see Fig. 39).

An excellent verification of charge exchange via an Auger spectroscopy experiment has also been presented by Woods and co-workers.[34] They demonstrate in Fig. 48 that Auger hypersatellite lines (emission from states having two K vacancies) are observed when using F^{9+} bombardment. Furthermore, Fig. 49 shows the energy dependence of the production probability for these lines as compared to the total Auger yield. The fast decrease in yield with bombarding energy agrees with calculations

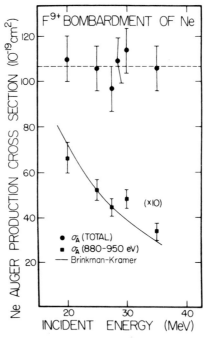

FIG. 49. Total K-shell Auger electron cross section (●) and partial cross section for exciting the hypersatellites centered at 915 eV (■). The latter is multiplied by 10 in the figure for clarity. The absolute accuracy is about 30% and the energy dependence is accurate to 10%, as shown. The solid curve is the Brinkman–Kramer calculation for electron capture into the K shell of the projectile from the K shell of the target, normalized at 25 MeV. The dashed line indicates the trend of the total K-shell Auger production cross section. From C. W. Woods, R. L. Kauffman, K. Jamison, N. Stolterfoht and P. Richard, *Phys. Rev. A* **12**, 1393 (1975) (courtesy of The Physical Review).

9.3. STUDIES OF ION–ATOM COLLISION PHENOMENA

using the Brinkman–Kramers approximation to charge exchange. Obviously, many more IAES studies of charge exchange by highly stripped ions needs to be performed.

9.3.3. Selected Experimental Topics

9.3.3.1. Measurement of Fluorescence Yields and Ionization Probabilities for Multiply Ionized Neon.

The fact that fluorescence yield increases with the degree of mutliple ionization was introduced in Section 9.3.1.4. Even though this phenomenon has been known theoretically for some time the actual measurement of this effect has only been recently performed. The first experimental indication of the effect came from the work of Burch et al.[48] For neon, they simultaneously measured the ratio of total x-ray to total Auger yield for proton and oxygen bombardment. Their measurements indicated that the fluorescence yield for the neon K shell was 2.4 times greater following O^{5+} rather than H^+ bombardment. Since we know from Figs. 2, 29, and 45 that O^{5+} inflicts more multiple ionization of neon than H^+, then we can conclude that the K-shell fluorescence yield is increasing with the extent of outer-shell ionization.

The only drawback to the method used by Burch et al.[48] was their assumption that $\sigma_X \ll \sigma_A$, and hence $\tilde{\omega} \simeq \sigma_X/\sigma_A$. Normally this approximation is valid but when using chlorine (or xenon) bombardment the states formed have high fluorescence yields, hence the proper expression $\tilde{\omega} \simeq \sigma_X/(\sigma_X + \sigma_A)$ must be used. In another measurement, Burch and colleagues[128] determined $\tilde{\omega}_K$ for neon following 50-MeV $Cl^{8+} \to Ne$ collisions. Figure 50 shows their results for $\tilde{\omega}_K$ variation as a function of projectile charge state. The rapid variation can be attributed to enhanced production of two- and three-electron neon excited states (large ω value) when using highly stripped chlorine ions. Fluorescence yield values for the different charge states of single K-multiple L ionization of neon have also been determined by experiment (see Stolterfoht et al.,[113] Matthews et al.,[136] Schneider and colleagues,[54] and Moore and Matthews[137]). The method is semiempirical in that measured values of x-ray (R_n) and Auger (Q_n) satellite strengths and $\tilde{\omega}_K$ values are combined with the binomial distribution for satellite formation (see Section 9.3.2.2). In this case the ionization probability of charge state n is

$$q_n^{\text{emp}} = \binom{8}{n} P_{K,L}^n (1 - P_{K,L})^{n-8}, \qquad (9.3.14)$$

---

[136] D. L. Matthews, B. M. Johnson, L. E. Smith, J. J. Mackey, and C. F. Moore, *Phys. Lett.* **48A**, 93 (1974).

[137] C. F. Moore and D. L. Matthews, *Proc. Third Conf. Sci. Industr. Appl. Small Comp.*, p. 419. NTIS Rept. #CONF-741040-P1, 1974.

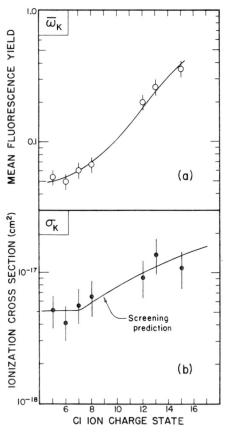

FIG. 50. Projectile charge state dependence of the mean neon K-fluorescence yield and neon K-excitation cross section in 50-MeV $Cl^{n+} \to$ Ne collisions. Solid curve in (b) is obtained assuming screening of the projectile nuclear charge. From D. Burch, N. Stolterfoht, D. Schneider, H. Wieman, and J. S. Risley, *Phys. Rev. Lett.* **32**, 1151 (1974) (courtesy of The Physical Review).

where $P_{K,L}$ is the probability of simultaneous K- and L-shell ionization. q_n^{ex} values can also be determined from experiment using formula $q_n^{ex} = \bar{a}Q_n + \bar{\omega}R_n$, where $\bar{a} = 1 - \bar{\omega}$ is the total probability of Auger decay. Values of the x-ray relative intensities were determined for all n by Kauffman et al.[71] and Matthews and co-workers.[138] Auger satellite strengths Q_n could only be determined for $N = 0, 1, 6$ because of overlapping lines. Therefore q_n^{ex} values were used to determine $P_{K,L}$ in expression (9.3.14), thus allowing q_n to be generated for the remaining charge states. Individ-

[138] D. L. Matthews, B. M. Johnson, and C. F. Moore, *Phys. Rev. A* **10**, 451 (1974).

TABLE VI. Measured X-Ray Ratios R_n, etc., Corrected for Multiplet Splitting: Experimental Ionization Probabilities q_n^{ex}; Semiempirical Ionization Probabilities q_n^{em}; Fluorescence Yields ω_n [a]

Initial ionization state KL^n n	30-MeV O^{5+} ($\bar\omega = 0.038$)					35-MeV O^{8+} ($\bar\omega = 0.076$)				
	R_n	q_n^{ex}	q_n^{em}	$\omega(n)$	$\omega(n)$ [b]	R_n	q_n^{ex}	q_n^{em}	$\omega(n)$	$\omega(n)$ [b]
0	—	0.028	0.028	—	0.018	—	0.0046	0.0046	—	—
1	0.061	0.127	0.127	0.018	0.019	0.008	0.037	0.035	0.017	0.018
2	0.138	—	0.250	0.021	0.022	0.031	—	0.018	0.020	0.021
3	0.159	—	0.280	0.022	0.031	0.074	—	0.227	0.025	0.027
4	0.229	—	0.195	0.045	0.054	0.172	—	0.273	0.048	0.053
5	0.117	—	0.088	0.051	0.099	0.129	—	0.210	0.047	0.125
6	0.239	0.02	0.024	0.378	0.161	0.451	0.117	0.101	0.339	0.201
				(0.450)[c]					(0.293)[c]	
7	0.058	—	0.004	0.551	1.000	0.142	—	0.028	0.385	1.000

[a] From Ref. 113.
[b] See Ref. 3.
[c] The $\omega(n)$ values in parentheses are obtained by using q_n^{ex} as opposed to q_n^{em}.

ual charge state fluorescence yields are then obtained from the expression $\tilde{\omega}_n = \bar{\omega} R_n/q_n$. Table VI from the work of Stolterfoht et al.[113] summarizes the semiempirically determined multiple-ionization probabilities q_n and fluorescence yields ω_n for $O^{5+,8+}$ bombardment of neon. Similar values were later obtained by Moore and Matthews.[138] Comparison of ω_n values from experiment and theory was shown in Fig. 34. Remarkably good agreement is shown especially in view of the semiempirical method used. The mean charge state can also be derived from the q_n values, namely $\bar{n} = \sum_{n=0}^{\infty} n q_n = 8 P_{K,L}$; $\bar{n}$ determined in this fashion was used to generate the $\bar{E}_L$ vs. n curve of Fig. 31. The statistical assumption that 2s and 2p ionizations are independent is deeply embodied in the binomial distribution formulation. In Section 9.3.3.2 we describe an experiment that tests this assumption.

9.3.3.2. Excited-State Populations in Lithiumlike Atoms. In this subsection we discuss some experiments that study the intensities of K-Auger lines of lithiumlike atoms found both in the projectile and target atom.

First we consider the statistical nature of populating lithiumlike neon configurations following bombardment by 45-MeV Cl^{13+} ions. Matthews and co-workers[83] have shown that both lithiumlike x-ray and Auger line strengths deviate from the theoretical prediction assuming a statistical population of configurations. Basically, they observed an overpopulation of $1s2s^2$ and $1s2s2p$ configurations relative to the $1s2p^2$. Since McGuire and Richard[109] have shown that statistical populations usually result in collisions governed by direct Coulomb ionization, Matthews and co-workers[83] concluded that other types of ionization processes must be present to explain the nonstatistical behavior.

A more detailed discussion of the results of the Cl → Ne measurement plus inclusion of new data from Xe → Ne measurements is instructive. Matthews and co-workers[83] define

$$I(^{2s+1}L) = \frac{P_c P_m [1 - \omega(LS)] I'(^{2s'+1}L')}{P'_c P'_m [1 - \omega'(LS')]} \quad (9.3.15)$$

as the relative Auger intensity of a lithiumlike state having quantum number L,S and fluorescence yield $\omega(LS)$ compared to the state $L'S'$. Also they define P_c as the probability for forming a given lithiumlike configuration and P_m as the probability of forming the multiplet state described by $L_1 S$. If we assume that the theoretical fluorescence yields $\omega(LS)$ are correct and that P_m are statistical [i.e., populated in proportion to $(2L + 1)(2S + 1)$] then we can test whether or not 2s or 2p vacancies are populated statistically. Of course, the binding energies of 2s and 2p electrons are different so that their ionization probabilities should not be

TABLE VII. Lithiumlike Neon Auger Electron Line Strengths, Experimental and Theoretical[a,b]

Transition to $1s^2(^1S_0)$	Energy (eV)[c] (theor.)	Energy (eV)[d] (expt.)	$(1-\omega_k)$[e]	P_c	P_m	I/I_0 (expt.)[d]	$I/I_0 = (1-\omega_k)P_c P_m$ (theor.)
$1s2s^2(^2S)$	653.2	652.2	1.000	2/56	1	0.20 ± 0.03	0.22
$1s2s2p(^4P^o)$	656.7	656.5	0.775	21/56	12/24	1.00 ± 0.15	$1.0 = I_0$
$(^2P_-)$[f]	668.1	668.9	0.512	24/56	6/24	0.29 ± 0.05	0.33
$(^2P_+)$[f]	674.7	674.5	0.988	24/56	6/24	Unresolved (1.1 ± 0.2)	0.64 } 1.84
$1s2p^2(^4P^o)$	673.7	673.5	0.930	30/56	12/30		1.20
(^2D)	681.3	681.0	0.970	30/56	10/30	Unresolved (~ 0.5)	1.04
(^2P)	684.0	—	0.028	30/56	6/30	—	0.02
(^2S)	688.5	—	0.920	30/56	2/30	—	0.20

[a] From Ref. 83.
[b] Assuming a statistical population of multiplets P_m and configurations P_c.
[c] Refs. 66, 73.   [d] Ref. 54.   [e] Ref. 66, 78.
[f] The $^2P_+$ ($^2P_-$) notation refers to the higher (lower) energy-valued term as explained by C. P. Bhalla in Ref. 2.

equal [unlike what is assumed in expression (9.3.15)]. However, for lithiumlike neon the binding energies only differ by 15 eV out of ~250 eV total, so that the difference in ionization cross section (~12%) is not important compared to other uncertainties.

The statistical values for P_c (assuming 2s and 2p ionization probabilities are the same) along with a comparison of theoretical and experimental line strengths and energies are shown in Table VII. A theoretical Auger spectrum derived from the quantities shown in Table VII and assuming the experimental lineshape is shown in Fig. 51. Comparison of the inten-

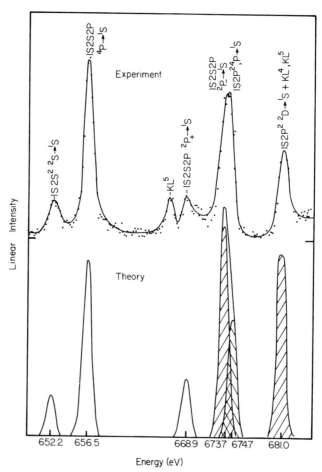

FIG. 51. Lithiumlike neon Auger spectrum produced by Cl^{13+} bombardment from D. L. Matthews, R. J. Fortner, D. Schneider, and C. F. Moore, *Phys. Rev. A* **14**, 1561 (1976) (courtesy of The Physical Review).

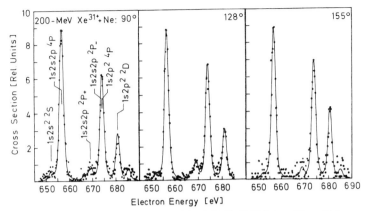

FIG. 52. Neon K-Auger spectra produced in 200-MeV $Xe^{31+} \rightarrow$ Ne collisions. Note growth of ^2D lines at backward angles. From N. Stolterfoht, D. Schneider, R. Mann, and F. Folkmann, *J. Phys. B* **10**, L281 (1977) (with permission, The Institute of Physics).

sity of theoretical and experimental peak height at 675 and 681 eV with the $^4P^0$ line at 656 eV should demonstrate the apparent underpopulation of $1s2p^2$ configurations. The data shown in Fig. 51 are plagued by some overlapping line from boronlike configurations, so that derivation of some of the line strengths is subject to large uncertainties. The more recent $Xe^{31+} \rightarrow$ Ne measurements[139] demonstrate that only lithiumlike configurations remain following the collision. The spectra of Stolterfoht and co-workers for three different observation angles are shown in Fig. 52. They measure some anisotropy in the relative line strengths, which will be discussed later. More importantly, the spectral peak heights corroborate the results of Matthews and co-workers. Namely, theory would give relative peak heights of 10:13:10 for the three most prominent lines at 656, 674, and 681 eV. However, estimating their spectral strengths for these lines gives values of 10:7:3, 10:7.8:2.9, and 10:7.8:4.2 at 90, 128, and 155°, respectively. Therefore, an underpopulation of the $1s2p^2$ is definitely observed. In addition, correction for the anisotropy definitely cannot improve the agreement.

The exact mechanism for producing this nonstatistical population of configurations is not known. Matthews *et al.*[83] suggested that a cascade-feeding model produced the surplus, although further theoretical consideration of the ideas enbodied in this model is necessary. Another unexplored possibility is two-step ionization by a combination of both Coulomb ionization and charge exchange. The Coulomb ionization contribution would supposedly be statistical, whereas further ionization by

[139] N. Stolterfoht, D. Schneider, R. Mann, and F. Folkmann, *J. Phys. B* **10**, L281 (1977).

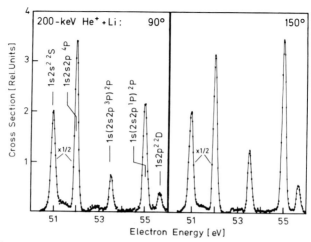

FIG. 53. Auger spectra of lithium following 200-keV He$^+$ → Li collisions. Lab observation angles are 90 and 150° as indicated. From P. Ziem, R. Bruch, and N. Stolterfoht, *J. Phys. B* **8**, L480 (1976) (with permission, The Institute of Physics).

electron capture may be nonstatistical. The S, L, M_L dependence of states formed by electron capture has not been explored.

Calculation of theoretical Auger intensities for other lithiumlike atoms has yet to receive attention. Nonetheless, interesting experiments have been performed. For example, Ziem *et al.*[140] have studied the Z_1 dependence of Auger decay for states formed by excitation from the ground state of lithium target atoms. They were able to relate the production of high spin states to the availability of electrons for exchange by the projectile. For instance, Fig. 53 shows the lithium Auger spectrum following excitation by 200-keV He$^+$ bombardment. The high spin value $^4P^0$ state (52 eV) is only formed when the He$^+$ ion is used for a projectile. Formation of these states using H$^+$ bombardment is unlikely since excitation from the lithium ground state ($1^2S_{1/2}$) to the 2^4P^0 states requires a spin–flip excitation that is a second-order effect in the Coulomb interaction matrix element (dipole term dominates). On the contrary, a significant probability for electron exchange exists when using He$^+$ bombardment. The electron exchange excitation mechanism is not spin selective.

Obviously, much theoretical work should be done on the lithium atom because of its simplicity. It will be essential to know the fluorescence yields of excited states as well as the L, S dependence of the e^- exchange cross section in order to form a theoretical spectrum.

[140] P. Ziem, R. Bruch, and N. Stolterfoht, *J. Phys. B* **8**, L480 (1976); N. Stolterfoht, *Proc. Second Int. Conf. Inner-Shell Ionization Phenom.*, p. 184. Invited Papers, Freiberg, 1976.

TABLE VIII. Lithiumlike KLL-Auger Transitions of Neon[a]

Configuration		Ref. (73)	This work			Ref. (65)
Initial	Final	E_{theor} (eV)	E_{theor} (eV)	E_{expt} (eV)	I^{rel}_{expt} (%)	I^{rel}_{theor} (%)
$1s(2s, 2p)^2$	$^2S^e/1s^2$	656.3	652.7	652 ± 1	7.1 ± 0.6	6.8
$1s(2s, 2p)^2$	$^4P^o/1s^2$	655.6	656.3	656 ± 1	1.4 ± 0.5	0.9[b]
$1s(2s, 2p)^2$	$^2P^e/1s^2$	668.6	668.9	669 ± 1	8.4 ± 0.7	10.4
$1s(2s, 2p)^2$	$^2P^o/1s^2$	672.3	673.1	674 ± 1	43.5 ± 1.2	20.1 ⎫ 42.4
$1s(2s, 2p)^2$	$^4P^o/1s^2$	672.6	673.5	—	—	22.3[b] ⎭
$1s(2s, 2p)^2$	$^2D^e/1s^2$	681.5	681.8	682 ± 1	32.7 ± 1	32.8 + 0.02 = 32.8
$1s(2s, 2p)^2$	$^2D^e/1s^2$	682.8	683.2	—	—	
$1s(2s, 2p)^2$	$^2S^e/1s^2$	688.5	693.3	693 ± 1	6.5 ± 0.5	6.3

[a] From Ref. 141.
[b] For the metastable states the incomplete decay during the passage of the electron-emitting projectiles through the spectrometer focal region is taken into account.

Auger electron intensity measurements following the decay of projectile lithiumlike states have also been performed. Curiously, for the case of neon ions stripped and excited by a thin carbon foil, Schumann et al.[141] have shown that the lithiumlike states are statistically populated. This may indicate a fundamental difference in the ionization process for projectiles excited in multiple collisions in a solid foil as compared to atoms excited in a single collision.

Table VIII summarizes the results of Schumann et al. and demonstrates their observation of a statistical population. Notice that the relative intensities for the $^4P^{e,0}$ lines, which are metastable states, are much less than those observed for the neon target atom. This phenomenon occurs because some of the moving ions can escape the viewing length of electron spectrometer prior to decaying. Schumann and co-workers have properly taken account of this effect in their analysis.

The beam–foil Auger work of Schumann et al. represents an interesting experimental method, which was originated by Groeneveld et al.[35] They study beam–Auger electrons in coincidence with a defined (by electrostatic analysis) final charge state of the beam. As an example, to look at lithiumlike neon initial states they gated the total Auger spectrum in coincidence with a Ne^{8+} beam. Figure 54 shows Groeneveld et al.'s noncoincidence and coincidence spectra. Their interesting spectroscopic phenomena are apparent from the data: (1) the centroid energy of the Auger lines decreases with increasing final charge state in agreement with previous observations and calculations, (2) the spectral complexity or number of Auger lines decreases with increasing final charge state, and (3) the yield of K-Auger electrons per unit ions increases with increasing final charge state. Phenomenon (3) illustrates that even though an ion emerges from the foil in a given high charge state it does not necessarily contain a K vacancy that Auger decays. Evidently, for this case, the highest probability for emerging from the foil with a K vacancy that Auger decays is for lithiumlike initial states ($q_f = 8^+$).

Other beam–Auger measurements have also studied lithiumlike transitions. Work by Sellin et al. for fluorine ions,[47] Schneider et al. for carbon ions[142] and Johnson et al. for oxygen ions[52] are some of the experiments that have determined intensities, energies, and some lifetimes of Auger lines from the decay of lithiumlike states. Some of the more intriguing results have come from studying the lithiumlike decays of low-Z ions such as lithium, beryllium, and boron (see Leithäuser et al.,[143] Pegg et al.,[144]

---

[141] S. Schumann, K. O. Groeneveld, K. D. Sevier, and B. Fricke, *Phys. Lett.* **60A**, 289 (1977).

[142] D. Schneider, unpublished.

[143] U. Leithäuser, D. Schneider, and N. Stolterfoht, in conference proceedings, Reference 140, page 65.

9.3. STUDIES OF ION–ATOM COLLISION PHENOMENA

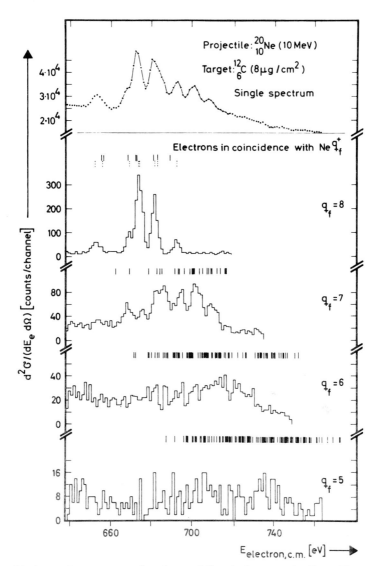

FIG. 54. Auger electron spectra from beam–foil excited neon projectiles, without coincidence (top spectrum) and in coincidence with final charge state $q_f = 5$ to 8^+ (bottom spectra). From K. O. Groeneveld, G. Nolte, S. Schumann, and K. D. Sevier, *Phys. Lett.* **56A,** 29 (1976) (courtesy of North-Holland, Amsterdam).

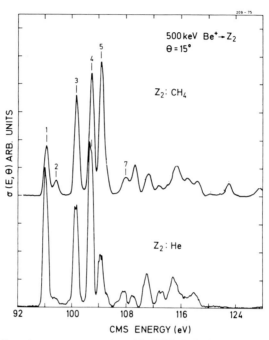

FIG. 55. Beryllium Auger spectra produced in 500-keV $Be^+ \to CH_4$ and $Be^+ \to He$ collisions. Peaks numbered 1–5 and 7 correspond to $n = 2$ states of lithiumlike beryllium. From M. Rødbro, R. Bruch, and P. Bisgaard, *J. Phys. B* **10**, L275 (1977) (with permission, The Institute of Physics).

and Rødbro and co-workers[145]). Figure 55 demonstrates the beryllium K-Auger measurements of Rødbro and colleagues[145] for $Be^+ \to CH_4$, He gas exciter targets. Notice the large variation in the $n = 2$ lithiumlike Auger lines (1–7) with the Z of the exciter gas. At these collision energies, the variations are perhaps caused by the molecular orbital available for e^- promotion. As of now, details of these collisions or effects have not been formulated.

9.3.3.3. Anisotropies in Ion-Induced Auger Electron Yields. The measurement of nonisotropic Auger electron emission from states other than those formed by electron bombardment (see Section 9.3.1.5) has only recently been performed. The measurements were performed by Ziem *et al.*[140] and Stolterfoht *et al.*[139] and involve emission from lithiumlike atoms. An interesting finding in both of these measurements is that

---

[144] D. J. Pegg, H. H. Haselton, R. S. Thoe, P. M. Griffin, M. D. Brown, and I. A. Sellin, *Phys. Lett.* **50A**, 447 (1974); *Phys. Rev. A* **12**, 1330 (1975).

[145] M. Rødbro, R. Bruch, and P. Bisgaard, *J. Phys. B* **10**, L275 (1977).

the high-spin states do not yield nonisotropic Auger decay, whereas the normal states do. In Section 9.3.3.2, we observed that it took different types of collision mechanisms to produce high- and low-spin states. Perhaps the isotropy and anisotropy of Auger yields from certain states is another clue to the infliction of separate ionization mechanisms.

The first collision system studied[140] was 200-keV He$^+$ bombardment of lithium. The ion velocity in this case was approximately equal to the orbital electron velocity of the 1s electron of lithium (i.e., $v_1 \sim v_2$). Therefore, direct Coulomb ionization perhaps with some electron exchange should be the dominant ionization mechanism. The results of Ziem and co-workers demonstrated the change in spectral linestrength for lithium Auger lines observed at 90 and 150°. The 1s2s^2(^{2}S) line should exhibit no anisotropy since it has no M_L dependence. Relative to the 1s2s^2(^{2}S) line there is little change in the high-spin 1s2s2p(^{4}P) line between 90 and 150°. However, the low-spin ^{2}P, ^{2}D lines exhibit nearly 100% more intensity at 150° than observed at 90°. Since the exchange mechanism is responsible for forming the ^{4}P state and that state appears isotropic, one may conclude that the exchange process does not selectively populate M_L values. On the other hand, the ^{2}P, ^{2}D states can be easily formed by direct excitation and at least for He$^+$ bombardment they show strong anistropy. More detailed studies of these collision systems will be necessary to make any further conclusions. For instance, is there any anistropy with H$^+$ bombardment and what is the possible energy dependence of the effect? Answering both of these questions would enable theoretical consideration. It is possible that the electron promotion mechanism enters in this velocity region. Although a theoretical description is necessary to substantiate this statement it is possible with molecular collision to selectively populate magnetic substates. This would occur since σ_{ROT} automatically selects a certain orientation of the collision system for maximum vacancy production.

As opposed to the lithium case just discussed, which depends on e$^-$ excitation for initial-state formation, anisotropy data are also available for the violent collision system 200-MeV Xe$^{31+} \rightarrow$ Ne. In this case, by a combination of ionization processes a ground-state neutral neon atom is stripped and excited into a three-electron atom. Recalling Fig. 52 from Stolterfoht et al.[139] we see the lithiumlike neon Auger spectrum viewed at three laboratory angles: 90, 128, 156°. For this experiment the 1s2s^2(^{2}S) line is not clearly observed because of poor statistics. However, comparison of the ^{2}P, ^{4}P^e, ^{2}D lines to the 1s2s2p^{4}p line at 656 MeV shows some anisotropy. In particular, the ^{2}D line increases by 60% in going from 90 to 155°.

Obviously more anisotropy measurements must be performed to draw reasonable conclusions from this measurement. We can conclude, though, that energetic heavy-ion bombardment does lead to alignment of magnetic sublevels. The exact collision mechanisms responsible for this alignment have yet to be determined.

9.3.3.4. Auger Linewidth Measurements. Earlier we mentioned the utility of measuring characteristic Auger linewidths to determine total transition rates for an inner-shell vacancy. IAES measurements of linewidths are primarily restricted to proton bombardment because of the large kinematical broadening present with heavier ions. The earliest determination of linewidths using proton bombardment was done by Volz and Rudd.[20] They measured the L_2 and L_3 level widths for argon. Most of the newer measurements have been done by Ridder et al.[32] and Schneider and co-workers.[146] They have studied K,L level widths for singly ionized neon and for singly and doubly ionized argon. In principle, the direct measurement of Auger linewidths is a near excellent method for determining lifetimes in the range $\sim 10^{-15}$–10^{-14} sec when there is negligible instrumental broadening. It is only limited by the Doppler broadening due to thermal motion of target (see Dahl et al.[29]). However, we show that even the best available instrumental energy resolution is still large by comparison to widths of some of the states. In addition, the large number of overlapping lines in Auger spectra can limit the number of lines that can be studied. Third, even the Doppler width sometimes dominates the widths of certain excited states. Even so, the technique is powerful and at present is the only practical method for obtaining decay rates of certain states (mainly those with very small fluorescence yields).

Two general problems face the experimenter before taking a high-resolution spectrum, namely, thermal broadening, due to target motion, and unfolding the instrument resolution function from observed spectral lineshapes. Rudd and Macek[7] initially formulated the expression for thermal contributions to linewidth

$$\Delta E(\text{eV}, \text{FWHM}) = 4 \left(E_0 \frac{mkT}{M} \ln 2 \right)^{1/2}, \qquad (9.3.16)$$

where E_0 is the Auger transition energy, M the electron mass, m the target mass, and T the target temperature. For room temperature ($T = 300$ K) the thermal broadening for argon L-Auger lines at 200 eV is 28 meV. Some L-shell states in argon have widths $\Delta E < 9$ meV, while the single-ionized K-vacancy state in neon has a width of 80 meV. Thus, we see that the thermal broadening is nonnegligible in comparison to natural

---

[146] D. Schneider, K. Roberts, W. Hodge, and C. F. Moore, *Phys. Rev. Lett.* **36**, 1065 (1976).

TABLE IX. Calculated Linewidths for $L_{2,3}M_{2,3}$ Argon Auger Lines

Initial state	Final state: transition energy (eV) (Ref. 28)		Width Γ (meV) (Ref. 78)
	^2D	^2P	
$^3D_3$	192.44	190.72	⟨Γ⟩ = 9.25
$^3D_2$	193.34	191.62	⟨Γ⟩ = 9.25
$^3D_1$	194.74	193.02	⟨Γ⟩ = 9.25
$^3P_2$	194.64	192.92	⟨Γ⟩ = 298.5
$^3P_1$	195.94	194.22	⟨Γ⟩ = 298.5
$^3P_0$	195.44	193.72	⟨Γ⟩ = 298.5
$^3S_1$	193.84	192.12	3.8
$^1P_1$	192.14	190.42	12.8
$^1D_2$	196.74	195.02	276.8
$^1S_0$	198.64	196.92	266.7

widths. We must also deconvolute the spectrometer resolution function or lineshape from the measured line profile. Usually the spectrometer function is determined to be Gaussian while the natural lineshape is, of course, Lorentzian. A Lorentzian lineshape convoluted into a Gaussian can be expressed by the Voigt function (see Gelius et al.[80]). Deconvolution of the two out of an spectral line can only reach a reliable degree of accuracy for spectrometer linewidths less than or equal to natural linewidth. This requires the maximum (and sometimes more) resolving power achievable.

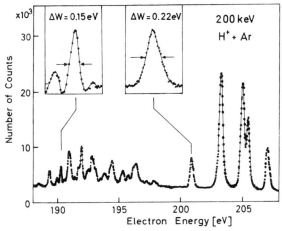

FIG. 56. Argon L-Auger spectrum produced by 200-keV H^+ projectiles. Δω indicates the measured linewidth. From D. Ridder, J. Dieringer, and N. Stolterfoht, Ref. (147). (Courtesy of University of Washington Press.)

TABLE X. Measured Argon L-Linewidths (in meV)[a]

Transition	Proton impact		Electron impact		Theory	
	This work	Other	This work	Other	Ref. (78)	Ref. (63)
$L_{23}M_{23}(^3D)-M_{23}^3$ [b]	≤15	—	≤20	(90)	9.4	9.4
$L_{23}M_{23}(^1P)-M_{23}^3$	80 ± 30	—	90 ± 40	—	13	19
$L_{23}M_{23}(^1D)-M_{23}^3$	150 ± 50	—	140 ± 50	—	278	250

[a] From Ref. 147.
[b] For this case the experimental value and McGuire's value (Ref. 63) correspond to that of the $^3D_3$ initial state, whereas the Chen and Crasemann value is averaged over j.

Assuming Doppler widths and spectrometer resolution forms are properly accounted for, the measurement of L,S,J dependence on linewidth provides a fascinating and rigid test of theory. No Auger measurements of the J dependence on linewidth have been successfully performed because for the main case of interest, i.e., the LM (single L, single M) vacancy lines of argon, lines of different J but the same L,S, drastically overlap with other lines (see Werme et al.[28]) and are therefore unresolvable. Table IX gives an indication of the experimental line energies and theoretical widths (L,S dependence) (Chen and Crasemann[65]) for $L_{2,3}M_{2,3} \rightarrow M_{2,3}^3$ ($^2D,^2P$) lines of doubly ionized Argon. Note that ^3P lines should be nearly 85 times wider than those from the 3s state. Ridder et al.[32] have published the best resolution so far for e⁻ bombardment of argon. Their instrumental width of $\Delta E = 90 \pm 15$ MeV allowed them to measure the widths of the argon $L_{2,3}$ vacancy states and some of the $L_{2,3}M_{2,3}$ multiple vacancy states. Their measurements using proton bombardment yielded larger instrumental widths, as can be seen in Fig. 56. The left inset in the figure shows the $L_{2,3}M_{2,3}(^3D_3) \rightarrow M_{2,3}^3(^2P)$ satellite whose natural width of ~9 meV (from Table IX) is negligible. Therefore, the entire width $\Delta W = 0.15$ eV is due to instrumental, kinematical, and Doppler effects. The right inset shows the $L_3-M_{2,3}^2(^1S)$ line, whose natural width of ~0.1 eV clearly broadens the observed lineshape.

Table X from the work of Ridder et al.[147] summarizes the known measurements for argon L- and LM-level widths. The L,S dependence of argon LM widths is demonstrated by comparison of the $L_{2,3}M_{2,3}(^3D,^1P,^1D)$ measured linewidths. Comparison with various theoretical values demonstrates the need for further consideration of this problem.

Table XI gives the final compilation of known linewidth measurements

[147] D. Ridder, J. Dieringer, and N. Stolterfoht in Reference 127, page 419 and private communication.

9.3. STUDIES OF ION–ATOM COLLISION PHENOMENA

TABLE XI. Natural Linewidth Γ (FWHM) of Various Neon and Argon Auger Peaks Derived from Experiment and Theory[a]

Atom line	Γ (eV) (Expt.)	Γ (eV) (Theor.)
Neon		
KLL (^{1}D)	0.23[b]	0.22[e]
	0.21[c]	
Argon		
$L_2M_{23}M_{23}$(^{1}S)	0.10[c]	0.16[f]
		0.13[g]
KL_1L_1(^{1}S)	7.5[d]	—
KL_1L_2(^{1}P)	3.6[d]	—
KL_2L_3(^{1}D)	1.0[d]	—

[a] From Ref. 74. [b] Ref. 80. [c] Ref. 32. [d] Ref. 146.
[e] Ref. 79. [f] Ref. 65. [g] Ref. 62.

for characteristic Auger lines. In particular, the KLL linewidths for argon were determined from proton bombardment of argon by Schneider et al.[146] The contribution of large L_1 final-state width, whose decay is by Coster–Kronig transitions evident in the KL_1L_1 and $KL_1L_{2,3}$ argon Auger widths. The absence of theoretical values for these states emphasizes the necessity of further theoretical work.

9.3.3.5. Inner-Shell Vacancy Fractions from Beam Auger Measurements. Studies of the population of excited states in projectiles moving through solids has been a subject of great interest since Lassen[148] first observed a higher mean charge state from ions emerging from solids rather than gases. Two different arguments have historically been used to describe the gas–solid discrepancy. The first argument, the so-called Bohr–Linhard (BL) model,[149] proposes that the most loosely bound electrons on the projectile are continually excited or prompted toward the continuum because of multiple collisions inside the solid. The ion therefore never relaxes to the ground state, as would be the case in single-collision conditions at gas target densities. Thus, the ion traversing a solid assumes its charge distribution primarily inside the solid. To the contrary, the Betz–Grodzins (BG) model[150] (on the basis of experimental data on ion collisions with dense-gas targets) suggests that the degree of electron loss is essentially independent of the excitation of one electron. Therefore, the electron loss is distributed among all outer-shell electrons. This means the ion may have many electrons in highly excited states upon

[148] N. O. Lassen, K. Danske, *Vidensk. Selsk. Mat.-Fys. Meddr.* **26**, 12 (1951).
[149] N. Bohr and J. Linhard, *K. Danske Vidensk. Selsk. Mat.-Fys. Meddr.* **28**, No. 7, (1954).
[150] H. Betz and L. Grodzins, *Phys. Rev. Lett.* **25**, 211 (1970).

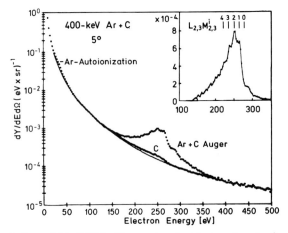

FIG. 57. The electron yield of 400-keV argon ions incident on 2 μg/cm² carbon foils observed at 5°. The carbon K-emission at 140° is added to the background. The insert shows the Auger yield after background subtraction. Also given are theoretical transition energies for different multiply ionized initial states. From P. Ziem, R. Baragiola, and N. Stolterfoht, Ref. 153.

emergence from the foil. The increased mean charge state over gas excitation is then assumed by extensive Auger decay upon emergence from the solid. Thus, with the BG model the mean charge distribution inside a solid is similar to that when an ion is transversing a gas target. The primary difference then between the BL and BG models is the location where the ion assumes its increased charge over that of gas excitation (i.e., inside the solid for BL or outside for BG).

Admittedly, from a collisions standpoint these models are naive but they serve as an introduction to the basic concepts. Until recently, the only way to study the states of ions inside a solid or immediately upon emergence from a foil was by x-ray emission (e.g., see Fortner and Garcia[151]). For instance, if one desired to measure the probability that the beam has an inner-shell vacancy (called the inner-shell vacancy fraction) then the x-ray method is difficult since fluorescence yields for states present both inside and outside the foil must be known. Measurement of Auger yields can be a simpler and more accurate solution to this problem. We can accurately determine the inner-shell vacancy fraction upon emergency from the foil provided $\tilde{\omega} \ll 1$ (usually the case for low-Z atoms having small degrees of multiple ionization). Furthermore, once the vacancy fraction is measured, then the contribution to mean charge state formation can be determined since the Auger process increases the mean charge by one charge unit. If the BG model is correct, then the

[151] R. J. Fortner and J. D. Garcia, *Phys. Rev. A* **12**, 856 (1975).

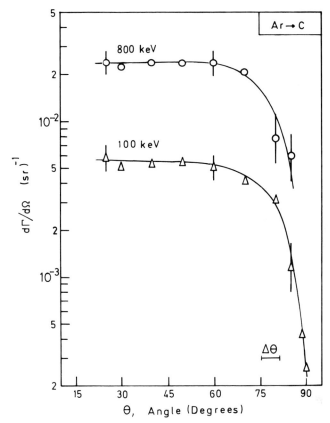

FIG. 58. The angular distributions of forward-emitted argon L-Auger electrons for 100- and 800-keV incident argon ions normal to the foil, corrected for kinematic effects. $\Delta\theta = 6°$ is the angular acceptance of the spectrometer. From R. Baragiola, P. Ziem, and N. Stolterfoht, *J. Phys.* **139**, L447 (1976) (with permission, The Institute of Physics, Great Britain).

measured contribution to mean charge state due to Auger processes should equal the increase in nuclear charge assumed by an ion traversing a solid rather than a gas.

The experimental method for determining vacancy fractions by Auger measurement was developed by Baragiola *et al.*[152] The first case they studied was argon L-vacancy fractions following argon ions traversing thin carbon foils. In essence, they measured the absolute yield of Auger electrons produced in collisions of 100–800-keV $Ar^{+,2+}$ ions with carbon foils. From simultaneously measuring the absolute beam flux they determine the Auger production probability (Auger yield per unit ion) $d\Gamma/d\Omega$ at a given lab observation angle. Figure 57 shows a sample electron spec-

[152] R. A. Baragiola, P. Ziem, and N. Stolterfoht, *J. Phys. B* **9**, L447 (1976).

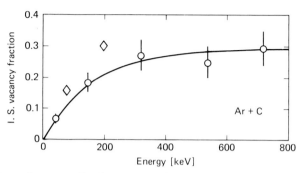

FIG. 59. Argon L-vacancy fraction vs. argon ion bombarding energies. From P. Ziem, R. Baragiola, and N. Stolterfoht, Ref. 153.

trum from their work. Their electron spectrometer views the foil directly, but the majority of electrons detected come from outside the solid since the mean free path for escape of a 200 eV electron out of a solid is only 8 Å as compared to a mean free path for decay $v_1\tau$ of at least 30–50 Å. As Fig. 58 indicates, they verify that the emission is essentially isotropic (the large drop in emission about $\theta_{Lab} = 75°$ is attributed to absorption of electrons due to surface irregularities). Since the yield is constant with angle then $d\Gamma/d\Omega$ can be integrated over the forward 2π hemisphere to give Γ, the Auger emission probability in the forward direction. When measuring argon L-Auger yields, this then is related to the inner-shell vacancy fraction for the L shell, f_L, through the expression $f_L = 2\Gamma$, where we have assumed $\tilde{\omega} \ll 1$ and Auger emission equal in backward and forward directions.

Figure 59 shows the variation in f_L with argon ion energy. The saturation of the curve demonstrates that the maximum probability of having an argon L vacancy is $\sim 30\%$. These values are substantially smaller than those deduced in the most complicated x-ray measurements.[151]

Multiplication of f_L by unity gives the contribution to charge increase outside the foil due to the L-Auger process. Figure 60 shows the variation in this contribution with argon in energy as compared to the solid–gas charge state differences. Obviously the L-Auger process can account for only part of the difference in mean charge between solids and gases. However, Ziem et al.[153] have determined that the probability of autoionization (M-Auger process) Γ_M is approximately equal to Γ_L at 400 keV. This implies that $f_L \simeq f_M$ (400 keV) $= 0.27$. Defining $\bar{q}_s - \bar{q}_q \equiv f_L + f_M$, which is the requirement for Auger processes accounting for solid–gas

[153] P. Ziem, R. Baragiola, and N. Stolterfoht, "Program and Abstracts of International Conference on the Physics of X-Ray Spectra," p. 278. Nat. Bur. Stds., 1976.

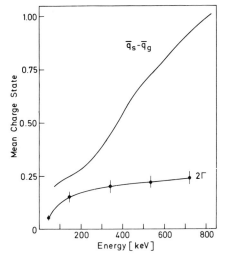

FIG. 60. The increase in mean charge due to L-Auger processes, 2Γ, and $\bar{q}_s - \bar{q}_g$, the difference in the mean charge of argon ions after traversing carbon foils and nitrogen gas. From R. A. Baragiola, P. Ziem, and N. Stolterfoht, *J. Phys.* **89,** L447 (1976) (with permission, The Institute of Physics, Great Britain).

charge difference, then we see from Fig. 60 that $\bar{q}_s - \bar{q}_q = 0.51$ in close agreement with the experimental value of $f_L + f_M = 0.54$. Obviously this result must be tested for other bombarding energies. However, if the agreement is nonfortuitous it demonstrates that the BG picture of ion charge state production in solids is more appropriate than the BL model.

There are some other useful and interesting measurements that have been or should be performed using this technique. If we assume $f_L \neq f_L(x)$, that is, an equilibrium thickness of foil has been reached, then the inner-shell vacancy lifetime can be obtained from the formula

$$f_L = n\sigma_p v\tau/(1 + n\sigma' v\tau), \qquad (9.3.17)$$

where n is target density, σ_p the argon L vacancy production cross section, v the ion velocity, τ the vacancy lifetime, and $\sigma' = \sigma_p + \sigma_Q$, where σ_Q is the probability for quenching the vacancy inside the solid by electron capture or other mechanisms. Solving for τ in expression (9.3.17) and assuming $\sigma_p > \sigma_Q$ (reasonable for this collision system) allows one to estimate upper and lower bounds on τ and σ_p (see Schartner et al.[154]). For 400-keV argon projectiles, Ziem and co-workers[153] used this technique to obtain $\sigma_p \geq 3.6 \times 10^{-18}$ cm^2 and $\tau \leq 4.3 \times 10^{-15}$ sec. Dif-

[154] K. H. Schartner, Th. P. Hoogkamer, P. Woerlee, and F. W. Saris, *Nucl. Inst. Meth.* **132,** 35 (1976).

ferences between τ_{expt} and τ_{theor} can most likely be ascribed to neglecting the importance of the quenching cross section (see Schartner et al.[154]).

Another important result derived from this method involves the measurement of argon beam and carbon target Auger yield from the front and back surface of the foil. Ziem and co-workers[153] observed no argon L emission from the back side of the foil ($\theta_L = 140°$) but C_K target Auger electrons were still observed. It is easy to understand why no argon L-Auger electrons were seen since the mean free path for production $l = 1/n\sigma_p \simeq 500$ Å is much larger than the electron escape depth (~ 8 Å). The equal C_K Auger yields for both front and back sides were also explained. Argon ions require multiple M-shell ionization before the production of C_K vacancies via electron promotion (MO level swapping). Therefore multiple M-vacancy production must equilibrate in much less than 8 Å since the same degree of multiply charged argon ions will be necessary near the front surface and back surface of foil in order to produce the same number of C_K vacancies.

Other useful measurements with this technique that have not been reported include measurement of f_L vs. target thickness and atomic number Z, and also measurements of $f_{K,L,M}$ vs. bombarding energy to understand the effects of inner-shell vacancy production on outer-shell vacancy probabilities.

9.3.3.6. Auger Spectra Following Ion–Molecule Collisions. For some time it has been known that heavy-ion bombardment of a molecular target tends to dissociate it into free atoms. However, IAES measurements indicate that Auger lines from states having prompt lifetimes ($\tau \leq 10^{-12}$ sec) still reflect the characteristics of decay from a molecular state. Figure 61 from the work of Johnson[155] demonstrates that the well-known band structure associated with Auger decay from molecular states is still present even when using O^{6+} bombardment of N_2. Recently, Mann et al.[156] have observed K-Auger emission from metastable "lithiumlike" states of highly ionized carbon and nitrogen ions present in various gaseous molecular components. Their data are shown in Fig. 62 along with the atomic spectrum of neon all formed by 56-MeV Ar^{12+} bombardment. The principal observation is that the width of the ^4P line from the 1s2s2p (atomic configuration) varies depending upon the chemical species of the target. They explained this as a Doppler broadening of the lines due to the emitting atoms nonzero velocity, which was obtained from a Coulomb explosion of the atoms normally bound in a molecular configuration. The amount of broadening depends upon the length of the

[155] B. Johnson, unpublished.
[156] R. Mann, F. Folkmann, and K. O. Groeneveld, *Phys. Rev. Lett.* **37**, 1674 (1976).

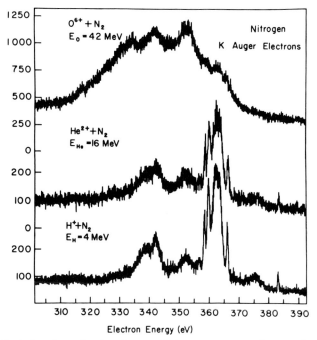

FIG. 61. Variations in molecular nitrogen K-Auger spectrum when produced by O^{6+}, He^{2+}, and H^+ bombardment. From B. Johnson, unpublished.

initial chemical bond (which, of course, depends on molecular specie). The line-broadening formula for effects due to a Coulomb explosion is given by

$$\Delta E_{\text{bond}} = 4 \left(\frac{m_1 m_2}{m_1 + m_2} E_c E_{\text{Auger}} \right)^{1/2}, \quad (9.3.17)$$

where E_c is the electrostatic potential (Coulomb) at bond length R, i.e., $E_c = Z_1 Z_2 e^2/R$, m_1 and m_2 are the dissociation atomic masses, atom 2 is assumed to eject the Auger electron, and Z_1, Z_2 are the effective nuclear changes of the molecular fragments, which are assumed to be $Z_1 = 1$ and $Z_2 = 1$ to account for screening of one K-shell electron in the configuration of question.

The calculated linewidths also must include the spectrometer resolution ΔE_{sp} and the effects of collisional broadening (Dahl et al.[29]) ΔE_{coll}. All widths are assumed to be Gaussian; hence they sum in Gaussian quadrature to give $\Delta E_T = (\Delta E_{\text{sp}}^2 + \Delta E_{\text{coll}}^2 + \Delta E_{\text{bond}}^2)^{1/2}$. Comparison with experimental linewidths is shown in Table XII. Excellent agreement is noted for a wide variety of targets.

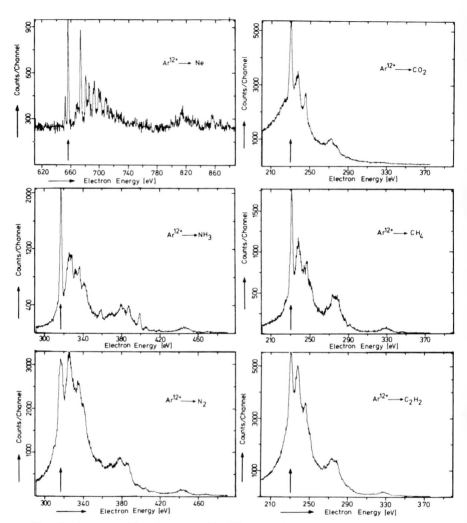

FIG. 62. Auger transitions in neon and in different carbon- and nitrogen-containing molecules from Ar^{12+} (56 MeV) impact. The arrow indicates the 1s2s2p → $1s^2$ transition. From R. Mann, F. Folkmann, and K. O. Groeneveld, *Phys. Rev. Lett.* **37**, 1674 (1976) (courtesy of The Physical Review).

This is a unique set of measurements since in one spectrum, depending upon the lifetime of the initial state, one can see Auger emission from a molecule that is either disassociating or intact. Since the lifetimes of the metastable ^4P state are on the few nanoseconds time scale, we can conclude that this is the characteristic dissociation time for a molecule ionized by heavy-ion bombardment.

TABLE XII. Measured and Calculated Linewidths ΔE in Auger Transitions in Atoms with Different Molecular Environments[a]

Target	ΔE_{coll} (eV)	ΔE_{bond} (eV)	ΔE_{result} (eV)	ΔE_{meas} (eV)	Molecular bond length R (Å)
Ne	1.39	—	2.2	2.0 ± 0.2	—
NH_3	1.37	1.0	3.0	2.5 ± 0.3	1.014
CH_4	1.11	0.9	2.9	2.5 ± 0.3	1.093
CO_2	1.11	3.36	4.3	3.7 ± 0.4	1.162
C_2H_2	1.11	4.8	5.5	5.6 ± 0.5	1.21
N_2	1.37	6.85	7.4	7.2 ± 0.5	1.1

[a] From Ref. 156.

Other uses of this technique that have not been explored include (1) measurement of inner-shell vacancy lifetimes via extent of line broadening, (2) measurement of metastable Auger linestrength as a function of bond length (to determine where the molecular regime exists), and (3) a study of the effects of screening on the Coulomb explosion energy. All these experiments could enhance our understanding of Auger decay in molecular environments.*

Acknowledgments

I would like to thank Drs. Nico Stolterfoht, Dieter Schneider, Peter Ziem, and Detluf Ridder of the Hahn-Meitner Institut in Berlin for reviewing this chapter.

---

* Reference to a company or product name in this chapter does not imply approval or recommendation of the product by the University of California or the U.S. Energy Research and Development Administration to the exclusion of others that may be suitable.

10. RADIATIVE AND AUGER BEAM–FOIL MEASUREMENTS

By David J. Pegg

10.1. Introduction

Since the early 1960s the beam–foil technique has been used to provide information on the structure and decay rates of atoms and positive ions of many elements in varying stages of ionization. Over 60 elements have been studied in beam–foil experiments to date. The method was developed from tools borrowed from two disparate disciplines, nuclear physics with its associated accelerator technology and pulse-handling systems and classical spectroscopy with its sophisticated wavelength-selecting equipment. The emphasis in nuclear physics at this time was shifting to higher-energy regimes and heavier-ion projectiles, which left many low-energy accelerators and associated facilities available for use on atomic and solid-state problems. Classical spectroscopy reached maturity in the latter part of the 1920s but ever since then spectroscopists have sought new and more refined light sources, particularly sources able to produce spectra of highly ionized species such as the beam–foil source. Even though beam excitation techniques involving gaseous targets (which are somewhat similar to the beam–foil method) were used in the 1920s,[1] the first reports of the successful mating of an accelerated ion beam, a thin *solid* excitation medium, and spectroscopic equipment to produce spectra of atoms and ions were not made until the period 1963–1964, when both Kay[2] and Bashkin and his collaborators[3] published the first "beam–foil" papers. Time of flight (tof) lifetime measurements by Bashkin and colleagues[4] were soon to follow these early spectral studies, and the potential for studying time-resolved decay process in virtually any element in any

---

[1] W. Wien, *Ann. Physik* **60**, 597 (1919); **66**, 229 (1921); **70**, 1 (1923); **73**, 483 (1924); **83**, 1 (1927).

[2] L. Kay, *Phys. Lett.* **5**, 36 (1963).

[3] S. Bashkin, *Nucl. Inst. Meth.* **28**, 88 (1964); S. Bashkin and A. B. Meinel, *Astrophys. J.* **139**, 413 (1964); S. Bashkin, L. Heroux, and J. Shaw, *Phys. Lett.* **13**, 229 (1964).

[4] S. Bashkin, A. B. Meinel, P. R. Malmberg, and S. B. Tilford, *Phys. Lett.* **10**, 63 (1964).

degree of ionization was quickly realized. Later, beam–foil quantum beat phenomena were observed[5] and attributed to interference effects arising from the radiative decay of two or more coherently excited and close-lying levels of a foil-excited atom to a common lower level. The coherent excitation is brought about by the impulsive nature of the foil excitation process. During the early 1970s, spectral and lifetime measurements were made by the University of Tennessee–Oak Ridge National Laboratory (hereafter referred to as UT-ORNL) group[6] on foil-excited ions that decayed via an Auger emission process (autoionization) rather than radiative deexcitation.

In this part we stress the practical aspects of beam–foil research, which is in the theme of this treatise. Hopefully, this information will be of some value to readers wishing perhaps to initiate a program of beam–foil research. The text is illustrated throughout with the work of many beam–foil investigators. Several good reviews of the entire field have been recently published.[7,8,9] Reviews of more specific areas are referred to at the appropriate place in the text.

10.2. Experimental Arrangements

10.2.1. Introduction

The beam–foil method is one aspect of a more general research area of atomic physics, often called accelerator-based atomic physics, in which accelerated ion beams interact with either solid or gaseous targets. This wider-based research topic is the subject matter of the present volume. In these fast-beam experiments many important structure and collision parameters associated with both the excited projectiles and target atoms are measurable. More recently laser beams, used both in the CW and pulsed mode, have also been employed to selectively excite (most other

---

[5] S. Bashkin, W. S. Bickel, D. Fink, and R. K. Wangsness, *Phys. Rev. Lett.* **15**, 284 (1965).

[6] B. Donnally, W. W. Smith, D. J. Pegg, M. D. Brown, and I. A. Sellin, *Phys. Rev. A* **4**, 122 (1971); I. A. Sellin, D. J. Pegg, M. D. Brown, W. W. Smith, and B. Donnally, *Phys. Rev. Lett.* **27**, 1108 (1971); D. J. Pegg, I. A. Sellin, P. M. Griffin, and W. W. Smith, *Phys. Rev. Lett.* **28**, 1615 (1972); H. H. Haselton, R. S. Thoe, J. R. Mowat, P. M. Griffin, D. J. Pegg, and I. A. Sellin, *Phys. Rev. A* **8**, 468 (1975).

[7] I. Martinson and A. Gaupp, *Phys. Rep.* **15**, 113 (1974).

[8] C. L. Cocke, *in* "Methods of Experimental Physics" (D. Williams, ed.), Vol. 13B, p. 213. Academic Press, New York, 1976.

[9] "Beam-Foil Spectroscopy" (S. Bashkin, ed.). Springer-Verlag, Heidelberg, 1976. The nine chapters of this book are addressed to different facets of beam-foil research.

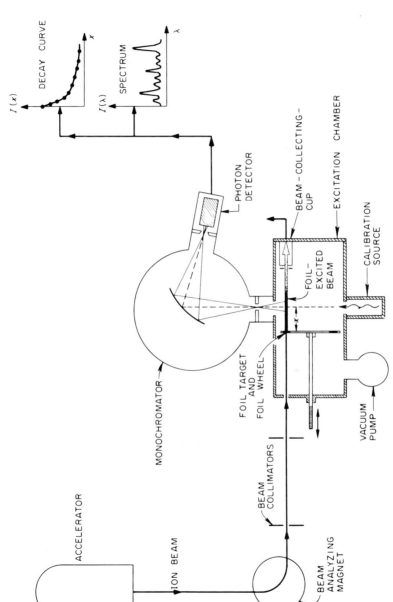

Fig. 1. Schematic diagram of a typical beam–foil arrangement used for spectral and lifetime measurements.

methods including the beam–foil technique do not selectively excite states) the atoms and ions of the fast-moving beam into specific energy states. Such laser–ion beam techniques are developing very rapidly and with the advent of the more powerful and versatile lasers of the future we can expect an even greater amount of activity in this field.

The beam–foil method is rather simple in both conception and execution. Figure 1 shows a schematic of the essential experimental arrangement necessary for radiative decay studies (for Auger decay studies the photon spectrometer is replaced by an electron analyzer, usually of the cylindrical mirror design). A fast-moving beam of ions produced by a particle accelerator is first magnetically analyzed in order to both define the beam energy and ensure chemical and isotopic source purity. The monoenergetic beam is then passed, after collimation, through a thin solid target placed in the path of the beam inside a well-evacuated chamber. The solid target, typically a thin self-supporting, amorphous carbon foil a few hundred atomic layers thick, defines rather precisely the location of the interaction region between the beam ions and the target atoms in both space and time. All projectile–target interactions (such as charge exchange and inelastic collision processes) occur during the rapid transit of the fast-beam ions through the foil, i.e., in approximately 10^{-14} sec or less. The foil thus serves to further strip and excite the remaining electrons of the beam ions as they traverse it. Target thicknesses are usually chosen so that equilibrium distributions of the various states of ionization and excitation are produced in the emerging ions. The spontaneous deexcitation of the individual foil-excited atoms and ions of the beam occurs in flight downstream from the foil. The resulting emitted radiation constitutes the beam–foil source.

One important property of this rather unique radiation source is the excellent time resolution that can be attained. This property permits excited-state decay studies to be made using a tof technique (the well-defined point of initial excitation, i.e., the foil, means that the $t = 0$ origin is known extremely well). We return to this important application in Chapter 10.5. From the spectroscopic viewpoint, the beam–foil light source must be considered as one of only moderate resolution (as compared with more traditional high-resolution sources). The beam–foil source does, however, have certain definite advantages over more traditional spectroscopic light sources. For example, the distribution of ionization states in the source can easily be varied by merely changing the beam energy incident upon the target foil, allowing one to study in principle the spectra and lifetimes of a particular heavy element from its atomic form all the way to the limit of the highly stripped hydrogenlike ion.

10.2.2. Particle Accelerators and Ion Sources

Several different varieties of ion accelerators, each covering a particular energy regime and using different types of ion sources, have been employed in beam–foil research. At low beam energies say less than about 500 keV) it is common to use Cockcroft–Walton-type machines, small single-ended Van de Graaffs, or isotope separators. Nowadays many of these low-energy machines are fitted with "universal-type" ion sources capable of producing ions of virtually any element. Such sources are compact enough to be installed in the terminal of the accelerator and can be used interchangeably with gaseous or solid feeds. The solid feed existing either in its elemental form or as a more volatile compound is heated in a source furnace to a temperature for which its vapor pressure is about 1 μm. Metal chlorides, which can be produced by introducing some CCl_4 into the source, have high vapor pressures at relatively low temperatures. The source vapor is then ionized usually by electron impact. Sources such as this are commercially available (for example, Physicon Corp.[10] manufactures a universal ion source based upon the design of Nielson[11]). For higher-energy beam–foil work, say incident beam energies greater than 10 MeV, common machines used are tandem Van de Graaff accelerators, cyclotrons, and linear accelerators. Most of these machines are equipped with some sort of heavy-ion source capable of producing a variety of species. For example, a "universal" negative-ion source (UNIS) recently introduced by Middleton[12] and available commercially,[13] has greatly extended the variety of heavy-ion beams that can be accelerated by tandem Van de Graaff machines. In the UNIS source a low-energy beam of cesium ions is directed onto a hollow cone, which is fabricated from the material whose negative ions you wish to extract. The role of the cesium beam seems to be twofold: to sputter positive ions from the cone material and to form a donor layer on the cone surface. This cesium layer is the source of electrons captured by the sputtered ions. Usable yields of negative ions for more than 20 elements have been extracted and focused into a beam for injection into the tandem. At ORNL for instance, such a source, which has been in operation at the EN tandem facility for several years, has produced useful beams of copper, gold, iron, nickel, aluminum, silicon, carbon, etc., in addition to the usual

---

[10] Physicon Corp., P. O. Box 232, Boston, Massachusetts, 02114.
[11] K. O. Nielson, *Nucl. Instr. Meth.* **1**, 289 (1957).
[12] R. Middleton and C. T. Adams, *Nucl. Instr. Meth.* **118**, 329 (1974).
[13] For example, Extrion Corp., Gloucester, Massachusetts, 01930 and General Ionex Corp., Ipswich, Massachusetts, 01938.

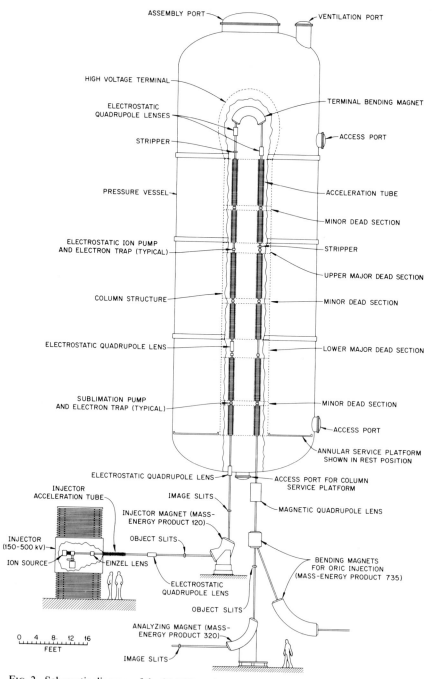

FIG. 2. Schematic diagram of the 25-MV tandem accelerator presently under construction at the Holifield National Heavy Ion Facility at the Oak Ridge National Laboratory.

electronegative species. A particularly interesting article by Alton[14] reviews this and other types of heavy-ion sources used to extract both negative and positive ions. A tandem accelerator has many desirable features for beam–foil work. For example, it is rather easy to change the beam energy during an experiment and the beam energy spread is quite small. In addition, with the advent of the UNIS-type source, it is also a simple matter to change the accelerated specie in a rather short period of time during the course of a run. The most energetic beams currently available from tandem accelerators are produced at the BNL three-stage tandem facility[15] where two MP tandems are arranged in line. The MP-6 tandem contains a negative-ion source in the terminal, which allows it to be used as a negative-ion injector for the MP-7 tandem accelerator. Using this arrangement 255-MeV ^{63}Cu ion beams (a few nanoamperes after analyzing magnet), for example, have been accelerated. A new 25-MV tandem is currently being built as part of the Holifield National Heavy Ion Facility at ORNL. Figure 2 shows a schematic of this accelerator, which involves a folded-beam design. The beam from this machine will eventually be injected into the cyclotron (ORIC) at ORNL. Details of the design and energy capabilities of this accelerator are given by Stelson[16] in a recent article. The highest beam energies employed in beam–foil experiments up to the present time involve linear accelerators such as the Super-HILAC at Berkeley, which is capable of accelerating heavy ions to energies of about 8 MeV/amu. For example, a recent beam–foil study of radiative transitions in Kr(XXXV) by Gould and Marrus[17] involved the use of a beam energy of 714 MeV. Beam–foil research in the intermediate-energy regime with beam energy from about 500 keV up to approximately 10 MeV is done primarily with single-ended Van de Graaff accelerators, dynamitrons, and perhaps tandems at the high end of this regime.

10.2.3. Beam Analysis and Source Purity

In conventional spectroscopic light sources impurities can pose problems. Chemical and isotopic purity of the beam–foil source is usually attained by magnetically analyzing the accelerated ion beam be-

---

[14] G. D. Alton, in "The Use of Small Accelerators in Research and Technology" (J. L. Duggan and I. L. Morgan, eds.), Vol. 1, p. 11. ERDA Technical Publication CONF-741040-P1, 1975.

[15] P. Thieberger and H. E. Wegner, *Nucl. Inst. Meth.* **122**, 205 (1974).

[16] P. H. Stelson, in "Beam-Foil Spectroscopy" (I. A. Sellin and D. J. Pegg, eds.), Vol. 1, p. 401. Plenum Press, New York, 1976.

[17] H. Gould and R. Marrus, in "Beam-Foil Spectroscopy" (I. A. Sellin and D. J. Pegg, eds.), Vol. 1, p. 305. Plenum Press, New York, 1976.

fore it enters the target chamber. In addition to this function, the calibrated magnet also serves to precisely define the energy of the ion beam before it interacts with the exciter foil. The energy calibration of the magnet frequently involves the use of well-known threshold and resonance energies of certain nuclear reactions.[18] The energy-calibrated magnetic analyzer is employed in a typical beam–foil arrangement to deflect ions of a certain selected magnetic rigidity (i.e., momentum-to-charge ratio) into the foil chamber and exclude all other ions. In most cases this ensures chemical and isotopic source purity, but occasionally contamination can fortuitously occur if different ions of the same mass and charge are present in the ion source (carbon ions, thought to be ejected from the foil by the beam, have also been observed in some cases to contaminate the postfoil source). For example, one could not separate beams of CO^+ and N_2^+ by magnetic analysis alone. The figures of merit of a magnetic analyzer are the mass resolving power and the mass–energy product, given by ME/q^2, where M is the ion mass (amu), E the energy (MeV), and q the charge on the ion. For low-energy beam–foil studies, Wien filters or electrostatic deflectors are sometimes used to deflect selected beams into the target chamber.

10.2.4. Foil Targets for Excitation

The unsupported thin-foil targets that are used as ionization–excitation media in beam–foil work are usually made of amorphous carbon and are typically several hundred atomic layers thick. Foil thicknesses are most often quoted in terms of areal densities in units of $\mu g/cm^2$. At low-to-intermediate beam energies, foils of thicknesses from 5 to 30 $\mu g/cm^2$ are usually sufficient to attain ionization and excitation equilibrium conditions, while at high energies thicker foils may be necessary. Carbon is the most frequently used target material because it is easy to work with, is rather strong mechanically, and is of sufficiently low Z to keep Rutherford scattering processes to a minimum. In addition, the "lifetime" of carbon foils under ion bombardment is longer than for most other materials. Carbon foils can be made in a laboratory with a vacuum evaporator or they can be obtained commercially.[19] Foil thicknesses can be determined by elastic scattering or energy loss determinations of α particles or other ions incident upon carbon. Stoner[20] describes a process in which the absolute amount of carbon in a piece of foil is determined by burning the foil

---

[18] J. B. Marion, *Rev. Mod. Phys.* **38**, 660 (1966).

[19] For example, The Arizona Carbon Foil Co., Tucson, Arizona and Yissum Carbon Foil Co., Jerusalem, Israel.

[20] J. Stoner, *J. Appl. Phys.* **40**, 707 (1969).

in pure oxygen, the pressure of the resulting CO_2 being measured accurately. Such a technique can then be used to calibrate optical transmission measurements, which are far easier to apply. Foil thicknesses measured in this manner are estimated to be accurate to from 5 to 10%. Foils are usually mounted on their target holders and are typically self-supporting over a diameter of about 5–10 mm. The exact cause of the breakage of carbon foils under ion bombardment is unknown (it is thought to be related to foil heating). The lifetime of a typical foil target depends upon the incident beam species and the beam intensity and energy. Foil breakage is rarely a problem at high energies but can be a serious problem at low energies when working with heavy elements, since it is then frequently necessary to normalize data taken with several exciter foils to accumulate a complete data set. (A fresh foil typically has slightly different ionization–excitation characteristics initially.)

It is now a quite well-established fact that foil characteristics change with time of bombardment due to foil-thickening effects and that these changes can cause intensity normalization problems as well as increasing energy loss in the foil. Chupp et al.[21] observed foil-thickening effects by studying a time-dependent change in the decay length of Ly α radiation from atomic hydrogen. In more quantitative studies by Bickel and Buchta[22] and Dumont et al.,[23] these foil-thickening effects were confirmed. Bickel and Buchta[22] studied the time dependence of energy loss and straggling of an 80 keV Li^+ beam after passage through a thin carbon foil by electrostatically analyzing the emergent beam. A carbon thickness increase rate of 2.7 $\mu g/cm^2/hr$ was determined from this experiment. Dumont et al.[23] found that the source of the deposited carbon was apparently carbon-containing molecules in the residual gas of the vacuum system. They found that the rate for carbon deposition depended upon the energy and current density of bombarding ions, as well as upon the conditions of the vacuum. Elimination of carbon deposition was achieved by surrounding the foil with a cooled baffle.

10.2.5. Normalization Procedures

It is essential, particularly for lifetime and quantum beat measurements, that any fluctuations in the beam intensity are not interpreted as being associated with "real" intensity changes of a foil-excited transition. Thus the signal corresponding to the intensity of a foil-excited spectral

[21] E. L. Chupp, L. W. Dotchin, and D. J. Pegg, Phys. Rev. **175**, 44 (1968).
[22] W. S. Bickel and R. Buchta, Phys. Scripta **9**, 148 (1974).
[23] P. D. Dumont, A. E. Livingston, Y. Baudinet-Robinet, G. Weber, and L. Quaglia, Phys. Scripta **13**, 122 (1976).

line must be normalized to some quantity that is proportional to the number of ions per second traversing the foil in order to take into account possible beam current variations. This is most conveniently done by normalizing the line intensity signal to the amount of beam charge collected in a shielded Faraday cup placed downstream from the foil (this cup must not receive or lose any electrons in the measurement and so it is desirable to use a shielded metal ring biased negatively by a few hundred volts to suppress secondary electrons). Care must be taken to demonstrate that the collected charge does not vary due to foil–cup distance changes, which are necessary in lifetime and quantum beam measurements. This geometrical effect is more likely to pose a problem when studying low-energy heavy ions since increased multiple-scattering effects in the foil under these conditions can cause considerable divergence in the postfoil beam. Another potential problem associated with this beam charge integration method of normalization might be due to changes in charge state and excitation distributions brought about by the time-dependent foil-thickening effects mentioned in the previous section. This method of normalization does, however, appear to be reliable in high-energy beam–foil measurements. A more consistent method of normalization (somewhat more difficult to apply in some experimental arrangements) is to use a second photon detector set at a fixed distance downstream from the foil. Either a specific foil-excited line or the total integrated radiation can be monitored by this normalization detector. Light pipes have been utilized in the visible region to transfer the foil-excited radiation to a detector positioned in a more convenient part of the target chamber. This type of normalization overcomes the potential difficulties that could be encountered in the aforementioned beam charge integration method since the foil–monitor detector geometry remains constant as the foil is translated and in addition any variation in the efficiency of production of a particular ionization–excitation state due to changing foil properties such as time-dependent thickening will be reflected in the intensity of the transition or transitions being monitored for normalization purposes.

10.2.6. Filters, Spectrometers, Spectrographs, and Detectors

The foil excitation process is nonselective and in general many different radiative and Auger transitions will occur at any instant in the beam–foil source. Thus some sort of energy-selective detection device must be used in beam–foil measurements to isolate the transition that "signatures" the decay of some particular excited state in the source. A simple and optically fast device for radiative transitions in the visible and near ultraviolet consists of a narrow-band interference filter–photoelectric detector arrangement. The use of such a device is limited, however, in practice to only those spectral features sufficiently well resolved to be isolated by the bandpass of the filter. Far more versatile instruments, with variable bandpasses, are spectrometers (monochromator plus a photoelectric detector) and spectrographs (with a photographic plate as a detector). Both of these types of dispersive instruments have been employed to spectroscopically analyze the radiation emitted by beam–foil sources. For lifetime studies the only spectroscopic objective is to isolate a particular line associated with a transition. Virtually all lifetime measurements made nowadays employ photoelectric detection since it involves the direct reading of a signal that is the linear response of the detector to the photon flux (the response remains linear over a wide dynamic range). Some of the early survey work in beam–foil spectroscopy was done with spectrographs that permitted qualitative studies of the intensity distribution of the source in terms of both wavelength and distance along the foil-excited beam (in an arrangement with the spectrograph entrance slit set parallel to the beam axis). The drawbacks associated with the use of photographic recording are numerous, however (for example, nonlinearity of response and problems associated with translating plate densities to line intensities), and spectrographs are rarely used for quantitative intensity work any more. In the future it is likely that a matrix of very small photoelectric detectors (placed perpendicular to the plane of dispersion) will entirely replace the exit slit of an instrument such as an ultraviolet grazing incidence spectrometer. This would, of course, give the spectrometer multichannel capabilities while still retaining the essential advantages of photoelectric recording.

Of course, no single spectrometer or detector can be used over the entire wavelength range encompassed by radiations from the beam–foil source. Volume 13 of this series contains some useful information on classical spectroscopic methods and different types of spectrometers. In particular, Kauffman and Richard[24] describe different types of monochromators (bent crystal and Doppler-tuned) as well as Si(Li) detectors, which have been used in beam–foil investigations in the x-ray region of the spectrum. In the visible and near-ultraviolet region, a favorite instrument in beam–foil work (because of its high optical speed, constant dispersion, low aberration, and production of stigmatic images) is a Czerny–Turner monochromator or a modifield version of it, for example, the McPherson model 218 ($f/3.5$, 0.33-m instrument). Rather than describe many different types of spectrometers and duplicate some of the

---

[24] R. Kauffman and P. Richard, in "Methods of Experimental Physics" (D. Williams, ed.), Vol. 13A, p. 148. Academic Press, New York, 1976.

material of Volume 13 of this series, we concentrate on only two types of instruments, both of which are used extensively in beam–foil work by the UT-ORNL group. First, we describe a grazing incidence monochromator, which can be used best to disperse radiation in the soft x-ray and extreme ultraviolet region, and second, we briefly discuss a cylindrical-mirror electron energy analyzer that has been used for beam–foil Auger (autoionization) work.

Grazing incidence monochromators are commercially available from several manufacturers.[25] A good description of grazing incidence spectrometers and far ultraviolet spectroscopy has been given by Sampson.[26] Heroux[27] was the first to study beam–foil transitions in the EUV with grazing incidence instrument. Figure 3 is a schematic of the beam–foil arrangement currently used by the UT-ORNL group for EUV measurements. Foil-excited radiation emitted at a nominal angle of 90° to the beam axis is collected and dispersed by a 2.2 m grazing incidence spectrometer (McPherson model 247). The dispersed radiation is detected behind the exit slit of the instrument with a low-noise (~ 0.2 Hz) channel electron multiplier (CEM). A negatively biased screen ($\sim 90\%$ transmission) is situated directly in front of the grounded end of the CEM detector to exclude stray electrons. The angle of incidence of the foil-excited radiation upon the gold-coated grating (blazed at 2°4′) can be varied from 82 to 88°. The optimum angle depends upon the wavelength region of interest (grating efficiency at a particular wavelength depends strongly on the angle of incidence at short wavelengths). The practical wavelength range from ~ 30 to 500 Å (other dispersive devices such as bent crystal x-ray spectrometers and normal incidence UV grating spectrometers adequately cover the region below and above this range respectively) is covered by the use of two interchangeable gratings with identical radii of curvature but different groove spacings. The gratings (300 and 1200 grooves/mm) are mounted in the instrument with their grooves parallel to the entrance slit and perpendicular to the beam axis. Various-sized masks can be used in front of the grating to limit the amount of surface area illuminated. The grating mask also serves another purpose in our experimental arrangement—it is the dominant aperture defining the length of beam viewed by the spectrometer. In the spectral region of re-

[25] For example, GCA/McPherson Co., 530 Main St., Acton, Massachusetts, 01720; λ Minutemen Labs, Inc., 916 Main St., Acton, Massachusetts, 01720; Jobin-Yvon Optical Systems, 20 Highland Avenue, Metuchen, New Jersey, 08840.
[26] J. R. Sampson, in "Techniques of Vacuum Ultraviolet Spectroscopy," J. Wiley and Sons, New York, 1976; and in "Methods of Experimental Physics" (D. Williams, ed.), Vol. 13A, p. 204. Academic Press, New York, 1976.
[27] L. Heroux, *Phys. Rev.* **153**, 156 (1967).

10.2. EXPERIMENTAL ARRANGEMENTS

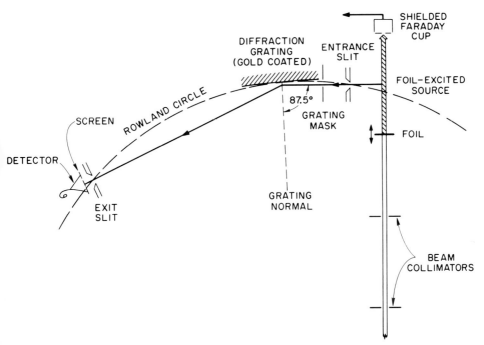

FIG. 3. Schematic diagram of the experimental arrangement used by the UT-ORNL beam–foil group for studies of highly charged ions that emit extreme ultraviolet or soft x-ray radiation in radiative deexcitation (Ref. 96).

cent interest to the UT-ORNL group (~100–400 Å) the acceptance angle (angular field of view in the plane of dispersion) has been 5.8 mrad (as determined by the grating mask, which corresponds to the optimum illuminated grating width with respect to both spectral and temporal resolution). This acceptance angle corresponds to an effective source length along the beam of 0.22 mm (average of the umbra and penumbra source lengths—the beam source profile being trapezoidal in shape) for the present beam-spectrometer separation. In spite of the inherent low sensitivity of grazing incidence monochromators (small acceptance angles, low grating reflectivities, large astigmatism, and other spherical grating aberrations) worthwhile work can still be accomplished in the soft x-ray and EUV regions (~30–500 Å) primarily because of a compensating factor—the availability of extremely low noise, and moderately high-quantum-efficiency detectors [for example, channel electron multipliers (CEM)]. In addition, lines in this spectral region are usually intrinsically strong (large f values). The size of the time windows (temporal resolution) attained with the aforementioned beam source viewing length is suf-

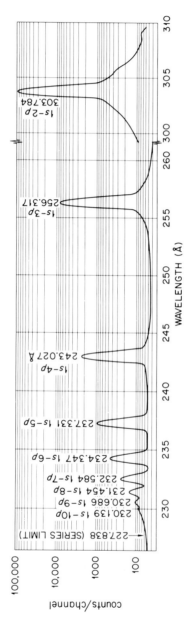

FIG. 4. Spectrum of the Lyman series of He(II) taken with the monochromator shown in Fig. 3 and a hollow-cathode light source.

10.2. EXPERIMENTAL ARRANGEMENTS

ficient to study lifetimes as low as tens of picoseconds. It might be necessary, however, to deconvolute the "time window" functions from the data if shorter lifetimes are to be measured. Both Barrette et al.[28] and Berry et al.[29] have achieved slightly better spatial resolution along the beam with similar grazing incidence spectrometers by moving the entrance slit of the instrument closer to the beam while still keeping it on the Rowland circle.

Wavelength "scanning" is accomplished in this type of instrument by setting the exit slit at discrete positions along the Rowland circle. A convenient and measurable quantity, which is related to the observed wavelength of the dispersed radiation, is the chord length between the middle of the grating and the exit slit. A usable form of the grating equation, expressed in terms of this quantity, is

$$m\lambda = \frac{10^7}{f} \left\{ \left[1 - \left(\frac{x_0 - k}{R}\right)^2\right]^{1/2} - \left[1 - \left(\frac{x - k}{R}\right)^2\right]^{1/2} \right\}, \quad (10.2.1)$$

where m is the order of interference, λ the wavelength of the dispersed radiation (Å), f the number of grooves/mm of the grating, R the diameter of the Rowland circle, and $x_0 - k$ and $x - k$ the chord lengths for the central image and the observed line, respectively (the grating equation is valid only when the normal at the center of the grating passes through the center of curvature of the machined ways upon which the slits travel). The values for the parameters shown in the wavelength equation can be obtained by curve-fitting methods involving measurements on spectral lines of well-established wavelengths from light sources such as a hollow cathode. In the UT-ORNL beam–foil arrangement, a windowless, differentially pumped hollow-cathode source is mounted on the target chamber directly opposite the entrance slits of the monochromator (this source is not shown in Fig. 3). Figure 4 is an example of a spectrum [Lyman series in He(II)] taken with the UT-ORNL hollow-cathode grazing incidence spectrometer system.[30] In addition to the aforementioned hollow-cathode calibration lines, certain well-known beam–foil lines from one- and two-electron ions have also been used as in-board reference lines to establish the proper Doppler shift corrections. Figure 5 is included to show a representative beam–foil lineshape obtained with the grazing incidence instrument. A schematic diagram of the data acquisition system used in the UT-ORNL beam–foil arrangement is shown in Fig. 6.

[28] L. Barrette, Université Laval, Doctoral Thesis (1975).
[29] H. G. Berry and C. H. Batson, in "Beam-Foil Spectroscopy" (I. A. Sellin and D. J. Pegg, eds.), Vol. 1, p. 367. Plenum Press, New York, 1976.
[30] P. M. Griffin, ORNL, unpublished data.

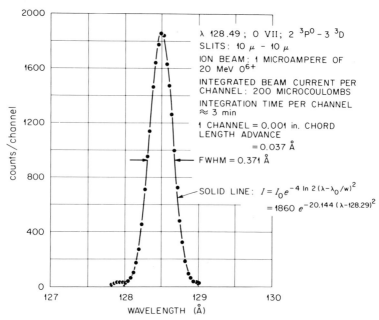

FIG. 5. Lineshape of a foil-excited transition in the heliumlike ion O(VII) taken with the experimental arrangement shown in Fig. 3.

Figure 7 shows a schematic diagram of an electron spectrometer that has been employed by the UT-ORNL group to energy-analyze Auger (autoionization) electrons emitted by foil-excited beams. The instrument is of a cylindrical mirror design,[31] which is very suitable for both spectroscopic and lifetime measurements involving projectile Auger electrons. This instrument has inner- and outer-cylinder radii of 5.71 and 12.22 cm, respectively, and is housed in a cylindrical high-vacuum chamber. Since electron trajectories can be affected by magnetic fields, all the material used in fabrication of the instrument is nonmagnetic and the whole spectrometer is shielded against external magnetic fields by a triple layer of shielding material (annealed Conetic). In an analyzer of this design, electrons leaving the beam ions at a mean polar angle of 42.3° (for second-order focusing properties) with respect to the beam direction are energy-selected by the electric field between the cylinders and detected by a CEM counter placed behind the exit slit. In the UT-ORNL instrument the spread in the polar angle is $\sim 0.07°$ and the azimuthal angle used is $\sim 120°$. The UT-ORNL electron spectrometer has been used success-

[31] V. V. Zashkvara, M. I. Korsunskii, O. S. Kosmachev, *Zh. Tekhn. Fiz.* **36**, 132 (1966) [*Sov. Phys.-Tech. Phys.* **11**, 96 (1966).].

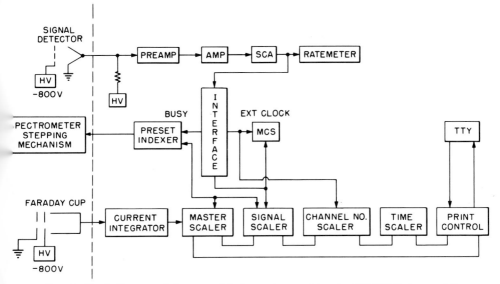

Fig. 6. Block diagram of the essential electronics used by the UT-ORNL beam–foil group.

fully for projectile studies with both solid (carbon foil) and gaseous excitation media. Lines in projectile electron spectra will be Doppler shifted and broadened in a similar manner to lines associated with radiative transitions. Line broadening can also result because of the postfoil beam divergence introduced by multiple-scattering processes in the foil.

Several different types of photoelectric detectors are necessary to span the different wavelength regions of the electromagnetic spectrum. In most beam–foil work it is common practice to operate a detector in a pulse-counting mode. The pulses can then be amplified, shaped if necessary, pulse height selected, and counted using the rather sophisticated electronic equipment "borrowed" from nuclear physics. Photomultiplier tubes (either with windows, windowless, or with fluorescent screens) can be used as detectors in the visible and ultraviolet. The "dark count" (thermionic) noise of such detectors is frequently quite high (especially for red-sensitive tubes) but it can be reduced to tolerable rates by cooling the photocathode with either liquid nitrogen or dry ice, or by using commercially built Peltier effect cooling housings.[32] A relatively low noise photomultiplier that finds a lot of use in beam–foil work in the visible is the EMI 6256S tube. In the EUV and soft x-ray regions of the spectrum, CEMs or continuous-strip magnetic electron multipliers are

[32] For example, Products for Research, Danvers, Massachusetts.

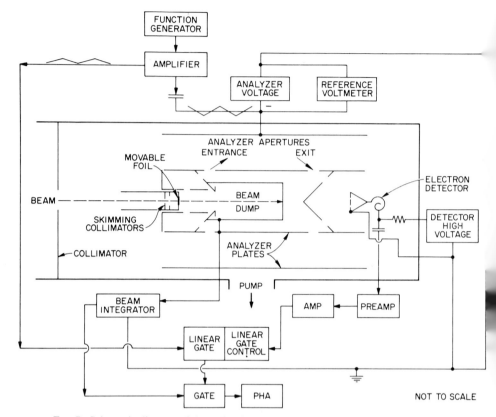

FIG. 7. Schematic diagram of the cylindrical-mirror electron energy analyzer used by the UT-ORNL beam–foil group for studies of Auger-emitting projectiles excited by either foil or gaseous targets [D. J. Pegg, I. A. Sellin, R. S. Peterson, J. R. Mowat, W. W. Smith, M. D. Brown, and J. R. Macdonald, *Phys. Rev. A* **8**, 1350 (1973)].

relatively efficient and -low noise detectors. Heroux[33] has described pulse-counting methods with these types of detectors. In a CEM a photoelectron, which is emitted from the inner surface of the cathode, is accelerated toward the high voltage output end of the electron multiplier producing secondary electrons on the way. The electron multiplier is usually a small curved tube whose inner surface is made of a high-resistance semiconductor. The CEM is usually operated in a pulse saturated mode. Figure 8 shows the quantum efficiency of such a CEM (Galileo Optics design) as a function of incident photon wavelength as measured by Mack *et al.*[34] It is of interest to note the rather sharp discontin-

[33] L. Heroux, *Nucl. Instr. Meth.* **90**, 1973 (1970).
[34] J. E. Mack, F. Paresce, S. Bowyer, *Appl. Optics,* **15**, 861 (1976).

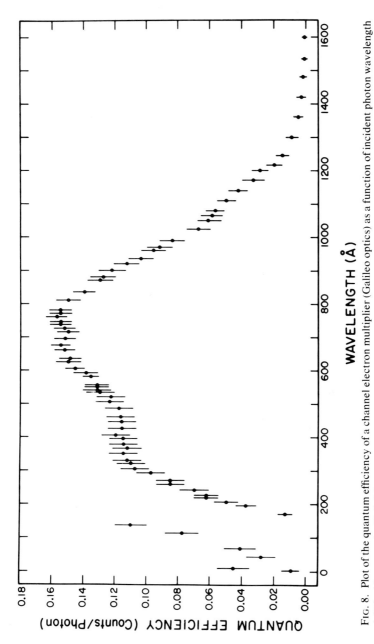

FIG. 8. Plot of the quantum efficiency of a channel electron multiplier (Galileo optics) as a function of incident photon wavelength and 0° incidence angle (Ref. 34).

uities existing in the 1–200 Å region, which are attributed to the absorption edges of elements present in the glass constituting the sensitive surface of the detector.

10.3. Source Characteristics

10.3.1. General Properties

Table I summarizes the approximate magnitudes and ranges of some quantities associated with the beam–foil method.

10.3.2. Comparison of the Beam–Foil Source with Other Spectroscopic Sources

There have been many creations and refinements of spectroscopic sources ranging from the traditional ones such as simple flames, arcs, sparks, and glow discharges to the more recently developed sources of highly ionized atoms such as plasma pinch discharges, laser-induced plasmas, and the beam–foil radiation source with high incident-beam energies. Differences between the beam–foil source and these other spectroscopic sources stem primarily from the fact that the beam–foil excitation process occurs in a high-density medium (a foil), while the deexcitation takes place spontaneously in a comparatively low density environment, i.e., a well-evacuated target chamber with residual gas particle densities $\sim 10^{10}/cm^3$. In most other sources both the excitation and deexcitation events occur in the same medium, which is often restricted to a gas at some intermediate particle density. This situation is a compromise between the optimum conditions for excitation and those for deexcitation.

TABLE I. Approximate Values of Some Beam–Foil Parameters

Beam energies	$\sim$20 keV to over 700 MeV
Ion velocities	$\sim$0.005c to 0.1c
Lifetimes measured	$\sim 2 \times 10^{-7}$ to 1×10^{-12} sec
Charge states encountered	H(I) to Kr(XXXV)
Exciter foil material	usually carbon
Foil aerial density	$\sim$5 to 30 $\mu g/cm^2$ (thicker for very high energy beams)
Ion beam currents	a few nanoamperes to 10 μA
Beam particle densities	$\sim 10^5$ ions/cm^3
Particle densities of residual gas in target chamber	$\sim 10^{10}$ particles/cm^3
Energy range of emitted:	
Radiation	$\sim$1.8 eV (7000 Å) to 3 keV (4 Å)
Auger electrons	$\sim$50 eV to 3 keV

10.3. SOURCE CHARACTERISTICS

Whereas it is obviously desirable to maintain low-density conditions in such a source to minimize perturbations to the spontaneous radiative process, it is of course equally desirable to operate the source at as high a density as possible to maximize the excitation (and sometimes ionization) efficiency. In the beam–foil arrangement, where different media are involved in excitation and deexcitation, these conditions can both be satisfied without compromise. Source perturbations (which may plague other sources) such as interionic fields, radiative recombination of beam ions, and stray electrons, resonance reabsorption of emitted radiation, and the stimulated emission of radiation are all negligibly small in the beam–foil source due to the low beam particle densities (typically $\sim 10^5/\text{cm}^3$) and residual gas particle densities (typically $\sim 10^{10}/\text{cm}^3$). In most other sources encountered, excitation and deexcitation occur in the same medium and as a consequence the particle densities are at least several orders of magnitude higher.

Although the details of the beam–foil interaction mechanism are far from being known at this time, it does appear that two distinct types of collision processes play a role in the production of the ionization–excitation states characteristic of the postfoil source. Multiple collisions involving such processes as charge exchange and stepwise inelastic excitation of both outer- and inner-shell electrons of the projectile ion occur in the high-density foil material (because of the high target density the frequency of collisions is such that excited states of the projectile ions do not have time to radiatively relax between collisions, resulting in stepwise excitation). Equilibrium conditions for both excitation and ionization processes are established if the target is sufficiently thick. In addition to the type of collisions that occur within the foil, there is considerable evidence for electron capture occurring as the emergent ion exits the foil. The capture of these surface electrons into high n and l states can account for the copious production of hydrogenic (Rydberg) states that are too fragile to survive if they were "born" inside the foil. Doubly excited states, which are also abundantly populated in the beam–foil source, can be explained by the capture of a single surface electron into an already excited state of the emerging projectile. Veje[35] describes an interesting independent-electron model of the foil interaction process along these lines, and the predictions are in reasonable agreement with charge state distributions and excitation function data. Also of interest with respect to understanding the foil interaction is the observation by Schnopper et al.[36] and

---

[35] E. Veje, *Phys. Rev. A* **14**, 2077 (1976).
[36] H. W. Schnopper, H. D. Betz, J. B. Delvaille, K. Kalata, and A. R. Sohval, *Phys. Rev. Lett.* **29**, 898 (1972).

Lindskog et al.[37] of radiative electron capture by highly ionized systems with foil-excited K vacancies.

Many plasma light sources (including astrophysical ones) involve the ionization and excitation of atoms and ions by inelastic collisions with electrons. In some cases a Maxwellian distribution of electron velocities is established, allowing an electron temperature to be assigned to characterize the source. To produce highly ionized species in such a source, a high electron temperature is needed. For example, the lines from highly ionized iron [~Fe(XI)–Fe(XV)] observed in spectra of the solar corona are produced at electron temperatures of $\sim 10^6$ K. It is, of course, impossible to assign a temperature to the beam–foil in the same sense as for plasma sources, since the excitation process is so different. One could, however, be tempted strictly for comparison purposes, to talk of an "equivalent temperature" for the beam–foil source that is related to the relative velocity between the electrons of the foil target and the ions of the unidirectional beam, a "temperature" easily changed by varying the incident-beam energy. A beam–foil spectrum of highly stripped iron taken by Bashkin et al.[38] using a 35-MeV beam produced many of the lines previously mentioned in connection with the solar corona spectrum. Thus the beam–foil source in this case can be called "hot" in the sense that it produces the same lines that are present in the coronal plasma of temperature $\sim 10^6$ K.

10.4. Beam–Foil Spectra

10.4.1. Introduction

The beam–foil source is capable of producing spectra of moderate resolution. A combination of high beam velocities and large acceptance angles (necessitated by low light levels) prohibit the high-resolution work characteristic of other more traditional spectroscopic sources. If, however, one is willing to accept moderate resolution spectra a great deal of useful work can be done with the beam–foil source that does not overlap the output of other sources. For example, one unique ability of the beam–foil source is to investigate the spectra of atoms and ions of an element over a very wide range of charge states due to the ease of controlling the postfoil charge state distribution by varying the incident-beam energy. This enables studies to be made, for example, of the spectra of many dif-

---

[37] J. Lindskog, J. Pihl, R. Sjödin, A. Marelius, K. Sharma, and R. Hallin, *Phys. Scripta* **14**, 100 (1976).

[38] S. Bashkin, K. W. Jones, D. J. Pisano, P. M. Griffin, D. J. Pegg, I. A. Sellin, and T. H. Kruse, *Nucl. Instr. Meth.* **154**, 169 (1978).

10.4. BEAM–FOIL SPECTRA

TABLE II. Excited Configurations of the Lithium Atom

Type	Example
Singly excited	$1s^2n\ell$ ($n \geq 2$)
Hydrogenic (or Rydberg)	$1s^2n\lambda$ (n, ℓ large)
Core-excited (K shell)	$1s2sn\ell$ ($n \geq 2$) or $2sn\ell n'\ell'$ (n, $n' \geq 2$)
Doubly excited	$1sn\ell n'\ell'$ (n, $n' \geq 2$, $\ell \geq 1$)

ferent members of an isoelectronic sequence allowing regularities to be investigated. Another interesting application of the beam–foil source to spectral measurements is associated with the study of states such as doubly excited states, core-excited states, hydrogenic (Rydberg) states, and displaced terms, which are not as commonly observed in other terrestrial sources but are abundantly produced in the foil interaction due to the rather unusual excitation–deexcitation conditions. Table II lists four types of excited configurations met in beam–foil work and to which we refer here. Core-excited configurations are frequently indistinguishable from doubly excited configurations (multiply excited configurations are those in which two or more electrons occupy orbitals differing from the ground state configuration) and so it is common practice to refer to both types of states as being "doubly excited."

10.4.2. Postfoil Charge Distributions

Due to the statistical nature of charge exchange processes in the foil material, the ion beam emerges from the foil in a distribution of charge states (ionization stages) that is approximately gaussian in shape. The mean of this distribution increases with incident beam energy and the width of the distribution tends to become greater for heavy ions. The beam–foil investigator requires knowledge of the energy dependence of these postfoil charge distributions in order to choose the optimum incident beam energy to maximize a particular ionization stage. A survey of the available experimentally measured charge fractions for various ions has been made by Betz[39] and by Wittkower and Betz.[40] Part 3 of this volume is devoted to the subject of charge equilibration of high-velocity ions in matter. When experimental values are not available one can use the semiempirical equation of Nikolaev and Dmitriev[41] to estimate the mean charge states at any beam energy. This equation, which is applicable for

---

[39] H. D. Betz, *Rev. Mod. Phys.* **44**, 465 (1972).
[40] A. B. Wittkower and H. D. Betz, *Atom. Data* **5**, 130 (1973).
[41] V. Nikolaev and I. Dmitriev, *Phys. Lett.* **28 A**, 277 (1968).

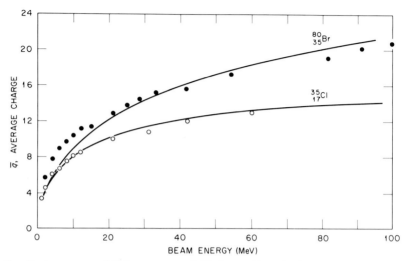

FIG. 9. Average postfoil charge states of chlorine and bromine ions as a function of incident beam surgery. The solid curves are from a semiempirical formula (Ref. 41). Experimental points, C. D. Moak, L. B. Bridwell, H. O. Lutz, and S. Datz, *in* "Beam-Foil Spectroscopy" (S. Bashikin, ed.), Springer-Verlag, Heidelberg, 1976.

$Z \geq 16$, is given in the convenient form

$$E = (0.067)MZ^{0.9} [(Z/\bar{q})^{1.67} - 1]^{-1.2}, \qquad (10.4.1)$$

where E is the incident beam energy (MeV), M the projectile mass (amu), Z the atomic number of the projectile ion, and $\bar{q}$ the mean charge of the postfoil distribution. That this equation yields reasonably good estimates can be seen from Fig. 9, where the calculated values are compared to experimental values for incident chlorine and bromine ions over a wide range of energies.

10.4.3. Charge State Identifications

It was shown in Section 10.4.2 that ions of several different charge states exist in the postfoil source for any given beam energy. Frequently spectral lines appear in a beam–foil spectrum that have not been previously classified and the first step in identifying these lines is to determine the charge state of the emitting ion. Several different techniques have been developed for this purpose. The simplest method, although not necessarily always reliable, is that used originally by Kay.[42] The excitation function (line intensity vs. incident beam energy) of a line is in-

---

[42] L. Kay, *Proc. Phys. Soc.* **85**, 1963 (1965).

10.4. BEAM–FOIL SPECTRA

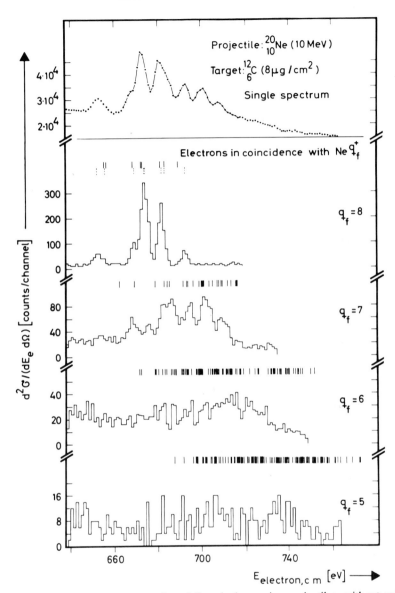

FIG. 10. Auger electron spectra from foil-excited neon ion projectiles, without coincidences (top spectrum) and in coincidence with the final charge state 5^+–8^+ (bottom spectra) (Ref. 46).

vestigated and compared to the energy dependence of the production of various ionization states under the assumption that changes in the line intensity with beam energy are primarily due to changes in the production of the ion. This technique, which has the advantage that no auxiliary equipment is necessary, seems to work quite well for singly excited states but, as may be expected, is less reliable for multiply excited states.

Other methods of charge state identification involve the application of external electric (or magnetic) fields to the beam–foil source. If the lifetime of the excited state is sufficiently long, ions of different charge states in the source can be spatially separated by an electric or magnetic field before deexcitation occurs. In this case ions of different charge states can be isolated and radiative transitions studied separately. Such a technique was employed by Lennard et al.[43] If deexcitation of the ion occurs in the electric field being used for separation, a Doppler shift technique described by Carriveau and Bashkin[44] can be used. The deflecting field gives an ion a transverse velocity component, the magnitude of which is dependent upon the ionic charge of the emitter. As a consequence of this the spectral line will exhibit a measurable Doppler shift.

Coincidence techniques have also been used to identify the charge state of emitting-beam ions. For example, Cocke[45] has described experiments involving coincidence counting between x-ray photons and scattered ions. Groeneveld et al.[46] and Schumann[47] describe similar coincidence measurements between Auger electrons emitted in the decay of doubly excited states and the charge state of the various ions in the beam. Figure 10 shows these coincidence spectra as well as a singles spectrum.

10.4.4. Doppler Broadening and Shifts

The fact that beam–foil radiation is emitted from a fast-moving unidirectional beam of ions precludes high-resolution spectroscopic measurements, at least in the visible and near ultraviolet, because of the Doppler broadening and shifts of the spectral features. The wavelength of a spectral line as measured in the laboratory frame will be Doppler-shifted with respect to the rest frame wavelength under these conditions. The magnitude of the first-order fractional shift depends upon the beam

---

[43] W. N. Lennard, R. M. Sills, and W. Whaling, *Phys. Rev. A* **6**, 884 (1972).

[44] G. W. Carriveau and S. Bashkin, *Nucl. Instr. Meth.* **90**, 203 (1970).

[45] C. L. Cocke, in "Beam-Foil Spectroscopy" (I. A. Sellin and D. J. Pegg, eds.), Vol. 1, p. 283. Plenum Press, New York, 1976.

[46] K.-O. Groeneveld, G. Nolte, S. Schumann, and K. D. Sevier, *Phys. Lett.* **56A**, 29 (1976).

[47] S. Schumann, Univ. of Frankfurt, Ph.D. Thesis (1976).

velocity and the angle of emission of the radiation with respect to the beam direction. Thus a spectral feature can be either red-shifted or blue-shifted according to whether the source ion has a velocity component along the line of sight of the spectrometer approaching (blue-shifted) or receding (red-shifted) from the spectrometer. The second-order Doppler shift (relativistic shift) is independent of the angle of emission and is always a red shift. This is summarized in the Doppler shift equation

$$\Delta\lambda/\lambda_0 \simeq \beta \cos\theta + \tfrac{1}{2}\beta^2 + \cdots, \qquad (10.4.2)$$

where the first term to the right corresponds to the first-order (in β) Doppler shift and the second term corresponds to the second-order shift, λ_0 is the rest frame wavelength, and $\beta = v/c$. The mean angle of observation for the most popular viewing arrangement (the side-on configuration) is nominally set at 90° in order to minimize the first-order shift (a small residual first-order shift remains because it is difficult to fix the emission angle exactly at 90°). The second-order shift can become appreciable at high beam energies. For example, Fig. 11 shows a spectrum

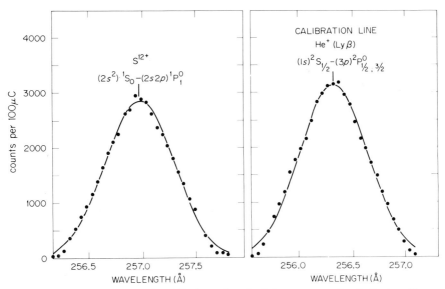

FIG. 11. An accurate Doppler-shifted wavelength of the resonance line in the beryllium-like ion S(XIII), determined by calibrating the wavelength scale in this region with the Lyβ line of He(II) obtained with a hollow-cathode source [D. J. Pegg, J. P. Forester, C. R. Vane, S. B. Elston, P. M. Griffin, K.-O. Groeneveld, R. S. Peterson, R. S. Thoe, and I. A. Sellin, UT-ORNL group, unpublished data].

taken by Pegg et al.[48] in which a beam–foil line from S(XIII) is compared to a He(II) line in the same wavelength region from a hollow-cathode source. The total (first- and second-order) Doppler shifts could be determined by comparing the measured wavelength (the spectrometer wavelength scale was calibrated in this region with the He(II) line shown) of the beam–foil S(XIII) line with the well-established rest frame wavelength measured by Behring et al.[49] In this manner it was determined that the total shift was 0.283 Å (~0.1%), which corresponds to a second-order red shift ($\beta = 0.0552$) contribution of 0.391 Å and a first-order blue shift contribution of 0.108 Å (the mean emission angle viewed was actually 89.56° and an observation angle of 90.44°).

Doppler-broadening effects occur as a result of the finite acceptance angle of the spectrometer. Radiation emitted from the source volume with some range of emission angles will enter the spectrometer and this leads to a symmetric line broadening due to the corresponding range of blue and red shifts associated with the radiation. Lines can be asymmetrically broadened if the decay length is comparable to or small compared to the length of the beam observed. The fractional Doppler broadening $\Delta\lambda/\lambda_0$, is proportional to this acceptance angle α of the spectrometer and to $\beta = v/c$, i.e.,

$$\Delta\lambda/\lambda_0 \simeq \beta\alpha. \qquad (10.4.3)$$

Thus in the visible and near-ultraviolet regions of the spectrum Doppler-broadening contributions to the linewidth can be as much as a few angstroms. In this region wavelengths are quite long, beam energies correspond to $\beta \sim 0.01$, and the acceptance angle α is usually relatively large since the rather weak beam–foil source demands fast optical systems. Stoner and Leavitt,[50] however, have demonstrated that this contribution can be substantially reduced by refocusing the spectrometer such that the illumination of the grating is no longer collimated. For example, the linewidths of lines from foil-excited N(IV) ions were reduced from 3 to 0.9 Å in this manner. In the far-ultraviolet and x-ray regions of the spectrum Doppler broadening becomes less of a problem as can be seen from Fig. 11. The wavelengths are relatively short, beam energies correspond to $\beta \sim 0.05$–0.1 (necessary to strip heavy ions), and the acceptance angle α can be reduced because many of the spectral lines are intrinsically stronger (large f values). Thus in the work by Pegg et al.,[48] shown in Fig. 11, the acceptance angle $\alpha = 5.83$ mrad, $\beta = 0.055$, and

---

[48] D. J. Pegg, J. P. Forester, C. R. Vane, S. B. Elston, P. M. Griffin, K.-O. Groeneveld, R. S. Peterson, R. S. Thoe, and I. A. Sellin, UT-ORNL group, unpublished data, 1976.

[49] W. E. Behring, L. Cohen, U. Teldman, and G. A. Doschek, Ap. J. **203**, 521 (1976).

[50] J. O. Stoner, Jr. and J. A. Leavitt, Appl. Phys. Lett. **18**, 477 (1971).

$\lambda_0 = 256.686$ Å, which yields a calculated Doppler broadening of $\Delta\lambda \sim 0.083$ Å (0.03%). This is rather small compared to the instrumental linewidth (fwhm) of 0.75 Å associated with the 50 μm slits employed in the experiment. If one reduces this instrumental linewidth (by narrowing the entrance and exit slits) to a minimum of ~ 0.25 Å (for the 300 grooves/mm grating used), the Doppler contribution to the overall linewidth will increase to some 30%, however (of course, the contribution would become even larger for the improved instrumental linewidth associated with the 1200 grooves/mm grating).

Line-broadening contributions associated with the divergence of the postfoil beam have been discussed by Stoner and Leavitt.[51] Again, this source of broadening will be most serious for low-energy heavy-ion beams (due to increased multiple-scattering processes) but will be practically negligible for high-energy beams of moderately heavy ions.

10.4.5. Singly Excited States

As a consequence of the resolution limitations and line-shift problems mentioned earlier, beam–foil experiments are less frequently designed to solely determine the wavelength of radiative transitions than they are to measure lifetimes of excited states. There are important exceptions, however. Thus beam–foil spectroscopic studies have supplied very useful information in cases when it is evident that there exists either a total or a partial lack of knowledge of the spectrum of a particular ionization stage of an element (a situation that becomes more probably as an element becomes more highly ionized.). Another particularly profitable spectroscopic application of the beam–foil source is related to doubly excited or Rydberg states, both types being populated with high efficiency in the beam–foil source (these types of states are found to be much less commonly produced in other spectroscopic sources.). Low-to-moderate-resolution survey spectral scans are practically always made prior to lifetime measurements in order to gain information on the relative intensities of lines associated with potential blend or cascade transitions.

A particularly useful review of the status of the spectroscopic aspect of beam–foil research has been recently provided by Martinson[52]. It is perhaps of considerable interest to note that many of the new lines that have been reported by beam–foil investigators have been found to originate in comparatively simple ionic systems containing only a few active electrons. This is particularly true for dipole transitions associated with

---

[51] J. O. Stoner, Jr. and J. A. Leavitt, *Optica Acta* **20**, 435 (1973).
[52] I. Martinson, *in* "Beam-Foil Spectroscopy" (S. Bashkin, ed.), p. 33. Springer-Verlag, Heidelberg, 1976.

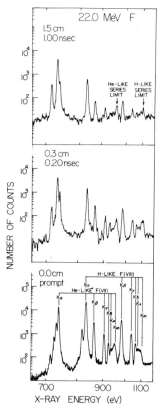

FIG. 12. X rays from foil-excited fluorine taken with a beam energy of 22 MeV. The bottom spectrum is taken looking straight at the foil target while the other spectra are accumulated at the indicated time decays (Ref. 62).

moderately high n. For example, Berry et al.[53] reported on many new transitions with $n = 6–12$ in the simple system Li(II). Several different investigators[54] have reported on the observation of new transitions in the heliumlike ion boron IV. Buchet et al.[55] produced new spectroscopic information in their studies of few-electron ions of carbon, nitrogen, ox-

---

[53] H. G. Berry, E. H. Pinnington, and J. L. Subtil, *J. Opt. Soc. Amer.* **62**, 767 (1972).

[54] For example, I. Martinson, W. S. Bickel, and A. Ölme, *J. Opt. Soc. Amer.* **60**, 1213 (1970); J. P. Buchet and M. C. Buchet-Poulizac, *J. Opt. Soc. Amer.* **63**, 243 (1973); H. G. Berry and J. L. Subtil, *Phys. Scripta* **9**, 217 (1974).

[55] J. P. Buchet, M. C. Buchet-Poulizac, G. DoCao, and J. Desesquelles, *Nucl. Instr. and Meth.* **110**, 19 (1973); J. P. Buchet, A. Denis, J. Desesquelles, M. Druetta, and J. Subtil, in "Beam-Foil Spectroscopy" (I. A. Sellin and D. J. Pegg, eds.), Vol. 1, p. 355. Plenum Press, New York, 1976.

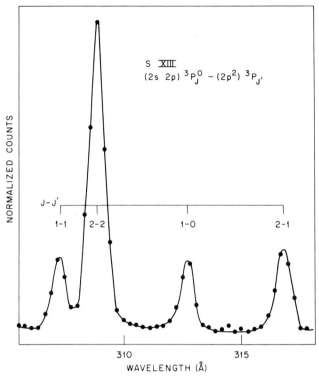

FIG. 13. Section of the EUV spectrum of foil-excited sulfur ions taken at an incident-beam energy of 38 MeV. The spectrum shows part of the $2s2p\ ^3P_J - 2p^2\ ^3P_{J'}$ multiplet in the berylliumlike ion S(XIII) (Ref. 63).

ygen, and neon. Pegg et al.[56] confirmed several of these observations for transitions in O(VII) and extended the work to fluorine observing several new transitions in the heliumlike ion F(VIII). Griffin et al.[57] reported on many unidentified lines in their spectral studies of highly ionized silicon [Si(VI)–Si(XI)] in the ~100–400 Å region alone. Many new lines were also observed in beam–foil studies of sodium atoms and ions by both Brown et al.[58] and Dufay et al.[59] Beam–foil spectra have also contributed to our knowledge of displaced terms in alkaline earth systems. For

[56] D. J. Pegg, P. M. Griffin, H. H. Haselton, R. Laubert, J. R. Mowat, R. S. Thoe, R. S. Peterson, and I. A. Sellin, *Phys. Rev. A* **10**, 745 (1974).

[57] P. M. Griffin, D. J. Pegg, I. A. Sellin, K. W. Jones, D. J. Pisano, and T. H. Kruse, in "Beam-Foil Spectroscopy" (I. A. Sellin and D. J. Pegg, eds.), Vol. 1, p. 321. Plenum Press, New York, 1976.

[58] L. Brown, K. Ford, V. Rubin, and W. Trachslin, in "Beam-Foil Spectroscopy" (S. Bashkin, ed.), Vol. 1, p. 45. Gordon and Breach, New York, 1968.

[59] M. Dufay, M. Gaillard, and M. Carre, *Phys. Rev. A* **3**, 1367 (1971).

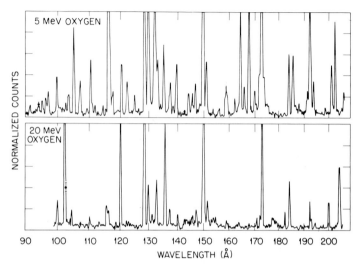

FIG. 14. EUV spectra of foil-excited oxygen ions accelerated to 5 MeV (top) and 20 MeV (bottom). Note the change in the spectra with incident beam energy due to changes in the postfoil charge state distributions (Ref. 56).

example, the study of Be (I) by Hontzeas et al.[60] and B(II) by Berry and Subtil[61] have yielded new information on the energies of displaced terms.

Figures 12-14 are included to illustrate beam-foil spectra taken with different types of spectrometers in different spectral regions. Figure 12 shows an X-ray spectrum of highly stripped fluorine ions obtain by Kauffman et al.[62] using a curved crystal spectrometer with a RAP crystal and a proportional counter. The EUV spectra shown in Figs. 13 and 14 were taken by Pegg et al.[56,63] using a grazing incidence spectrometer equipped with a CEM detector.

10.4.6. Doubly Excited States

Perhaps one of the most interesting and useful spectroscopic applications of the beam-foil method lies in the investigation of the wavelengths of radiative transitions involving an upper doubly excited (including core-excited) state. A doubly excited configuration is often character-

[60] S. Hontzeas, I. Martinson, P. Erman, and R. Buchta, Phys. Scripta **6**, 55 (1972).
[61] H. G. Berry and J. B. Subtil, Phys. Scripta **9**, 217 (1974).
[62] R. L. Kauffman, C. W. Woods, F. F. Hopkins, D. O. Elliott, K. A. Jamison, and P. Richard, J. Phys. B **6**, 2197 (1973).
[63] D. J. Pegg, J. P. Forester, C. R. Vane, S. B. Elston, P. M. Griffin, K.-O. Groeneveld, R. S. Peterson, R. S. Thoe, and I. A. Sellin, Phys. Rev. **15**, 1958 (1977); D. J. Pegg, P. M. Griffin, G. D. Alton, S. B. Elston, J. P. Forester, M. Suter, R. S. Thoe, and C. R. Vane,

ized by at least one vacancy in an inner (nonvalence) shell of the system, an arrangement that is potentially highly unstable due to the large amount of internal excitation energy involved. For light and near neutral systems (for which the fluorescence yield is small) this inner-shell vacancy is very likely to be rapidly filled in a nonradiative deexcitation process called Auger emission (analogous to autoionization) in which one electron drops into the vacancy and another (usually a valence electron) is ejected from the system with kinetic energy equal to the energy balance between the initial highly excited state and the final state of the residual ion. For example, consider the 1s2snl or 1s2pnl states of lithiumlike ions that decay via Auger emission (autoionization) when an electron falls into the K hole. These electrons can be coupled (LS coupling scheme) to produce two doublet states and one quartet state. The quartet state is metastable against the allowed Auger process because it would require a spin–flip transition to fill the K vacancy in a manner demanded by the exclusion principle. The allowed Auger process, which is induced by the electrostatic (Coulomb) interaction between the electrons, is governed by a set of selection rules in a manner similar to radiative decay. Forbidden Auger emission (autoionizing) processes can occur even if the allowed process is prohibited although with considerably less probability (the selection rules on forbidden transitions are more relaxed). The decay rates or probabilities for forbidden Auger transitions induced by the weaker magnetic interactions between the electrons are much smaller than for the allowed Auger process but can become appreciable at high Z owing to the increase in strength of the magnetic interactions themselves. In this chapter we concentrate on the radiative decay of doubly excited states that are metastable against Auger emission (the state lives long enough that radiative deexcitation can compete with or perhaps dominate forbidden Auger decay modes). Here it becomes convenient to distinguish between two different types of radiative transitions involving doubly excited states, both of which have been observed in beam–foil studies. Thus, allowed electric dipole (E1) transitions take place between two different doubly excited states of opposite parity. Bickel et al.,[64] Buchet et al.,[65] and Berry et al.[53] have investigated such transitions in Li(I). Analogous transitions in other members of the lithium-sequence have been reported, for

---

Phys. Scripta **18**, 18 (1978); J. P. Forester, D. J. Pegg, P. M. Griffin, G. D. Alton, S. B. Elston, H. C. Hayden, R. S. Thoe, C. R. Vane, and J. J. Wright, *Phys. Rev. A* **18**, 1476 (1978); B. Denne, D. J. Pegg, E. Alvarez, R. Hallin, K. Ishii, J. Pihl. R. Sjödin, *J. Phys.* **40**, C1-183 (1979).

[64] W. S. Bickel, I. Bergström, R. Buchta, L. Lundin, and I. Martinson, *Phys. Rev.* **178**, 118 (1969).

[65] J. P. Buchet, A. Denis, J. Desesquelles, and M. Dufay, *Phys. Lett.* **28A**, 529 (1969).

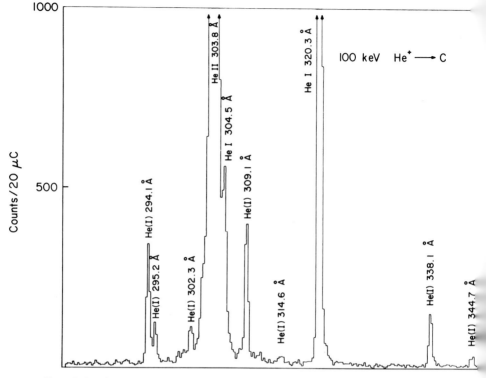

FIG. 15. Portion of the EUV spectrum of foil-excited helium. Satellite lines are shown in the vicinity of the resonance (1s–2p) line in He(II) [E. J. Knystautas and R. Drouin, *Nucl. Inst. Meth.* **110**, 95 (1973)].

example, by Hontzeas et al.,[60] Berry et al.,[66] Buchet et al.,[67] Matthews et al.,[68] Pegg et al.,[56] Irwin et al.,[69] To et al.,[70] and Knystautas et al.[71] Both allowed and forbidden radiative transitions have been observed in the beam–foil decay of a doubly excited state to a lower-lying singly excited state. For example, the allowed $1s^22p\ ^2P \to 1s2p^2\ ^2P$ transition in Li(I) was reported by Buchet et al.[65] and the forbidden transition (intermulti-

[66] H. G. Berry, M. C. Buchet-Poulizac, and J. P. Buchet, *J. Opt. Soc. Amer.* **63**, 240 (1973).

[67] J. P. Buchet and M. G. Buchet-Poulizac, *J. Opt. Soc. Amer.* **64**, 1011 (1974).

[68] D. L. Matthews, W. J. Braithwaite, H. W. Wolter, and C. F. Moore, *Phys. Rev. A* **8**, 1397 (1973).

[69] D. J. G. Irwin and R. Drouin, in "Beam-Foil Spectroscopy" (I. A. Sellin and D. J. Pegg, eds.), Vol. 1, p. 347. Plenum Press, New York, 1976.

[70] K. X. To, E. J. Knystautas, R. Drouin, and H. G. Berry, in "Beam-Foil Spectroscopy" (I. A. Sellin and D. J. Pegg, eds.), Vol. 1, p. 385. Plenum Press, New York, 1976.

[71] E. J. Knystautas and R. Drouin, in "Beam-Foil Spectroscopy" (I. A. Sellin and D. J. Pegg, eds.), Vol. 1, p. 393. Plenum Press, New York, 1976.

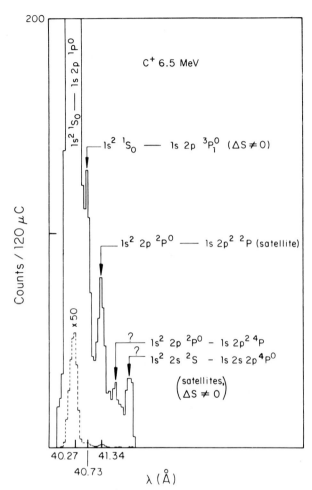

FIG. 16. Spectrum of foil-excited carbon ions at an incident beam energy of 6.5 MeV showing satellite and intercombination lines in the vicinity of the C(V) resonance line (Ref. 74).

plet M2) $1s^22s\ ^2S \rightarrow 1s2s2p\ ^4P$ was observed to occur in highly stripped lithiumlike ions by Cocke et al.[72].

Studies of analogous transitions involving doubly excited states in He(I) and the helium sequence have been recently reviewed by Berry.[73] Allowed transitions between doubly excited states of He(I) and singly excited states fall in the EUV and are shown in Fig. 15. These latter lines,

---

[72] C. L. Cocke, B. Burnutte, and J. R. Macdonald, *Nucl. Instr. Meth.* **110**, 493 (1973).
[73] H. G. Berry, *Phys. Scripta* **12**, 5 (1975).

which are close in wavelength to the resonance line (1s–2p) of He(II), are often called satellite lines and have been observed in other terrestrial and astrophysical sources. Knystautas and Drouin[74] have reported on similar lines in lithiumlike ions and Fig. 16 illustrates their results for C(V). Of course, the reason the wavelengths of satellite lines are often close to the resonance line wavelength is due to the fact that the initial- and final-state configurations only differ from those of the resonance transition by a single "spectator" electron. Interest in these satellite lines stems from the plasma source study of Jordan and Gabriel,[75] who noted that satellite line intensities could be used as an indicator of the electron density in the plasma. These satellite lines are often produced in a plasma by the process of dielectronic recombination in which an electron is captured into a doubly excited state (inverse of autoionization) and the resulting state subsequently relaxes via a stabilizing radiative transition to a lower-lying singly excited state, producing the observed satellite line.

10.4.7. Hydrogenic (Rydberg) States

A foil-excited beam is clearly visible to the naked (but dark-adjusted) eye even, rather suprisingly, in the case of high incident-beam energies, when the characteristic postfoil radiation is expected to fall primarily in the EUV or x-ray region of the spectrum. The origin of the visible radiation under such conditions can be traced to E1 transitions between singly excited hydrogenic (Rydberg) states. Such states are characterized by a tightly bound, highly charged ion core and a single electron in an orbit of much larger radius (perhaps several tens of angstroms). The electron under these conditions is labeled by high n and l values (transitions involving levels as high as $n = 14$ have been reported), and thus the binding energy for such a nonpenetrating orbit can be determined by a simple hydrogenic expression with account taken for the polarization of the core by the electron. Edlén[76] has listed both dipole and quadrupole polarizabilities of the core. It is obvious that when core polarization is small the positions of spectral lines are practically independent of the element, depending only upon the charge state of the core and the change in the quantum number n. From the theoretical point of view, studies of weakly bound states such as these may be of interest because of their similarity to the unbound state consisting of the scattering of electrons from highly ionized systems, a problem of considerable interest in astrophysics

---

[74] E. J. Knystautas and R. Drouin, in "Beam-Foil Spectroscopy" (I. A. Sellin and D. J. Pegg, eds.), Vol. 1, p. 393. Plenum Press, New York, 1976.

[75] A. H. Gabriel and C. Jordan, *Nature* **221**, 947 (1969).

[76] B. Edlén, in "Handbuch der Physik" (S. Flugge, ed.), Vol. 27, p. 80. Springer-Verlag, Berlin, 1964.

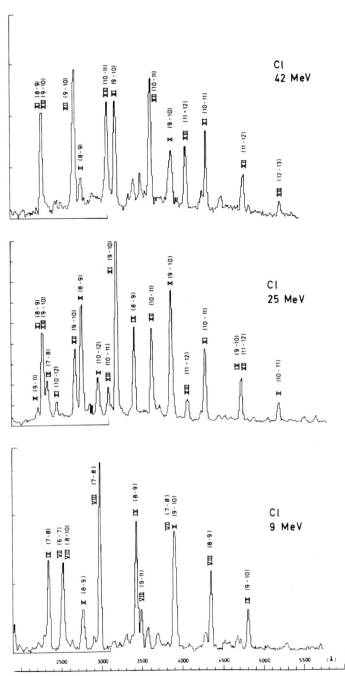

FIG. 17. Portion of a foil-excited spectrum of highly charged chlorine ions taken in the visible and near UV region. The spectral features are due to allowed transitions between hydrogenic (Rydberg) levels of high n (Ref. 77).

and thermonuclear fusion research. It is interesting to note that many of the unidentified features observed in early beam–foil spectra in the visible have been subsequently understood in terms of transitions between these hydrogenic states. Hallin et al.[77] observed many visible hydrogenic transitions associated with the oxygen ions O(V)–O(VIII). Similar studies in chlorine ions by Bashkin and Martinson[78] produced many new lines associated with hydrogenic transitions and this work was extended by Hallin et al.[77] Figure 17 shows typical spectra with prominent hydrogenic lines. Lennard et al.[43] observed hydrogenic lines in the spectra of both Fe(IV)–Fe(VIII) and Ni(IV)–Ni(VIII), and in their work on argon Buchet et al.[79] reported on numerous hydrogenic transitions originating in Ar(X)–Ar(XV). Hydrogenic lines in the visible spectrum of highly ionized fluorine, silicon, and copper were observed by McIntyre et al.[80]

Hydrogenic (Rydberg) states such as we have just described are rather fragile and it is difficult to visualize these states surviving if they were produced within the high-density foil where atoms spacings are typically a few angstroms. It is thought that hydrogenic states may be produced when an emerging ion captures an electron at the exit surface of the foil. Lennard and Cocke[81] have measured cross sections for this process and found an expected n^{-3} dependence. Once outside the foil, the tenuous environment of the beam–foil source is ideal for the hydrogenic states to survive and subsequently relax radiatively without being collisionally deexcited. Lifetimes of such states have also been measured and as expected are found to be relatively long especially for high-l states. Externally applied electric fields have been used to identify these hydrogenic levels[82] (the lifetime of such a level changes when it is a Stark-mixed with close-lying levels of opposite parity) and Stark-induced line intensity changes are an indication that a particular line originates on a hydrogenic level.

10.4.8. Auger-Emitting States

In Section 10.4.6 we discussed beam–foil studies of radiative transitions involving doubly excited states. In order to have time to radiate,

---

[77] R. Hallin, J. Lindskog, A. Marelius, J. Pihl, and R. Sjödin, *Phys. Scripta* **8**, 209 (1973).

[78] S. Bashkin and I. Martinson, *J. Opt. Soc. Amer.* **61**, 1686 (1971).

[79] J. P. Buchet, M. C. Buchet-Poulizac, A. Denis, J. Desesquelles, and G. DoCao, *Phys. Scripta* **9**, 221 (1974).

[80] L. C. McIntyre, J. D. Silver, and N. A. Jelley, in "Beam-Foil Spectroscopy" (I. A. Sellin and D. J. Pegg, eds.), Vol. 1, p. 331. Plenum Press, New York, 1976.

[81] W. N. Lennard and C. L. Cocke, *Nucl. Instr. Meth.* **110**, 137 (1973).

[82] S. Bashkin, in "Beam-Foil Spectroscopy" (I. A. Sellin and D. J. Pegg, eds.), Vol. 1, p. 129. Plenum Press, New York, 1976.

10.4. BEAM–FOIL SPECTRA

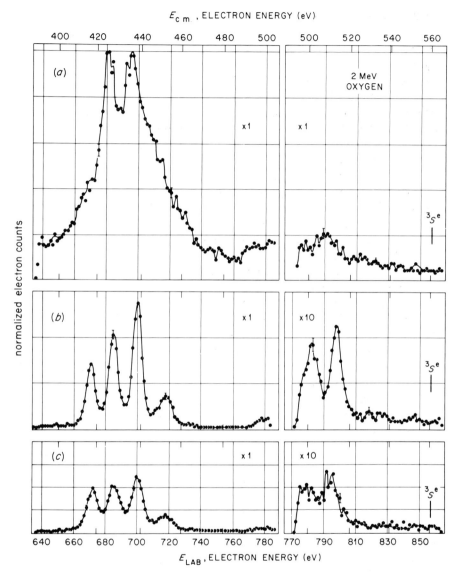

FIG. 18. Spectra of Auger (autoionization) electrons emitted by 2-MeV foil-excited oxygen ions undergoing decay in flight. The energy scale is divided, separating the "low"- and "high"-energy group of peaks. The expansion factors shown normalize the intensity scales to that of the spectrum in the top left-hand corner. (a) "Foil-zero" position; (b,c) time delays of 0.7 and 1.0 nsec, respectively, relative to (a) [D. J. Pegg, I. A. Sellin, R. S. Peterson, J. R. Mowat, W. W. Smith, M. D. Brown, and J. R. Macdonald, *Phys. Rev. A* **8**, 1350 (1973)].

such states must be metastable against allowed Auger emission (autoionization). In this section we investigate beam–foil studies of the autoionizing decay of doubly excited states via forbidden decay channels induced by magnetic interactions between the electrons such as spin–orbit, spin–other orbit, and spin–spin. The UT-ORNL group applied electron spectroscopic techniques to the beam–foil source in order to determine binding energies and lifetimes of doubly excited states in the lithium sequence with K-shell excited configurations of the type $1s2snl$ and $1s2pnl$. Figure 18 shows a spectrum of Auger or autoionizing electrons emitted in flight by foil-excited oxygen ions. The spectrum was taken with a cylindrical-mirror electron analyzer similar to the one described in Section 10.2.6. The special features in this type are associated with doubly excited (K shell excited) states of lithium-, beryllium-, and boronlike oxygen ions. Figure 19 shows a similar Auger spectrum taken by Pegg *et al.*,[83] which involves lines from sodiumlike chlorine ions. Subsequent beam–foil work by Bruch *et al.* on the lithium atom[84] and beryllium, boron, and carbon ions[85] and by Groeneveld *et al.*[86] on nitrogen, oxygen, and neon ions extended the earlier measurements on oxygen, fluorine, aluminum, silicon, sulfur, chlorine, and argon ions made by the UT-ORNL group[6]. Coincidence studies on neon ions by Groeneveld *et al.*[46] clearly showed (see Fig. 10) the charge state of the emitting ion associated with the various features of the electron spectra. Such measurements are extremely time consuming, however, and almost all other data are accumulated in the singles mode.

It should be clear from what has been said up until now that many ions emerging from a foil are left in doubly excited states, which in turn will decay in flight either rapidly via an allowed Auger event or, if this is forbidden, more slowly via either a forbidden Auger transitions or an allowed or forbidden radiative transition. In general, of course, forbidden processes become more probable as the nuclear charge on the ion increases. From the practical point of view, whether electron spectroscopic studies can be made using the beam–foil source depends upon both the lifetime of the states against both radiative and Auger emission (allowed Auger events are usually too rapid to be easily detected downstream from the foil) and the energy of the electron ejected in the Auger process (if the electrons are very low in energy it becomes a difficult prac-

---

[83] D. J. Pegg, I. A. Sellin, P. M. Griffin, and W. W. Smith, *Phys. Rev. Lett.* **28**, 1615 (1972).
[84] R. Bruch, G. Paul, J. Andrä, and L. Lipsky, *Phys. Rev. A* **12**, 1808 (1975).
[85] R. Bruch, G. Paul, and H. J. Andrä, *J. Phys. B* **8**, L253 (1975).
[86] K.-O. Groeneveld, R. Mann, G. Nolte, S. Schumann, and R. Spohr, *Phys. Lett.* **54A**, 335 (1975).

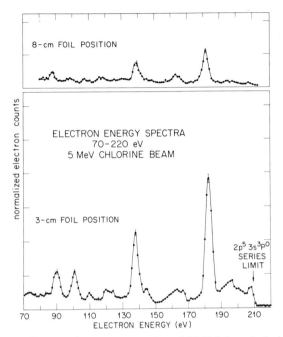

FIG. 19. Section of Auger electron spectra from 5-MeV foil-excited chlorine ions undergoing decay in flight, plotted in the ionic rest frame. Data shown are for foil–spectrometer separations of 3 and 8 cm (Ref. 83).

tical problem to analyze and detect them). The energies of electrons emitted in the Auger decay of doubly excited states of ions with alkalilike or alkaline-earth-like structures are usually relatively large due to the rather big energy difference between the initial state and the final state of the tightly bound inert gas core.

As in the case of the emission of radiation from the moving foil-excited source, one can expect the electron lines to be Doppler-shifted and broadened. A simple nonrelativistic expression, based upon the conservation of linear momentum in the emission process, relates the Auger emission energies T_L in the laboratory frame to those in the rest frame of the emitting ion T_c in the following manner:

$$T_c = T_L + \frac{m}{M} T_i - 2 \left(T_L T_i \frac{m}{M} \right)^{1/2} \cos \theta, \qquad (10.4.4)$$

where T_i is the kinetic energy of the ion beam, θ the angle of emission in the laboratory frame, and m and M the masses of the electron and the emitting ion, respectively.

Finally, it should be mentioned that not all beam–foil electron investigations have been associated with discrete energy electrons. Meckbach[87] and Groeneveld[88] for example, have both recently described studies of the electron energies and angular distributions of the continuous energy electrons that are always found to accompany the passage of heavy ions through foils.

10.5. Lifetimes

10.5.1. Introduction

The single most useful application of the beam–foil method involves the study of time-dependent decay processes by a tof process that utilizes the excellent time resolution produced by fast unidirectional beams and which also scale short decay times into small but measurable decay lengths. The lifetime of a particular state of an ion in the beam is measured by studying the decay in intensity of a line associated with a transition out of a state (the wavelength of the line "signatures" the decay of the particular state) as a function of the distance between the foil and the spectrometer viewing region. Spatial coordinates are converted to temporal coordinates by dividing by the constant postfoil beam velocity. The relative distance between the foil and the viewing region is usually varied by translating the foil in a direction parallel to the beam axis. This technique is indeed very versatile in that, in principle, it is applicable to any ionization stage of any element (subject, of course, to the availability of an appropriate accelerator and ion source). Other lifetime measurement methods such as the Hanle effect[89], electron-impact[90] and ion-impact[91] delayed coincidence techniques, electron-impact phase shift,[92] and laser absorption[93] are in general restricted to neutral atoms or near neutral ions of an element. For highly stripped ions no alternative to the beam–foil method exists at the present time. The precision of individual beam–foil lifetime measurements varies somewhat depending upon the conditions pertaining in the experiment (i.e., presence or absence of com-

---

[87] W. Meckbach, in "Beam-Foil Spectroscopy" (I. A. Sellin and D. J. Pegg, eds.), Vol. 2, p. 577. Plenum Press, New York, 1976.

[88] K.-O. Groeneveld, in "Beam-Foil Spectroscopy" (I. A. Sellin and D. J. Pegg, eds.), Vol. 2, p. 593. Plenum Press, New York, 1976.

[89] For example, W. W. Smith and A. Gallagher, Phys. Rev. **145**, 26 (1966).

[90] For example, W. R. Bennett and P. J. Kindlemann, Phys. Rev. **149**, 38 (1966).

[91] For example, L. W. Dotchin, E. L. Chupp, and D. J. Pegg, J. Chem. Phys. **59**, 3960 (1973).

[92] For example, G. M. Lawrence, Phys. Rev. **175**, 40 (1968).

[93] For example, H. J. Andrä, A. Gaupp, and W. Wittmann, Phys. Rev. Lett. **31**, 501 (1973).

10.5. LIFETIMES

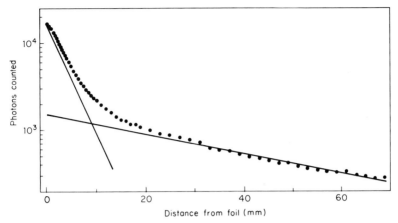

FIG. 20. Intensity decay of the foil-excited Be(I) resonance line (2s² ¹S–2s2p ¹P) measured at a 180-keV incident beam energy. A decomposition into two exponentials that yields lifetimes $\tau_1 = 1.83$ nsec (2s2p ¹P) and $\tau_2 = 20$ nsec (weighted average of the cascading states) [I. Martinson, A. Gaupp, and L. J. Curtis, *J. Phys. B* **7**, L463 (1974)].

plicating cascades, strong or weak lines, line-blending problems, knowledge of the postfoil velocity). Thus the precision of the measurement may be ~10–15%, for example, when cascading complicates the data analysis, or closer to 5% under more ideal conditions where cascading is shown to be unimportant, the line is relatively strong and unblended, and the energy loss suffered by beam ions in the foil is either well known or is negligibly small (for example, as is often the case in high-energy experiments). A particularly well-designed beam–foil lifetime experiment was performed by Astner *et al.*[94] recently at relatively low beam energies in order to estimate how precise a measurement could be made with this technique under optimum conditions. Astner *et al.*[94] quoted a value for the lifetime of the 1s3p¹P₁ level in He(I) of 1.7225 ± 0.0046 nsec (0.26%). Details of this experiment are discussed in Section 10.5.3.

Due to the relative ease of application of the beam–foil technique to a large variety of species, the method has become the primary experimental source of lifetimes, transition probabilities, and oscillator strengths (*f* values). These fundamentally important quantities find frequent application in such areas as astrophysics, laboratory plasma physics, laser physics, and the atomic structure theory. The large amount of relatively accurate beam–foil *f* value data that have been accummulated over the past decade has been instrumental in the establishment of certain systematic *f* value

---

[94] G. Astner, L. J. Curtis, L. Liljeby, S. Mannervik, and I. Martinson, *Z. Physik A* **279**, 1 (1976).

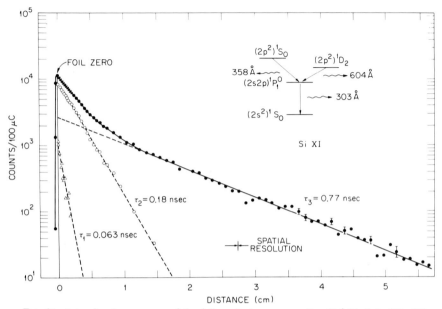

FIG. 21. Intensity decay curve of the foil-excited resonance line (2s² ¹S–2s2p ¹P) of the berylliumlike ion Si(XI) taken at an incident beam energy of 28 MeV. The shape of this curve is similar to that of the corresponding transition in Be(I) (shown in Fig. 20). A decomposition into three exponentials yields the lifetimes $\tau_1 = 63$ psec (2p² ¹S), $\tau_2 = 180$ psec (2s2p ¹P), and $\tau_3 = 770$ psec (2p² ¹D) (Ref. 63).

trends that have been found to exist along isoelectronic sequences. The beam–foil method is, of course, ideally suited for such studies along isoelectronic sequences since the degree of ionization of any element constituting the source can be easily varied by changing the incident beam energy. Figures 20 and 21 illustrate the fact that decay curves associated with a particular transition (in this case the 2s² ¹S → 2s2p ¹P transition in berylliumlike ions) are similar in general shape for the two well-separated members of the same isoelectronic sequence. More is said concerning f value systematics in Section 10.7.2.

10.5.2. Population Changes and Cascading

In this section we consider the effect of cascading (the repopulation of a state by states of higher energy) on the shape of radiative decay curves. Cascading is a consequence of the nonselective nature of the beam–foil collisional excitation process. A recent review of this problem has been given by Curtis.[95] Consider an initial state $|i\rangle$ whose lifetime τ_1 is to be

---

[95] L. J. Curtis, in "Beam-Foil Spectroscopy" (S. Bashkin, ed.), p. 63. Springer-Verlag, Heidelberg, 1976.

10.5. LIFETIMES

measured by studying a radiative transition (wavelength λ_{if}) to some final state $|f\rangle$. If no cascading is present the population of the initial state n_i will be depleted in an exponential manner according to

$$n_i(t) = n_i(0) \exp(-t/\tau_i), \tag{10.5.1}$$

where $\tau_i = (\Sigma_j A_{ij})^{-1}$ (A_{ij} is the transition probability for spontaneous decay to some lower state $|j\rangle$), $n_i(t)$ and $n_i(0)$ are the populations of state $|i\rangle$ at any instantaneous time t and at $t = 0$, respectively. (The origin of time, $t = 0$, in a beam–foil experiment is, of course, the foil.) Let us now consider the simplest cascade situation that can arise. Assume that a single upper level $|k\rangle$ cascades into the state $|i\rangle$ with a transition probability A_{ki} (also assume for simplicity that $A_{ki} = \tau_k^{-1}$, that is, unity branching ratio). Under these conditions the rate equation describing the time dependence of the population of state $|i\rangle$ is given by

$$\frac{dn_i(t)}{dt} = -\frac{n_i(t)}{\tau_i} + \frac{n_k(t)}{\tau_k}. \tag{10.5.2}$$

The solution to this differential equation is

$$n_i(t) = \left[n_i(0) - n_k(0) \frac{\tau_i}{\tau_k - \tau_i} \right] \exp\left(\frac{-t}{\tau_i}\right)$$
$$+ \left[n_k(0) \frac{\tau_i}{\tau_k - \tau_i} \right] \exp\left(\frac{-t}{\tau_k}\right). \tag{10.5.3}$$

Curtis[95] has expressed the population equation in terms of more general cascade situations. The intensity of the emitted radiation $I_{if}(t)$ at any instant of time t is given by $A_{if} n_i(t)$. It is not possible, however, to observe at any one instant of time following excitation since the finite acceptance angle of the apparatus defines a short-beam-source segment of length $\Delta x = v \Delta t$ (Δt is the time "window" of the apparatus). We must then integrate the expression for $n_i(t)$ [Eq. (10.5.2)] over the symmetric time window from $t - \Delta t/2$ to $t + \Delta t/2$ or after converting to spatial coordinates (which are the quantities actually measured in a beam–foil lifetime measurement) from $x - \Delta x/2$ to $x + \Delta x/2$. Then we obtain an expression for the intensity $I_{if}(x, \Delta x)$ that can be written

$$I_{if}(x, \Delta x) = \epsilon_{if} A_{if} \frac{\Delta x}{v} n_i(0) \left\{ \left[1 - \frac{1}{N}\left(\frac{T}{1-T}\right) \right] \exp\left(-\frac{x}{v\tau_i}\right) \right.$$
$$\left. + \left[\frac{1}{N}\left(\frac{T}{1-T}\right) \right] \exp\left(-\frac{xT}{v\tau_i}\right) \right\}, \tag{10.5.4}$$

where ϵ_{if} is the detection efficiency of the apparatus for radiation of wavelength λ_{if}, Δx the length of the beam "viewed" by the apparatus, v the postfoil beam velocity, $n_i(0)$ the initial population of the $|i\rangle$ state following foil excitation, $N = n_i(0)/n_k(0)$ the ratio of the initial populations of the state $|i\rangle$ to the cascade state $|k\rangle$, and $T = \tau_i/\tau_k$ the ratio of the lifetimes of the state $|i\rangle$ to that of the cascade state $|k\rangle$. This equation is cast in a particularly easy form to simulate artifical decay curves for various combinations of N and T. For example, Figure 22 displays two

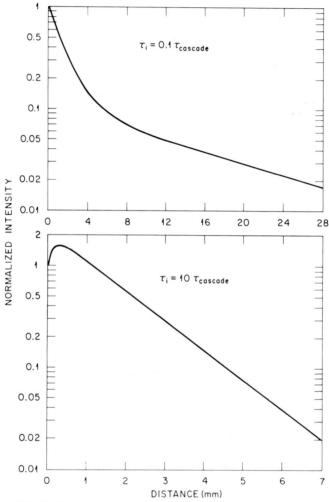

FIG. 22. Artificially simulated intensity decay curves showing two types of cascading effects observed in beam–foil experiments. The upper curve shows a cascade component with a lifetime longer than the primary level and the lower curve shows a cascade component with a lifetime shorter than the primary level resulting in a "growing-in" effect [D. J. Pegg, unpublished data].

different shapes of decay curves involving cascades obtained by programming Eq. (10.5.2) on a simple hand calculator. The upper curve ($N = 1$, $T = 0.1$) illustrates the most commonly met type of cascade situation, i.e., when the cascade lifetime is longer than that of the primary level. The lower curve ($N = 1$, $T = 10$) shows the less frequently observed cascade situation of "growing-in." It is comparatively simple to extract the decay constant of interest from such decay curves if the decay constant associated with the cascade component is sufficiently different (as is the case in Fig. 22). The condition $N \simeq 1$ is physically realistic for the case of in-shell cascades, which were the dominant cascade contributors observed by Pegg et al.[63, 96] in their work on $\Delta n = 0$ dipole transitions within the L shell of highly stripped sulfur ions. This condition will in general not prevail for the most frequently met cascade situation, i.e., out-of-shell cascading. It is very instructive to generate artificial decay curves such as those shown in Fig. 22 since they make one aware of what to look for under real experimental conditions.

In general, a beam–foil decay curve is a multiexponential sum that includes terms representing many cascade levels in addition to the one describing the level of interest. In practice, however, it is frequently possible to approximate the cascade contributions by one or perhaps two dominant components. The first step in any data analysis of decay curves is the graphical method (using semilogarithmic paper). This technique can provide useful input estimates for a more sophisticated data analysis procedure involving a least-squares fit computer program.[97] Curtis[95] includes a useful description of various curve-fitting techniques in his review article.

10.5.3. Postfoil Beam Velocity

The postfoil beam velocity is an extremely important parameter in any tof technique since it is, of course, used to connect spatial decay measurements to the time domain. Via this constant scaling factor very small time intervals are translated into easily measurable spatial intervals. In fact, it is certainly worthwhile, if possible, to perform a lifetime measurement at several different beam energies (consistent, of course, with efficiently producing the charge state of interest) in order to check the aforementioned scaling with beam velocity. An example of such a measurement is shown in Fig. 23.

---

[96] D. J. Pegg, S. B. Elston, P. M. Griffin, H. C. Hayden, J. P. Forester, R. S. Thoe, R. S. Peterson, and I. A. Sellin, *Phys. Rev. A* **14**, 1036 (1976).

[97] For example, Program HOMER written by D. J. G. Irwin and A. E. Livingston, *Comput.-Phys. Comm.* **7**, 95 (1974).

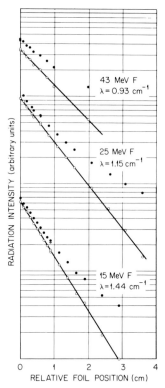

FIG. 23. Decay curves for the $1s^2\ ^1S_0 \to 1s2p\ ^3P_1$ intercombination transition in the heliumlike ion F(VIII) taken at three different beam energies to check the velocity scaling. The open circles represent data after subtraction of a background component [J. R. Mowat, I. A. Sellin, R. S. Peterson, D. J. Pegg, M. D. Brown, and J. R. Macdonald, *Phys. Rev. A* **8**, 145 (1973)].

At low incident-beam energies, in particular, a correction must be made for the rather appreciable fractional energy loss (often $\geq 10\%$) suffered by the beam ions in transit through the foil. In this situation it is preferable to directly measure the postfoil velocity of the beam using a calibrated charged-particle analyzer such as an electrostatic device. Andrä[98] describes a rather novel calibration technique involving the use of the zero-field quantum beat pattern associated with the well-established fine-structure interval between the 2^3P_J levels of He(I). Astner *et al.*[94] went one step further in their precise measurement of the 1s3p $^1P_1$ level in He(I) by using two spectrometers simultaneously, one to record the decay

---

[98] H. J. Andrä, *in* "Beam-Foil Spectroscopy" (I. A. Sellin and D. J. Pegg, eds.), Vol. 2, p. 835. Plenum Press, New York, 1976.

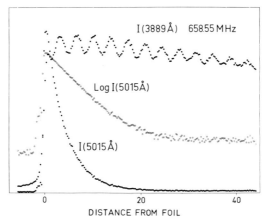

FIG. 24. A decay curve for the 1s3p ^{1}P level of He(I). The data are velocity-normalized by a simultaneous quantum beat measurement using the 3889 Å line (Ref. 94).

curve of the 1s3p $^1P_1$ state and the other to monitor the aforementioned quantum beat pattern in order to obtain an accurate value for the postfoil velocity. Figure 24 shows the results obtained in this experiment.

At higher beam energies, the fractional energy loss in the foil target is often negligibly small enabling, with little error, the postfoil velocity to be equated directly to the incident-beam velocity as determined by the calibrated magnetic analyzer associated with the accelerator. An energy loss correction can be made if necessary by using specific energy loss tables[99] and a value for the foil thickness (one can accept the manufacturer's estimate or measure it using α particle ranging for example).

It is also important, particularly in beam–foil lifetime measurements, to avoid uncertainties associated with the potential problem of foil thickening, which could steadily decrease the post-foil beam velocity throughout the measurement. Foil thickening effects were discussed in detail in Section 10.2.4. Other effects that could potentially lead to problems in lifetime measurements are associated with the dispersion of the ion beam velocity by the foil. Up until this point we have assumed that all the beam ions are emerging with equal velocities in a direction perpendicular to the line of sight of the spectrometer. (This situation, in fact, never really pertains since even before passage through the foil the beam is usually best represented by a Gaussian-shaped velocity profile and beam collimation is never perfect.) The foil interaction introduces a velocity dispersion to the beam. Transverse components to the beam velocity vector are introduced by multiple-scattering processes in the foil

[99] For example, L. C. Northcliffe, and R. F. Schilling, *Nucl. Data Tables A* **7**, 233 (1970).

and these produce a spatial divergence in the previously collimated beam. Additional longitudinal velocity components are also introduced in the foil interaction, causing energy straggling in the postfoil beam. If the beam divergence is severe (as may be the case for low-energy heavy ions) some of the beam ions may travel outside the viewing region of the spectrometer, causing incorrect intensity measurements. Fortunately, this is rarely a problem since most experimental arrangements employ spectrometer entrance slits that are perpendicular to the beam axis (the same beam divergence problem can, however, effect the normalization procedure as was discussed in Section 10.2.5). It can also be shown that the effect of Gaussian beam velocity profile on the semilogarithmic slope of a beam–foil decay curve is given approximately by the expression

$$\text{slope} \simeq \frac{1}{\tau}\left[1 - \frac{1}{5.6}\left(\frac{\Delta v}{v_0}\right)^2 \frac{t}{\tau}\right], \quad (10.5.5)$$

where τ is the lifetime of the decay state, v_0 and Δv the mean and full-width-half-maximum (FWHM) spread of the assumed gaussian beam velocity profile, and t the time following excitation. In all cases likely to be met in beam–foil work, the correction factor will be small enough to be neglected in the decay curve analysis.

Other methods have also been employed to determine the postfoil beam velocity accurately. For example. Sørensen[100] describes a method suited to low-energy, heavy-ion beams. In this technique the change in the Doppler shift of a spectral line from the beam (measured at 30° to the beam axis) is determined when the beam is excited first by a dilute gas and then by a foil target.

Relativistic time dilation corrections to lifetime measurements are negligibly small (compared to other sources of error) for most beam velocities employed in beam–foil experiments (it amounts to a $\sim 2\%$ correction at $\beta = 0.2$).

10.5.4. Limits on Beam–Foil Lifetimes

The beam–foil method is not suited to measurements of very short ($\tau \leqslant 10^{-12}$ sec) or very long ($\tau \geqslant 10^{-7}$ sec) lifetimes. In this section we discuss the factors determining these limits. It was shown in Section 10.5.2 [Eq. (10.5.4)] that the recorded intensity of a particular spectral line at any position downstream from the foil depends not only upon the intrinsic strength of the transition (f value), the "direct" population of the upper

---

[100] G. Sørensen, in "Beam-Foil Spectroscopy" (I. A. Sellin and D. J. Pegg, eds.), Vol. 1, p. 165. Plenum Press, New York, 1976.

state of the transition in the foil excitation process, and of course the efficiency of the apparatus, but also upon such factors as the amount of cascade repopulation of the upper state and the instrumental "window" attenuation factor ($\Delta x/v\tau$) for the transition (Δx is the length of beam that is viewed). Because of this window factor, a transition will be recorded most strongly in a beam–foil experiment when the upper-state lifetime (τ) approximately matches the time "window" ($\Delta x/v$) of the apparatus. Thus for long-lived states $\tau \gg \Delta x/v$, the probability of decays occurring within the window is so small that the corresponding intensity measurements are too weak to be useful for lifetime determinations. The long-lifetime limit of the method thus depends upon both the length of beam view (Δx) and the beam velocity (the smaller the better for long lifetime work). Consider a beam velocity of $\sim 1 \times 10^7$ cm/sec and a spatial resolution of ~ 1 cm (remember this cannot be made indefinitely large since it will produce increased Doppler broadening and spectral bandpass with the potential for line broadening). Thus it is doubtful whether the beam–foil technique can yield meaningful lifetime results below $\tau \sim 1 \times 10^{-7}$ sec. For very short lifetimes, $\tau \ll \Delta x/v$, the decay process occurs so close to the foil that it becomes exceedingly difficult to track spatially. The short-lifetime limit of the method thus depends upon how small the time window $\Delta x/v$, of the apparatus can be made. Since the beam velocity is somewhat fixed in order to maximize a particular charge state in the source, the beam length viewed, Δx, must be made small to measure short lifetimes (this again cannot be made indefinitely small since the amount of radiation entering the spectrometer is proportional to the length of the beam that is viewed). In the recent investigation of lifetimes in highly ionized sulfur by Pegg et al.,[63] made using a grazing incidence spectrometer, a spatial resolution $\Delta x \simeq 0.2$ mm was employed with a beam velocity of $\sim 1.66 \times 10^9$ cm/sec (chosen to maximize the S(XIII) charge component of the beam). These factors correspond to a time window of ~ 13 psec. Spatial resolutions of only ~ 0.05 mm have been obtained using similar grazing incidence spectrometers by Barrette[28] and Berry and Batson[29] in a manner described in Section 10.2.6. This enabled lifetime measurements to be extended down to $\sim 10^{-12}$ sec.

10.5.5. Allowed Radiative Transitions

A radiative transition is said to be allowed if the selection rules on the electric dipole (E1) decay process are satisfied. It is convenient to further distinguish between different types of allowed transitions: in-shell ($\Delta n = 0$) and out-of-shell ($\Delta n \neq 0$). The decay rates for the two processes scale quite differently with Z along an isoelectronic sequence for the reasons to be discussed below. The E1 transition rate A_{E1} is given by

$$A_{E1} \propto (\Delta E)^3 \langle r \rangle^2, \qquad (10.5.6)$$

where ΔE is the energy separation between the upper and lower states of the transition, and $\langle r \rangle$ indicates the electric dipole matrix element connecting the upper and lower states. For out-of-shell transitions, ΔE is determined primarily by the electron–nucleus interaction, which scales as Z^2, while for in-shell transitions ΔE is the electrostatic fine structure and is determined primarily by the electron–electron interaction, which scales approximately linearly with Z. In the realm of a simple Bohr model, $\langle r \rangle \propto 1/Z$. Thus combining these results yields $A_{E1} \propto Z^4$ for out-of-shell processes while $A_{E1} \propto Z$ for in-shell transitions. It is easy to see that states decaying primarily via out-of-shell transitions rapidly become too short-lived to be studied by the beam–foil tof method, whereas states that can only deexcite via in-shell transitions remain amenable to beam–foil measurements to far higher Z due, of course, to the less-severe Z-scaling dependence. Figures 25, 26, and 27 show beam–foil decay curves for in-shell ($\Delta n = 0$) transitions in highly ionized systems. Of particular interest is the presence in Fig. 27 of the long-lived cascade compo-

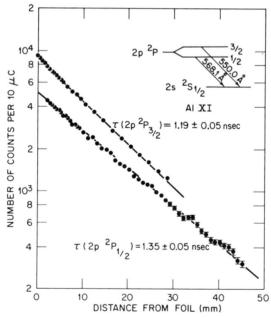

FIG. 25. Decay curve for the $1s^2 2p$ ^2P levels in the foil-excited lithiumlike ion AL(XI) obtained with a 45-MeV incident beam energy [Ref. (63)].

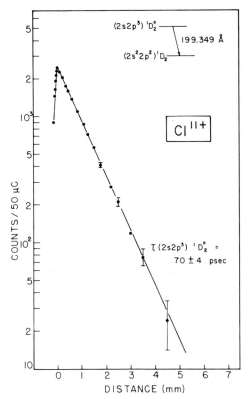

FIG. 26. Decay curve for the $(2s2p^3)^1D_2$ level in the foil-excited carbonlike ion S(XIII) obtained with a 38-MeV beam energy (J. P. Forester et al., Ref. 63).

nent (the decay curve was investigated out to more than ten decay lengths of the primary level), which is probably attributable[101] to cascading from long-lived yrast levels (the term "yrast level," borrowed from nuclear physics, is used here to indicate a Rydberg level of maximum l for a given value of n). This tail seems to be always present in decay curves associated with $\Delta n = 0$ transitions in M and N shells. It was also observed in this experiment that if the rather strong cascade tail was neglected and the data analysis was based upon only three decay lengths of the primary level, the measured "lifetime" would be raised by some 40%!

Of particular importance in lifetime work are those transitions for which there exists an approximately unity branching ratio out of the upper state since under these conditions the reciprocal of the measured lifetime yields the probability for the transition (A value), a fundamental atomic

[101] R. J. S. Crossley, L. J. Curtis, and C. Froese-Fischer, Phys. Lett. **57A**, 220 (1976).

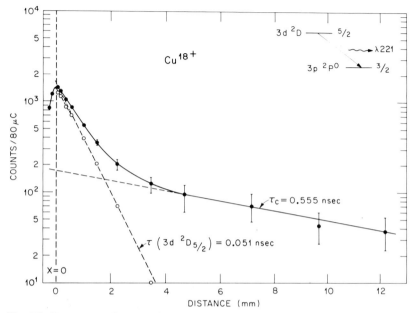

FIG. 27. Decay curve for the 3d $^2D_{5/2}$ level in the foil-excited sodiumlike ion Cu(XIX) obtained using a 70-MeV beam energy. The long-lived cascade tail is thought to be associated with highly excited Rydberg levels (Ref. 102).

quantity that finds frequent application in many areas of research. The absorption oscillator strength or f value for dipole transitions is a quantity related to the A value by

$$f_{fi} = 1.499 \times 10^{-16} (g_i/g_f) \lambda_{if}^2 A_{if}, \quad (10.5.7)$$

where g_i and g_f are statistical weight factors of the initial and final states, respectively, λ_{if} the wavelength (Å) of the transition i → f, and A_{if} the atomic transition probability (sec^{-1}). Another useful expression relates the f value to the line strength S of a transition (in a.u.):

$$f_{fi} = (303.7/g_f \lambda_{if}) S_{if}. \quad (10.5.8)$$

Frequently f values are the bridge between theory and experiment, i.e., one usually compares f values. Thus the experimentalist arrives at an f value via Eq. (10.5.7) after having measured τ_i and derived A_{if} from the measurement. The theorist calculates an **f** value [using Eq. (10.5.8)] by combining a calculated line strength S with either a calculated or a measured wavelength λ_{if}. It is preferable to use the same measured λ (if

[102] D. J. Pegg, P. M. Griffin, B. M. Johnson, K. W. Jones, J. L. Cecchi, and T. H. Kruse, Phys. Rev. A **16**, 2008 (1977).

10.5. LIFETIMES

available) in both Eqs. (10.5.7) and (10.5.8) so that the experiment becomes a way of checking the line strength factor alone. This in turn tests the appropriateness of the wavefunctions used in the calculation. As one goes to highly stripped, heavy-ion systems where wavelength values are not so readily available, it may become necessary to accurately measure the wavelength of the transition as well as the lifetime to obtain an accurate f value.

At this point we refer to several sets of tables of both wavelengths and transition probabilities (and f values) that are very useful to beam–foil investigators: The wavelength tables of Moore,[103] Kelly and Palumbo,[104] and Striganov and Sventitskii[105] aid in line identifications. The tables of Wiese et al.[106] and Smith and Wiese[107] are important sources of A and f values.

10.5.6. Forbidden Radiative Transitions

A radiative transition is said to be forbidden if the allowed E1 process is prohibited by any one of the well-known selection rules on E1 transitions. If this is the case the state may still radiate, of course, but via one of the weaker, forbidden decay channels such as M1 (magnetic dipole), E2 (electric quadrupole), M2 (magnetic quadrupole), or 2E1 (two-photon electric dipole) for which the selection rules are less rigorous. The transition rates for all such processes are considerably smaller than for the allowed process and are very difficult to measure for atoms or near neutral ions. As a heavy ion becomes more highly stripped, however, the forbidden decay rates, which scale strongly with Z, become appreciable and eventually even dominate over competing E1 decays. A state that can only radiate via forbidden processes is said to be metastable against radiation. Thus metastable states are long-lived for near neutral species but become short enough for beam–foil tof measurements in highly stripped, high-Z ions. Most atoms and their isoelectronic ions have at least one meta-

---

[103] C. E. Moore, "Atomic Energy Levels," Vol. 1, National Bureau of Standards (NSRDS-NBS 35) Reference Data Series, Washington, D.C., 1971.

[104] R. L. Kelly and L. J. Palumbo, "Atomic and Ionic Emission Lines Below 2000 Angstroms," Naval Research Laboratory Report 7599, Washington, D.C., 1973.

[105] A. R. Striganov and N. S. Sventitskii, "Tables of Spectral Lines of Neutral and Ionized Atoms," Plenum Press, New York, 1968.

[106] W. L. Wiese, M. W. Smith, and B. M. Glennon, "Atomic Transition Probabilities," Vol. 1, National Bureau of Standards (NSRDS-NBS 4) Reference Data Series, Washington, D.C., 1966; W. L. Wiese, M. W. Smith, and B. M. Miles, "Atomic Transition Probabilities," Vol. 2, National Bureau of Standards (NSRDS-NBS 22) Reference Data Series, Washington, D.C., 1968.

[107] M. W. Smith and W. L. Wiese, *Astrophys. J.* **23**, 103 (1971).

stable state, but up until now beam–foil studies have concentrated on metastability in one-, two-, and three-electron ions. A good review of the topic has been given by Marrus[108].

A partial energy level diagram (the $n = 2$ manifold) for a two-electron system is shown in Fig. 28 to illustrate many of the forbidden decay modes previously mentioned (for ions with $Z \geq 6$, all the 2^3P levels lie below the 2^1S level). All of the 2^3P_J levels can decay by E1 transitions to the 2^3S level, an in-shell process that scales as Z in the limit of large Z. A competing decay mode for the depopulation of the 2^3P_1 level is the spin-forbidden (intercombination or intermultiplet) transition to the 1^1S_0 ground state. Singlet–triplet mixing, bought about by the spin–orbit interaction, increases rapidly with Z and the $1^1S_0 \to 2^3P_1$ rate increases initially as Z^{10} (and ultimately as Z^4). For ions with $Z \geq 6$ this spin-forbidden process dominates the E1 decay channel to the 2^3S_1 level. The beam–foil results, for example, of Sellin et al.[109] ($Z = 7,8$), Mowat et al.[110] ($Z = 9$), Richard et al.[111] ($Z = 9$), and Varghese et al.[112] ($Z = 14,16$) are all in essential agreement with theoretical predictions for this transition. The 2^3P_2 level has a spin-forbidden (intermultiplet) M2 decay channel to the ground state, which also competes with the allowed E1 decay to the 2^3S_1 level. The rate for this M2 process scales as Z^8 and so eventually ($Z \geq 17$) it dominates the in-shell E1 process. There are no possible E1 transitions out of the 2^1S_0 or the 2^3S_1 levels. The 2^1S_0 cannot decay by any single-photon process due to the rigorous selection rules on J ($0 \leftrightarrow 0$). A two-photon decay process is possible (2E1) and this gives rise to continuum emission. Rates for the 2E1 process, which scale as Z^6, become sufficiently rapid for high-Z ions that the emission continuum is observable in beam–foil experiments. For example, Marrus and Schmieder[113] studied this continuum using coincidence techniques in several ions including heliumlike argon. The 2^3S_1 level can also radiate by a 2E1 process similar to that just described for the 2^1S_0 level. However, it has also been quite recently established that a competing M1 transition, producing a discrete line, is also permitted and in addition since

[108] R. Marrus, in "Beam-Foil Spectroscopy" (S. Bashkin, ed.), p. 209. Springer-Verlag, Heidelberg, 1976; R. Marrus, in "Atomic Physics 3" (S. J. Smith and G. K. Walters, eds.), p. 291. Plenum Press, New York, 1973.

[109] I. A. Sellin, B. L. Donnally, C. Y. Fan, Phys. Rev. Lett. 21, 717 (1968); I. A. Sellin, M. Brown, W. W. Smith, and B. Donnally, Phys. Rev. A 2, 1189 (1970).

[110] J. R. Mowat, I. A. Sellin, R. S. Peterson, D. J. Pegg, M. D. Brown, and J. R. Macdonald, Phys. Rev. A 8, 145 (1973).

[111] P. Richard, R. L. Kauffman, F. F. Hopkins, C. W. Woods, and K. A. Jamison, Phys. Rev. Lett. 30, 888 (1973).

[112] S. L. Varghese, C. L. Cocke, and B. Curnutte, Phys. Rev. A 14, 1729 (1976).

[113] R. Marrus and R. W. Schmieder, Phys. Rev. A 5, 1160 (1972).

10.5. LIFETIMES

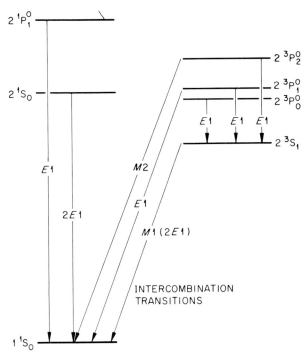

FIG. 28. Partial energy level diagram (low-lying levels) of a heliumlike ion showing the various forbidden radiative decay modes.

this relativistically induced M1 process scales as $\sim Z^{10}$ it dominates the 2E1 process at high Z. An experimental laboratory confirmation that this type of transition occurs was first made by Marrus and Schmeider[114] in heliumlike silicon, sulfur, and argon. Lifetime measurements on other highly stripped heliumlike ions were later made by Gould et al.,[115] Cocke et al.,[116] and Bednar et al.[117] At first there appeared to be a discrepancy between the experimental measurements and the theory of Drake[118] and Johnson and Lin.[119] Later, however, it was found that experimental results approached the theoretical values if measurements were made downstream from the foil at distances corresponding to more

[114] R. Marrus and R. W. Schmieder, Phys. Lett. **32A**, 431 (1970).
[115] H. Gould, R. Marrus, R. W. Schmieder, Phys. Rev. Lett. **31**, 504 (1973); H. Gould, R. Marrus, and P. Mohr, Phys. Rev. Lett. **33**, 676 (1974).
[116] C. L. Cocke, B. Curnutte, and R. Randall, Phys. Rev. Lett. **31**, 507 (1973).
[117] J. A. Bednar, C. L. Cocke, B. Curnutte, and R. Randall, Phys. Rev. A **11**, 460 (1975).
[118] G. W. F. Drake, Phys. Rev. A **3**, 908 (1971).
[119] W. R. Johnson and C. Lin, Phys. Rev. A **9**, 1486 (1974).

than one decay length of the 2^3S level. Transition probabilities for the spin-forbidden $(1s^2 2s^2)^1 S_0 - (1s^2 2s 2p)^3 P_1$ transition in Be-like Fe and Kr have been measured by Dietrich et al.[120]

10.5.7. Forbidden Auger Transitions

In Section 10.4.8 we discussed forbidden Auger deexcitation processes, which often become observable when allowed Auger events are prohibited. Electron spectra resulting from the Auger (autoionizing) decay of these metastable autoionizing states are shown in Figs. 18 and 19. Here we describe tof lifetime measurements that have been made on some of these spectral features. The $1s2s2p\ ^4P_{5/2}$ level of the lithium sequence is particularly interesting since being the lowest-lying quartet level it is metastable against both radiation and Auger emission. The UT-ORNL group has measured[121] the lifetime of this metastable level in lithiumlike ions of oxygen, fluorine, silicon, sulfur, chlorine, and argon (this incidentally is a good example of how the beam–foil method can be used to study many different members of an isoelectronic sequence). Figure 29 shows a typical decay curve for Cl(XV). Groeneveld et al.[86] have made similar measurements in lithiumlike nitrogen, oxygen, and neon and Bruch et al.[85] have reported measurements on near neutral lithiumlike ions. The $(1s2s2p)^4P_{5/2}$ level has two primary deexcitation channels both of which are forbidden but which scale quite differently with Z. Thus the level can relax via an Auger process (autoionization) induced by the tensor part of the spin–spin interaction between the electrons, a process that appears to scale as $\sim Z^{3-4}$. Radiative deexcitation of this level via a spin-forbidden M2 transition to the ground state has also been observed[72,122] and this process scales $\sim Z^8$ for large Z. Following the suggestion by Cocke et al.[72] that the rate for this M2 process in lithiumlike ions would be approximately equal to the known rates for the analogous transition in heliumlike ions, Pegg et al.[123] estimated from the lifetime results that the fluorescence yield (branching ratio for the M2 radiative process) was less than 1% for lithiumlike ions with $Z \leq 10$. This yield increases, however, to $\sim 20\%$ for $Z = 17$. These estimates were later confirmed and extended by the Dirac–Hartree–Fock calcula-

---

[120] D. D. Dietrich, J. A. Leavitt, S. Bashkin, J. G. Conway, D. MacDonald, R. Marrus, B. M. Johnson, and D. J. Pegg, *Phys. Rev. A* **18**, 208 (1978); D. D. Dietrich, J. A. Leavitt, H. Gould, and R. Marrus, *J. Physique* **40**, C1–215 (1979).

[121] A summary of results in H. H. Haselton, R. S. Thoe, J. R. Mowat, P. M. Griffin, D. J. Pegg, and I. A. Sellin, *Phys. Rev. A* **11**, 468 (1975).

[122] J. R. Mowat, K. W. Jones, and B. M. Johnson, *Phys. Rev. A* **14**, 1109 (1976).

[123] D. J. Pegg, H. H. Haselton, P. M. Griffin, R. Laubert, J. R. Mowat, R. Peterson, and I. A. Sellin, *Phys. Rev. A* **9**, 1112 (1974).

10.5. LIFETIMES

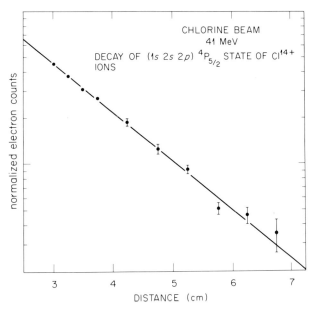

FIG. 29. Auger decay in flight of the 1s2s2p $^4P_{5/2}$ foil-excited level in the lithiumlike ion Cl(XV) obtained using a 41-MeV beam energy [D. J. Pegg, P. M. Griffin, I. A. Sellin, and W. W. Smith, *Nucl. Instr. Meth.* **110**, 489 (1973)].

tions of Cheng *et al.*,[124] who were the first to calculate the transition rates for both the Auger and radiative channels. It is apparent that while these theoretical results agree quite well with the experimental measurements, there is still a small yet significant difference for most of the ions of the lithium sequence [the ratio of experimental-to-theoretical lifetime values seems to be constant at ~1.14 for ions from O(VI) to Cl(XV)].

10.5.8. Lamb Shift Determinations in Heavy Ions

Radiative or Lamb shifts in atoms and ions are quantum-electrodynamical (QED) effects caused by the interaction of electrons with the quantized electromagnetic field of free space. The Dirac relativistic theory of an electron in a pure Coulomb field predicts degeneracy between the $2s_{1/2}$ and $2p_{1/2}$ levels of one-electron systems but this degeneracy (Lamb shift) is lifted primarily by QED effects such as electron self-energy and vacuum polarization. Accurate measurements of Lamb shifts provide a convenient way of testing the validity and limitations of

---

[124] K. T. Cheng, C. P. Lin, and W. R. Johnson, *Phys. Lett. A* **48**, 437 (1974).

QED, our most sophisticated theory of atomic structure and radiative processes. Experimental measurements of the Lamb shift in the $n = 2$ states of highly ionized one-electron heavy ions have been made using the beam–foil or beam–gas technique. Such measurements are important since any shortcomings of QED theory are expected to be more apparent for highly ionized systems where higher-order corrections to the established theory (already well established for atomic hydrogen and near neutral one-electron ions) become increasingly important. Lamb shift measurements on highly stripped one-electron ions are usually made using a field-quenching technique. This method is based upon the fact that the s

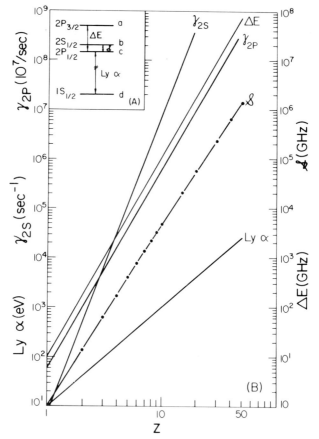

FIG. 30. Plot of the Z dependence of the energy separations and widths associated with the low-lying levels of a hydrogenlike atom (nuclear spin 0) [D. E. Murnick, *in* "Beam-Foil Spectroscopy" (I. A. Sellin and D. J. Pegg, eds.), Vol. 2, p. 815. Plenum Press, New York, 1976].

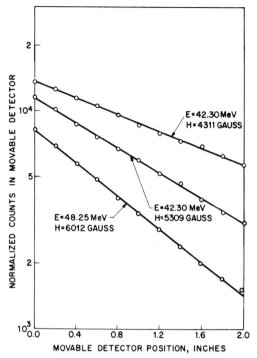

FIG. 31. Semilog plot of normalized decay curves associated with the Stark-mixed $n = 2$ states of O(VIII). These results yield a Lamb shift value for the hydrogenlike ion O(VIII) (Ref. 125).

and p levels have very different field-free lifetimes and that these opposite parity levels can be "mixed" by either an external electric field or a motional electric field produced by passing the ion beam through the field of a magnet. Figure 30 shows the Z dependence of the energy intervals and decay rates of the $n = 2$ levels of a hydrogenlike ion of zero nuclear spin. The magnitude of the field-dependent lifetime is related to the Lamb shift separation between the two mixed components by (the Bethe–Lamb formula)

$$\gamma(E) = \gamma(2p) \frac{|V|^2}{\hbar^2\{\mathscr{S}^2 + [\gamma^2(2p)/4]\}} + \gamma(2s), \quad (10.5.9)$$

where $\gamma(E)$, $\gamma(2p)$, and $\gamma(2s)$ are the decay rates associated with mixed states in the electric field and the 2p and 2s states separately without the electric field, respectively. The Stark matrix element V connects the $2p_{1/2}$ and $2s_{1/2}$ levels and $\mathscr{S}$ corresponds to the Lamb shift interval between the two levels. Figure 31 shows three field-dependent decay curves obtained by Leventhal et al.[125] in their measurement of the $n = 2$ Lamb shift in O(XIII). Marrus[108] summarizes both experimental measurements and theoretical calculations for one-electron systems from hydrogen to fluorine. The F(IX) measurement described by Kugel et al.[126] did not involve the usual Stark-mixing method but rather a laser absorption technique (using a pulsed HBr laser). Extension of both techniques to higher-Z ions is planned.

Berry and Schectman[127] have used the beam–foil technique to measure Lamb shifts in the two-electron ion N(VI) by accurately determining the wavelengths of the $3^3S_1 \rightarrow 3^3P_J$ transitions in the foil-excited source.

10.6. Quantum Beat Phenomena

10.6.1. Introduction

It was previously stated that from the spectroscopic point of view the beam–foil source is one of only low-to-moderate resolution. However, we now discuss certain interesting source properties that sometimes enables high-resolution measurements to be made on small-energy separations (such as fine and hyperfine intervals), which are themselves unresolvable by normal spectroscopic means. The quantum beat method described in this section is based upon an interference effect arising from the radiative decay of two or more coherently excited, aligned (or oriented), and closely spaced levels to a common lower energy level. Excellent reviews of this subject exist in the literature. For example, on the theoretical side the papers of Macek,[128] Fano and Macek,[129] and Macek and Burns[130] provide a good introduction to the subject. The papers by Andrä,[131] Berry,[132] and Church[133] review the many experimental measurements.

[125] M. Leventhal, D. E. Murnick, and H. W. Kugel, Phys. Rev. Lett. **28**, 1609 (1972).
[126] H. W. Kugel, M. Leventhal, D. E. Murnick, C. K. N. Patel, and O. R. Wood, Phys. Rev. Lett. **35**, 647 (1975).
[127] H. G. Berry and R. M. Schectman, Phys. Rev. A **9**, 2345 (1974).
[128] J. Macek, in "Physics of Electronic and Atomic Collisions" (J. S. Risely and R. Geballe, eds.), p. 627. Univ. of Washington Press, Seattle, 1976.
[129] U. Fano and J. H. Macek, Rev. Mod. Phys. **45**, 553 (1973).
[130] J. Macek and D. Burns, in "Beam-Foil Spectroscopy" (S. Bashkin, ed.), p. 237. Springer-Verlag, Heidelberg, 1976.
[131] H. J. Andrä, in "Atomic Physics 4" (G. Zu Putlitz, E. W. Weber, and A. Winnacker, eds.), p. 625. Plenum Press, New York, 1976; ibid. Phys. Scripta **9**, 257 (1974).
[132] H. G. Berry, in "Beam-Foil Spectroscopy" (I. A. Sellin and D. J. Pegg, eds.), Vol. 2, p. 755. Plenum Press, New York, 1976.
[133] D. A. Church, in "Physics of Electronic and Atomic Collisions" (J. S. Risely and R. Geballe, eds.), p. 660. Univ. of Washington Press, Seattle, 1976.

10.6.2. Alignment, Orientation, and Coherence

The postfoil beam source consists of ions in a distribution of charge states and a variety of different excited states. For each state characterized by a particular angular momentum quantum number l there will exist a population distribution among the $2l + 1$ magnetic substates m_l. If the excitation conditions are such that there is a preferential probability for populating a given magnetic substate within the l manifold, either a net alignment (magnetic substates of opposite sign are equally populated but this population is different for different values of $|m_l|$) or orientation (a single magnetic substate is overpopulated relative to the equilibrium value) of this manifold exists. Such a state is often referred to as an anisotropic state since radiation emitted in the decay of this state to a lower state will be polarized and exhibit an anisotropic angular distribution as a result of the preferential population of the magnetic substates. The radiation intensity per unit solid angle observed at an angle θ with respect to the beam axis is then given by

$$I(\theta) = \frac{I}{4\pi}\left(\frac{1 - P\cos^2\theta}{1 - \frac{1}{3}P}\right), \qquad (10.6.1)$$

where I is the total intensity emitted from the source and P the polarization of radiation emitted at 90°, which is given by

$$P = (I_\parallel - I_\perp)/(I_\parallel + I_\perp), \qquad (10.6.2)$$

where $I_\parallel$ and $I_\perp$ are the intensities of the radiation observed in a direction perpendicular to the beam, polarized parallel and perpendicular to the beam axis, respectively. It is, in fact, the disturbance of this radiation anisotropy by internal magnetic interactions (for example, fine structure and hyperfine structure) or externally applied electric or magnetic fields that leads to the quantum beat pattern observed in certain beam–foil experiments.

Sources in which a unique spatial axis is defined are generally aligned. For example, consider a beam of particles that is either excited itself (beam–foil excitation) or is used to excite target atoms (electron impact excitation). The beam axis then defines a preferred direction in space (a quantization axis) and the excited species are then generally left in anisotropic states due to the lack of spherical symmetry. Such a source still has cylindrical symmetry and the associated alignment produces linearly polarized radiation in a given direction. In the beam–foil source this cylindrical symmetry can be destroyed by tilting the foil target so that its surface normal is no longer coincident with the beam axis, and then the source can also become oriented.[132] The decay of an oriented state produces circularly polarized radiation.

In the foil excitation process, beam ions are collisionally excited into aligned states that are eigenstates of some Coulombic collision Hamiltonian (it is assumed that magnetic interactions play a negligible role in the excitation process). A nonstationary coherence may be induced in the wavefunction following the collision by internal and/or external interactions. For example, outside the foil the emergent beam ions are described by some different Hamiltonian, which includes internal magnetic interactions such as fine and hyperfine interactions (if present) and any interactions that are the result of externally applied electric or magnetic fields. The rapid transition between these two conditions (the transit time through the foil is typically 10^{-14} sec or less, which corresponds to a frequency uncertainty of some 10^{14} Hz) causes the initial aligned state to be described in terms of a coherent and nonstationary superposition of eigenstates of the emergent ion Hamiltonian. For example, if we assume these eigenstates are of the JM_J representation we can write

$$\psi|(LSM_LM_S, t)\rangle = \sum_{JM_J} C(JM_J|LM_LSM_S)\, U^{LS}_{JM_J} \exp\left[-\left(\frac{\gamma_J}{2} + i\omega_J\right)t\right], \quad (10.6.3)$$

where C represents a Clebsch–Gordon coefficient, U^{LS}_{JM} time-independent eigenfunctions of the JM_J representation, γ_J the decay probability of the J state, and ω_J its angular frequency. The light intensity $I(t)$ can then be found from

$$I(t) \propto |\langle f|\mathbf{D}|\psi(t)\rangle|^2, \quad (10.6.4)$$

where $|f\rangle$ is the common final state of the E1 transition, and $\mathbf{D}$ the dipole operator for the transition. If we now simplify things by assuming that there are only two states, each with the same decay constant γ, in the coherent sum we get

$$I(t) \propto e^{-\gamma t}[A + B \cos \omega_{JJ'} t], \quad (10.6.5)$$

where $\omega_{JJ'} = E_{JJ'}/\hbar$ is the separation between the two coherently excited states. Equation (10.6.5) shows that as long as $B \neq 0$ the otherwise exponential decrease in intensity will be modulated cosinusoidally with a frequency $\omega_{JJ'}$, the separation of the two coherently excited states.

10.6.3. Quantum Beat Experiments

Fine and hyperfine intervals and Stark and Zeeman splittings have been measured for several different atoms and near neutral ions using the beam–foil quantum beat technique. Details of this work can be found in the review papers.[131,132,133] It is, of course, only possible to make these

10.6. QUANTUM BEAT PHENOMENA

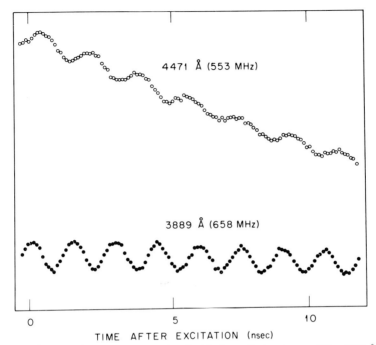

FIG. 32. Quantum beat pattern associated with the radiative decay of the 4471 Å (1s2p ^{3}P–1s4d ^{3}D) and 3889 Å (1s2s ^{3}S–1s3p ^{3}P) lines in He(I) (Ref. 134).

measurements if the temporal resolution of the apparatus is made approximately equal to or less than the period of the beat pattern. At present it is quite possible to obtain temporal resolutions of ~0.1–0.01 nsec with beam–foil arrangements allowing high-resolution structure studies to be made on intervals with frequencies up to ~10–100 GHz. As in beam–foil lifetime measurements at relatively low beam energies, one of the major sources of precision limitation in quantum beat experiments lies in the determination of the postfoil beam velocity, which is used to convert the spatial beat pattern into the time and frequency domain. Beam spreading, velocity straggling, and foil-thickening effects can introduce errors in the time calibration process and hence in the frequency determinations. Of the many transitions studied in hydrogen-, helium-, and lithiumlike atoms and ions, we choose two recent zero-field experiments to illustrate the quantum beat method. Andrä[131] and Church[133] both give a good description of Stark and Zeeman beat experiments.

The work of Astner et al.,[134] who measured the fine-structure separa-

[134] G. Astner, L. J. Curtis, L. Liljeby, S. Mannervik, and I. Martinson, *J. Phys. B* **9**, L345 (1976).

tions n^3D_1–n^3D_2 ($n = 3$–8) in ^{4}HeI, is particularly innovative. In this experiment the serious problems associated with accurately determining the postfoil beam velocity were overcome by using simultaneously two optical spectrometers in the experimental arrangement, one tuned to observe the decay of the n^3D_J levels and the other set to monitor the decay of the 1s3p ^3P_J levels for which the fine-structure intervals are well

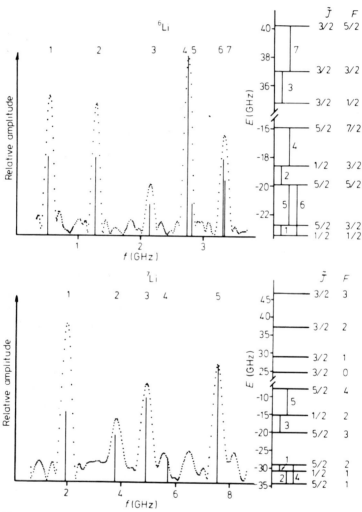

FIG. 33. Fourier transform of the quantum beat pattern associated with the radiative decay of the doubly excited 1s2p^2 ^4P_J levels of Li(I). A study was made of both isotopes ^{6}Li (upper) and ^{7}Li (lower) (Ref. 136).

known. In particular, the $3^3P_1-3^3P_2$ interval, measured to be 658.55 ± 0.15 MHz by Wittmann et al.,[135] was easily observed in the decay curves of the 1s2s $^3S \rightarrow$ 2s3p ^3P_J transition. The spatial beat pattern that was recorded simultaneously by Astner et al.[134] in their second spectrometer was used to accurately determine the postfoil velocity and thus the time scale calibration for the less well known beat pattern observed with the first spectrometer, i.e., associated with the $n^3D_1-n^3D_2$ ($n = 3-8$) intervals. Figure 32 shows the intensity modulations associated with the $4^3D_1-4^3D_2$ interval as well as the $3^3P_1-3^3P_2$ calibration interval. The $4^3D_1-4^3D_2$ fine structure splitting was measured to be 553.0 ± 0.7 MHz in this experiment.

Figure 33 shows the Fourier transform of a rather complex quantum beat signal obtained recently by Gaupp et al.[136] The fine- and hyperfine-structure intervals measured in this experiment are associated with the doubly excited levels 1s2p^2 ^4P_J in the ^{6}Li and ^{7}Li isotopes. The state is metastable against Auger emission (autoionization) but can radiate to the 1s2s2p ^4P_J levels, giving rise to a multiplet at ~3714 Å. The postfoil velocity of the ion beam was determined to better than 0.5% in this experiment by using a 90° cylindrical electrostatic energy analyzer. As is the procedure in most quantum beat measurements, the exponential decay dependence was subtracted from the data, leaving a beat signal that oscillates about zero. The beat frequencies and amplitudes were then determined from this oscillatory signal by a Fourier transform program.

10.7. Applications of Beam–Foil Results

10.7.1. Introduction

Atomic and ionic structure measurements (for example, binding energies, fine and hyperfine splittings, and radiative energy shifts) and the determination of decay process parameters (for example, transition probabilities or f values), as measured by the beam–foil process, find frequent practical application in other areas of research such as astrophysics, laboratory plasma physics, and laser physics. In addition, many-particle approximations of atomic structure (nonrelativistic and relativistic) can be sensitively tested by comparing the calculated results with accurately measured values of these quantities. In this section we discuss some of the applications of beam–foil results.

---

[135] W. Wittmann, K. Tillman, H. J. Andrä, and P. Doberstein, Z. Physik **257**, 279 (1972).
[136] A Gaupp, M. Dufay, and J. L. Subtil, J. Phys. B **9**, 2365 (1976).

10.7.2. Atomic Structure Theory

Beam–foil f value measurements have been instrumental in establishing certain systematic f value trends that exist along isoelectronic sequences. A detailed account of the present status of these studies has been published recently by Wiese.[137] The theoretical background for such f value trends was discussed by Wiese and Weiss[138] and is based upon a nonrelativistic perturbation series expansion of the f value in inverse powers of the nuclear charge Z in the following manner:

$$f = f_0 + f_1 Z^{-1} + f_2 Z^{-2} + \cdots . \tag{10.7.1}$$

In the limit of large Z, the f value approaches asymptotically the hydrogenic term f_0. In Section 10.5.5 we distinguished between two types of dipole transitions: in-shell ($\Delta n = 0$) and out-of-shell ($\Delta n \neq 0$). As can be seen from Eq. (10.7.1) in the limit of high Z, the f value approaches asymptotically the hydrogenic term f_0, which is zero for in-shell transitions due to the degeneracy of levels of the same n in the hydrogenic situation. Relativistic effects, described later in this section, cause f values of high-Z ions to strongly deviate from nonrelativistic values, especially in the case of in-shell transitions. Equation (10.7.1) indicates that a good way to display f value data along an isoelectronic sequence is to plot f vs. $1/Z$, as shown in Fig. 34 for the 2s → 2p transition multiplet in the lithium sequence. Many such plots for various low-lying transitions in simple few-active-electron sequences have been published by Smith and Wiese.[107]

In the nonrelativistic realm, particularly for near neutral ions, rather sharp deviations from the normally smooth f vs. $1/Z$ plots can occur over a small range of Z. The origin of these anomalies can usually be traced to configurational level crossings or cancellations within the transition integrand as the radial nodes of the wavefunction shift toward the origin with increasing Z. Configurational level crossings occur as the energy levels of the ion adjust themselves with increasing Z so that they finally group together according to n in the high-Z limit. The f values depend upon the energy separation between the initial and final states of the transition and any anomalies in the position of either of these levels along a sequence will show up as an anomalous f value in the f vs. $1/Z$ plot. In the region of low to intermediate Z, electron correlation (configurational mixing) can become important particularly for in-shell transitions. These intrashell transitions are correlation sensitive owing to the interpenetration of the

---

[137] W. L. Wiese, in "Beam-Foil Spectroscopy" (S. Bashkin, ed.), p. 147. Springer-Verlag, Heidelberg, 1976.
[138] W. L. Wiese and A. W. Weiss, *Phys. Rev.* **175**, 50 (1968).

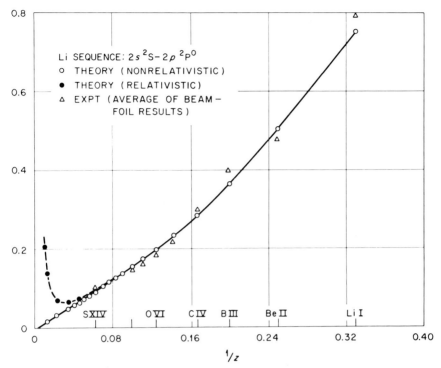

FIG. 34. A plot of multiplet f values vs. $1/Z$ for the 2s ^{2}S–2p ^{2}P resonance doublet in the lithium isoelectronic sequence (Ref. 63).

electrons of the same principal quantum number and, in general, many-particle atomic models that include configuration mixing effects are used to replace the simple independent-particle picture. Sinanoglu,[139] Sinanoglu and Luken,[140] and Nicolaides and Beck[141] have made nonrelativistic calculations of f values that include electron correlation effects. Comparisons of accurately measured f values with calculations of such quantities afford sensitive tests of the correctness of the wavefunctions of the upper and lower states of the transition. Configurational mixing effects, which may be present in the wavefunctions, strongly affect the calculation of the f value for the transition. Nonrelativistic calculations on the sodium sequence have been made by Froese-Fischer,[142] including corrections for

[139] O. Sinanoglu, *Nucl. Instr. Meth.* **110**, 193 (1973).
[140] O. Sinanoglu and W. Luken, *J. Chem. Phys.* **64**, 4197 (1976).
[141] C. A. Nicolaides and D. R. Beck, *J. Phys. B* **6**, 535 (1973).
[142] C. Froese-Fischer, *in* "Beam-Foil Spectroscopy" (I. A. Sellin and D. J. Pegg, eds.), Vol. 1, p. 69. Plenum Press, New York, 1976.

core polarization effects. Hibbert[143] has described configuration interaction calculations in the beryllium sequence.

Relativistic effects become appreciable for highly stripped heavy ions since the active electrons in these systems attain relativistic speeds in the strong field of the core. Two distinct types of relativistic effects can be important: configurational effects and orbital shrinkage due to mass increase of the fast moving electrons. The latter effect, which can cause changes in both orbital energies and transition integrals, is usually negligible for $Z \leq 50$. Relativistic configurational effects are caused by the increased strength of the spin–orbit interaction for high-Z ions. The LS coupling scheme must be replaced by intermediate coupling when the spin–orbit interaction becomes comparable in magnitude to the electrostatic interaction. These term mixings affect the line strength, the transition energy, and of course their product, the f value. For $20 \leq Z \leq 50$, the major effect on the f values for most in-shell transitions is associated with relativistic corrections to the transition energy alone. Even small changes in the energy separation of the initial and final states of this type of transition have large effects on the f value because the energy difference is a small quantity. Figure 35 shows an f vs. $1/Z$ plot for the $2^2S_{1/2} \to 2^2P_{1/2}$ and $2^1S_{1/2} \to 2^2P_{3/2}$ transitions in high-Z members of the lithium sequence. It can be seen from this figure that the f value for the $1/2 \to 3/2$ transition is affected strongly by relativistic effects on the transition energy, while the f value for the $1/2 \to 1/2$ transition remains close to the nonrelativistic value for most of the ions of the sequence. This was experimentally verified in measurements on several different highly ionized three-electron ions. The theorical results shown on this curve are due to Kim and Desclaux,[144] who employed relativistic self-consistent field wavefunctions. Similar relativistic calculations have been made by Weiss,[145] Armstrong *et al.*,[146] Lin *et al.*,[147] Sinanoglu and Luken,[148] and Aymar and Luc-Koenig[149] on resonance transitions in ions of simple structure such as those of the helium-, lithium-, beryllium-, boron-, sodium-, magnesium-, argon-, and copperlike sequences. Beam–foil results for very high Z heavy ions will be useful to confirm theoretical pre-

---

[143] A Hibbert, *in* "Beam-Foil Spectroscopy" (I. A. Sellin and D. J. Pegg, eds.), Vol. 1, p. 29. Plenum Press, New York, 1976.

[144] Y.-K. Kim and J. P. Desclaux, *Phys. Rev. Lett.* **36**, 139 (1976).

[145] A. W. Weiss, *in* "Beam–Foil Spectroscopy" (I. A. Sellin and D. J. Pegg, eds.), Vol. 1, p. 51. Plenum Press, New York, 1976; A. W. Weiss, J. Q. S. R. T. (to be published).

[146] L. Armstrong, Jr., W. R. Fielder, and D. L. Lin, *Phys. Rev.* **14**, 1114 (1976); D. L. Lin, W. Fielder, L. Armstrong, Jr., *Phys. Rev. A* **16**, 589 (1977).

[147] C. D. Lin, W. R. Johnson, and A. Dalgarno, *Phys. Rev. A* **15**, 154 (1977).

[148] O. Sinanoglu and W. Luken, *Chem. Phys. Lett.* **20**, 407 (1973).

[149] M. Aymar and E. Luc-Koenig, *Phys. Rev. A* **15**, 821 (1977).

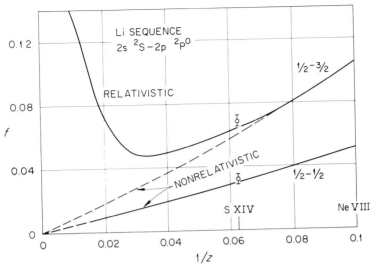

FIG. 35. High-Z region (relativistic regime) of the f vs. $1/Z$ plot for the $2^2S_{1/2} \to 2^2P_{1/2}$ and $2^2S_{1/2} \to 2^2P_{3/2}$ transitions in the lithium sequence. The solid lines are theoretical relativistic f values (Ref. 144).

dictions. The aim of the recent work in the relativistic regime is to be able to confidently extrapolate the existing f vs. $1/Z$ plots to highly stripped members of the sequences. Once this is accomplished it will be a simple matter to interpolate for specific ions of the sequence.

10.7.3. Astrophysical and Laboratory Plasmas

Accurate f values determined from beam–foil tof lifetime measurements are frequently used in studies of both astrophysical and laboratory plasmas. Absolute intensity measurements on spectral lines emitted by plasma sources can yield important information on the chemical composition (elemental identifications and abundances) and the physical conditions (temperature and densities) in the immediate vicinity of the emitter if such atomic parameters as radiative transition probabilities (f values) and excitation rate coefficients are independently known. Such plasma diagnostic techniques were developed on laboratory plasmas but the experience is frequently applied to astrophysical plasmas for which there exists no alternative to spectroscopic diagnosis.

Of high current interest in the study of magnetically confined thermonuclear fusion plasmas is the role played by small concentrations of highly charges heavy ions in the overall energy balance of the plasma. There is strong evidence[150] that these impurities cause unwanted radiation cooling

---

[150] E. Hinnov, Princeton Phys. Lab. Rep. MATT-1024 (1974).

that may be sufficient to prevent the attainment of ignition temperatures in future fusion reactors. A knowledge of f values and excitation rate coefficients for transitions associated with these radiative energy losses would enable more accurate predictions to be made concerning the extent of the problem. At the present time beam–foil measurements are the only experimental source of such f value data.

Astrophysical applications of beam–foil measurements cover, for example, elemental abundance determinations in the photosphere and atmosphere of the sun and other stars, and interstellar media. Spectroscopic diagnosis combined with laboratory-measured f values has yielded valuable information on electron temperatures and densities in the solar atmosphere, for example, both in the case of the quiet sun and under conditions of high solar activity. Astrophysical plasmas are usually less dense than laboratory plasmas and it is often possible to observe for-

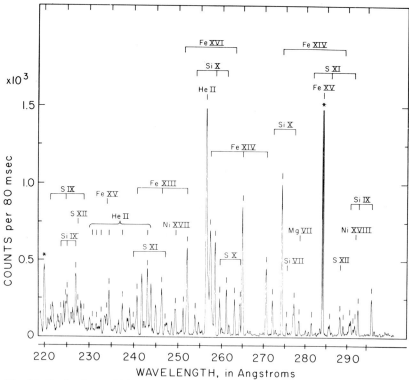

FIG. 36. Rocket-borne EUV spectrum of the solar chromosphere and corona in the region 220–295 Å (photoelectrically recorded). Reprinted courtesy of M. Malinovsky and L. Heroux[157] and *Astrophys. J.*, Univ. Chicago Press; © (1973) The American Astronomical Society.

10.7. APPLICATIONS OF BEAM–FOIL RESULTS

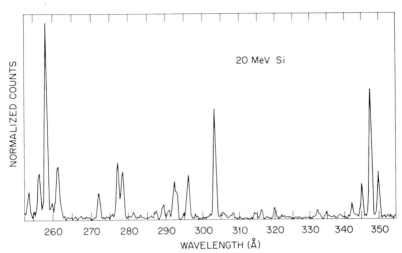

FIG. 37. Portion of EUV spectrum of foil-excited silicon ions accelerated to 20 MeV. In the overlap region of this spectrum and that of Fig. 36 many lines are common (Ref. 57).

bidden transitions due to the smaller collisional deexcitation rates. Transition rates for these forbidden lines can be particularly useful for electron density determinations.[151] The rates for forbidden decay modes become accessible to beam–foil studies for highly stripped ions (Section 10.5.6). Since many chromospheric and coronal emission lines arise from resonance transitions in highly charged ions they are not accessible to ground-based observations (the earth's atmosphere absorbs solar radiation below ~ 3000 Å). The advent of space vehicles (rockets and satellites) has enabled astrophysicists to actively study the chemical and physical properties of the solar atmosphere. Not all solar emission lines have yet been identified and it is expected that beam–foil measurements can aid in identifications. Good reviews of the applications of the beam–foil technique to astrophysical problems have recently been given by Whaling,[152] Heroux,[153] and Morton.[154] A review article by Smith[155] lists a number of beam–foil measurements of interest to astrophysicists.

Not long ago it appeared that inconsistencies existed between the photospheric, coronal, and meteoritic abundances of the iron group elements.

---

[151] A. H. Gabriel and C. Jordan, *Mon. Not. Roy. Astron. Soc.* **145**, 241 (1969).

[152] W. Whaling, in "Beam-Foil Spectroscopy" (S. Bashkin, ed.), p. 179. Springer-Verlag, Heidelberg, 1976.

[153] L. Heroux, in "Beam-Foil Spectroscopy" (S. Bashkin, ed.), p. 193. Springer-Verlag, Heidelberg, 1976.

[154] D. C. Morton, in "Beam-Foil Spectroscopy" (I. A. Sellin and D. J. Pegg, eds.), Vol. 2, p. 907. Plenum Press, New York, 1976.

[155] P. L. Smith, *Nucl. Instr. Meth.* **110**, 395 (1973).

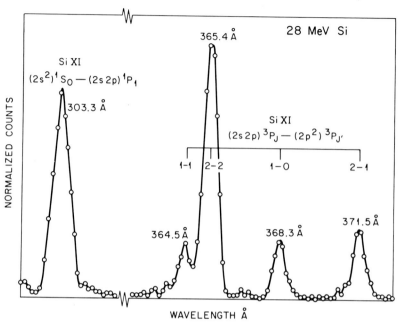

Fig. 38. A section of EUV spectrum emitted by 28-MeV foil-excited silicon ions (Ref. 63).

Whaling et al.[156] made beam–foil measurements on FeI and the results bring the solar photospheric abundance of iron into much better agreement with the coronal value. Since then most of the other elements of the iron group have been studied with the beam–foil method (often combined with separate branching ratio measurements) and their abundances have been revised upwards. Beam–foil measurements on titanium, vanadium, and the rare earth elements have also been used to determine solar abundances.

Figure 36 shows a portion of the EUV spectrum of the solar chromosphere and corona obtained photoelectrically by Malinovsky and Heroux.[157] A similar spectrum has recently been obtained photographically by Behring et al.[49] The lines are associated with resonance transitions of highly charged ions produced in the high-temperature ($\sim 10^4$–10^6 K) and low-density plasma of this region of the sun. Because of the low-density conditions prevailing, excitation and ionization processes are induced primarily by electron impact and deexcitation occurs

[156] W. Whaling, R. B. King, and M. Martinez-Garcia, Astrophys. J. **158**, 389 (1969).
[157] M. Malinovsky and L. Heroux, Astrophys. J. **181**, 1009 (1973).

10.7. APPLICATIONS OF BEAM–FOIL RESULTS

solely via radiative processes, i.e., nonlocal thermodynamic equilibrium conditions. Many of the lines of the solar EUV spectrum shown can be efficiently produced in beam–foil experiments with incident beam energies of ~1 MeV/amu. For example, the portion of the foil-excited spectrum of highly ionized silicon shown in Fig. 37 was taken by Griffin *et al.*[57] at 20 MeV. As can be seen from Fig. 36 most of the prominent lines in the overlap region are common to both spectra. Figure 38 shows a portion of the silicon foil-excited spectrum taken by Pegg *et al.*[63] at a somewhat higher beam energy of 28 MeV. Figure 39 shows the decay of three of the fine structure levels of the $(1s^2 2p^2)\,^3P$ term of Si(XI). Figure 40

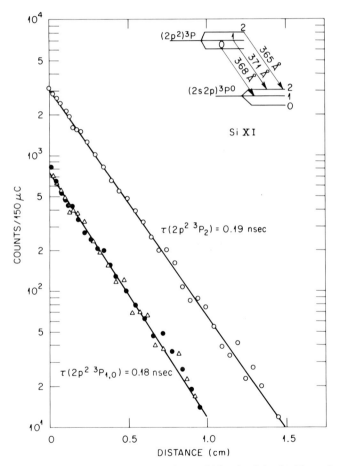

FIG. 39. Intensity decay curves for the $J = 2, 1,$ and 0 levels of the doubly excited $(2p^2)\,^3P$ term in Si(XI) (Ref. 63).

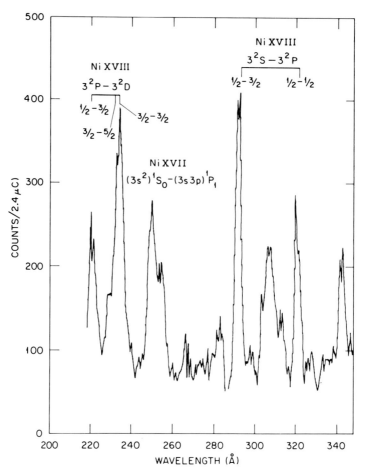

FIG. 40. A portion of EUV spectrum of foil-excited nickel ions accelerated to 65 MeV. These lines are prominent features of solar flare spectra. Reprinted courtesy of D. J. Pegg et al.[158] and Astrophys. J., Univ. Chicago Press; © (1978) The American Astronomical Society.

shows a beam–foil spectrum of nickel ions in which the sodium-like resonance lines are prominent features. These same lines are observed in solar flare events.[159]

Figure 41 shows the resonance doublet ($2s^2S-2p^2P$) associated with the lithium-like ion, S(XIV) taken in a beam–foil experiment by Pegg et al.[160]

[158] D. J. Pegg, P. M. Griffin, B. M. Johnson, K. W. Jones, and T. H. Kruse, Astrophys. J. **224**, 1056 (1978).

[159] K. G. Widing and J. D. Purcell, Astrophys. J. **204**, L151 (1976); G. D. Sandin, G. E. Brueckner, V. E. Scherrer, and R. Tousey, Astrophys. J. **205**, L50 (1976).

[160] D. J. Pegg, J. P. Forester, S. B. Elston, P. M. Griffin, K.-O. Groeneveld, R. S. Peterson, R. S. Thoe, C. R. Vane, and I. A. Sellin, Astrophys. J. **214**, 331 (1977).

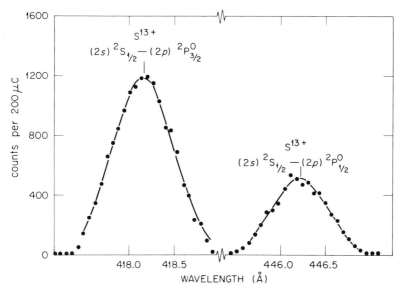

FIG. 41. The resonance doublet (2s²S–2p²P) of the lithiumlike ion, S(XIII) obtained using a 45-MeV foil-excited sulfur ion beam. The doublet splitting corresponds to a fine structure interval between the $J = \frac{1}{2}$ and $\frac{3}{2}$ levels of 15,056 cm^{-1}. The wavelength scale was calibrated using well-established reference lines from a hollow cathode choice. Reprinted courtesy D. J. Pegg et al.[160] and Astrophys. J., Univ. Chicago Press; © (1977) The American Astrophysical Society.

using a 45-MeV beam energy. Reports have been made[159] of the observations of this doublet in highly stripped lithium-like ions made aboard the Skylab staellite during solar flare events. In fact this doublet was among the strongest of the high-temperature lines observed and was employed as a diagnostic of the plasma temperature. The beam–foil experiment of Pegg et al.[160] determined the f values for both lines and also confirmed the doublet splitting measured using the astrophysical observations.

10.7.4. Recent Additions

The following list of references are related to significant material published since the first writing of this article. I hope it will be of some value to people interested in the present status of the field. A good general up-to-date review of the whole area of beam-foil spectroscopy has been given by Andrä.[161] In addition, the Proceedings of the Fifth International Conference on Fast Ion Beam Spectroscopy, which was held in Lyon, France in 1978, has been published.[162] This is an addition to the proceed-

---

[161] H. J. Andrä, in "Progress in Atomic Spectroscopy," (W. Hanle and H. Kleinpoppen, eds.), Chap. 20, p. 829, Plenum Press, New York (1979).

[162] Proc. Fifth Int. Conf. Fast Ion Beam Spectrosc. (J. Desequelles, ed.), publ. in J. Physique 40, C-1 (1979).

ings of the other four conferences.[163] A new set of wavelength tables and Grotian level diagrams, which are often useful in planning experiments, has been published by Bashkin and Stoner.[164] References 165–172 relate to some recent beam–foil measurements.

Acknowledgments

The author would like to thank Debbie Citraro and Meryl Montgomery for their valuable typing assistance. Preparation of this chapter was supported in part by the National Science Foundation, the Office of Naval Research, and Union Carbide Corporation under contract with the Energy Research and Development Administration.

[163] "Beam–Foil Spectroscopy," *Proc. Fourth Int. Conf. Beam–Foil Spectros.*, Plenum Press, New York (1976); *Proc. Third Int. Conf. Beam–Foil Spectrosc.*, North-Holland Publishing Co., Amsterdam (1973); *Proc. Second Int. Conf. Beam–Foil Spectros.* North-Holland Publishing Co., Amsterdam (1970); *Proc. First Int. Conf. Beam–Foil Spectrosc.* Gordon and Breach, New York (1968).

[164] "Atomic Energy Levels and Grotian Diagrams," (S. Bashkin and J. O. Stoner, Jr., eds.), North-Holland Publishing Co., Amsterdam (1975).

[165] H. G. Berry, R. DeSerio and A. E. Livingston, *Phys. Rev. Lett.* **41**, 1652 (1978).

[166] A. E. Livingston and H. G. Berry, *Phys. Rev. A* **17**, 1966 (1978).

[167] E. Träbert, I. A. Armour, S. Bashkin, N. A. Jelley, R. O'Brien, and J. D. Silver, *J. Phys. B* **12**, 1665 (1979).

[168] E. Träbert, P. H. Heckmann, and H. V. Buttlar, *Z. Phys.* **281**, 333 (1977).

[169] K. Ishii, E. Alvarez, R. Hallin, J. K. Lindskog, H. Marelius, J. Pihl, R. Sjödiu, B. Denne, L. Engström, S. Huldt, and I. Martinson, *Phys. Scripta* **18**, 57 (1978).

[170] K. W. Jones, *J. Physique* **40**, C1-197 (1979).

[171] P. Ceyzeriat, J. P. Buchet, M. C. Buchet-Poulizac, J. Desesquelles, and M. Druetta, *J. Physique* **40**, C1-171 (1979).

[172] Y. Baudinet-Robinet, P. D. Dumont, H. P. Garnier, E. Biemont, and N. Grevesse, *J. Physique* **40**, C1-175 (1979).

AUTHOR INDEX

Numbers in parentheses are footnote numbers. They are inserted to indicate that the reference to an author's work is cited with a footnote number and his name does not appear on that page.

A

Aashamar, O., 328
Abdulwahah, M., 145
Aberth, W., 157, 167(12), 173
Abignoli, M., 173
Abrath, F., 233(80, 139), 242, 278
Abrines, R., 93, 97(83)
Acerbi, E., 94
Adams, C. T., 84, 533
Adriamonje, S., 318(*bb*), 320, 328
Åberg, T., 397, 398, 401
Afrosimov, V. V., 22, 151, 155, 156(6), 158, 186, 187, 304, 331(1)
Agaard, B., 332, 333(76)
Aitken, D. W., 358(18), 359
Aksela, S., 455
Aladag, J. E., 183, 184(32)
Alessi, J., 117, 118(175)
Alkhazov, G. D., 357
Allison, S. K., 76, 82(21), 83(21, 22), 102, 104, 114(21, 22), 116(131), 118(21, 22), 257, 264, 267, 281, 287
Alpert, V. A., 281
Alton, G. D., 110, 114(145), 115, 116(161, 162), 117(165), 118, 127(179), 133, 210, 231(128), 278, 535, 559(63), 560, 561, 572(63), 575(63), 579(63), 580(63), 581(63), 597(63), 602(63), 603(63)
Alvarez, E., 561, 572(63), 575(63), 579(63), 580(63), 581(63), 597(63), 602(63), 603(63), 606
Alvarez, L. W., 9, 16
Ambros, R., 318(*s*), 320, 344
Amundsen, P., 330
Amundsen, P. A., 209(36), 210(37), 211, 318(*f*), 319, 324, 325, 327, 328(57)
Andersen, H. H., 26, 35, 37(25), 46, 48, 49

Anderson, J. U., 233(78), 241, 312, 313, 318 (*a, j, k, l, n*), 319(*ff*), 320, 324, 325(50), 328, 330, 337, 338(50), 339, 341, 342
Andrä, H. J., 568, 570, 576, 586, 590, 592(131), 593, 595, 605
Andrews, H. R., 39, 46(30)
Angert, N., 85, 99, 100(41), 102(117), 103, 115(41, 117), 116(41, 163), 131
Anholt, R., 211, 221, 227(55), 228(55), 336
Ankudinov, V. A., 85, 102
Annett, C., 310, 318(2), 320, 335
Antar, A., 115, 116(161), 133
Anttila, A., 53, 54
Appleton, B. R., 122, 139, 145, 257, 277(100)
Arianer, J., 21, 281
Armbruster, P., 22, 305, 314(7), 318(*r*), 320, 331(7)
Armour, I. A., 606
Armstrong, L., 350, 598
Asaad, W. N., 434, 439, 468, 470, 474, 475, 478, 505(3), 511(3)
Asgaard, B., 473, 490, 491(116) 492
Ashley, J. C., 71
Astner, G., 571, 576, 577(94), 593, 595
Auger, P., 433
Awaya, Y., 230(125), 278
Aymar, M., 598

B

Bacsó, J., 230(118), 278
Bakhru, H., 240, 403, 418
Ball, G. C., 39, 46(30)
Bambynek, W., 195, 197, 290, 393, 478

Bang, J., 110, 200, 202, 205, 321, 323, 327, 341(33), 486
Banyard, K. E., 94
Baragiola, R., 104, 520, 521, 522(151), 523(153), 524(153)
Barat, M., 158, 173, 181(13), 182(13), 190(32), 213, 490
Barkas, W. H., 129
Barnett, C. F., 121
Baron, E., 118, 122, 127(180), 130(180)
Barrette, L., 543, 579
Barros Leite, C. V., 234(85), 248
Barrus, D. M., 364
Barton, H. A., 193
Basbas, G., 111, 112, 113(150, 151), 203, 204, 206(24), 207, 219, 232(112), 266, 321, 338(37), 383, 384, 486, 495
Bashkin, S., 87, 88(52), 134(52), 529, 530, 550, 554, 566, 586, 606
Baskin, A. B., 232(79, 92), 242, 250
Bassel, R. H., 90
Bastide, R. R., 85
Basu, D., 77, 109
Bates, D. R., 80, 90, 92(34), 94, 110, 111(34), 140(95), 286, 296, 343
Batho, H. F., 73, 118
Batson, C. H., 543, 579
Batz, H. D., 431
Baudinet-Robinet, Y., 537, 606
Baudon, J., 173
Bayfield, J. E., 89, 107, 108, 281
Bearse, R. C., 231(131), 235(131), 278
Beck, D. R., 597
Beck, G., 73, 118(8)
Bederson, B., 21
Bednar, J. A., 291(27), 292, 311, 316, 317(31), 318(p), 320, 338(31), 340(31), 341(31), 342(31), 350, 393(84), 394, 585
Beeck, O., 152
Behring, W. E., 556, 602
Behrisch, R., 61
Belkic, Dz., 323, 354
Bell, F., 109, 110(142), 140, 141(236), 142, 143, 144(239, 240), 145(236), 147(236), 148, 232(101), 257, 260, 287, 428, 431
Bell, G. I., 95
Bell, K. L., 114
Beloshitsky, V. V., 323
Benka, O., 234(83), 246

Bennett, W. H., 9
Bennett, W. R., 570
Berényi, D., 230(118), 278
Berg, R. E., 14
Bergmark, T., 444, 518(28)
Bergström, I., 561
Beringer, R., 85
Berkner, K. H., 115, 116(157)
Berry, H. G., 87, 88(52), 134(52), 543, 558, 560, 561, 562, 579, 590, 591(132), 592(132), 606
Bertin, E. P., 88
Bertolini, G., 358
Besenbacher, F., 70
Bethe, H. A., 20, 28, 34(5-7), 131, 147(213)
Bethge, K., 64, 65, 85, 231(94), 234(146), 250, 252(94), 255(94), 278, 313, 314, 318(d, m), 319, 320, 329, 330, 338, 344
Betz, H., 519, 125(24), 126, 127(31, 195), 128, 129, 130(24, 195, 208), 132, 133(220,-221), 134(221), 135, 136, 137(24, 227, 228), 138, 139(24), 140, 141(236), 142, 143, 39, 77, 78, 81(24), 82(24, 35), 83(24), 87(24), 97(24), 99, 100(24, 51), 101, 102(35, 119), 104(35, 119), 105(35), 109, 110(142), 114(24), 115(35, 51, 118, 119), 116(35, 51, 118, 119), 118(24, 31), 119(24), 121, 122, 144(239, 240), 145(24, 236), 147(236), 148, 232(101), 257, 260, 272, 281, 287, 372, 397, 426, 428, 549, 551
Bhadha, H. J., 20
Bhalla, C. P., 291(33), 292, 293(29), 294(29), 295(33), 297(33), 347, 380, 389, 393(52, 74), 394, 395, 406(52), 409, 468, 470, 474(66), 475, 478, 479, 500, 507(66), 519(62)
Bichsel, H., 37
Bickel, W. S., 530, 537, 558, 561
Biermann, J. D., 173
Biermont, E., 606
Biersack, J. P., 44, 72(36)
Biggerstaff, J. A., 116, 117(165), 118, 127(179), 145
Bigham, C. B., 14
Bingham, F. W., 165
Bisgaard, P., 514
Bissinger, G., 230(72), 231(126, 127),

233(72, 76), 239, 241, 246, 278, 316, 327, 426, 427
Blackett, P. M. S., 20
Blake, R. L., 364, 377, 416(39)
Blasche, K., 120
Blatt, S. L., 233(81, 137), 234(141), 244, 248, 278
Blauth, E., 454
Blewett, J. P., 13
Bloch, F., 20, 35
Blokhin, M. A., 418
Blosser, H. G., 14
Bogdanov, D. D., 128
Bohr, N., 8, 20, 28, 31(3, 4), 33, 73(15), 74, 77, 82(17), 94, 96, 114, 125, 126(15), 129, 132, 164, 308, 309, 519
Bolger, J., 499
Bolger, J. E., 400
Bollinger, L. M., 16
Bolton, H. C., 439
Bonani, G., 396, 397, 398
Bonnelle, C., 361
Børgesen, P., 62
Borovik, E. S., 85
Bothe, W., 193
Bottcher, C., 102
Bourland, P. D., 52, 53
Bøving, E., 332, 333(76), 473, 490, 491(116), 492
Bowman, H., 211
Bowyer, S., 546, 547(34)
Brackmann, R. T., 98, 102, 116(111)
Bragg, W. H., 69
Braithwaite, W. J., 387, 562
Brandon, F. L.
Brandt, W., 71, 72, 110, 112, 113(150, 151), 134, 200, 203, 204(24), 205, 206(24, 25), 207(24), 219(24), 222, 231(130), 245, 253, 278, 288, 318(b, x, y), 319, 320, 321, 324, 337, 338, 383, 431, 486
Bransden, B. H., 94
Bratton, T. R., 314(124), 318(z), 320, 335, 353, 354
Braun, M., 69
Bray, C. W., 104
Breit, G., 8
Brenn, R., 230(75), 240, 286, 384
Brennan, J. G., 68
Brenot, J. C., 158, 181, 182, 190

Briand, J. P., 318(h), 319, 327, 397
Brice, D. K., 45, 60
Bridwell, L. B., 78, 86, 87, 100(51), 115(51), 116(51, 161), 117(165), 118, 120, 121(32), 122(32, 45), 127(32, 179), 133, 552
Briggs, G. H., 20
Briggs, J. S., 147, 158, 166, 213, 224, 225(59), 226(59), 227(59), 289, 306, 314(9), 317(9), 318(t), 320, 332, 333(74), 385, 484, 490(98)
Brinkman, H. C., 90, 107(61), 341
Broadhurst, J. H., 318(e), 319, 325, 326
Bromander, J., 87, 88(52), 134(52)
Brongersma, H. H., 110, 114(145)
Brooks, N. B., 85
Broude, C., 65, 66
Brown, L., 559
Brown, M. D., 38, 39, 46, 68, 96, 122, 137, 139, 140(103), 145, 206, 256, 268, 284(11), 285, 291(12), 293, 294(12), 298, 299(36), 341, 342(100), 374, 384, 414, 415, 512(44), 514, 530, 546, 567, 568(6), 576, 584
Browne, J. C., 107
Browning, R., 89, 104(59), 115
Bruch, R., 454, 510, 514(140), 515(140), 568, 586
Brueckner, G. E., 604
Brunnings, J. H., 73(14), 74
Bryant, H. W., 2
Buchet, J. P., 558, 561, 562, 566, 606
Buchet-Poulizac, M. C., 558, 562, 566, 606
Buchta, R., 537, 560, 561, 562(60)
Burch, D., 278, 284(10), 291(10), 292, 305, 306, 313, 318(q), 320, 327, 332, 333(76), 338, 339, 344, 374, 375, 377, 379, 393, 416(39), 434, 456, 458, 473, 490, 491(116), 492, 494, 495(132), 496, 499, 503, 504
Burek, A. J., 364
Burhop, E. H. S., 433, 434, 439, 468, 470, 474, 475, 478, 505(3), 511(3)
Burkhalter, P. G., 2, 376, 377, 397, 398, 416, 420(119), 428, 429
Burns, D., 590
Busch, C. E., 232(92), 233(79), 242, 250
Busol, F. I., 85

Buttlar, H. V., 606
Bygrave, W. D., 5

C

Cacak, R. K., 168, 332, 351, 368, 490
Cage, M. E., 318(e), 319, 325, 326
Cahill, T. A., 233(136), 278
Cairns, J. A., 224, 225(59), 226(59), 227(59)
Campbell, J. L., 216(a, d), 217
Carbone, R. J., 151, 152(1), 165, 173(1), 309
Carlson, T. A., 96, 434
Carlston, C. E., 102
Carre, M., 559
Carriveau, G. W., 554
Castiglioni, M., 94
Cauchois, Y., 361, 368
Cecchi, J. L., 581, 582(101)
Cernosek, R. W., 104
Ceyzeriat, P., 606
Chadwick, J., 193
Chamberlain, G. E., 442
Chan, F. T., 91
Chan, K. C., 117, 118(175)
Chang, C. N., 233(81, 137), 234(141), 244, 248, 278
Charles, M. W., 357
Chase, L. F., Jr., 221, 227(55), 228(55)
Chatterjee, G., 94
Chaturvedi, R. P., 230 (124), 231(129), 233(84), 234(144), 246, 247(84), 278, 286, 384
Chau, E. K. L., 69, 70
Chau, H. K., 78, 114(33), 116(33)
Chemin, J. F., 318(g, bb), 319(gg), 320, 328
Chen, F. K., 312(130), 278
Chen, J. R., 234(143, 149), 278
Chen, M. H., 197, 208, 292, 393, 394, 395, 418, 468, 475, 476, 477, 478, 479(77, 78), 507(78), 517(78), 518(78), 519(65)
Cheng, K. T., 587
Cheshire, I. M., 91, 94
Chevallier, P., 318(h), 319, 327
Chiao, T., 98, 99, 100(114, 115), 104(114, 115), 115(114, 115), 116(114, 115), 140, 206, 256, 268, 284(11), 286, 291(12), 294(12), 341, 342(100), 374, 381, 384, 405, 406(105), 419, 420, 421
Chiu, K. C., 110, 114(145)
Choi, B. H., 203, 210, 241, 250, 383
Chu, W. K., 35, 36, 45, 46, 52(49), 53, 58, 60(41), 61, 65(41), 69(57, 58), 70
Chupp, E. L., 537, 570
Church, D. A., 590, 592(133), 593
Ciocchetti, G., 328
Cipolla, S. J., 231(129), 278
Citrin, P. H., 420
Clark, D. J., 14
Clark, D. L., 234(147, 148), 248, 278, 318(e), 319, 325, 326
Clark, R. B., 118, 127(177)
Clausnitzer, G., 85
Cleff, B., 479, 480
Close, D. A., 231(131), 235(131), 278
Clover, M. R., 86
Coates, W. M., 16, 21(40)
Cocke, C. L., 260, 277, 291(27), 292, 311, 314(124), 316, 317(31), 318(o, p, v, w, z), 320, 334, 335, 337, 338, 340(31), 341(31), 342(31, 99), 348, 350, 351, 352(99), 353, 354, 369, 370, 371(28), 388(28), 393(84), 394, 414, 415, 495, 496, 530, 554, 563, 566, 584, 585, 586(72)
Cockroft, J. D., 4
Cohen, B. L., 118
Cohen, L., 556, 602
Coleman, J. P., 77, 90, 92, 94(27), 95(27)
Comeaux, A. R., 102
Conway, J. G., 586
Cooke, B. A., 357
Costa, G. J., 39, 46(30)
Cox, H. L., 69
Crandall, D. H., 2, 104, 107, 280
Crasemann, B., 112, 195, 197(9), 208, 290, 292, 393, 394, 395, 418, 468, 475, 476, 477, 478, 479(77, 78), 507(78), 517(78), 518(78), 519(65)
Crispin, A., 34
Crone, A., 170, 171(27), 173(27), 332
Crooks, G. B., 493
Cross, M. C., 134, 135(236)
Crossley, R. J. S., 581
Crumpton, D., 231(135), 278
Cue, N., 230(75), 240, 286, 384, 403, 418, 424, 425
Cuevas, J., 102

Cunningham, M. E., 221, 223(54), 224(54), 376
Curnutte, B., 260, 291(27), 292, 311, 314(124), 316, 317(31), 318(p, v, w, z), 320, 334, 335, 337(82), 338(31), 340(31), 341(31), 342(31), 348, 350, 353, 354, 369, 370, 371(28), 388(28), 393(84), 394, 563, 584, 585, 586(72)
Curran, S. C., 356
Curtis, L. J., 571, 572, 573, 575, 576, 577(94), 581, 593, 595(134)
Czuchlewski, S. J., 96, 140(103), 298, 299(36)

D

Dahl, O., 5
Dahl, P., 178, 191, 333, 446, 490, 516, 525
Dalgarno, A., 90
Danske, K., 519
Darby, P. F., 9
Datz, S., 78, 86, 87, 100, 115(51), 116(51), 120, 121(32), 122(32), 127(32), 139, 145, 233(78), 241, 257, 277, 426, 552
Davies, J. A., 69
Davies, W. G., 39, 46(30)
Dawton, R. H., 85
Dearnaley, G., 358
de Castro Faria, N. V., 234(85), 248
Dedrick, K. G., 157, 167(12)
Degener, L., 118
de Heer, F. J., 384
Dehmel, R. C., 78, 114(33), 116(33)
Delaunay, B., 118, 122, 123, 124, 127(180), 130(180)
Delaunay, J., 122, 123, 124
Delvaille, J. P., 145(260), 146, 549
de Meijer, R. J., 318(f), 319
Demkov, Y. N., 485
Denagbe, S., 318(bb), 320, 328
Denis, A., 558, 566, 561, 562(65)
Denne, B., 561, 572(63), 575(63), 575(63), 580(63), 581(63), 597(63), 602(63), 603(63), 606
de Pinho, A. G., 234(85), 248
Der, R. C., 221, 223(54), 224(54), 376, 379
Desclaux, J. P., 96, 598, 599(144)
Deslattes, R. D., 391

DeSerio, R., 606
Desesquelles, J., 558, 561, 562(65), 566, 606
Dettmann, K., 110, 147
DeVries, J. L., 356, 357(7), 361(7)
Devries, R. M., 86
Dewey, R. D., 372
Dhuieq, D., 158, 181(13), 182(13), 190(32)
Diamond, R. M., 145
Dick, G. J., 16
Dien, Q. T., 328
Dieringer, J., 453, 477(32), 516(32), 517(147), 518(32), 519(32)
Dietrich, D. D., 586
Dmitriev, I. S., 97, 102(107, 108), 104, 114, 115(107), 116, 122, 129(204), 130(204, 208), 131, 134, 551, 552(41)
Dobberstein, P., 467, 595
DoCao, G., 558, 566
Dolgov, A. S., 117
Donets, E. D., 281
Donets, W. M., 21
Donnally, B., 530, 568(6), 584
Doschek, G. A., 556, 602
Dose, V., 114
Dost, M., 319(ii), 320
Dotchin, L. W., 537, 570
Doughty, B. M., 104
Doyle, B. L., 231(127), 278, 288, 289, 296, 300 (20, 35), 316, 391
Dozier, C. M., 2
Drake, G. W. F., 585
Drisko, R. M., 92
Drouin, R., 562, 563(74), 564
Druetta, M., 558, 606
Drukarev, G. F., 102
Druse, T. H., 318(y), 320
Dufay, M., 559, 561, 562(65), 594(136), 595
Duggan, J. L., 210, 230(70, 124), 231(128, 129), 232(112), 233(78), 238, 255, 265, 266, 278
Dumont, P. D., 537, 606
Duncan, M. M., 459
Dunn, K. F., 89, 104(59), 115
Dunning, K. L., 376, 377, 416, 420(119)
Dupree, A. K., 2
Dutkiewicz, V., 230(75), 240, 286, 384, 403, 418, 424, 425
Dutto, G., 94

E

Easthan, D. A., 118, 127(177)
Eckardt, J. C., 47, 51
Edlén, B., 564
Edwards, A. K., 440
Egidi, A., 437
Eichler, J., 91, 484
Eidson, W. W., 341, 342(99), 352(99)
Eland, J. D., 439
Elliott, D. O., 380, 386, 387, 388(68), 393(52), 406(52), 409, 410, 558(62), 560
Ellis, D. J., 234(143), 278
Ellsworth, L. E., 87, 98, 99, 100(51, 114, 115), 104(114, 115), 115(51, 114, 115), 116(51, 114, 115), 140, 206, 256, 268, 284(11), 286, 288, 291(12), 294(12), 296, 299, 300(20, 35), 301(37), 316, 341, 342(100), 374, 384
Elston, S. B., 110, 114(145), 368, 541(96), 555, 556, 559(63), 560, 561, 572(63), 575(63), 579(63), 580(63), 581(63), 597(63), 602(63), 603(63), 604, 605(160)
Enge, H. A., 464
Englestein, P., 65, 66(67)
Engström, L., 606
Erman, P., 560, 562(60)
Evans, G. E., 121
Evans, R. D., 308
Everhart, E., 22, 117, 151, 152(1), 154, 155, 156, 157, 165, 173(1), 186, 188, 304, 309, 331(2, 5), 345(5), 490
Evers, D., 142, 143, 144(239), 260, 287

F

Fan, C. Y. 584
Fano, U., 22, 35, 117, 152, 153, 212, 306, 321(8), 330, 357, 384, 480, 482, 490(91), 590
Fastrup, B., 114, 116(155), 117(155), 157, 158, 167, 168(11), 170(11), 171, 173(11, 27), 176, 179, 188(11), 189, 190(14), 191(14), 192(14), 213, 222, 223(57), 266, 304, 331(4), 332, 333(76), 342, 355, 382(1), 446, 473, 484, 490, 491(116), 492, 516(29), 525(29)

Fateeva, L. N., 97, 102(107, 108), 104, 115(107), 116, 131
Fedorenko, N. V., 151, 155, 156(6), 186, 187(6), 304, 331(1)
Feldman, L. C., 233(78), 241, 263, 318(*a*), 319, 24, 325(50), 338(50), 368, 420(23)
Feng, J. S., -Y., 61, 69
Ferguson, S. M., 99, 100(114, 115), 104(114, 115), 115(114, 115), 116(114, 115)
Fermi, E., 8, 33
Feuerstein, A., 44, 72(37)
Fielder, W. R, 598
Fink, D., 530
Fink, R. W., 195, 197(9), 216(*b*), 217, 290, 356, 393, 478
Fintz, P., 344
Firsov, O. B., 28, 39, 40, 164
Fischbeck, H. J., 318(*i*), 319, 328
Fitchard, E., 384
Fite, W. L., 77, 97(30), 98, 100(30), 102, 114(30), 116(30, 111)
Fleischmann, H. H., 78, 114(33), 116(33)
Fleming, J., 8
Fock, V., 90
Fogel, I. M., 85, 102
Folkmann, F., 509, 514(139), 515(139), 524, 526, 527(156)
Folland, N. O., 394, 470, 478
Ford, J. L., 86, 384
Ford, K., 559
Forester, J. P., 368, 541(96), 555, 556, 559(63), 560, 561, 572(63), 575(63), 579(63), 580(63), 581(63), 597(63), 602(63), 603(63), 604, 605(160)
Forster, J. S., 39, 46(30), 289
Fortner, R. J., 77, 111(26), 114(26), 116(26), 117(26), 158, 194, 213(8), 221, 223(54), 224(54), 227, 263, 316, 331, 355, 376, 379, 382(2), 384(2), 415, 422, 426, 427, 428, 429, 430, 431, 434, 447(9), 479, 483(9), 492(9), 494, 497(83), 506(83), 507(83), 508, 509(83), 520
Fox, J. D., 118
Fowler, G. N., 34
Fraenkel, B. S., 2
Francois, P. E., 231(135), 278
Frank, L. A., 436
Frankel, R. S., 358(18), 359

AUTHOR INDEX

Franz, H., 193
Franzke, B., 85, 99, 100, 102(117), 103, 115(41, 117), 116(41, 163), 120, 121, 131
Fredorenko, N. V., 22
Freedman, M. S., 318(i), 319, 328
Freeman, W., 422
Freund, A., 478
Freund, H. U., 195, 197(9), 290, 393
Fricke, B., 350, 511(141), 512
Froese-Fischer, C., 581, 597
Fulbright, H. W., 86
Fulmer, C. B., 118
Fuls, N., 151, 152(1), 173(1)
Furuta, T., 374

G

Gabler, H., 449, 454(31)
Gabriel, A. H., 2, 564, 601
Gaillard, M., 559
Galejs, A., 84
Gallagher, A., 570
Gangadharan, S., 119
Garcia, J. D., 77, 111(26), 114(26), 116(26), 117(26), 158, 194, 201, 213, 221, 223(54), 224(54), 225, 227, 331, 347, 355, 376, 379, 382(2), 383, 384(2), 428, 429, 430, 431, 434, 447, 483, 486, 489, 492, 494, 520
Garcia-Munoz, M., 76, 83(22), 102, 114(22), 118(22)
Gardner, L. D., 281
Gardner, R. K., 143, 144(242), 194, 205, 213(8), 230(68, 70), 231(73), 232(109, 113), 234(82), 235(88), 237, 238, 239, 245, 246(82), 248(73), 263, 264(109), 265(109), 266(109), 267(109), 268, 270, 271, 272(88), 277, 287, 314(124), 353
Garfinkel, A. F., 48, 49(46)
Garnier, H. P., 606
Gaukler, G., 315
Gaupp, A., 530, 570, 571, 594(136), 595
Gauyacq, J. P., 190(32)
Gavrila, D. M., 398
Gedcke, D. A., 358, 360
Gehrke, R. J., 216(c), 217
Geiger, H., 3, 20

Geiger, J. S., 50, 51(47)
Gelius, U., 439, 460(17), 477, 517, 519
Genevese, F., 193
George, J. M., 183, 184(32)
Geretschläger, M., 234(83), 246
Gerjuoy, E., 80, 90, 92(34), 94, 111(34), 201, 483, 486, 489(94)
Gershon, S., 122, 124(198)
Gersten, J. I., 427
Gerthsen, C., 193
Gilbody, H. B., 89, 104(59), 115, 121(129)
Gilmore, B. J., 89, 104(59)
Glässel, P., 86
Glennon, B. M., 583
Gluckstern, R. L., 96
Glupe, G., 479
Goad, M. L., 104
Golden, J. E., 354
Goldstein, Ch., 21, 281
Gordeev, Y. S., 22, 155, 156(6), 158(10), 166(25), 167, 186, 187(6), 304, 331(1)
Gould, H., 535, 584, 586
Graboske, H. C., 134
Graham, R. L., 50, 51(47)
Grahmann, H., 44, 55, 56, 57, 72(37)
Grant, I. S., 118, 127(177)
Graw, G., 85
Gray, T. J., 143, 144(242), 194, 205, 213(8), 230(68, 70), 231(64, 73), 232(109, 113), 233(80, 84, 139), 234(82), 235(88), 237, 238, 239, 242, 245, 246, 247(84), 248(73), 263, 264, 265, 266, 267, 268(113), 278, 270(113), 271(113), 272(88), 277, 287, 302, 391, 395(76), 399, 400, 422(76), 423
Greenless, G. W., 234(147, 148), 248, 278, 318(e), 319, 325, 326
Grevesse, N., 606
Griffin, P. M., 110, 114(145), 284(8), 285, 368, 465, 512(144), 514, 530, 541(96), 543, 550, 555, 556, 559(63), 560(56), 561, 562(56), 568(6), 569(83), 572(63), 575(63), 579(63), 580(63), 581(63), 582(101), 586, 587, 597(63), 602(63), 603(63), 604, 605(160)
Griffing, G., 110
Grigorev, A. N., 102
Grodzins, L., 132, 133(221), 134(221), 135, 137(227), 519
Groeneveld, K. O., 231(107), 262, 349,

350, 454, 458, 502(35), 511(141), 512, 513, 524, 526, 527(156), 553(46), 554, 555, 556, 559(63), 560, 568, 570, 572(63), 573(63), 579(63), 580(63), 581(63), 586, 597(63), 602(63), 603(63), 604, 605(160)
Grosse, E., 145
Grün, N., 93, 95(85)
Grundinger, U., 99, 102(117), 115(117)
Gruppe, A., 309, 310
Gryzinski, M., 383, 486
Guertin, J., 231(73), 239, 248(73)
Günther, G., 85
Guffey, J. A., 140, 299, 301(37), 384
Guillaume, G., 318(y), 320, 344
Gur, D., 117, 118(175), 351

H

Hachiya, Y., 230(63), 237
Hafstad, L. R., 5, 8
Haggmark, L. G., 44, 72(36)
Hahn, O., 7
Hall, J. M., 143, 144(242), 232(109, 113), 263, 264(109), 265(109), 266(109), 267(109), 268(113), 270(113), 271(113), 287, 399, 400, 401
Hallin, R., 145, 550, 561, 565(77), 566, 572(63), 575(63), 579(63), 580(63), 581(63), 597(63), 602(63), 603(63), 606
Halpern, A. M., 140, 206, 256, 286, 321, 341(38), 384, 483, 489
Hamada, T., 230(125), 278
Hanke, C. C., 48, 49(46)
Hansen, J. S., 201, 210, 216(b), 217, 323, 379, 398, 417, 418, 487
Hansteen, J. M., 110, 200, 202, 205(15), 211, 318(f), 319, 321, 322, 323, 325, 327, 328(34, 57), 330, 337(36, 40), 341(33), 379, 383, 481, 486, 487(90), 498, 499
Hardt, T. L., 194, 213(8), 231(132), 234(145), 278
Harrison, K. G., 110
Harrower, G. A., 439
Harwood, D. J., 134
Haselton, H. H., 232(97), 255, 512(144), 514, 530, 559, 560(56), 562(56), 568(6), 586
Hashiyume, A., 230(125), 278

Hasse, R. W., 109, 110(142)
Hauser, U., 65, 66, 67(68), 68, 69(68)
Havas, P., 73, 118(8, 9)
Hayden, H. C., 368, 490, 541(96), 561, 572(63), 575(63), 579(63), 580(63), 581(63), 597(63), 602(63), 603(63)
Heckman, H. H., 129
Heckmann, P. H., 606
Heerden, I. V., 319(dd), 320
Heffner, R., 458, 503(48)
Hein, M. A., 394, 470, 478
Heinemeier, J., 70
Henderson, G. H., 20, 73, 74
Hendrichs, C., 118
Hendrie, D. L., 86
Henke, B. L., 358, 369(13), 467
Henneberg, W., 20, 486
Hermann, G., 157, 165, 166(22), 167(11), 168(11), 170(11), 171(22, 27), 173(11, 27), 175, 176(11), 179(11), 188(11), 189(11), 332
Heroux, L., 529, 540, 546, 600(157), 601, 602
Hesterman, V. W., 173
Heydenburg, N. P., 5
Hibbert, A., 598
Hill, K. W., 391, 397, 398
Hinnov, E., 599
Hiskes, J. R., 109
Hock, G., 230(118), 278
Hodge, W., 408, 435, 446, 454(30), 460(30), 461(50), 497(30), 498(30), 516, 519(146)
Hoffman, G. W., 413, 414
Holloway, T. C., 231(73), 239, 248(73)
Hontzeas, S., 560, 562
Hoogkamer, Th. P., 523, 524(154)
Hopkins, F. F., 88, 102(55), 141, 143(55, 238), 144(238, 241), 230(75), 232(105, 106), 240, 260, 261, 265, 266, 286, 292, 294(30), 379(52), 380, 384, 386, 387, 388(68), 393(52), 394, 406(52), 409, 410, 411, 412(85), 415, 416, 422, 424, 425, 428, 430(136), 465, 558(62), 560, 584
Hoppenau, S., 319(ii), 320
Hortig, G., 129, 130(208)
Hsu, C. N., 230(121), 233(121), 278
Hubbard, E. L., 129
Hughes, A. L., 462

AUTHOR INDEX

Huldt, S., 606
Hvelplund, P., 70, 102, 115, 121, 128(194)

I

Ilyushchenko, V. I., 281
Ingalls, W. B., 313, 318(q), 320, 327, 339, 458, 503(48)
Inokuti, M., 111, 114(149)
Ioannou-Yarnou, J. G., 211
Irwin, D. J. G., 562, 575
Ishii, K., 230(63, 121, 122), 233(121, 122, 140), 237, 249, 278, 561, 572(63), 575(63), 579(63), 580(63), 581(63), 597(63), 602(63), 603(63), 606
Ishiwari, R., 47, 48
Iyumo, K., 230(125), 278

J

Jackson, J. D., 90
Jacobs, W. W., 231(127), 278
Jakubassa, D. H., 147
Jamison, K. A., 143, 144(242), 218, 232(109, 113), 263, 264(109), 265(109), 266(109), 267(109), 268(113), 270(113), 271(113), 277, 287, 302, 386, 387, 388(68), 391, 395(76), 397, 398, 399, 400, 401, 406, 407, 409, 410, 412, 413, 422(76), 423, 454, 456(33), 470, 473(71), 480, 481, 482(87), 494(34), 495, 496, 501, 502(34), 504(71), 558(62), 560, 584
Jamnik, D., 210
Jarvis, O. N., 211
Jelley, N. A., 566, 606
Jenkins, R., 356, 357(7), 361(7)
Jensen, F. E., 296, 381, 405, 406(105), 419, 420, 421
Johann, H. H., 365
Johannson, T., 365
Johnsen, O. M., 321, 323(36), 330, 337(36)
Johnson, A., 318(h), 319, 327
Johnson, B. M., 318(y), 320, 335, 377, 379, 388, 394, 395, 412, 413, 414, 435, 446, 454(30), 460(30), 461(50), 462(51), 470, 472(73), 488, 497(30), 498(30), 499, 503(54), 504, 506(138), 507(54, 73), 512(52), 524, 581, 582(101), 586, 604

Johnson, W. R., 585, 587, 598
Jones, C. M., 115, 116(161, 162), 118, 127(179), 133
Jones, K. W., 145, 318(b, x, y), 319(ee), 320, 324, 335, 337, 344, 388, 550, 559, 581, 582(101), 586, 603(57), 604, 606
Jones, L. W., 13
Jones, J. L., 361, 366(19), 367, 368
Jones, P. R., 151, 152(1), 173(1)
Jordan, C., 2, 564, 601
Jundt, F. C., 318(y), 320, 344
Joy, T., 118, 127(177)
Joyce, J. M., 231(126, 127), 278

K

Kaji, H., 230(121, 122), 233(121, 122, 140), 249, 278
Kalata, K., 145(260), 146, 549
Kalbitzer, S., 44, 55, 56, 57, 72
Kalish, R., 85, 102(44), 139
Kalkoffen, G., 142, 143, 144(239), 260, 287
Kallmann, H., 73
Kaminker, D. M., 151
Kaminsky, A. K., 323
Kapitza, P., 20
Kaplan, S. N., 115, 116(157)
Kardontchik, A., 85, 102(44), 139
Karlowicz, R. R., 233(138), 237, 249, 278
Karnaukhov, V. A., 128
Katou, T., 230(125), 278
Kauffman, R. L., 88, 194, 213(8), 231(73), 239, 248(73), 277, 292, 294(30), 357, 368, 379, 381(49), 386, 387, 388(68), 390(12), 391, 392(12), 394, 395(76), 399, 400, 401, 402, 403(49), 404, 405, 406(49), 409, 410, 411, 412(85), 413, 414, 415, 416, 420(23), 422(76), 423, 454, 456(33), 465, 470, 472, 473, 488, 489(113), 493(113), 494(34), 495, 496, 497(113), 501, 502(34), 503(113), 504, 505(113), 506(113), 539, 558(62), 560, 584
Kauppila, W., 183
Kavanagh, T. M., 77, 111(26), 114(26), 116(26), 117(26), 158, 194, 213(8), 221, 223(54), 224, 331, 355, 376, 379, 382(2), 384(2), 434, 447(9), 483(9), 492(9)
Kay, L., 529, 552

Keller, R., 280
Kelley, C. H. P., 398
Kelly, E. L., 13
Kelly, H. P., 468, 475, 476(64), 477, 519(79)
Kelly, R. L., 583
Kerst, D. W., 13
Kessel, Q. C., 22, 114, 115, 116(155, 161), 117(155, 165), 118, 127(179), 133, 155, 156, 157, 158, 168, 170, 171(27), 173(27), 186, 188, 190(14), 191(14), 192(14), 213, 304, 331(4), 332, 347(3), 355, 382(1), 490
Khan, J. M., 221, 223(54), 224(54), 376, 379
Khandelwal, G. S., 203, 383
Khayrallah, G. A., 89, 107, 108(60)
Khelil, N., 237
Kienle, P., 145
Kim, H. J., 102, 103
Kim, Y.-K., 598, 599(144)
Kindlemann, P. J., 570
King, R., 118, 127(177)
King, R. B., 602
Kingston, A. E., 114
Kirby-Docken, K., 145(260), 146
Kiss, I., 230(118), 278
Kleber, M., 92, 145, 147
Kleeman, R., 69
Klinger, H., 104, 107(135)
Knaf, B., 230(93), 231(133, 134), 250, 266(93), 278
Kneis, W., 99, 102(117), 115(117)
Knipp, J., 73(14), 74
Knudsen, H., 70, 115, 116(162), 309
Knudson, A. R., 376, 377, 397, 398, 416, 420(119), 428, 429
Knystautas, E. J., 156, 490, 562, 563(74), 564
Kocbach, L., 211, 312, 313(21), 319(ff), 320, 323(36), 324, 325, 327, 328(21), 330, 337(36)
Koch, P. M., 89, 108(60)
Kocher, D. C., 104
Koenigsberger, J., 73
Koike, F., 442
Kolb, B., 231(107), 262
Koltay, E., 230(118), 278
Konecny, E., 118
Konopinski, E. J., 193
Korpf, A., 234(83), 246

Korsunskii, M. I., 454, 544
Kosmachev, O. S., 454, 544
Kostroun, V. O., 197, 208, 418
Kramers, H. A., 33, 90, 107(61), 341
Kraner, H. W., 318(b, x), 319, 320, 324, 337
Krause, H. F., 145
Krause, M. O., 434, 439
Kreutz, R., 69
Krupnik, L. I., 85
Kruse, T. H., 327, 550, 559, 581, 582(101), 603(57), 604
Kuehn, E., 65
Kuenhold, K. A., 231(129), 278
Kugel, H. W., 89, 231(126), 278, 589(125), 590
Kurzweg, L., 98, 116(111)
Kutschewski, J., 73
Kuyatt, C. E., 442

L

Laegsgaard, E., 121, 128(194), 191, 222, 223(57), 233(78), 241, 312, 313(21), 318(a, j, k, l, n), 319(ff), 320, 324, 325, 328(21), 330, 337, 338(93), 339, 341(93), 342(93)
Lamb, W. E., 73, 77, 125, 131
Lambert, J. M., 5
Land, D. J., 68
Langenberg, A., 229, 236(62)
Lapicki, G., 204, 205, 206(25), 207, 232(112), 245, 266, 286, 384
Larkins, F. P., 208, 347, 393, 470
Larsen, G. A., 333, 490
Larson, H., 231(66), 237
Laslett, L. J., 13
Lassen, N. O., 74, 75(16), 118(16), 131, 519
Latimer, C. J., 115
Latta, B. M., 65, 69
Laubert, R., 96, 111, 113(150), 139, 140(103), 144, 200, 203, 204(24), 206(24), 207(24), 219(24), 222, 231(95, 130), 232(97), 251, 253, 255(95), 257, 277(100), 278, 288, 298, 299(36), 321, 338(37), 383, 414, 486, 559, 560(56), 562(56), 586
Lauderi, P. C., 304, 331(5), 345(5)
Law, J., 140, 206, 256, 286, 321, 341(38), 384, 483, 489

AUTHOR INDEX

Lawrence, E. O., 12, 16, 21(39)
Lawrence, G. M., 570
Layne, C. B., 376
Layton, J. K., 102
Lazarus, S. M., 221, 227(55), 228(55)
Lear, R. D., 230(67, 70), 231(64), 233(84), 237, 238, 246, 247(84)
Leavitt, J. A., 556, 557, 586
Lee, N., 440
Leeper, A. K., 296, 420, 421
Leischner, E., 129, 130(208)
Leithauser, U., 449, 454(31), 486, 512
Leminen, E., 53, 54, 61, 62
Lennard, W. N., 289, 554, 566
Leventhal, M., 589(125), 590
Levy, G., 89, 104(59)
Levy, H., 114
Lewis, C. W., 218
Lewis, H. W., 110, 321, 372, 482
Li, T. K., 230(120), 234(147, 148), 238, 248, 278, 318(e), 319, 325, 326, 417, 418
Lichten, W., 22, 117, 152, 153, 212, 213, 306, 321(8), 330, 384, 482, 484, 490(91)
Lichtenberg, W., 309, 310, 316, 344(28)
Liebert, R. B., 231(66), 234(144), 237, 278
Liesen, D., 289, 319(hh), 320
Light, G. M., 205, 230(70), 231(73), 234(82), 238, 239, 245, 246(82), 248(73)
Liljeby, L., 571, 576, 577(94), 593, 595(134)
Lin, C. C., 94
Lin, C. D., 286, 343, 354(104)
Lin, C. P., 585, 587
Lin, D. L., 350, 598
Lin, J., 231(129), 278
Lin, W. K., 58, 59, 60, 62, 63, 69
Lindhard, J., 28, 33, 35, 39, 45, 71, 74, 82(17), 94, 125(17), 132, 164, 165
Lindhart, J., 308, 309(14)
Lindner, P., 145
Lindskog, J. K., 145, 550, 565(77), 566, 606
Linhard, J., 519
Lipsky, L., 568
Little, A., 141, 143(238), 144(238), 221, 227(55), 228(55), 232(106), 261, 266(106), 384, 422, 424, 425
Livingston, A. E., 537, 575, 606
Livingston, M. S., 12, 13, 20, 193

Lo, H. H., 77, 97(30), 98, 100(30), 114(30), 116(30, 111)
Lodge, J., 97
Lodhi, A. S., 69
Loftager, P., 165, 166(22), 171(22)
Lokken, R. A., 216(c), 217
Lorents, D. C., 157, 158, 167(12), 173
Losonsky, W., 204, 207, 266, 286, 384
Losonsky, William, 231(95), 251, 255(95)
Lu, C. C., 96
Lucas, M. W., 110
Luc-Koenig, E., 598
Luken, W., 597, 598
Lund, M., 312, 313(21), 318(c, j, k, l, n), 319(ff), 320, 324, 327, 328(21), 330, 337, 338(93), 339, 341(93), 342(93)
Lundin, L., 561
Lutz, H. O., 78, 86, 87, 100(51), 115(51), 116(51), 120, 121(32), 122(32, 45), 127(32), 192, 305, 306, 308, 314(7, 9), 317(9), 318(r, s, t, u), 319(cc), 320, 331(7), 332, 333, 337, 344, 426, 552
Luz, N., 318(u), 319(cc, dd), 320, 333, 337

M

McCarroll, R., 322
McCaughey, M. P., 117, 156, 490
McCrary, D. G., 378, 379
McDaniel, F. D., 210, 230(68), 231(128, 129), 232(112), 234(82), 237, 246(82), 255, 265, 266, 278
Macdonald, J. R., 96, 98, 99, 100(113, 114, 115), 104(113, 114, 115), 105, 115(113, 114, 115), 116(113, 114, 115), 140, 206, 256, 268, 284(11), 285, 286, 288, 289, 291(12), 293, 294(12), 296, 298, 299(36), 300(20, 35), 301(37), 314, 316, 319(hh), 320, 341, 342, 348, 351, 352, 353(26), 354, 374, 384, 388, 414, 415, 546, 563, 567, 576, 584, 586(72)
MacDonald, J. R., 546, 563, 586(72)
McDowell, M. R., 77, 91, 92, 94(27), 95(27)
Macek, J. H., 191, 213, 224, 225, 226(59), 227, 332, 385, 434, 463(7), 480, 484, 490(98), 492, 516, 590
McFarlane, S. C., 480
McGeorge, J. C., 216(b), 217

McGowan, J. W., 110, 114(145)
McGuire, E. J., 197, 209, 468, 475, 518(63)
McGuire, J. H., 201, 321, 323, 327, 330, 341(39), 354, 379, 381(49), 382(50), 384, 402, 403(49), 404, 406(49), 418, 483, 486, 487(109), 488, 489, 502, 506
McIntosh, A. I., 89, 104(59)
McIntyre, L. C., 566
McKnight, R. H., 233(138), 237, 249, 278, 440
Mack, J. E., 546, 547(34)
Mackey, J. J., 394, 395, 435, 446, 454(30), 460(30), 461(50), 462(51), 488, 497(30), 498(30), 499, 503
McMurray, R., 319(dd), 320
McNelles, L. A., 216(a, d), 217
McWherter, J., 400, 416, 420(120)
Madison, D. H., 112, 197, 202, 233(79), 242, 391, 483, 486
Maeda, K., 456
Magnuson, G. D., 102
Mahadevan, P., 102
Maier, M. R., 145
Malanify, J. J., 231(131), 235(131), 278
Malik, F. B., 96
Malinovsky, M., 600(157), 602
Mallory, M. L., 104, 280
Malmberg, P. R., 529
Mann, A., 122, 124(198)
Mann, R., 509, 514(139), 515(139), 524, 526, 527(156), 568, 586(86)
Mannervik, S., 571, 576, 577(94), 593, 595(134)
Mapes, R. S., 372
Mapleton, R. A., 77, 91, 92, 94, 140(95)
Marchi, P. R., 157, 167(12)
Marconero, R., 437
Marelius, A., 145, 550, 565(77), 566
Marelius, H., 606
Marion, J. B., 536
Mark, H., 195, 197(9), 290, 393, 478
Markham, R. G., 86
Marrus, R., 369, 535, 584, 585, 586, 590
Marsden, E., 3, 20
Martin, F. W., 98, 100(113), 104(113), 105, 115(113), 116(113), 122, 351, 368, 422
Martinez-Garcia, M., 602
Martinson, I., 87, 88(52), 134(52), 530, 557, 558, 560, 561, 562(60), 566, 571, 576, 577(94), 593, 595(134), 606
Maruhn-Rezwani, V., 93, 95(85)
Mason, E. V., 119
Massey, H. S., 80, 90, 92(34, 62), 102, 111(34)
Matsunami, N., 69
Matteson, S., 62, 63(61), 69, 70
Matthews, D. L., 88, 227, 292, 316, 387, 388, 394, 395, 412, 413, 414, 415, 422, 426, 427, 428, 430, 431, 435, 446, 454, 460(30), 461(50), 462(51), 470, 472, 474, 479, 488, 494, 497, 498(30), 499, 500, 503, 504, 506(138), 507(73, 83), 508, 509, 562
May, R. M., 90, 97, 109
Mayer, J. W., 62
Meckbach, W., 110, 114(145), 570
Mehlhorn, W., 454, 479, 480
Meijer, R. J., 325, 327, 328(57)
Meinel, A. B., 529
Menendez, M. G., 459
Menzel, D. H., 396, 431(91)
Mercereou, J. E., 16
Merzbacher, E., 110, 112, 197, 202, 203, 321, 372, 383, 482, 483, 486
Metz, W. D., 74
Meyer, F. W., 102
Meyer, L., 309
Meyer, R. C., 8
Meyerhof, W. E., 192, 221, 227, 228(55), 265, 332, 335, 336, 342, 385, 485, 490
Middleton, R., 84, 533
Mielczarek, S. R., 442
Milazzo, M., 234(142), 278
Miles, B. M., 583
Miljanic, D., 231(66), 234(144), 237, 278
Miller, P. D., 115, 116(161, 162), 117(165), 118, 127(179), 133, 210, 120(124), 231(128, 129), 232(112), 266, 278
Miller, T. E., 234(143), 278
Mitchell, I. V., 39, 46(30)
Mitchell, J. V., 289
Mittleman, M. H., 92
Moak, C. D., 38, 39, 46, 78, 86, 87, 100(51), 115(51), 116(51, 161, 162), 118, 120, 121(32), 122(32, 45), 127(32, 179), 133, 137, 139, 145, 312, 313(21), 318(j, k), 319, 328(21), 337, 338(93), 339, 341(93), 342(93), 426, 552

Möller, A., 99, 102(117), 103, 115(117), 116(163), 131
Möller, C., 20
Mohr, P., 585
Mokler, P. H., 289, 305, 314(7), 318(r), 319(hh), 320, 331(7), 375
Molinari, A., 328
Moliere, G., 309
Moore, C. F., 233(84), 246, 247(84), 377, 378, 379, 381(49), 387, 388, 394, 395, 400, 402, 403(49), 404, 405, 406(49), 408, 409, 412, 413, 414, 415, 416, 420(120), 435, 446, 454(30), 460(30), 461(50), 462(51), 470, 472(73), 479, 488, 497(30, 83), 498(30), 499, 500, 503(54), 504, 506(83, 138), 507(54, 73, 83), 508, 509(83), 516, 519(146), 562, 583
Morgan, G. H., 152, 154, 155, 304, 331(2)
Morgan, I. L., 374
Morgan, J. F., 233(81, 137), 234(181), 244, 248, 278
Morgenstern, R., 158
Morita, S., 230(63, 121, 122), 233(121, 122, 140), 237, 249, 278
Morse, P. M., 144(247), 145
Morton, D. C., 601
Moruzzi, V. L., 69
Mosebekk, O. P., 110, 200, 202, 205(15), 321, 323, 328(34), 379, 383, 481, 487(90), 498, 499
Moss, J. M., 318(e), 319, 325, 326
Mott, N. F., 20, 80, 92(34), 111(34)
Mourino, M., 253, 288
Mowat, J. R., 96, 139, 140(103), 232(97), 255, 257, 277(100), 284(8), 285, 293, 298, 299(36), 388, 414, 465, 530, 546, 559, 560(56), 562(56), 567, 568(6), 576, 584, 586
Müller, A., 104, 107(135)
Müller, M., 280
Münzer, H., 118, 127(178)
Muggleton, A. H., F., 358
Mukherjee, S. C., 77, 109(29)
Murnick, D. E., 89, 588, 589(125), 590

N

Nagamiya, S., 211
Nagarajan, M. A., 92
Nagel, D. J., 2, 376, 377, 397, 398, 416, 420(119), 428, 249
Natowitz, J. B., 218
Naylor, H., 9
Neelavathi, V. N., 134
Nelson, J. William, 230(70), 238
Nemirovsky, I. B., 104
Nestor, C. W., 96
Nettles, P. H., 230(72), 232(92), 233(72), 239, 244(72), 250
Neuwirth, W., 65, 66(65), 67(65), 68(65), 69
Nicholson, J. B., 361, 366(19), 367, 368
Nicolaides, C. A., 597
Nicolet, M.-A., 61, 62, 69(57, 58)
Nielsen, V., 164, 165(19), 308, 309(14)
Nielson, K. O., 533
Nikolaichuk, L. I., 102, 121, 131
Nikolaev, V. S., 76, 83(23), 94, 96(23, 94), 97(23), 100(23), 102(23, 107, 108), 104, 105(23), 114(23), 115(23, 107), 116(23), 118(23), 122(23), 129(204), 130(204, 208), 131, 134, 140(94), 207, 265, 323, 353, 489, 551, 552(41)
Nir, D., 122, 124(198)
Nishimura, F., 442
Nix, D., 216(b), 217
Noggle, T. S., 145
Nolte, G., 316, 344(28), 349, 350, 454, 502(35), 512(35), 513, 553(46), 554, 568(46), 586(86)
Northcliffe, L. C., 29, 30, 46, 78, 86, 120, 121(32), 122(32, 45), 127(32), 219, 577
Northrop, D. C., 358
Novikov, M. T, 98, 115(109), 117, 121, 130(192)
Nussbaum, G., 98, 104(110), 115(110), 116(110)

O

O'Brien, R., 606
Oda, N., 442
Oetzman, H., 44, 72(37)
Ogurtsov, G. N., 492
Ölme, A., 558
Okano, M., 230(125), 278
Olsen, D. K., 378, 379, 400, 402, 403, 405, 408, 409, 416, 420(120)
Olsen, R. E., 107

Olson, H. G., 58, 59(54), 60(54)
Oltjen, J., 399, 400, 401
Omidvar, K., 90, 109, 330, 354
Oona, H., 430, 431
Opower, H., 118
Oppenheimer, J. R., 75, 89, 92, 108, 144(20), 341
Ott, W. R., 102
Ovsyannikov, V. P., 21

P

Palumbo, L. J., 583
Panke, H., 142, 143, 144(239, 240), 145, 232(101), 257, 260, 287, 428, 431
Panov, M. N., 22, 155, 156(6), 186, 187(6)
Pape, A., 118
Papineau, A., 118, 129(185)
Paresce, F., 546, 547(34)
Parillis, E. C., 434
Park, J. T., 183, 184
Parratt, L. G., 418
Paschen, K. W., 361, 366(19), 367, 368
Patel, C. K. N., 590
Paul, G., 568, 586
Paul, H., 234(83), 246
Pauling, L., 90
Pavov, M. N., 304, 331(1)
Peacher, J. L., 183, 184(32)
Peacock, N. J., 2
Pedersen, E. H., 115, 121, 128(194), 314, 353(26)
Pegg, D. J., 87, 88(52, 53), 134(52), 284(8), 285, 368, 465, 512, 514, 530, 537, 541(96), 546, 550, 555, 556, 559(63), 560, 561, 562, 567, 568(6), 569(83), 570, 572(63), 574, 575, 576, 579, 580(63), 581(63), 582(101), 584, 586, 587, 597(63), 602(63), 603(57), 604, 605(160)
Pepper, G. H., 230(70), 233(84), 238, 244, 247
Percival, I. C., 93, 97(83), 480
Perfilov, N. A., 118
Perkins, B. L., 129
Pernot, A., 173
Peter, O., 193
Peterson, R. S., 139, 232(97), 255, 257, 277(100), 284(8), 285, 368, 465, 541(96), 546, 555, 556, 559(63), 560(56), 562(56), 567, 572(63), 575(63), 576, 579(63), 580(63), 581(63), 584, 586, 597(63), 602(63), 603(63), 604, 605(160)
Petrov, L. A., 128
Pettus, E. W., 140, 206, 268, 286, 291(12), 294(12), 341, 342(100), 374, 384
Phaneuf, R. A, 2, 102
Phillips, D., 289
Phillips, G. C., 231(66), 234(144), 237, 278
Piacentini, R. D., 213
Pietsch, W., 65, 66(65), 67(65, 68), 68(65), 69(68)
Pihl, J., 145, 550, 561, 565(77), 566, 572(63), 575(63), 579(63), 580(63), 581(63), 597(63), 602(63), 603(63), 606
Pilipenko, D. V., 102
Pinnington, E. H., 558, 561(53)
Pisano, D. J., 318(y), 320, 335, 550, 559, 603(57)
Pivovar, L. I., 98, 102, 115(109), 117, 121, 130(192), 131
Pizzella, G., 437
Podolsky, B., 90
Pol, V., 183
Pollack, E., 190(32)
Polyanskii, A. M., 156, 158(10)
Pommier, J., 158, 181(13), 182(13), 183, 190(32)
Popova, M. I., 323
Popp, M., 65, 66(67)
Povh, B., 145
Powers, D., 35, 36, 45, 52(49), 53, 58, 59(54), 60(41, 54), 62, 63(61), 65(41), 69, 70
Pradhan, T., 90
Preischkat, E., 119
Presser, G., 230(93), 231(133, 134), 250, 266(93), 278, 401
Pretorius, R., 319(dd), 320
Price, R. E., 195, 197(9), 290, 393, 478
Proetel, D., 145
Purcell, E. M., 458
Purcell, J. D., 604, 605(159)
Pyle, R. V., 13, 115, 116(157)

Q

Quaglia, L., 537

AUTHOR INDEX

R

Rai, D. K., 94
Rains, R. G., 440
Raisbeck, G., 94
Randall, R., 310, 317(19), 318(v), 320, 335, 337(82), 348, 350, 354, 369, 370, 371(28), 388(28), 585
Randall, R. R., 291(27), 292, 311, 316, 317(31), 318(o, p, w), 320, 334, 335, 338, 340, 341(31), 342, 394(84), 385
Rao, P. V., 290, 393, 478
Rashiduzzamankhan, M. D., 231(135), 278
Rasmussen, J. O., 211
Ratkowski, A., 383
Rauscher, E., 211
Rayburn, L. A., 231(129), 278
Reading, J. F., 384
Reber, J. D., 234(143), 278
Reenen, R. V., 319(dd), 320
Reeves, E. M., 2
Reichelt, K., 318(s), 320, 344
Resmini, F., 94
Restelli, G., 358
Reusse, W., 193
Reynolds, H. L., 98
Reynolds, T. W., 372
Richard, P., 88, 143, 144(242), 194, 201, 213(8), 218, 231(73), 232(109, 113), 235(88), 239, 248(73), 263, 264(109), 265(109), 266(109), 267(109), 268(113), 270(113), 271(113), 271(113), 272(88), 277, 287, 292, 293, 294(30), 302, 305, 323, 355, 357, 374, 375, 377, 378, 379, 380, 381(49), 382(50), 386, 387, 388(68), 390(12), 391, 392(12), 393(52), 394, 395(76), 397, 398, 399, 400, 401, 402, 403(49), 404, 406(49, 52), 407, 408, 409, 410, 411, 412(85), 413, 414, 415, 416(39), 422(76), 423, 434, 454, 456(33), 465, 470, 473(71), 480, 481, 482(87), 483, 486, 487(109), 488(10), 489(113), 492, 493(113), 494(34), 495, 496, 497(113), 501, 502(34), 503(113), 504(71), 505(113), 506(113), 539, 558(62), 560, 584
Richter, K., 65, 66(65), 67(65, 68), 68(65), 69(68)
Ricobono, C., 234(142), 278
Ricz, S., 230(118), 278
Ridder, D., 453, 477, 516, 517(147), 518, 519(32)
Risley, J. S., 284(10), 285, 291(10), 292(10), 393, 455, 458, 494, 495(128), 496(128), 503(48, 128), 504
Ritchie, R. H., 71, 72, 110, 112, 113(151), 134, 486
Roberts, K., 516, 519(146)
Roberts, R. B., 8
Robertson, R. G., 86
Robinson, M. T., 166
Robinson, R. L., 86
Rødbro, M., 191, 314, 353, 446, 514, 516(29), 525(29)
Roeckl, E., 22
Röhl, S., 319(ii), 320
Roentgen, W. C., 193
Rogers, F. J., 134
Rogers, S. R., 354
Roget, J. P., 318(h), 319, 327
Rojansky, V., 462
Roll, W., 85
Roos, M., 85
Rose, P. H., 84, 85, 98, 104(110), 115(110), 116(110), 131, 139
Rosenfeld, L., 8
Rosner, B., 117, 118(175), 351
Rothermel, J., 109, 110(142)
Rousseau, C. C., 45, 60, 65
Routurier, J., 318(g, bb), 319, 320, 328
Rubin, V., 559
Rudd, M. E., 168, 191, 332, 434, 440, 441, 446, 463(7), 490, 492, 493, 499(20), 516(29), 525(29)
Rudnick, P., 73
Russek, A., 77, 83, 92(25), 93(25), 94(25), 97(25), 114(25), 117, 188
Rutherford, E., 3, 20(9, 11), 73, 74(3)
Rutledge, C. H., 194, 213(8)
Ryding, G., 98, 99, 100, 101, 102(119), 104(110, 119), 115(110, 119), 116(110, 119), 119, 122, 130, 131, 132, 133(221), 134(221), 139

S

Saboya, B., 318(g, bb), 319, 320, 328
Sackmann, S., 306, 314(9), 317(9), 318(t, u), 319(cc), 320, 332, 333, 337
Salin, A., 213, 322, 323, 336, 354

Salop, A., 93, 97(84)
Salpeter, E. E., 131, 147(213)
Salzborn, E., 104, 107(135)
Sampson, J. R., 540
Sanders, J. T., 230(67), 237
Sandlin, G. D., 604
Sandner, P., 64(63), 65
Sandström, A. E., 356
Sar-el, H., 455
Saris, F. W., 523, 524(154)
Sauls, J., 422
Savin, W., 327
Savoy, S. A., 98, 99, 100(114, 115), 104(114, 115), 115(114, 115), 116(114, 115)
Saxman, A., 98, 104(110), 115(110), 116(110)
Sayer, R. O., 115, 116(161, 162), 133
Saylor, T. K., 221, 227(55), 228(55), 281, 314(124), 353
Scanlon, P. J., 65, 69
Scharfer, U., 118
Scharff, M., 28, 35, 39, 164, 165(19), 308, 309(14)
Schader, J., 231(107), 262
Schartner, K. H., 523, 524
Schectman, R. M., 590
Scheid, W., 93, 95(85)
Scherrer, V. E., 604
Scherzer, B. M. U., 61, 62
Schiebel, U., 235(88), 248, 272, 288, 289, 296, 300(20, 35), 316
Schiff, H., 90, 92, 107(81), 322
Schilling, R. F., 29, 30, 46, 219, 577
Schiokawa, T., 230(121, 122), 233(121, 122, 140), 249, 278
Schiøtt, H. E., 28, 39
Schmeider, R. W., 369, 584, 585
Schmelzer, Ch., 85, 99, 100(41), 102(117), 103, 115(41, 117), 116(41, 163), 129, 130(208), 131
Schmidt, G. B., 347
Schmidt, H., 64(63), 65
Schmidt, K. C., 437
Schmidt-Böcking, H., 231(94), 234(146), 250, 252(94), 255(94), 278, 309, 310, 313, 314, 315, 316, 318(*d, m, aa*), 319, 320, 329, 330, 336, 337(89), 338, 344(28)
Schmidt-Ott, W. D., 216(*b*), 217

Schmiedekamp, C, 232(113), 266, 268(113), 270(113), 271(113), 287
Schneider, D., 284(10), 285, 291(10), 292(10), 332, 333(76), 393, 394, 395, 412, 415, 460, 467, 473, 479, 488, 489(113), 490, 491(116), 492, 493, 494, 495(128, 132), 496(128), 497(83, 113), 499, 503(113, 128), 504, 505(113), 506(83, 113), 507(54, 83), 508, 509(83), 512, 514(139), 515(139), 516, 518(127), 519
Schneider, H. R., 14
Schnopper, H. W., 145(260), 146, 549
Schowengerdt, F. D., 183
Schuch, R., 315, 316, 318(*aa*), 320, 336, 337(89), 344
Schulé, R., 231(94), 234(146), 250, 252(94), 255(94), 278, 315, 316, 318(*d, m*), 319, 320, 329, 330, 336, 337(89), 338, 344(28)
Schumann, S., 349, 350, 454, 458, 470, 502(35), 511(141), 512(35), 513, 553(46), 554, 568(46), 586(86)
Schwarzschild, A., 253, 288, 383
Schwob, J. L, 2
Scofield, J. H., 197, 356
Scott, H. A., 115, 116(161), 133
Scott, W. T., 309
Seaman, G., 388
Seaton, M. J., 480
Sellin, I. A., 87, 88(52), 96, 110, 114(145), 134(52), 139, 140(103), 200, 232(97), 255, 257, 277(100), 284(8), 285, 293, 298, 299(36), 321, 338(37), 368, 414, 415, 454, 456, 458, 465, 512(144), 514, 530, 541(96), 546, 550, 555, 556, 559(63), 560(56), 562(56), 567, 568(6), 569(83), 572(63), 575(63), 579(63), 580(63), 581(63), 584, 586, 587, 597(63), 602(63), 603(57, 63), 604, 605(160)
Semrad, D., 234(83), 246
Sen, P., 240, 403
Senashenko, V. S., 114
Senglaub, M., 377, 378, 379
Sens, J. C., 118
Septier, A., 173
Sevier, K. D., 231(107), 262, 349, 350, 434, 436, 438, 439, 444(8), 454, 463(8), 470, 478, 502(35), 511(141), 512(35), 513, 553(46), 554, 568(46)

Shabason, L., 117, 118(175)
Shafroth, S. M., 230(72), 231(127), 232(76, 92), 233(72, 79), 239, 241, 242, 246(72), 278, 391
Shank, C., 205
Shapira, D., 86
Sharma, K., 145, 550
Shaw, J., 529
Shepard, K. W, 16
Shergin, A. P., 156, 158(10)
Shevelko, V. P., 94, 109, 140
Shiomi, N., 47, 48(45)
Shipsey, E. J., 107
Shirai, S., 47, 48(45)
Shirley, D. A., 469
Shore, B. W., 396, 431(91)
Sibenko, E. I., 85
Sidis, V., 190(32)
Siegbahn, K., 439, 444, 460(17), 518(28)
Siegert, G., 118
Sigmund, P., 26, 30, 42, 70, 85, 191, 222, 224(56), 309
Sills, R. M., 554, 566(43)
Silver, J. D., 566, 606
Silverman, P. J., 263, 368, 420(23)
Simon, W. G., 129
Simons, D. G., 68
Sinanoglu, O., 597, 598
Sistemich, K., 305, 314(7), 318(r), 320, 331(7)
Sizmann, R., 134
Sjödin, R., 145, 550, 561, 565(77), 566, 572(63), 575(63), 579(63), 580(63), 581(63), 597(63), 602(63), 603(63), 606
Slater, J. C., 470, 473
Sloane, D. H., 16, 21(39, 40)
Smith, F. M., 129
Smith, F. T., 157, 167(12)
Smith, K., 114
Smith, K. J., 157, 167(11), 168(11), 170(11), 173(11), 176(11), 179(11), 188(11), 188(11)
Smith, L., 16
Smith, L. E., 394, 395, 435, 446, 454(30), 456(30), 461(50), 462(51), 488, 497(30), 498(30), 499, 503
Smith, M. W., 583, 596
Smith, P. L., 601
Smith, R. A., 90, 92(62)
Smith, W. W., 347, 530, 546, 567, 568(6), 569(83), 570, 584, 586
Smythe, R., 115, 116(158)
Soares, C. G., 230(67), 237
Sørensen, G., 578
Sørensen, H., 35, 37(25), 48, 49(46)
Softky, S. D., 70
Sohval, A. R., 145(260), 146, 549
Sokolov, J., 141, 143(238), 144(238), 232(106), 261, 266(106), 422
Solz, M., 260
Sonobe, B. I., 296, 420, 421
Soong, S. C., 286, 343, 354
Sorbo, R. A., 119
Spangenberg, K. R., 442, 443(24)
Specht, H. J., 22, 315, 344, 374
Sperli, F., 437
Spindler, E., 140, 142, 144(240), 145(236), 148, 428
Spivak, M. A., 358
Spohr, R., 568, 586(86)
Stadler, B., 129, 130(208)
Stähler, J., 230(93), 231(133), 250, 266(93), 278
Staub, H. A., 314
Stebbings, R. F., 102
Stehling, W., 142, 144(240), 145, 428, 431
Stein, H. J., 305, 314, 318(r), 320, 331, 426
Steiner, R., 120
Stelson, P. H., 86, 102, 535
Stephens, F. S., 145
Stephenson, R., 281
Stiebing, K. E., 234(146), 278, 318(d, m), 319, 320, 329, 330, 338
Stier, P. M., 121
Stobbe, B. M., 144(246), 145
Stöckli, M., 396, 397, 398
Stoller, Ch., 396, 397, 398
Stolterfoht, N., 284(10), 285, 291(10), 292(10), 332, 333, 393, 394, 395, 434(13), 435, 440(13), 449, 453, 454(31), 456(33), 467, 473, 474, 477(32), 483, 485, 486, 488, 489(113), 490, 491, 492, 493, 494(34), 495(128, 132), 496(128), 497, 499, 501, 502(34), 503(128), 504, 505(113), 506, 509, 510, 512, 514(140), 515(140), 516(32), 517(147), 518(32, 127), 519(32, 74), 520, 521, 522(151), 523(153), 524(153)

Stone, G., 151, 152(1), 165, 173(1), 309
Stoner, J. O., Jr., 536, 556, 557, 606
Strassmann, F., 7
Striganov, A. R., 583
Stuckelberg, E. C., 144(247), 145
Subtil, J. L., 558, 561(53), 594(136), 595
Succi, C., 94
Sural, D. P., 77, 109(29)
Suter, M., 110, 114(145), 396, 397, 398, 559(63), 560, 572(63), 575(63), 579(63), 580(63), 581(63), 597(63), 602(63), 603(63)
Sventitskii, N. S., 583
Swift, C. D., 195, 197(9), 290, 393, 478
Symon, K. R., 13
Szuster, B. J., 94

T

Tagliaferri, G., 94
Tahira, S., 442
Takahuhi, T., 230(125), 278
Tandon, P. N., 65, 66(67)
Tashaev, Y. A., 102
Taulbjerg, K., 166, 191, 213, 222, 223(57), 224, 266, 289, 305, 322, 332, 333, 342, 343, 344, 385, 484
Tavernier, M., 318(h), 319, 327
Tawara, H., 77, 83, 92(25), 93(25), 94(25), 97(25), 114(25), 230(63, 121, 122), 233(121, 122, 140), 237, 249, 278, 302
Taylor, P. O., 2
Teldman, U., 556, 602
Teller, E., 73(14), 74
Tendow, Y., 230(125), 278
Teplova, Y. A., 97, 102(107, 108), 104, 115(107), 116, 131, 134
Terwilliger, K. M., 13
Thein, Q. T., 318(g), 319
Theisen, R., 215
Thibaud, J. P., 318(g, bb), 319, 320, 328
Thieberger, P., 535
Thoe, R. S., 110, 114(145), 139, 257, 277(100), 347, 368, 512(144), 514, 530, 541(96), 555, 556, 559(63), 560(56), 561, 562(56), 568(6), 572(63), 575(63), 579(63), 580(63), 581(63), 586, 597(63), 602(63), 603(63), 604, 605(160)
Thomas, L. H., 13
Thompson, D. A., 69

Thomson, G. M., 304, 331, 345, 346, 347
Thornton, R. L., 13
Thornton, S. T., 233(138), 237, 249, 278
Tilford, S. B., 529
Tillman, K., 595
To, K. X., 562
Toburen, L. H., 436, 437, 457, 458, 493
Toevs, J. W., 115, 116(158)
Toke, J., 86
Touati, A., 318(h), 319, 327
Tousey, R., 604
Trachslin, W., 559
Träbert, E., 606
Treado, P. A., 5
Tricomi, J., 230(124), 231(129), 278
Tripathy, D. N., 90, 94
Trubchaninov, F. M., 131
Tserruya, I., 231(94), 250, 252(94), 255, 313, 314, 315, 316, 318(d, m, aa), 319, 320, 329, 330, 336, 337, 338, 344(28)
Tuan, T. F., 94
Tubaev, V. M., 98, 115(109), 121, 130(192)
Tucker, T. C., 96
Tunnell, L. N., 286, 343, 354
Turkenburg, W. C., 173
Turner, J. E., 35
Tuve, M. A., 5, 6, 8

U

Uemura, Y., 47, 48(45)
Ulbricht, J., 85
Umbarger, C. J., 231(131), 235(131), 278
Unus, I., 216(b), 217
Urata, N., 90
Urnov, A. M., 109

V

Vaaben, J., 266, 289, 332, 333(74), 342, 343, 344
Vader, R. J., 318(f), 319, 325, 327, 328
Vajda, P., 35, 37(25)
Valenzuela, A., 47
Valkovic, V., 231(66), 237
Van de Graaff, R. J., 5, 6, 9
Vandenbosch, R., 119, 313, 318(q), 320, 327, 339

AUTHOR INDEX 625

van der Woude, A., 318(f), 319, 325, 327, 328(57)
Vane, C. R., 110, 114(145), 555, 556, 559(63), 560, 561, 572(63), 575(63), 579(63), 580(63), 581(63), 597(63), 602(63), 603(63), 604, 605(160)
van Eck, J., 229, 236(62)
Van Rinsvelt, H. A., 230(67, 70), 237, 238
Vanugopala Rao, P., 195, 197(9)
Varghese, S. L., 260, 318(w), 320, 334, 335, 354, 388, 584
Veje, E., 122, 549
Vinogradov, A. V., 94, 109, 140
Vinti, A. J. P., 398
Volkar, M. A., 119
Vollath, D., 215
Vollmer, A., 22
Volodyagin, Y. S., 102
Volz, D. J., 440, 441, 493, 499(20), 516
von Brentano, P., 118
Vriens, L., 201, 486

W

Walgate, R., 74
Walske, M. C., 37
Walters, D. L., 347, 468, 475, 519(62)
Waltner, A. W., 230(72), 232(76), 233(72), 239, 241, 246(72)
Walton, E. T. S., 4
Wangsness, R. K., 530
Wannberg, B., 439, 460
Warczak, A., 289, 319(hh), 320
Ward, D., 39, 46, 50, 51
Warren, B. E., 373
Warters, W. D., 58
Watanabe, T., 90
Watson, R. L., 194, 213(8), 218, 230(120), 231(132), 234(145), 238, 278, 296, 381, 405, 406(105), 417, 418, 419, 420, 421, 493
Weaver, O. L., 354
Weber, G., 537
Wegner, H. E., 145, 535
Wehring, B. W., 116, 117(165), 118, 127(179)
Weihrauch, J., 129, 130(208)
Weiss, A. W., 596, 598
Weiz, M., 142, 143, 144(239), 287

Weizel, W., 152
Welker, J., 201, 483, 489(94)
Wentzel, G., 474
Wenzel, W. A., 61
Werme, L. O., 444, 518
Wessel, W., 144(248), 145
Whaling, W., 61, 554, 566(43), 601, 602
Wheeler, R. M., 230(124), 231(129), 234(144), 278
White, M. G., 12
Whitehead, C., 211
Whittemore, A. R., 230(75), 240, 286, 384
Wideröe, R., 14
Widing, K. G., 604, 605(159)
Widson, W. W., 351
Wieman, H., 284(10), 285, 291(10), 292(10), 313, 318(q), 320, 327, 339, 393, 494, 495(128), 496(128), 503(128), 504
Wien, W., 73, 529
Wiese, W. L., 583, 596
Wille, U., 484
Williams, E. J., 20
Williams, M. E., 327
Wilson, W. D., 44, 72
Winter, T. G., 94
Winterbon, K. B., 85, 309
Winters, L. M., 96, 140(103), 256, 268, 284(11), 286, 291(12), 294, 298, 299(36), 206, 341, 342(100), 374, 384
Winther, A., 45
Wittkower, A. B., 78, 81, 82(35), 85, 98, 99, 100, 101, 102(35), 119), 104(35, 110, 119), 105(35), 115(35, 110, 118, 119), 116(35, 110, 118, 119), 118(31), 119, 121(129), 122, 127(31, 195), 128, 130(195), 131, 132, 133(221), 134(221, 139), 372, 551
Wittmann, W., 570, 595
Wölfli, W., 396, 397, 398
Woerlee, P., 523, 524(154)
Wolke, R. L., 119
Wolter, H. H., 387, 400, 416, 420(120), 562
Wood, O. R., 590
Woods, C. W., 292, 294(30), 386, 387, 388(68), 394, 399, 400, 401, 409, 410, 411, 412(85), 413, 415, 416, 454, 456, 465, 470, 473(71), 494, 495, 496, 501, 502, 504(71), 558(62), 560, 584

Wright, J. J., 561, 572(63), 575(63), 579(63), 580(63), 581(63), 597(63), 602(63), 603(63)
Wyly, L. D., 98
Wyrick, R. K., 233(136), 278

Y

Yadav, H. N., 90
Yarnold, G. D., 439
Yiou, F., 94
Young, K. M., 2
Yuferov, V. B., 85

Z

Zabel, T., 231(66), 234(144), 237, 278
Zaharis, E. J., 221, 223(54), 224(54), 376
Zander, A. R., 230(70), 238
Zashkvara, V. V., 454, 544
Zekl, H., 313, 314, 318(d), 319, 338
Zhileikin, Y. M., 114
Ziegler, J. F., 39, 45, 46, 69
Ziem, P., 493, 510, 514, 515(140), 518(127), 520, 521, 522(151), 523, 524
Ziemba, F. P., 151, 152(1), 173(1)
Zucker, A., 98
Zupančič, C., 210

SUBJECT INDEX

A

Accelerating-decelerating system, 183
Accelerating lens system, of parallel-plate analyzer, 440, 442, 450
Accelerator, 4–21; *see also* specific accelerators
 beam-foil experiments, 530, 533–535
 for charge state fraction measurements, 83–84
Accelerator-decelerator stripping technique, 281–282
Air-insulated pressure tank Van de Graaff accelerator, 7
Allotropic effect, on stopping cross sections, 70
Allowed radiative transition, 579–583
Alpha-particle projectile, 402–403
Anisotropy, 191
Auger electron emission, 479–482, 514–516
 in beam-foil research, 591
Astrophysical studies, and beam-foil experiments, 599–605
Asymmetric collision, 384–385
 energy spread, 178
 inner-shell vacancy production, 485
 noncoincidence technique in study of, 157
 SCA calculations, 341
 x-ray production cross section, 285–288
Atomic number of projectile, and degree of target ionization, 379
Atomic structure, beam-foil measurements of, 596–599
Atomic transitions, in solids, 139–144
Auger decay, 524
Auger effect, 433–434
Auger electron
 Coulomb ionization, 492–500
 energy, 401
 analysis, 544
 measurement of, 88
 coincidence measurements, 318–319
 production and detection, 436–467
 by electron promotion, 490–491
 spectral identification, 347–350
Auger electron detector, 314
Auger electron emission, 191, 468–482, 561, 566–570
 angular distribution, 479–482
 anisotropies in, 514–516
 in coincidence experiment, 345
Auger electron spectroscopy, *see* Ion—induced Auger electron spectroscopy
Auger line, 468–469
 of lithiumlike atoms, 506–514
Auger linewidths, 524–527
Auger measurement, of Auger yield, 292–294, 393, 395
Auger production cross section, 479
Auger transition, 470, 472–477
 forbidden, 586–587
Autoionizing electron, 191; *see also* Auger electron
AVF cyclotron, 13–14

B

Backscattering energy loss, 58–65
Bare projectile nucleus, 299
Barkas effect, on stopping cross section, 71
BEA, *see* Binary encounter approximation
Beam energy, fluctuations in, 187
Beam-foil decay curve, 572–578, 580–582
Beam-foil experiments, 371–373, 387, 529–606
 allowed radiative transitions, 579–583
 applications of, 595–606
 Auger measurements

627

cylindrical-mirror analyzer, 458
lithiumlike atoms, 512–514
cascading, 572–575
energy-loss measurements, 47–51
experimental arrangements, 530–548
  accelerators and ion sources, 533–535
  beam analysis and source purity, 535–536
  filters, spectrometers, spectrographs, detectors, 538–548
  foil targets, 536–537
  normalization procedures, 537–538
forbidden Auger transitions, 586–587
forbidden radiative transitions, 583–586
inner-shell vacancy measurements, 260, 519–524
Lamb shift in heavy ions, 587–590
lifetime measurements, 570–590
for measurement of excited states, 88
population changes, 572–575
postfoil beam velocity, 575–578
postfoil charge distribution, 551–552
quantum beat phenomena, 590–595
source characteristics, 548–550
spectra, 550–570
  Auger-emitting states, 566–570
  charge state identifications, 552–554
  Doppler broadening and shifts, 554–557
  doubly excited states, 560–564
  hydrogenic (Rydberg) states, 564–566
  singly-excited states, 557–560
Beam impurities, 84, 535–536
Beam transport system, 464–465
Bethe formula, for electronic stopping, 34–37, 45
Betz–Grodzins model, 136, 519–520, 523
Binary encounter, 159–173
Binary encounter approximation, 200–202, 229, 237, 323, 328, 383
  Coulomb ionization, 486, 492
  K-shell ionization, 326
  L-shell ionization, 240–242, 244, 329–330
  relativistic electron wavefunctions, 210
Binding energy, 204–207, 236–237, 385
  Auger spectroscopy, 470, 473–474
  Coulomb ionization, 487
  and equilibrium charge-state distributions, 121
  lithiumlike atoms, 506, 508

BK approximation, *see* Brinkman–Kramers approximation
Bohr–Lindhard model, 136, 519–520, 523
Bohr potential, 164
Born approximation, 89, 92, 110–112, 114; *see also* Plane-wave Born approximation
Bragg crystal reflection spectrometer, 361–365
Bragg's rule, for stopping cross sections, 69
Bremsstrahlung radiation, 458
Brinkman–Kramers approximation, 489
Broadening effects, 188; *see also* Doppler broadening; thermal broadening
  inelastic energy-loss measurements, 168–173

C

Calorimetric method, of energy-loss measurements, 48–49
Carbon foil, in beam-foil experiments, 532, 536–537
Cascading, effect of, on decay curves, 572–575, 581
CBEA, *see* Constrained binary encounter approximation
CEM, *see* Channel electron multiplier
Center-of-mass system, in inelastic kinematics, 159, 161–163, 174
  scattering cross section, 164, 166
Channel electron multiplier, 540–541, 544–547
Charge-analyzed reaction products, coincidence experiments involving, 345–354
Charge-changing cross section, 79–82; *see also* Electron capture cross section; Electron loss cross section
  measurement of, 83–84
Charge-changing processes, in ion–atom collisions, 303–354
Charge exchange, 384–385; *see also* Electron capture; Electron loss
  Auger spectroscopy, 482, 488–489, 500–503
  fundamental processes, 78–83
  lithiumlike atoms, 509–510
  probabilities, 108
Charge state, 189–190

SUBJECT INDEX

heavy-ion-atom collision, 151, 253, 256-257, 272-278
high-velocity ions in matter, 73-148
  atomic transitions and charge equilibrium in solids, 139-144
  density effects, 131-135
  electron capture, 89-110
  electron loss, 110-118
  equilibrium charge-state distribution, 118-124
  experimental techniques and data analysis, 83-89
  historical background and review, 73-78
  radiative electron capture, 144-148
  identification, 552-554
  postfoil charge distributions in beam-foil interactions, 551-552
Charge-state dependence
  conditions for measuring, 280-284
  in Coulomb ionization, 495-496
  in electron promotion, 490-491
  inner-shell cross sections, 279-302
  projectile cross sections, 297-302
  target x-ray production, 284-289
Charge-state fractions, 118-119, 121
  density effect, 132-133
  measurement of, 83-89
  nonequilibrium and equilibrium conditions, 80-83
CMA, *see* Cylindrical-mirror analyzer
CM system, *see* Center-of-mass system
Cockcroft-Walton accelerator, 4-5, 533
Coherence, in beam-foil experiments, 530, 591-592
Coincidence experiments, in ion-atom collision study, 188, 303-354
  broadening effects, 170-171
  charge-analyzed reaction products, 345-354
  detectors, 313-316
  inelastic energy-loss determination, 150-153, 155-159, 162, 173, 184-188
  inner-shell vacancy production, 304-344
  intermediate regions, 337-344
  molecular regime, 330-337
  point projectiles, 321-330
  single-encounter experiment, 306-310
  target-detector arrangements, 316-317

timing considerations, 311-313
Configurational effects, in beam-foil experiments, 596-598
Constant-frequency cyclotron, 13
Constrained binary encounter approximation, 201-202, 242, 244
Contamination of source, in beam-foil experiments, 84, 536
Continuous-channel electron multiplier, 436-439
Continuous-strip magnetic electron multiplier, 545
Core-excited state, beam-foil interactions, 551, 560
Coulomb approach, to inner-shell vacancy production 321-330
Coulomb deflection, 487
  low projectile velocities, 327-328
  in target ionization, 205-207, 236-237
Coulomb electron capture, 92
Coulomb ionization or excitation, 200-207, 211, 253, 384-386, 482-483, 486-488
  alignment mechanism 481
  Auger electron measurements, 492-500
  lithiumlike atoms, 509
  relativistic electron wavefunctions, 211
Coulomb potential, 163-164
Coulomb scattering, 31-33
Coupled accelerator, *see* Multiaccelerator
Crystal, energy loss in, 70
Crystal spectrometer, 372-374, 376-377, 386-388
  single cylindrical curved-crystal, 365-369
  single-plane, 361-365
Curved crystal spectrometer, 387
Cyclotron, 12-14, 533, 535
Cylindrical analyzer, 462-463
Cylindrical-mirror analyzer, 454-458, 467, 544, 546
Cylindrical-sector field analyzer, 178, 180
Czerny-Turner monochromator, 539

D

Dark count noise, of photoelectric detectors, 545
DCE, *see* Direct Coulomb excitation
Decay process, 296, 299, 311
  beam-foil studies, 529-530, 532, 570-590

doubly-excited states, 561
  autoionizing, 568–570
Deceleration lens system, of parallel-plate analyzer, 440–449
Deexcitation process, 149
Deflection analyzer, see Electrostatic deflection analyzer
Density effect, 131–139
  in gaseous targets, 131–134
  in heavy ion-atom collisions, 75–76
  in solid targets, 134–139
Detector, 175, 311–317, 436–438
  time-of-flight energy analysis, 180–181
Dielectric description, of energy-loss problem, 33–34
Differential pumping, 52–53
  in scattering chamber, 175
  in x-ray spectroscopy, 373
Differential-scattering cross section, 165, 167, 174
Dirac relativistic theory, 587
Direct Coulomb excitation, 482–483, 486–488
Direct Coulomb ionization, 200–207, 482–483, 486–487
  Auger electron measurements, 492–500
Dispersion of beam, in beam-foil experiments, 577
Dispersive counter, 358–359
Dispersive detector, 376
Distorted-wave method, for electron capture, 90, 92
Doppler broadening, 67, 516, 518, 524
  beam-foil interactions, 545, 554–557, 569
Doppler shift
  beam-foil interaction, 545, 554–557, 569
  of photons emitted by fast-moving ions, 369, 373
Doppler-shift attenuation method, of lifetime measurements, 65
Doppler-tuned spectrometer, 369–371, 388
Double-electron transition, 344
Doubly-excited state, 107
  autoionizing decay, 568–570
  beam-foil interactions, 549, 551, 560–564
DSA method, see Doppler-shift attenuation method

DTS, see Doppler-tuned spectrometer
Dynamic screening model, of heavy-ion incident on solid target, 253–255, 266

E

Elastic energy loss, see Nuclear energy loss
Electric dipole decay process, 579, 583–584
Electric dipole selection rules, in multielectron transitions, 396–397
Electric dipole transition, in doubly-excited states, 561–564
Electric quadrupole decay process, 583
Electron, Auger, see Auger electron
Electron capture, 89–110, 139–142, 384, 482, 488–489; see also charge exchange; Radiative electron capture
  beam-foil interactions, 549–550
  charge exchange probabilities, 108
  in complex systems, 94–97
  into continuum states, 109–110
  in heavy-collision systems, 75, 78
  into high quantum states, 108–109
  hydrogenic states, 566
  from K-shell of heavy targets, 350–354
  multiple, 104–107
  in simple collision system, 89–94
Electron capture cross section, 75, 79, 96, 143–144
  density effects and excited states, 132–133
  experimental results on, 97–104
  K shell, 140
  multiple capture, 104–105
  probabilities, 108
  simple collision systems, 91
Electron detector, 314
  continuous-channel electron multiplier, 436–439
Electronic energy loss, see Inelastic energy loss
Electron loss, 75, 78, 110–118
  multiple, 116–118
  projectile ionization, 113–114
  target ionization, 110–113
Electron loss cross section, 75, 79, 98–100
  density effects and excited states, 132–133

SUBJECT INDEX

experimental results, 114–115
multiple loss, 105–106
probabilities, 108
Electron multiplier, 436–439
Electron promotion, 482–486
  Auger electron production, 490–492
Electron temperature, and beam–foil interactions, 550
Electron transfer, 321, 341
Electrostatic analyzer, 175, 178–180, 183–185, 439–463
Electrostatic focusing lens, 442
Emission–yield studies, 191–192
Energy–calibrated magnetic analyzer, 536
Energy loss, of high-velocity ions in matter, 25–72
  experimental methods, 46–68
  mechanisms of, 26–30
  stopping cross sections, 68–72
  theories of, 30–46
Energy–loss measurements, 46–68
  backscattering, 58–65
  gaseous target transmissions, 51–53
  inelastic, in single collisions, 149–192
  inverted Doppler–shift attenuation method, 65–68
  supported films 53–54
  thin foil transmissions, 47–51
  thin layer on solid–state detector, 54–58
Energy straggling, 70–71
Energy transfer, in projectile–atom collision, 197–198
Equilibrium charge state, average, 125–131
  experimental, 127–128
  Lamb–Bohr criterion, 125–127
  semiempirical relationships, 129–131
Equilibrium charge state distributions, 80–83, 118–124
  experimental results, 118–122
  semiempirical descriptions of, 122–124
Equilibrium excitation, of fast ions in solids, 142–144
Excitation, 78, 87, 149, 188–191, 384–385
  after electron capture, 107
  density effect and, 131–139
  detection of excited states, 88
  direct Coulomb, 482–483, 486–488
  energy loss of high-velocity ions, 26
  fast ions in solids, 142–144
  inner shell, 116–117

mean excitation energy, 35–37
multiple, 135–139

F

Fast collision, 31
Fast–ion accelerator, 464–467
Field–dependent lifetime, and Lamb shift, 589–590
Field–quenching technique, Lamb shift measurements, 588–590
Filter, in beam–foil experiments, 538–539
Fine interaction, in foil excitation process, 592, 595
Firsov theory, of electronic energy loss, 39–40
Fission fragments, 73–74, 118
Flow proportional counter, 357–358
Fluorescence yield, 191, 380, 382, 389, 392–395, 477–479
  argon L–shell, 347
  in gases, 422
  in high–charge collisions, 289–297, 300
  hypersatellites, 409
  K–shell ionization, 195
  lithiumlike atoms, 506, 586
  L–shell ionization, 197
  in multiple ionization, 207–209, 238–239, 503–506
  for thick target, 227–228
Foil target, 371–372, 387, 532, 536–537; see also Beam–foil experiments
  charge state fraction measurements, 84–86
  energy–loss measurements, 47–51
  spectral identification of emitters, 348–350
  thickening effects, 537, 577
Folded tandem accelerator, 9–10, 12
Forbidden Auger transition, 561, 586–587
Forbidden radiative transition 583–586
Fourier transform, of quantum beat pattern, 594–595

G

Gas–filled detector, 356–358
Gas projectile, 415–416
Gas proportional counter, 86
Gas target, 316–317, 373–374, 379

angle resolution effects, 310
average equilibrium charges, 125–130
charge state fraction measurements, 85, 87
collisional quenching, 426–427
density effect in, 131–134
energy-loss measurements, 51–53
energy straggling, 70–71
isotopic composition of, and broadening effects, 170–171
K-shell vacancy production, 335, 337, 342
multielectron ions passing through, 136
neon, 410–411
outer-shell rearrangements in, 424–425
scattering, 306
spectra, 422–423
x-ray production rate, 228
Gaussian beam velocity profile, 578
Germanium (lithium) detector, 67, 313, 358–361
Grazing incidence spectrometer, 539–544
Gridded lens, of parallel-plate analyzer, 443

H

Half-cylindrical mirror analyzer, 457–458, 467
Hartree-Fock-Slater charge distribution, 41, 45
Heavy ion
  beam transport, 464
  charge exchange phenomena, 74–75
  charge-state fractions in solids, 272–278
  Coulomb excitation, 384
  ionization cross sections, 384
  K-shell ionization, 250–272
  Lamb shift, 587–590
  multiple inner-shell vacancy production, 374–375, 377
  projectile, 403–408
  sources, in beam-foil experiments, 533, 535
  target considerations in inner-shell ionization, 221–228
Heavy-ion accelerator, 16–19, 22, 74, 84
Heavy-ion-atom collision, 151–155, 500
  charge dependence of atomic inner-shell cross sections, 279–302
  electron loss, 117

Heavy target, electron capture from, 350–354
High-charge collision, fluorescence yields, 289–297
High-density gas jet target, for charge state fraction measurements, 85
High-energy beam-foil experiments, 533, 535–537
  lifetime measurements, 577
High-resolution energy spectrometry, 183–184
High-spin state, 510, 515
High-velocity ion
  charge equilibrium in matter, 73–148
  energy loss in matter, 25–72
High-voltage ion accelerator, 466
Hydrogenic state, in beam-foil interactions, 549, 551, 564–566
Hyperfine interaction, in foil excitation process, 592, 595
Hypersatellite Auger electron, 501–502
Hypersatellite production, 408–410

I

IAES, see Ion-induced Auger electron spectroscopy
IDSA method, see Inverted Doppler-shift attenuation method
IEL, see Inelastic energy loss
Impact-parameter dependence, in ion-atom collisions, 303–344
  charge exchange processes, 79–80
  multiple ionization, 487
  projectile ionization, 113
  small, and electron loss, 117–118
  target ionization, 113
  vacancy production, 380–382
Impulse approximation, for electron capture, 90–92
Inelastic collision, 22, 31
  beam-foil interactions, 549–550
  inner-shell cross sections, 279
Inelastic energy loss (electronic energy loss), 26–27, 32–33, 38, 42–44, 150, 303–304
  binary encounters, 159–173
  broadening effects, 168–173
  errors in analysis, 186–188
  experimental determination of, 173–188
  historical overview, 151–158

SUBJECT INDEX

K-shell vacancy production, 332
  at low velocities, 39–41
  measurements of, in single collisions, 149–192
  at medium velocities, 45–46
Inner-shell collision, 149, 163, 165
Inner-shell corrections to stopping power, 36–37
Inner-shell electron capture, 97
Inner-shell electron loss, 116–117
Inner-shell excitation, 22–23, 190
  excitation energy, 149
  heavy-ion atom collision, 152
Inner-shell ionization, 384–385; see also K-shell ionization; L-shell ionization; M-shell ionization
  binary encounter approximation, 200–202
  electron loss processes, 75
  experimental arrangements, 213–228
  with heavy-ion projectiles, 203, 205–207, 221–228
  molecular orbital excitation, 212–213
  multiple ionization effects, 207–209
  plane-wave Born approximation, 202–203
  relativistic electron wavefunctions, 209–211
  by relativistic projectiles, 211–212
  semiclassical approximation, 202
  and target thickness, 255–257
    thick targets, 218–219
    thin targets, 218–221
  theoretical models, 197–213
  and x-ray cross sections, 195–197
Inner-shell vacancy, 287, 383; see also K-shell vacancy; L-shell vacancy
  Auger lines, 468–470, 516–517
  collisional quenching, 426–428
  decay time, 311
  doubly-excited state, 561
  electron capture and, 139–142
  electron promotion process, 484–485
  fluorescence yield, 292–294, 297
  heavy ion within solid, 257, 273–274, 277
  impact-parameter dependence, 304–344
  lifetime, 523
  from multiple collisions, 224–227
  multiple production, 374–386
  two- and three-component models for target x-ray production, 257–262, 264–267, 270–272
Inner-shell vacancy cross section
  charge dependence, 279–302
  for heavy ions on thick targets, 224
Inner-shell vacancy fraction, beam Auger measurements 519–524
In-shell transition, 579–580, 584, 596, 598
Interaction potential
  inelastic energy loss, 163–167
  nuclear energy loss, 72
Intermediate-energy beam-foil experiments, 535–536
Intrinsic germanium detector, 358
Inverted Doppler-shift attenuation method, of energy-loss measurement, 65–68
Ion accelerator, 464–467
Ion-atom collision, 1–23; see also Beam-foil experiments
  Auger spectroscopic studies of, 468–527
  charge equilibrium of high-velocity ions in matter, 73–148
  direct Coulomb ionization or excitation, 486–488
  electron capture, 89–110, 488–489
  electron promotion, 484–486
  energy loss of high-velocity ions, 25–72
  heavy ions, 279
  impact-parameter dependent processes, 303–344
  mechanisms, 382–386, 482–503
  multiple-collision phenomena, 426–431
  single collision, 306–310, 401–416
  inelastic energy loss in, 149–192
Ion clusters effect, in stopping cross sections, 71–72
Ion-induced Auger electron spectroscopy, 433–527
  ion-atom collision phenomena, 468–527
  emission-yield studies, 191–192
  techniques, 436–467
Ion-induced x-ray spectroscopy, 355–431
  chemical effects, 416–426
  detectors, 356–371
  fluorescence yields 392–395
  line energies, 386–392
  multielectron transitions, 395–401
  multiple-collision phenomena, 426–431
  multiple inner-shell vacancy production, 374–386

single-collision phenomena, 401–416
targets, 371–374
Ionization, 383
  energy, 35–37
  energy loss of high-velocity ions, 26
  probabilities, for multiply ionized neon, 503–506
  projectile, 113–114
  solid targets and x-ray production, 193–278
  of target, due to light particles, 110–113
Ionization cross section, 195, 383–385
  heavy ion incident on solid target, 253, 255
  K-shell ionization, 111–113, 195, 201, 210
  L-subshell ionization, 196–197
Ion-molecule collision, Auger spectra following, 524–527
Ion-solid interaction, 428–431
Iron-group elements, solar abundances, 601–602
Isotope separator, beam-foil experiment, 533
Isotropic yield of target x rays, 218

J

Johann curved-crystal spectrometer, 365–369
Johannson curved-crystal spectrometer, 365–369

K

Kinetic energy of scattered particles, 159–161
KLL-Auger transition, 469–470, 472, 474–475, 498, 519
K shell
  corrections to stopping power, 37
  electron capture, 140, 350–354
  multielectron transitions, 396–401
  orbital velocity, 483
  proton capture from, 353–354
K-shell ionization, 195
  binary encounter approximation, 201
  fluorescence yield, 478–479
    in multiple ionization, 207–209
  by heavy ions, 203, 250–272
  by light ions, 204

  low-velocity effects, 204–206
  plane-wave Born approximation, 206–207, 323
  projectile charge dependence, 290–292
  relativistic electron wavefunctions, 209–211
  by relativistic projectiles, 211
  semiclassical approximation, 323–328
  x-ray measurements, 229–240
K-shell ionization cross section, 296–297, 380–381
K-shell vacancy, 305, 375–377, 417
  Auger spectroscopy, 468–469, 476, 481, 516
  charge state dependence, 292–297
  Coulomb ionization, 492–499
  fluorescence yields, 392–395
  heavy ions in solids, 273–277
  hypersatellite production, 408–410
  intermediate region, 338–344
  ions passing through solid, 141–142
  molecular orbital model, 331–337
  multiple ionization, 379–382
  $Ne^+ \rightarrow Ne$ collisions, 490–491
  sharing ratios, 491, 493
  in three-component model for target x-ray production, 270–272
  in two-component model for target x-ray production, 257–267
K-shell vacancy cross section, 385
K-shell x-ray yields, 140, 143, 287–292, 300–301, 386–392

L

Lamb-Bohr criterion, 125–127, 142
Lamb shift, in heavy ions, 587–590
Laser beam, in beam-foil experiments, 530, 532
Lenz-Jensen potential, 164, 166
Level crossing, in collision systems, 385, 409, 596
Lifetime measurements, see also Time-of-flight measurements
  beam-foil experiments, 529, 537, 570–590
  and density effects, 133–134
  Doppler-shift attenuation method, 65
  hydrogenic states, 566
  photoelectric detection, 539

SUBJECT INDEX

Light ion
  Coulomb excitation, 384
  ionization of inner shells, 383
  projectiles, 402–403
  target ionization due to, 110–113
Light pipes in beam–foil experiments, 538
Linear accelerator (linac), 14–17
  beam–foil experiments, 533, 535
Line energies
  ion–induced Auger electron spectroscopy, 469–474
  ion–induced x-ray spectroscopy, 386–392
Linewidth, spectral, see Spectral linewidth
Liquid target, charge state fraction measurements, 85
Lithium–drifted germanium diode, 358; see also Germanium (lithium) detector
Lithium–drifted silicon diode, 358; see also Silicon (lithium) detector
Lithiumlike atom
  anisotropies, 514–515
  Auger decay, 586–587
  excited–state populations, 506–514
  K–Auger emission, 524–526
LMM–Auger decay, 475–476
Low–energy beam–foil experiments, 533, 536–537
  lifetime measurements, 571, 576
Low–spin state, 515
Low–velocity effects
  electron capture, 100–101, 104
  target ionization, 113, 204–205, 236
Low–velocity ion
  electronic energy loss, 39–41
  nuclear energy loss, 26, 28, 30
L shell
  corrections to stopping power, 37
  fluorescence yields, 393, 395
  multielectron transitions, 397–398, 401
  population probability, in ion–solid interactions, 428–430
L–shell excitation, 190
L–shell ionization, 195–197, 232–235
  binary encounter approximation, 329–330
  fluorescence yield, 479
  with heavy ion projectiles, 203
  multiple, 401–408
  neon, 412–413

plane–wave Born approximation, 323
semiclassical approximation, 323, 329–330
x–ray measurements, 240–249, 271–272
L–shell ionization cross section, 380, 382
L–shell vacancy, 375–376, 378–379, 416–418
  Auger spectroscopy, 474, 476, 481, 516–518
  charge–analyzed reaction products related to production of, 345–347
  Coulomb ionization, 493–495, 499
  filling, 420, 422, 424–425
  fluorescence yield, 292–293
  hypersatellite production, 409–410
  impact–parameter dependence, 304–305
  molecular orbital model, 330–331
  multiple ionization, 379–382
L–shell vacancy fraction, 521–522
LSS theory, of electronic energy loss, 39–44, 72
L–subshell ionization, 195–197, 240–246
  binary encounter approximation, 201
  constrained binary encounter approximation, 202
  with light ion projectiles, 204
  low–velocity effects, 204–206
L–subshell vacancy, 380, 382

M

Magnetically confined thermonuclear fusion plasma, 599–600
Magnetic analyzer, in beam–foil experiments, 536
Magnetic dipole decay process, 583–585
Magnetic momentum analysis, 464–465
Magnetic quadrupole decay process, 583–584, 586
Mean charge state, 189–190
Mean excitation energy, 35–37
Mean inelastic energy loss, 153, 155, 189–190
Medium–velocity region, electronic energy loss, 45–46
Merzbacher–Lewis relationship, 218–219, 221, 227
Metastable state, 427–428, 583–584
  detection of, 89
  forbidden Auger transition, 586

K–Auger emissions, 524–526
lithiumlike atoms, 511–512
ML relationship, see Merzbacher–Lewis relationship
Molecular–orbital excitation, 190, 212–213, 330–337, 384–386
Molecular–orbital model, of ion–atom collision, 112, 152, 155, 482
  K–shell ionization, 483
  L–shell vacancy production, 345–347
Momentum transfer, 198
  and energy loss, 33, 40
Monochromator, 539–544
MO model, of ion–atom collision, see Molecular–orbital model
M–shell excitation, 190
M–shell ionization
  semiclassical approximation, 323
  x–ray measurements, 249–250
M–shell vacancy, 375, 391, 416, 418, 424
  Auger spectroscopy, 476, 517–518
  beam–foil interaction, 524
M–shell x–ray production, 344
Multiaccelerator, 17–19
Multielectron transitions, 395–401
Multiple–collision effects, 224–227
  beam–foil experiments, 549
  charge–changing probabilities, 79
  collisional quenching, 426–428
Multiple electron capture, 104–107
Multiple electron capture cross section, 98, 104–105
Multiple–electron loss, 116–118
Multiple–electron loss cross section, 98, 100, 105–106
Multiple excitation
  and charge–state density effect, 135–139
  shell effect, 107
Multiple inner–shell vacancy production, 374–386
Multiple ionization, 379–382, 417
  Auger decay, 468–482
  Coulomb ionization, 487, 496–500
  fluorescence yields, 207–209, 393–395, 503
  L shells, 247–248, 401–408
  molecular orbital collisions, 484
  neon, 411
Multiple scattering, in ion–atom collisions, 309–310
  beam–foil experiments, 538, 577–578

N

Narrow–band interference filter, in beam–foil experiments, 538–539
Neon target, 410–416
Neutral particle, energy analysis, 180
Noise, in photoelectric detectors, 545
Noncoincidence method of inelastic energy–loss measurement, 173–184, 188
Nondispersive detector, 358, 372
Nonequilibrium charge state distributions, 80–83
Nonradiative electron capture, 89
Nonrelativistic stopping formula, 34
Nuclear collision, energy loss, 26, 28–29
Nuclear energy loss, 26–28, 32, 41–44, 72
Nuclear reaction, energy loss, 26

O

OBK approximation, see Oppenheimer–Brinkman–Kramers approximation
One–electron system
  capture cross section, 299
  electron capture, 96
  Lamb shift, 587–588
  line energies, 387–388
  target x–ray cross section, 286–287
  x–ray decay, 398
Oppenheimer–Brinkman–Kramers approximation, 90–92, 97, 286, 341–342, 353–354
Orbital shrinkage, in beam–foil experiments, 598
Outer–shell collision, 149, 163, 165
Outer–shell electron capture, 299
Outer–shell electron loss, 116
Outer–shell excitation, 190, 376
  Auger lines, 470
  excitation cross section, 284, 287
  excitation energy, 149
  inelastic energy–loss measurements, 157
Outer–shell ionization, 384–385
Outer–shell rearrangement, in gaseous molecules, 424–425
Outer–shell relaxation, effect on x–ray spectra, 416–418
Outer–shell vacancy
  Auger spectroscopy, 473–474
  production at higher energies, 490
Out–of–shell transition, 579–580

P

Parallel–plate electrostatic deflection analyzer, 178, 180, 439–454, 467–468
Partial ionization cross section, 480
Particle detector, *see* Detector
Penning ionization source, 280
Perturbation method, of energy–loss calculation, 33–34
Perturbation theory
  electron capture, 91, 95
  inner–shell vacancy production, 321, 338, 340–341
  semiclassical approximation calculations, 322
Perturbed–stationary–state approximation, 112
PHA, *see* Pulse height analyzer
Photoelectric detector, in beam–foil experiments, 538–539, 545
Photoionization, 481
Photomultiplier, 545
Photon
  emission, 191
  generation of, 26
  relative intensities for calibrated sources, 216–218
Photon detector, 538
Photon–induced x rays, 375
Plane crystal spectrometer, 387
Plane–mirror electrostatic deflection analyzer, *see* Parallel–plate electrostatic deflection analyzer
Plane–wave Born approximation, 111–112, 199–200, 202–206, 210–211, 229, 236–239, 241–252, 321, 323, 383, 482–483, 486
Plasma light source, 550
Plasma studies, and beam–foil experiments, 599–605
PMA, *see* Parallel–plate electrostatic deflection analyzer
Point projectile, in inner–shell vacancy production, 321–330
Polarized x rays, 373
Position–sensitive avalanche detector, 315–316
Position–sensitive solid–state detector, 86
Postacceleration stripping technique, 281–282
Postfoil beam velocity 575–578, 593–595
Postfoil charge distribution, in beam–foil interactions, 551–552
Potential energy, of Ne–O system, 166
Pressurized tank accelerator 7–8
Probe–layer technique, for x–ray yield measurements, 260–262
Projectile charge dependence, 297–302
  target fluorescence yields, 290–297
  target inner–shell cross sections, 284–289
Projectile ion, charge equilibrium in matter, 73–148
Projectile ionization, 113–114
Projectile velocity, and energy loss, 37–39
Proportional counter, 86, 356–358, 374
Proton, 402–403
  K–shell ionization, 229, 236–237
  L–subshell ionization, 242
  passage through solids, 134–135
Proton linac, 16–17
Proton stopping cross section 28–30, 61
Pulse height analyzer, 181
Pulse height spectrum, 181–182
PWBA, *see* Plane–wave Born approximation

Q

QED, *see* Quantum–electrodynamical effects
Quadrupole doublet lens, 464–465
Quantum beat phenomena, 590–595
  beam–foil research, 530, 537
Quantum–electrodynamical effects, 587–588
Quasi–adiabatic collision, 151–152
Quasi–molecular treatment, *see* Molecular orbital excitation

R

Radiative Auger effect, 401
Radiative beam–foil measurements, 532
Radiative electron capture, 94, 136, 144–148
Radiative electron capture cross section, 146–148
Radiative electron rearrangement, 400–401
Radiative recombination, *see* Radiative electron capture

Radiative transition
  allowed, 579–583
  in doubly-excited state, 561–564
  forbidden, 583–586
  measurement of excited states, 88–89
Rare-gas ion-atom collision, 155–156
REC, see Radiative electron capture
Recoil effect, for heavy ions on solid targets, 222–224, 227
Recoil energy spectrum, 152, 154, 167, 180
Recoil-particle method, of inelastic energy-loss determination, 159–161, 163, 173–174, 177, 185–187, 189
  broadening effects, 170–173
Relativistic effects, in beam-foil experiments, 598
Relativistic electron wavefunctions, in target ionization, 209–211
Relativistic projectiles, in inner-shell ionization, 211–212
Relativistic semiclassical Coulomb approximation, in K-shell ionization, 324–327
RER, see Radiative electron rearrangement
Retarding lens system, of parallel-plate analyzer, 440–449
RSCA, see Relativistic semiclassical Coulomb approximation
Rutherford scattering, 30–31
Rydberg state, see Hydrogenic state

S

Satellite Auger decay, 469–474
Satellite line, 564
Satellite production, Coulomb ionization, 496–497, 499
SCA theory, see Semiclassical approximation theory
Scattered particle
  energy analysis, 185–187, 189
  energy spectra, 180
  kinetic energy, 159–161
  relative energy spread, 177
Scattered-particle method, of inelastic energy-loss determination, 159, 173–174, 177, 188
  beam energy fluctuations, 187
  broadening effects, 170–172
Scattering, 30–31, 306–310; see also Coulomb scattering
Scattering angle, 159–160, 162, 166–167
  deflection effects, 328
  and electron loss, 117
  uncertainty of, 187
Scattering chamber, 175–176, 436–439
  for recoil energy spectrum studies, 154
Scattering cross section, 164–165, 167–168, 174
Schrödinger equation, in electron capture problems, 93, 95
Screened hydrogenic wavefunction, in ion-atom collision, 198–202
Screening, 163–166
  and projectile velocities, 495–496
  protons in solids, 134–135
Screening distance, 164–165
Screening function, 163–165
Sector-focusing cyclotron, see AVF cyclotron
Semiclassical approximation theory, 110, 112, 202, 229, 237, 241, 321–330, 337–342, 383
  Coulomb ionization, 486
  low-velocity effect, 205
  relativistic electron wavefunctions, 211
Semiconductor detector, 358–361; see also Germanium (lithium) detector; Silicon (lithium) detector
Shell correction, in energy-loss calculations, 36–37
Shell effect, 102, 107
  on equilibrium charge-state distributions, 119, 121, 124, 128
Silicon detector, with thin layer, energy loss measurements on, 55–57
Silicon (lithium) detector, 88, 214, 218, 313, 358–361, 374–375
Single collision, 306–310
  charge-changing probabilities, 79
  electron capture in 89–94, 351
  multiple, 104–107
  single, 105
  heavy ions in gases, 272
  hypersatellite production, 408–410
  inelastic energy-loss in, 149–192
  multiple-electron loss, 116–118
  multiple L-shell ionization, 401–408
  neon, 410–416
Single cylindrical curved-crystal spectrometer, 365–369

SUBJECT INDEX 639

Single-plane crystal spectrometer, 361–365
Single electron capture cross section, 98, 103
Single electron loss cross section, 98, 106
Singly-excited state, beam-foil interactions, 551, 557–560
Slab target, in ion-atom collision, 371
Slow ion-atom collision, 484–485
  charge-changing cross sections, 80
  electron capture, 102, 104
  excitation cross sections, 168
  multiple capture cross section, 107
Sodium iodide x-ray detector, 313
Solar studies, beam-foil measurements, 600–605
Solid-state detector, 54–57, 86–87; see also Germanium (lithium) detector; Silicon (lithium) detector
Solid-state effect
  on stopping cross section, 69–70
  in target x-ray production, 257
Solid target, 191, 371–374, 379
  angle resolution effects, 310
  atomic transitions and charge equilibrium in, 139–144
  average equilibrium charge, 127–128, 130–131
  backscattering energy loss, 58–65
  beam-foil experiments, 532
  charge-state fractions for heavy ions in, 272–278
  collisional quenching, 426–428
  density effects in, 134–139
  ionization and x-ray production, 193–278
  K-shell vacancy production, 335, 337
  mean charge state of ions emerging from, 519–524
  multiple scattering, 309
  spectra, 418–424
Spectral linewidths, 524–527
  Auger spectroscopy, 476, 497–498, 516–519
  Doppler broadening, in beam-foil interactions, 556–557
  measurement of, with parallel-plane analyzer, 452–454
Spectrograph, in beam-foil experiments, 539
Spectrometer
  in beam-foil experiments, 539–544
  in quantum beat experiments, 594–595
Spectrometer resolution function, 516–517
Spectroscopy, 529; see also Ion-induced Auger electron spectroscopy; Ion-induced x-ray spectroscopy
  coincident charge state analysis for spectral identification, 347–350
  heavy-ion-atom collisions, 151–152
  inelastic energy-loss measurements, 191–192
Spherical-sector analyzer, 458–462, 467
Spin-forbidden transition, 584, 586
Spin-orbit interaction, 598
Spin-spin interaction, 586
Sputtering ion source, 84
Stepwise excitation, in beam-foil interactions, 549
Stopping cross section, 27–30, 68–72
  backscattering method of determining, 58–65
  electronic, 42
  inverted Doppler-shift attenuation method of measuring, 66–68
  measurements on thin layer on solid-state detector, 56–57
  nuclear, 42
Straggling effect, for heavy ions on solid targets, 222–224
Structure effect, on stopping cross section, 70
Subshell transfer cross section, 96–97
Superconducting cyclotron, 14
Superconducting linac, 16–17
Supported film, energy-loss measurements, 53–54
Symmetric collision
  energy spread, 178
  K-shell vacancy production, 332–334, 336
  orbital transitions, 321
  scattering, 306
  target x-ray cross section, 288

T

TAC, see Time-to-amplitude converter
Tandem accelerator, 8–12, 84, 465–467, 533–535
Target, in ion-atom collisions, 371–374; see also Gas target; Solid target;

Thick target; Thin target
  ionization
  due to light particles, 110–113
  and x-ray production, 193–278
  density, and charge-changing probabilities, 131–139
  medium, and energy loss, 37
  thickness of, 47, 51
    in beam–foil experiments, 532, 549
    charge state fraction measurements, 84–85, 87
    equilibrium charge state distribution, 81
    and x-ray production, 255–257, 287
Target-detector arrangements, 316–317
Target mass per unit area, 47
Thermal broadening, 156, 168–170, 516–517
Thermionic noise, in photoelectric detectors, 545
Thick target, in ion–atom collision, 371–372
  backscattering energy loss, 61–65
  energy-loss measurements, 47
  equilibrium charge state distribution, 81
  heavy incident ions, 221–228
  for inner-shell ionization, 218–219
Thin film, backscattering energy loss, 58–59
Thin foil, energy-loss measurements, 47–51
Thin layer on solid-state detector, energy-loss measurements, 54–58
Thin target, in ion–atom collisions, 214, 371–373
  beam–foil experiments, 536–537
  charge-dependent cross section measurement, 283
  for inner-shell ionization, 218–221
  K-shell ionization cross sections, 250–252
  K-shell vacancy production, 335
  multiple scattering, 309
Thomas–Fermi atomic model
  electronic energy loss, 40–41
  potential energy, 166
  scattering cross sections, 42, 164–165
Three-component model, for target x-ray production 257–272
Three-electron ion, target x-ray cross section, 288

Three-electron line energies, 387–388
Time-dependent Schrödinger equation, in electron capture problems, 93, 95
Time-of-flight measurements, 180–183, 529, 532, 570
  forbidden Auger transitions, 586
  postfoil beam velocity, 575
Time-resolved decay process, beam–foil studies, 529, 570
Time-to-amplitude converter, 181
  dead time, 311
  spectrum, 311–313
TOF measurements, see Time-of-flight measurements
Total cross section, emission-yield data, 192
Transition energy, 386–392, 470–474
Two-component model for target x-ray production, 257–278
Two-electron ion
  Lamb shifts, 590
  partial energy level diagram 584–585
  projectile cross section, 299
  target x-ray cross section, 288
Two-electron line energies, 388
Two-electron-one-photon transitions, 344, 400–401
Two-electron transition, 398
Two-photon electric dipole decay process, 583–585

U

Universal negative-ion source (UNIS), 533, 535

V

Vacancy configuration effect, in target x-ray production, 257
Vacancy-sharing, 385, 485–486, 491, 493
Van de Graaff accelerator, 5–12, 84, 312–313, 466, 533, 535
Velocity dependence, of electron capture and loss cross sections, 100–101

W

Weakly bound state, beam–foil interactions, 564
Window-attenuation factor, in beam–foil lifetime measurements, 573, 579

SUBJECT INDEX

X

X-ray detector, 88, 213–218, 311–317, 356–371
X-ray electron, spectral identification, 347–350
X-ray–ion coincidence measurements, 305–307
X-ray production, 140–144, 284–289
  bombardment of solid targets by high-velocity ions, 193–278
  coincidence measurements, 318–319
  single-encounter experiment, 306–307
  and target thickness, 255–257
  two- and three-component models, 257–272
X-ray production cross section, 285–288
  for heavy incident ions, 221–228
  inner-shell ionization and, 195–197
  measurements, 228–278
    and detector efficiency, 214–218
    K-shell, 229–240
    L-shell, 240–249
    M-shell, 249–250
  multiple-collision effects, 225–227
  thick targets, 218–219
  thin targets, 219–221
X-ray spectroscopy, 191–192; *see also* Ion-induced x-ray spectroscopy
  for fluorescence yield calculations, 292–294
X-ray spectrum, 375–379, 386–387, 396, 399–400
  dependence of solid target spectra on environment, 418–424
  effects of outer-shell relaxation on, 416–418
  neon, 413–414
  spectral lines and transitions to L shell, 196
  and vacancy production cross section, 380

Y

Yrast level, 581

Z

Zero-field experiment, in quantum beam method, 593–595